Lecture Notes in Computer Science 12096

More information about this series at http://www.springer.com/series/7407

Ilias S. Kotsireas · Panos M. Pardalos (Eds.)

Learning and Intelligent Optimization

14th International Conference, LION 14
Athens, Greece, May 24–28, 2020
Revised Selected Papers

 Springer

Editors
Ilias S. Kotsireas (iD)
CARGO Lab.
Wilfrid Laurier University
Waterloo, ON, Canada

Panos M. Pardalos (iD)
Center for Applied Optimization
University of Florida
Gainesville, FL, USA

ISSN 0302-9743   ISSN 1611-3349 (electronic)
Lecture Notes in Computer Science
ISBN 978-3-030-53551-3   ISBN 978-3-030-53552-0 (eBook)
https://doi.org/10.1007/978-3-030-53552-0

LNCS Sublibrary: SL1 – Theoretical Computer Science and General Issues

This Springer imprint is published by the registered company Springer Nature Switzerland AG
The registered company address is: Gewerbestrasse 11, 6330 Cham, Switzerland

Guest Editorial

The 14th International Conference on Learning and Intelligent Optimization (LION 2020) was scheduled to be held in Athens, Greece, May 24–28, 2020, but regrettably it was canceled, due to travel restrictions imposed world-wide by the COVID-19 pandemic. However, we felt it was important to publish the proceedings of the conference, in order to minimize disruption to the participant's careers and especially the potentially devastating effects in the careers of PhD students, post-doctoral fellows, and young scholars. An additional reason for us to undertake the publication of these LNCS proceedings, was to ensure the continuity of the LION conference series.

LION 2020 originally featured two invited talks and a tutorial talk:

- "UberAir: Optimization Problems in the Sky," by Youssef Hamadi, Uber AI, France
- "Combinatorial Methods and Optimization Algorithms for Testing and Explainable AI," by Dimitris E. Simos, SBA Research, Austria (joint work with Rick Kuhn (NIST, USA) and Raghu Kacker (NIST, USA))
- "A Tutorial in Robust Machine Learning and AI with Applications," by Theodore B. Trafalis, University of Oklahoma, USA

We wish to express our heartfelt thanks to the organizers of the eight LION 2020 special sessions:

- Automatic Solver Configuration
- Massively Parallel Methods for Search and Optimization
- DC Learning: Theory, Algorithms and Applications
- Intractable Problems of Combinatorial Optimization, Computational Geometry, and Machine Learning: Algorithms and Theoretical Bounds
- Intelligent Optimization in Health, e-Health, Bioinformatics, Biomedicine and Neurosciences
- Scientific Models, Machine Learning and Optimization Methods in Tourism and Hospitality
- Nature Inspired Algorithms for Combinatorial Optimization Problems
- Intractable problems of combinatorial optimization, computational geometry, and machine learning: algorithms and theoretical bounds

We would like to thank the authors for contributing their work and the reviewers whose tireless efforts resulted in keeping the quality of the contributions at the highest standards. A special thank you goes to the Technical Program Committee Chair, Professor Roberto Battiti, Director of the LION Lab (machine Learning and Intelligent OptimizatioN) for prescriptive analytics.

The editors express their gratitude to the organizers and sponsors of the LION 2020 international conference:

- Center for Applied Optimization at the University of Florida University of Florida, USA
- CARGO Lab, Wilfrid Laurier University, Canada
- APM Institute for the Advancement of Physics and Mathematics

Even though organization of all physical conferences is currently on hiatus, we are very pleased to be able to deliver this LNCS proceedings volume for LION 2020, in keeping with the tradition of the two most recent LION conferences [1] and [2]. We sincerely hope we will be able to reconnect with the members of the vibrant LION community next year.

References

1. Roberto Battiti, Mauro Brunato, Ilias S. Kotsireas, Panos M. Pardalos: Learning and Intelligent Optimization – 12th International Conference, LION 2018, Kalamata, Greece, June 10–15, 2018, Revised Selected Papers, *Lecture Notes in Computer Science*, LNCS 11353, Springer (2018)
2. Nikolaos F. Matsatsinis, Yannis Marinakis, Panos M. Pardalos: Learning and Intelligent Optimization – 13th International Conference, LION 2019, Chania, Crete, Greece, May 27–31, 2019, Revised Selected Papers, *Lecture Notes in Computer Science*, LNCS 11968, Springer (2019).

June 2020   Ilias S. Kotsireas
                                                                                    Panos M. Pardalos

Organization

General Chairs

Ilias S. Kotsireas   CARGO Lab, Wilfrid Laurier University, Canada
Panos M. Pardalos   Center for Applied Optimization, University of Florida, USA

Program Committee Chair

Roberto Battiti   LION Lab, University of Trento, Italy

Program Committee

Francesco Archetti	Consorzio Milano Ricerche, Italy
Annabella Astorino	ICAR-CNR, Italy
Amir Atiya	Cairo University, Egypt
Rodolfo Baggio	Bocconi University, Italy
Roberto Battiti	University of Trento, Italy
Christian Blum	Spanish National Research Council (CSIC), Spain
Juergen Branke	University of Warwick, UK
Mauro Brunato	University of Trento, Italy
Dimitrios Buhalis	Bournemouth University, UK
Sonia Cafieri	Ecole Nationale de l'Aviation Civile, France
Antonio Candelieri	University of Milano-Bicocca, Italy
John Chinneck	Carleton University, Canada
Kostas Chrisagis	City University London, UK
Andre Augusto Cire	University of Toronto, Canada
Patrick De Causmaecker	Katholieke Universiteit Leuven, Belgium
Renato De Leone	University of Camerino, Italy
Luca Di Gaspero	DPIA, University of Udine, Italy
Ciprian Dobre	University Politehnica of Bucharest, Romania
Adil Erzin	Sobolev Institute of Mathematics, Russia
Giovanni Fasano	University Ca'Foscari of Venice, Italy
Paola Festa	University of Napoli Federico II, Italy
Antonio Fuduli	Università della Calabria, Italy
Martin Golumbic	University of Haifa, Israel
Vladimir Grishagin	Nizhni Novgorod State University, Russia
Mario Guarracino	ICAR-CNR, Italy
Youssef Hamadi	Uber AI, France
Cindy Heo	Ecole hôtelière de Lausanne, Switzerland
Laetitia Jourdan	INRIA, LIFL, CNRS, France
Valeriy Kalyagin	Higher School of Economics, Russia

Alexander Kelmanov	Sobolev Institute of Mathematics, Russia
Marie-Eleonore Kessaci	Université de Lille, France
Michael Khachay	Krasovsky Institute of Mathematics and Mechanics, Russia
Oleg Khamisov	Melentiev Institute of Energy Systems, Russia
Zeynep Kiziltan	University of Bologna, Italy
Yury Kochetov	Sobolev Institute of Mathematics, Russia
Ilias Kotsireas	Wilfrid Laurier University, Canada
Dmitri Kvasov	DIMES, University of Calabria, Italy
Dario Landa-Silva	University of Nottingham, UK
Hoai An Le Thi	Université de Lorraine, France
Daniela Lera	University of Cagliari, Italy
Vittorio Maniezzo	University of Bologna, Italy
Silvano Martello	University of Bologna, Italy
Francesco Masulli	University of Genova, Italy
Nikolaos Matsatsinis	Technical University of Crete, Greece
Kaisa Miettinen	Jyväskylä University, Finland
Serafeim Moustakidis	AiDEAS OU, Greece
Evgeni Nurminski	FEFU, Russia
Panos Pardalos	University of Florida, USA
Konstantinos Parsopoulos	University of Ioannina, Greece
Marcello Pelillo	University of Venice, Italy
Ioannis Pitas	Aristotle University of Thessaloniki, Greece
Vincenzo Piuri	Università degli Studi of Milano, Italy
Mikhail Posypkin	Dorodnicyn Computing Centre, FRC CSC RAS, Russia
Oleg Prokopyev	University of Pittsburgh, USA
Helena Ramalhinho	Universitat Pompeu Fabra, Spain
Mauricio Resende	Amazon, USA
Andrea Roli	University of Bologna, Italy
Massimo Roma	Sapienza Universita of Roma, Italy
Valeria Ruggiero	University of Ferrara, Italy
Frédéric Saubion	University of Angers, France
Andrea Schaerf	University of Udine, Italy
Marc Schoenauer	INRIA Saclay Ile-de-France, France
Meinolf Sellmann	GE Research, USA
Yaroslav Sergeyev	University of Calabria, Italy
Marc Sevaux	Lab-STICC, Université de Bretagne-Sud, France
Thomas Stützle	Université Libre de Bruxelles (ULB), Belgium
Tatiana Tchemisova	University of Aveiro, Portugal
Gerardo Toraldo	University of Naples Federico II, Italy
Michael Trick	Carnegie Mellon University, USA
Toby Walsh	The University of New South Wales, Australia
David Woodruff	University of California, Davis, USA
Dachuan Xu	Beijing University of Technology, China
Luca Zanni	University of Modena and Reggio Emilia, Italy

Qingfu Zhang   University of Essex, UK and City University of
                        Hong Kong, Hong Kong
Anatoly Zhigljavsky   Cardiff University, UK
Antanas Zilinskas   Vilnius University, Lithuania
Andre de Carvalho   University of São Paulo, Brazil
Julius Zilinskas   Vilnius University, Lithuania

Contents

Optimization for Urban Air Mobility

Youssef Hamadi[✉]

Uber Elevate, Paris, France
youssefh@uber.com

Abstract. Urban Air Mobility (UAM) has the potential to revolutionize urban transportation. It will exploit the third dimension to help smooth ground traffic in densely populated areas. To be successful, it will require an organized and integrated approach able to balance efficiency and safety while harnessing common airspace resources. We believe that mathematical optimization will play an essential role to support the development of Urban Air Mobility. In this paper, we describe two important problems from this domain, operators 4D volume deconfliction, and air taxi trajectory deconfliction.

1 Urban Air Mobility

Urban Air Mobility (UAM), designates urban air transport systems that will move people and goods by air within and around dense city areas. Its purpose and objective is to help smooth urban ground traffic despite the increasing population density. The vast majority of urban air mobility aircraft designs, will share two main characteristics. Vertical Take-Off and Landing (VTOL) to operate in relatively small areas, e.g., rooftops, and distributed electric propulsion, which will exploit multiple small rotors to minimize noise (due to rotational speed) while providing high system redundancy. Two classes of vehicles are distinguished in UAM. Small drones, typically 55 lbs and below, will be used to carry cargo, e.g., parcel delivery. This category is generally referenced as Unmanned Aircraft System (UAS). Larger aircraft able to carry important cargo and passengers, e.g., air taxi.

Operating these new aircraft over large and densely populated areas will require an organized approach able to balance efficiency, and safety. The Urban air mobility Traffic Management (UTM) research initiative [9] has produced a general architecture which leverages fundamental ideas from large-scale air-traffic control, and adjust them to the key differences that provide for UAM (maneuverability, method of control, function, range, and operational constraints).

Over time, this architecture has been refined and adopted by the US Federal Aviation Agency (FAA) [3]. It is presented in Fig. 1 which exposes, at a high level, the various actors and components, their contextual relationships, as well as high-level functions and information flows. This architecture is grounded on layers of information sharing and data exchange - from operator to operator, vehicle to vehicle, and operator to the FAA - to achieve safe operations.

© Springer Nature Switzerland AG 2020
I. S. Kotsireas and P. M. Pardalos (Eds.): LION 14 2020, LNCS 12096, pp. 1–8, 2020.
https://doi.org/10.1007/978-3-030-53552-0_1

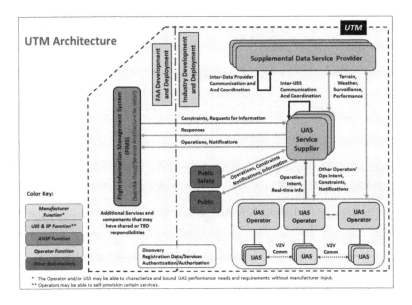

Fig. 1. Urban traffic management, notional architecture 2.0

Operators share their flight intent with each other and coordinate to de-conflict and safely separate trajectories. Through this architecture, the FAA makes real-time airspace constraints available to operators, who are responsible for managing their own operations safely within these constraints. Operators may choose to use third party UAS/UTM Service Suppliers (USSs) to support their operations, or they may choose to provision their own set of services. USSs provide services in the context of an UTM platform. Services providers can be accessed to support operations e.g., weather forecast, terrain limitations. All these interactions - between the regulator (FAA), operators and service providers - represent the UTM ecosystem, made of services, capabilities, and information flows between participants collaborating following the 'rules of the road' defined by the regulator.

Actors of this ecosystem need to efficiently cooperate, following high level regulator constraints, to get a fair and efficient access to restricted airspace resources. In the following, we propose to consider the previous from a resource optimization perspective, and present two important optimization problems in Urban Air Mobility.

2   Unmanned Aircraft System Service Providers Deconfliction

UTM operations need to be strategically deconflicted, i.e., reserved 4D volumes of airspace within which an operation is expected to occur should not intersect.

The sharing of intent through these volumes allow operators to deconflict naturally, avoiding operations traversing a pre-allocated volume. This implements a first-come first-served resource allocation, traditional in general aviation. However, since demands could be addressed simultaneously or conflict with a more recent but more important operation, negotiation between USSs is required for deconfliction. In the following, we illustrate this through an example and give guidance for the application of optimization to this problem.

Example. In this scenario, multiple operations performed by independent operators are scheduled in the morning, including construction and rail inspections, package delivery, photography, agriculture spraying, and training. See Fig. 2. Operators participate in UTM using the services of USSs to meet the requirements of their operation, including, but not limited to, sharing operation intent for situational awareness, strategically deconflicting to avoid 4D overlap of operations, obtaining airspace access authorizations, and receiving airspace notifications [3].

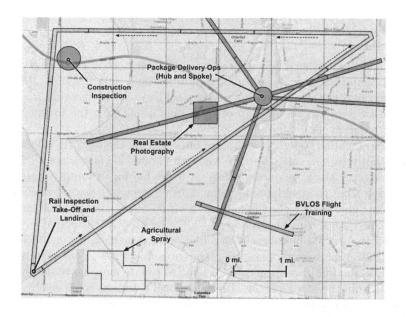

Fig. 2. Multi-operators interaction scenario

We will focus on a medical emergency which necessitates patient transport to a nearby medical facility; a MedEvac helicopter is dispatched. Flight operations personnel from the MedEvac company subscribed to the services of a USS that supports public safety operations. See Fig. 3.

This operator generates a 4D Volume Reservation that adheres to the constraints (defined spatial and temporal boundaries) of the request, and distributes it to the USS Network.

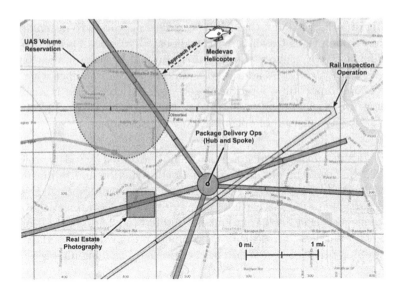

Fig. 3. MedEvac operational overview

Once informed, other USSs with subscribed operators in the vicinity of the MedEvac operation provide automated notifications. For instance, in Fig. 3, the delivery and rail inspection operators would receive notification from their respective USS due to the overlap of their operations with the urgent patient transport requirements.

Upon notification, operators who are impacted by the new 4D volume evaluate whether they can safely operate within its bounds. They adapt their operation as appropriate to maintain safety of flight by, for example, strategically deconflicting from the overlapping volume, using detect-and-avoid technologies to maintain separation from the helicopter while not changing their intent, or landing their UAS during the period in which the MedEvac is active.

The previous could be executed in an automated fashion, for instance at the USS level, using high level operators preferences and rules to adequately react. For instance, the rail inspection could be shifted to late morning to deconflict its 4D volume from the MedEvac one. However, local reactions could easily cascade into upstream conflicts which pairwise resolution are likely to result in globally less efficient operations.

For instance, the rail operator deconfliction could now conflict with parcel delivery operations. See Fig. 4, red circles. Once informed these two operators could coordinate and deconflict locally, again through automated rules. For instance, the delivery operations could be shifted earlier in time. In this simple example, a new UTM operation has successively created two more conflicts, which were successively deconflicted through basic automation. Overall, this has resulted in a less efficient use of airspace and time resources.

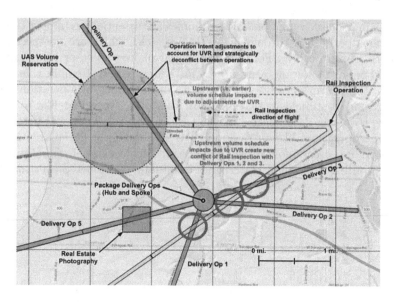

Fig. 4. Upstream conflicts (Color figure online)

For this problem, we propose to consider a general optimization approach, integrating, general rules of the road established by FAA, hard 4D volume constraints, and operator preferences. Solving this global problem should bring the best use of common airspace resources. There are two ways to practically solve it.

A centralized approach collecting global and individual input, to adjust volumes, while minimizing overall disruption and resource usage. This would require the full knowledge of inner operator preferences or cost functions, and therefore, might be implemented by the regulator or some third party authority. This could be supported by any mathematical programming formalism, e.g., MILP.

A distributed negotiation approach, for example, using the distributed constraint based optimization framework would allow direct USS to USS negotiation without requiring any centralizing point. This paradigm preserves locality of decision, and privacy. Several algorithms have been devised to solve problems expressed in this formalism [5–8, 10].

In our example, a global view on the problem, solved centrally or in a p2p way, would have resulted in the railway operator shifting inspection to early morning to deconflict with the MedEvac volume, without impacting package delivery operations.

3 Deconfliction for Trajectory-Based Operations

In the previous section, we have seen how UTM 4D volumes could be deconflicted through time and space adjustments in order to safely reconcile diverse operations in a shared airspace. These volumes are essential to support the kind of operations performed through light UAS vehicles which are not necessarily using a straight trajectory. For instance, aerial photography, or bridge inspection, could require loitering over an area for a long period. Remark that, as we have seen, this feature gives some flexibility in deconfliction, operations could be paused and continued to deconflict. In this part, we would like to isolate operations requiring deconfliction at the trajectory level. They will better support heavier operations like air taxi mobility. This time, the granularity is finer; all operators use a common 4D volume reserved for a class of operations, and operate according to straight trajectories eventually deconflicted over some key areas.

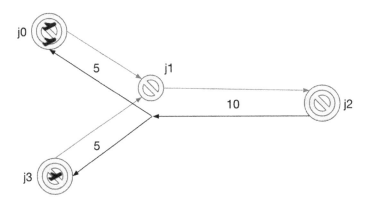

Fig. 5. Air taxi network using 4 exclusive airspace junctions (j0 to j4), 3 vertiports (associated to j0, j2, and j3), and 3 air taxi operations ready for departure (2 on j0, and 1 on j3)

The Fig. 5, presents an air taxi mobility scenario. It corresponds to 4 exclusive airspace junctions (j0 to j4), 3 vertiports for vehicle landing and take-off (associated to j0, j2, and j3), and 3 air taxi operations ready for departure (2 on j0, and 1 on j3). We will assume that the whole network uses a single 4D volume deconflicted from any other UAM operation.

The problem here, is to organize the traffic. Three flights have to be synchronized to travel from j0, and j3 to reach the same vertiport, figured at j2. They share common critical airspace resources (junctions) that can only be traversed by one vehicle at a time, along sufficient separation provision with other vehicles. In strategic conflict management, a "conflict" occurs whenever there is a competing demand for the airspace resource. This is the case here, since vehicles at j0 might want to depart at the same time while jointly constrained by the critical junction j0. The same problem happen during the flight with critical junction j1 and j2.

This problem is equivalent to the more general job-shop scheduling problem, known to be NP-hard [4]. In this problem jobs are made of successive tasks mutually competing for processing over a given set of common machines. The usual optimization criterion corresponds to the overall makespan for the whole set of jobs, which has to be minimized. The mapping is straightforward. Jobs represent flights which have to travel through a predefined ordered set of critical junctions, similar to machines with exclusive processing capacities. The makespan is equivalent to the arrival time of the latest flight, a criterion consistent with a good use of airspace resources.

There are multiple optimization approaches to tackle job-shop problems, with large instances successfully solved through complete or incomplete approaches. Reusing and adapting these methods and algorithms would be highly beneficial for efficient trajectory-based deconfliction in UTM.

4 Conclusion

Urban Air Mobility (UAM) has the potential to revolutionize urban transportation. It will exploit the third dimension to help smooth ground traffic in densely populated areas. To be successful, it will require an organized and integrated approach able to balance efficiency and safety while harnessing common airspace resources. Inspired by traditional air traffic management, the research in UAM has produced a general traffic management (UTM) architecture to organize airspace access through information sharing around precisely defined rules and regulations. See Fig. 1. We have described, at a high level, USSs 4D volume deconfliction, and air taxi trajectory deconfliction. For each of them, we have crafted resolution approaches, centralized and decentralized. In the following, in order to conclude and give perspectives, we are going to characterize good solutions to the above problems.

Equity. Within the cooperative rules and processes for the shared UTM platform, there is no assumption of a priority scheme that would diminish equity of access for users. The solutions to the above problems produced by mathematical optimization modeling and algorithms should be fair and preserve an equitable access to the resource. There are several ways to apply general fairness principles while deciding for resource usage in optimization [1].

Robustness. Weather changes, upcoming no-fly zone, synchronization with other transport modes make the above problems highly dynamic. This could result into unfeasible operational solutions if the underlying models are too rigid. Optimization under uncertainty explicitly takes into account uncertainties involved in the data or the model. It computes robust solutions which can tolerate approximation in the input data [2].

We believe that fair and robust optimization will play an essential role to support the development of Urban Air Mobility, and we hope that this research community will actively contribute to this important application domain.

References

1. Bertsimas, D., Farias, V.F., Trichakis, N.: The price of fairness. Oper. Res. **59**(1), 17–31 (2011). https://doi.org/10.1287/opre.1100.0865
2. Bertsimas, D., Sim, M.: The price of robustness. Oper. Res. **52**(1), 35–53 (2004). https://doi.org/10.1287/opre.1030.0065
3. Federal Aviation Agency: UTM concept of operations version 2.0. Technical report, Federal Aviation Agency (2020)
4. Garey, M.R., Johnson, D.S.: Computers and Intractability: A Guide to the Theory of NP-Completeness (Series of Books in the Mathematical Sciences), 1st edn. W. H. Freeman (1979)
5. Hamadi, Y.: Optimal distributed arc-consistency. In: Jaffar, J. (ed.) CP 1999. LNCS, vol. 1713, pp. 219–233. Springer, Heidelberg (1999). https://doi.org/10.1007/978-3-540-48085-3_16
6. Hamadi, Y.: Interleaved backtracking in distributed constraint networks. Int. J. Artif. Intell. Tools **11**(2), 167–188 (2002). https://doi.org/10.1142/S0218213002000836
7. Hamadi, Y.: Optimal distributed arc-consistency. Constraints Int. J. **7**(3-4), 367–385 (2002). https://doi.org/10.1023/A:1020594125144
8. Modi, P.J., Shen, W., Tambe, M., Yokoo, M.: An asynchronous complete method for distributed constraint optimization. In: Proceedings of the Second International Joint Conference on Autonomous Agents & Multiagent Systems, AAMAS 2003, 14–18 July 2003, Melbourne, Victoria, Australia, pp. 161–168. ACM (2003). https://doi.org/10.1145/860575.860602
9. Prevot, T., Rios, J., Kopardekar, P., III, J.E.R., Johnson, M., Jung, J.: UAS traffic management (UTM) concept of operations to safely enable low altitude flight operations (2016). https://doi.org/10.2514/6.2016-3292
10. Ringwelski, G., Hamadi, Y.: Boosting distributed constraint satisfaction. In: van Beek, P. (ed.) CP 2005. LNCS, vol. 3709, pp. 549–562. Springer, Heidelberg (2005). https://doi.org/10.1007/11564751_41

A Matheuristic Algorithm for Solving the Vehicle Routing Problem with Cross-Docking

Aldy Gunawan[1(✉)], Audrey Tedja Widjaja[1], Pieter Vansteenwegen[2], and Vincent F. Yu[3]

[1] Singapore Management University, 80 Stamford Road, Singapore, Singapore
{aldygunawan,audreyw}@smu.edu.sg
[2] Centre for Industrial Management/Traffic and Infrastructure, KU Leuven,
Celestijnenlaan 300, Box 2422, Leuven, Belgium
pieter.vansteenwegen@kuleuven.be
[3] Department of Industrial Management,
National Taiwan University of Science and Technology,
43, Sec. 4, Keelung Road, Taipei 106, Taiwan
vincent@mail.ntust.edu.tw

Abstract. This paper studies the integration of the vehicle routing problem with cross-docking, namely VRPCD. The aim is to find a set of routes to deliver single products from a set of suppliers to a set of customers through a cross-dock facility, such that the operational and transportation costs are minimized, without violating the vehicle capacity and time horizon constraints. A two-phase matheuristic approach that uses the routes of the local optima of an adaptive large neighborhood search (ALNS) as columns in a set-partitioning formulation of the VRPCD is designed. This matheuristic outperforms the state-of-the-art algorithms in solving a subset of benchmark instances.

Keywords: Vehicle routing problem · Cross-docking · Scheduling · Matheuristic

1 Introduction

Cross-docking is an intermediate activity within a supply chain network for enabling a transshipment process. The purpose is to consolidate different shipments for a particular destination in a full truckload (FTL), such that direct shipment with less than truckload (LTL) can be avoided, and thus the transportation cost is minimized [1]. The VRPCD as the integration of the vehicle routing problem (VRP) and cross-docking was first introduced by [5], which aims to construct a set of routes to deliver a single type of products from a set of suppliers to a set of customers through a cross-dock facility, such that the operational and transportation costs are minimized, with respect to vehicle capacity and time limitations.

© Springer Nature Switzerland AG 2020
I. S. Kotsireas and P. M. Pardalos (Eds.): LION 14 2020, LNCS 12096, pp. 9–15, 2020.
https://doi.org/10.1007/978-3-030-53552-0_2

The idea of combining metaheuristics with elements of exact mathematical programming algorithms, known as matheuristics, for solving the VRP was first introduced by [3]. [4] introduced matheuristic based on large neighborhood search for solving the VRPCD with resource constraints. In this study, we design a matheuristic which only requires a heuristic scheme to generate columns [2]. The column generation scheme is performed by an adaptive large neighborhood search (ALNS) and the set partitioning formulation is used to solve a subset of columns to find the final solution. The matheuristic is tested on one set of benchmark VRPCD instances, and the results are compared against those of the state-of-the-art algorithms. Preliminary experimental results show that our proposed matheuristic is able to obtain 29 out of 30 optimal solutions and outperform the state-of-the-art algorithms: tabu search (TS) [5], improved tabu search (imp-TS) [6], and simulated annealing (SA) [7].

2   Problem Description

The VRPCD network consists of a set of suppliers $S = \{1, 2, \ldots, |S|\}$ delivering a single product to a set of customers $C = \{1, 2, \ldots, |C|\}$ through a single cross-dock facility, denoted as node 0. Two major processes involved are: the pickup process at the suppliers and the delivery process to the customers. P_i products must be picked up from node i in S, and D_i products must be delivered to node i in C. Each pair of nodes (i, j) in S is connected by travel time t'_{ij} and transportation cost c'_{ij}. Each pair of nodes (i, j) in C is connected by travel time t''_{ij} and transportation cost c''_{ij}. The VRPCD network is illustrated in Fig. 1.

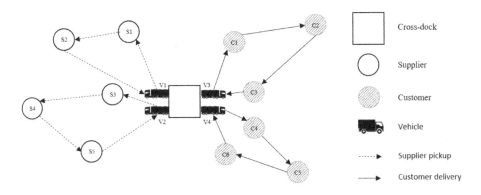

Fig. 1. VRPCD network

A fleet of homogeneous vehicles $V = \{1, 2, \ldots, |V|\}$ with capacity Q is available at the cross-dock facility to be utilized for shipments. Each vehicle can only perform either a pickup process or a delivery process, or neither. In the pickup process, vehicles depart from the cross-dock, visit one (or more) supplier(s) to pickup their products, and return to the cross-dock for consolidating products. After the products are consolidated according to customers' demand, vehicles depart from the cross-dock, visit one (or more) customer(s) to deliver their demand, and return to the cross-dock. For each vehicle used, an operational cost H will be charged. The VRPCD aims to determine the number of vehicles used and its corresponding routes, such that the operational and transportation costs are minimized. The constraints in the VRPCD are as follows:

- the total transportation time for the pickup and delivery processes together does not exceed T_{max}
- each supplier and customer can only be visited exactly once
- the number of vehicles utilized in both the pickup and delivery process together does not exceed $|V|$
- the amount of loads on the pickup route and on the delivery route in each vehicle does not exceed Q.

3   Proposed Algorithm

The matheuristic is decomposed into two phases: (i) adaptive large neighborhood search (ALNS) and (ii) the set partitioning formulation. The first phase aims to generate feasible candidate routes as many as possible, represented as columns. Those routes are then accommodated in two different pools, Ω_s and Ω_c, for pickup and delivery process respectively. In the second phase, a set partitioning formulation is solved over the set of routes stored in Ω_s and Ω_c to find a combination of routes that satisfies the VRPCD constraints. We define Sol_0 and Sol^* as the current and the best found solutions so far. An initial solution is constructed based on a greedy approach, where the node with the least additional transportation cost is inserted, such that each vehicle starts (ends) its route from (in) the cross-dock without violating the vehicle capacity and time horizon constraints. First, the Sol_0 and Sol^* are set to be the same as the initial solution. Then, the constructed routes (pickup routes and delivery routes) are added into Ω_s and Ω_c respectively. Let $R = \{1, 2, \ldots, |R|\}$ be a set of destroy operators, $I = \{1, 2, \ldots, |I|\}$ be a set of repair operators. The score s_j and weight w_j of each operator $j \in R \cup I$ is set such that its probability of choosing each operator j, p_j, in both R and I is equally likely in the beginning. At each iteration, a destroy operator R_i is randomly selected to remove π nodes from Sol_0. Consequently, a repair operator I_i is selected to reinsert the π removed nodes back to the Sol_0, resulting in a new neighborhood solution. In our implementation, $\pi = 5$.

Several destroy and repair operators that we use are **Random removal** (R_1): remove a randomly selected node from Sol_0, **Worst removal** (R_2): remove a node with a high removal cost (the difference in objective function values between including and excluding a particular node), **Route removal** (R_3): randomly select a vehicle and remove its visited nodes, **Greedy insertion** (I_1): insert a node to a position with the lowest insertion cost (the difference in objective function values between after and before inserting a node to a particular position), k-**regret insertion** (I_2, I_3, I_4): insert a node to a position with the largest regret value (the difference in objective function values when a node is inserted in the best position and in the k-best position). We use $k = 2, 3$, and 4.

Each of the removed nodes is only considered as a candidate to be inserted in a route of Sol_0 if it satisfies both vehicle capacity and time horizon constraints. Therefore, the feasibility of Sol_0 is guaranteed, unless some of the removed nodes cannot be inserted to any positions in Sol_0. If that happens, a high penalty value is added to the objective function value (total cost TC). Sol_0 is accepted if and only if it improves Sol^*. Otherwise, Sol_0 is set to be Sol^*, such that a new neighborhood solution is always explored from Sol^*. Each of the operators' score s_j is then updated by following Eq. (1), where $\delta_1 > \delta_2$. We implemented 0.5 and 0 for δ_1 and δ_2 respectively.

$$s_j = \begin{cases} s_j + \delta_1, & \text{if } Sol_0 < Sol^* \\ s_j + \delta_2, & \text{if } Sol_0 \geq Sol^* \end{cases} \quad \forall j \in R \cup I \tag{1}$$

After η_{ALNS} iterations, each of the operators' weight w_j is updated by following Eq. (2), where γ refers to the reaction factor $(0 < \gamma < 1)$ to control the influence of the recent success of an operator on its weight and χ_j is the frequency of using operator j. Consequently, each of the operators' probability p_j is updated by following Eq. (3). The ALNS is terminated when there is no solution improvement after $\eta \times \theta$ iterations. Upon this termination, the Sol^* constructed by ALNS becomes an upper bound of the VRPCD solution. It means that solving the following set partitioning formulation will only yield a lower (or at least the same) objective function value as the Sol^* constructed by ALNS.

$$w_j = \begin{cases} (1-\gamma)w_j + \gamma\frac{s_j}{\chi_j}, & \text{if } \chi_j > 0 \\ (1-\gamma)w_j, & \text{if } \chi_j = 0 \end{cases} \quad \forall j \in R \cup I \tag{2}$$

$$p_j = \begin{cases} \frac{w_j}{\sum_{k \in R} w_k} & \forall j \in R \\ \frac{w_j}{\sum_{k \in I} w_k} & \forall j \in I \end{cases} \tag{3}$$

Each candidate route r in Ω_s is associated to a transportation cost of c'_r and a transportation time of t'_r, while each candidate route r in Ω_c is associated to a transportation cost of c''_r and a transportation time of t''_r. Let a'_{ir} be a binary parameter equal to 1 if route r visits node i; 0 otherwise $(r \in \Omega_s, i \in S)$ and a''_{ir} be a binary parameter equal to 1 if route r visits node i; 0 otherwise $(r \in \Omega_c, i \in C)$. Several decision variables in the set partitioning formulation:

- $x'_r = 1$ if route r is selected; 0 otherwise $(r \in \Omega_s)$
- $x''_r = 1$ if route r is selected; 0 otherwise $(r \in \Omega_c)$
- Tp_{max} = the maximum transportation time for pickup process
- Td_{max} = the maximum transportation time for delivery process

The objective is to minimize the total of transportation and operational costs, as formulated in (4). All supplier and customer nodes must be visited, as required in (5) and (6) respectively. (7) limits the number of selected routes (i.e. does not exceed the number of available vehicles). (8) and (9) records the maximum transportation time in pickup and delivery process respectively. Finally, the two processes must be done within the time horizon, as expressed in (10).

$$Min \ \sum_{r \in \Omega_s} c'_r x'_r + \sum_{r \in \Omega_c} c''_r x''_r + H \left(\sum_{r \in \Omega_s} x'_r + \sum_{r \in \Omega_c} x''_r \right) \tag{4}$$

$$\sum_{r \in \Omega_s} a'_{ir} x'_r = 1 \ \forall i \in S \tag{5}$$

$$\sum_{r \in \Omega_c} a''_{ir} x''_r = 1 \ \forall i \in C \tag{6}$$

$$\sum_{r \in \Omega_s} x'_r + \sum_{r \in \Omega_c} x''_r \leq |V| \tag{7}$$

$$t'_r x'_r \leq Tp_{max} \ \forall r \in \Omega_s \tag{8}$$

$$t''_r x''_r \leq Td_{max} \ \forall r \in \Omega_c \tag{9}$$

$$Tp_{max} + Td_{max} \leq T_{max} \tag{10}$$

4   Computational Results

The matheuristic is tested on benchmark VRPCD instances with 10-nodes [5]. We report the average values found for all instances out of ten runs. Since the instances are small, we could use CPLEX and the mathematical model presented in [5] to obtain the optimal solution for all these instances. It should be noted, however, that these optimal solutions were not reported in the state of the art yet. In Table 1, we evaluate the performance of our approach and those of the state-of-the-art algorithms based on these optimal solutions. The matheuristic is implemented in C++ with CPLEX 12.9.0.0 to solve the set partitioning formulation. All experiments were performed on a computer with Intel Core i7-8700 CPU @ 3.20 GHz processor, 32.0 GB RAM. The parameter values are: γ: 0.7, θ: 20, η_{ALNS}: 200, η: $(|S| + |C|) \times 2$.

Our proposed matheuristic is able to obtain either the same or further improve the best known solutions (BKS) which are consolidated from the state-of-the-art algorithms. On average, we outperform the BKS with 1.5%. Moreover, we obtain the optimal solution for each instance. In terms of the average of CPU time, our proposed matheuristic spends 0.16 s while [5–7] use 2.02, 0.12 and 2.06 s respectively. The average calculation time for generating the optimal solutions with CPLEX takes 1.05 s.

Table 1. Total cost comparison of the matheuristic and state-of-the-art algorithms

Instance	[5]	[6]	[7]	BKS	Opt	Matheuristic	Gap BKS to Opt	Gap Matheuristic to Opt
1	7571.4	6847.6	6953.0	6847.6	6823.0	6823.0	0.4%	0.0%
2	7103.7	6816.8	6741.0	6741.0	6741.0	6741.0	0.0%	0.0%
3	9993.5	9615.6	9269.0	9269.0	9269.0	9269.0	0.0%	0.0%
4	8338.0	7289.7	7255.0	7255.0	7229.0	7229.0	0.4%	0.0%
5	8709.9	6599.0	6524.0	6524.0	6475.0	6475.0	0.8%	0.0%
6	9143.5	9324.6	7613.0	7613.0	7434.0	7434.0	2.4%	0.0%
7	12721.2	12083.0	11990.0	11990.0	11713.0	11713.0	2.4%	0.0%
8	9275.7	8719.6	8158.0	8158.0	8158.0	8158.0	0.0%	0.0%
9	8096.5	7362.2	7120.0	7120.0	6989.0	6989.0	1.9%	0.0%
10	7044.8	6204.5	6056.0	6056.0	5960.0	5960.0	1.6%	0.0%
11	8051.8	7635.3	7434.0	7434.0	6916.0	6916.0	7.5%	0.0%
12	8661.0	7867.2	7800.0	7800.0	7656.0	7656.0	1.9%	0.0%
13	7370.2	7097.9	6934.0	6934.0	6783.0	6783.0	2.2%	0.0%
14	7132.3	5208.0	4704.0	4704.0	4417.0	4417.0	6.5%	0.0%
15	7563.4	7103.2	7088.0	7088.0	7072.0	7072.0	0.2%	0.0%
16	9983.6	8768.7	8616.0	8616.0	8440.0	8440.0	2.1%	0.0%
17	9538.1	9003.0	9003.0	9003.0	9003.0	9003.0	0.0%	0.0%
18	8057.4	6887.5	6911.0	6887.5	6760.0	6760.0	1.9%	0.0%
19	9042.6	7123.0	7051.0	7051.0	7051.0	7051.0	0.0%	0.0%
20	10478.0	10471.0	10004.0	10004.0	9786.0	9786.0	2.2%	0.0%
21	8380.5	5431.4	4753.0	4753.0	4644.0	4646.0	2.3%	0.0%
22	9016.9	6908.0	6442.0	6442.0	6442.0	6442.0	0.0%	0.0%
23	9489.2	9224.1	9156.0	9156.0	9156.0	9156.0	0.0%	0.0%
24	12513.6	11976.0	11976.0	11976.0	11976.0	11976.0	0.0%	0.0%
25	7114.3	6638.0	6346.0	6346.0	6346.0	6346.0	0.0%	0.0%
26	8421.3	7216.9	6880.0	6880.0	6817.0	6817.0	0.9%	0.0%
27	10666.8	9709.8	9541.0	9541.0	9541.0	9541.0	0.0%	0.0%
28	10123.3	7408.0	7107.0	7107.0	6782.0	6782.0	4.8%	0.0%
29	7503.2	6748.5	6762.0	6748.5	6591.0	6591.0	2.4%	0.0%
30	7642.6	7304.4	6942.0	6942.0	6919.0	6919.0	0.3%	0.0%
Avg							1.5%	0.0%

5   Conclusion

We study the integration of vehicle routing problem with cross-docking (VRPCD). A matheuristic approach based on ALNS and set partitioning is proposed. Preliminary results show that the matheuristic outperforms the state-of-the-art algorithms in terms of both solution quality and computational time. Solving larger benchmark instances will be included in future work.

Acknowledgment. This research is supported by the Singapore Ministry of Education (MOE) Academic Research Fund (AcRF) Tier 1 grant.

References

1. Apte, U.M., Viswanathan, S.: Effective cross docking for improving distribution efficiencies. Int. J. Logist. **3**(3), 291–302 (2000)

2. Archetti, C., Speranza, M.G.: A survey on matheuristics for routing problems. EURO J. Comput. Optim. **2**(4), 223–246 (2014). https://doi.org/10.1007/s13675-014-0030-7
3. Foster, B.A., Ryan, D.M.: An integer programming approach to the vehicle scheduling problem. J. Oper. Res. Soc. **27**(2), 367–384 (1976)
4. Grangier, P., Gendreau, M., Lehuédé, F., Rousseau, L.M.: The vehicle routing problem with cross-docking and resource constraints. J. Heuristics (2019). https://doi.org/10.1007/s10732-019-09423-y
5. Lee, Y.H., Jung, J.W., Lee, K.M.: Vehicle routing scheduling for cross-docking in the supply chain. Comput. Ind. Eng. **51**(2), 247–256 (2006)
6. Liao, C.J., Lin, Y., Shih, S.C.: Vehicle routing with cross-docking in the supply chain. Expert Syst. Appl. **37**(10), 6868–6873 (2010)
7. Yu, V.F., Jewpanya, P., Redi, A.A.N.P.: Open vehicle routing problem with cross-docking. Comput. Ind. Eng. **94**, 6–17 (2016)

Physical Activity as a Risk Factor in the Progression of Osteoarthritis: A Machine Learning Perspective

Antonios Alexos[1], Serafeim Moustakidis[4(✉)], Christos Kokkotis[2,3], and Dimitrios Tsaopoulos[2]

[1] Department of Electrical and Computer Engineering,
University of Thessaly, Volos, Greece
[2] Center for Research and Technology Hellas,
Institute of Bio-Economy and Agri-Technology, Volos, Greece
[3] Department of Physical Education and Sport Science,
University of Thessaly, Trikala, Greece
[4] AIDEAS OÜ, Narva mnt 5, Tallinn, Harju maakond, Estonia
s.moustakidis@aideas.eu

Abstract. Knee osteoarthritis (KOA) comes with a variety of symptoms' intensity, frequency and pattern. Most of the current methods in KOA diagnosis are very expensive commonly measuring changes in joint morphology and function. So, it is very important to diagnose KOA early, which can be achieved with early identification of significant risk factors in clinical data. Our objective in this paper is to investigate the predictive capacity of physical activity measures as risk factors in the progression of KOA. In order to achieve this, a machine learning approach is proposed here for KOA prediction using features extracted from an accelerometer bracelet. Various ML models were explored for their suitability in implementing the learning task on different combinations of feature subsets. Results up to 74.5% were achieved indicating that physical activity measured by accelerometers may constitute an important risk factor for KOA progression prediction especially if it is combined with complementary data sources.

Keywords: Knee osteoarthritis prediction · Machine learning · Wearable activity data · Physical function · Knee joint

1 Introduction

Among the most chronic conditions of the joints is osteoarthritis (OA), while Knee osteoarthritis (KOA) is the most famous one being highly correlated with quality of life. KOA is called "wear-and-tear" type, because the cartilage in the knee joint progressively wears away. KOA is most often observed in people, that are over 55 years old, while the disease prevails in people over 65 years old. It may be a common disease in old people, but it is often diagnosed also in young

© Springer Nature Switzerland AG 2020
I. S. Kotsireas and P. M. Pardalos (Eds.): LION 14 2020, LNCS 12096, pp. 16–26, 2020.
https://doi.org/10.1007/978-3-030-53552-0_3

athletes after suffering an injury. An important uniqueness of this disease is that it has a variety of symptoms in intensity, frequency and pattern. Its complexity in combination with lack of sufficient data limit our understanding of the processes governing KOA progression.

KOA is not easy to define, predict or treat. Identification of risk factors for developing arthritis has been limited by a lack of longitudinal data, as well as an absence of reproducible, non-invasive methods to measure changes in joint morphology and function. As a result, the disease processes governing osteoarthritis progression are still poorly understood. Although most of the existing research has focused on factors associated with the disease, the lack of longitudinal data examining the factors associated with disease onset and progression has resulted in a lack of prevention and treatment interventions. Hence, there is a need for interventions that aim to treat the most appropriate modifiable risk factors and, therefore, prevent or delay the onset and/or progression of the disease. Medical risk factors known to influence development of KOA include advanced age, gender, hormonal status, body weight or size, usually quantified using body mass index (BMI), and a family history of disease. Additionally, there is now evidence supporting a strong genetic association. Other known risk factors for the onset and progression of OA include joint loading during occupational activity and sports participation, muscle weakness, a past history of knee injury, depression and reduced physical activity. Although many of the above factors are fixed, other risk factors such as body weight, physical activity are modifiable. Moreover, the rapid increase of large observational studies and the availability of big heterogeneous clinical databases bring new challenges as well as opportunities for enhanced diagnosis of OA through advanced data-driven approaches.

Despite being relatively slow adopting advanced analytics, OA field has recently made some progress on developing prediction models based on ML techniques. This occurred due to the increasing availability of data from a variety of sources that need to be included, so that the risk prediction model is based on all the known risk factors and their interactions. Due to the complexity of the data, there is a lack of bibliography in KOA prediction with accelerometers. By using kinematic data for the prediction of the KOA, in [17], the authors achieved an accuracy of 97.4% in knee OA detection and prediction of pain (83.3%) using ML approaches (Support Vector Machines). The authors in [11] used PCA combined with Support Vector Machines (SVM) in the task of KOA classification achieving accuracies of 98–100%. In [16] an AUC performance of 85% considering 3D knee kinematic parameters as diagnostic disease biomarkers of medial compartment KOA was achieved. KOA classification has been also attempted in [3,9,10] utilizing various data sources (including ground reaction forces and clinical data) and state-of-the-art machine and deep learning algorithms.

Moreover, the authors in [8] worked on gait biomechanical parameters and outcome scores for KOA prediction, achieving a low error rate (0.02). Finally, clinical data and independent predictors have been investigated as potential risk factors for KOA prediction resulting to moderate performance (0.66–0.88 AUC and AUC of 0.823 by using only 5 variables) in [17] and [7], respectively.

To the best of our knowledge, the level/intensity of physical activity has never been explored as the sole contributor in the development of KOA predictive models. An accelerometer is a great addition to KOA studies and has some major advantages (small weight, and convenience in observing physical activity of a wearer/patient). Data from Osteoarthritis Initiative (OAI) (available at http:// www.oai.ucsf.edu/) were utilised in our study with special emphasis on vital features extracted from accelerometers such as intensity, frequency and daily duration of physical activity. Decision trees, nearest neighbor classifiers, support vector machines along with boosting techniques were trained on the extracted features to predict whether a patient's KOA KL grade will progress or not. This means that KOA prediction was implemented here as a binary classification task grouping the pool of patients' data into two groups: (i) the patients whose KL grade will progress in the future and (ii) those whose KL grade remains stable. Different subsets of features were assessed and the best combination was used to produce the final results.

The rest of the paper is organised as follows. Section 2 provides a short description of the data used in our study, whereas Sect. 3 presents the proposed ML methodology for KOA prediction. Results are given in Sect. 4 and conclusions are drawn in the final section of the paper.

We release all code used for this work[1].

2   Data Description

Data was obtained from the OAI database. OAI was designed to (i) identify risk factors that cause KOA, (ii) boost the research in the area of KOA, and therefore contribute to a better quality of life for patients with KOA. It was launched in 2002, and contains data from patients in the ages 45–79 years old, with symptomatic KOA, or being on the verge of developing it, in at least one knee. The study has taken place in four US medical centers, where a total of 4796 participants were enrolled in a study that lasted for an 8-year period. It is very important to note that it has a follow-up rate of more than 90% for the first 4 years. The data that we used in this paper are features extracted from accelerometer bracelet data combined with Kellgren and Lawrence (KL) grade data from each patient. Table 1 below cites the features from OAI that were used in our experimentation.

3   Methodology

The proposed ML methodology comprises the following phases: (i) data preprocessing, (ii) learning and (iii) validation. More details about the aforementioned processing phases are given below.

3.1   Data Preprocessing

Data Selection. The OAI database has a significant number of missing values due to the fact that many subjects, that were participating in the study,

---

[1] https://github.com/antonyalexos/Accelerometer-Knee-Osteoarthritis-Prediction.

Table 1. Features of physical activity used in our experimentation.

Feature category	Number of features	Description
Ilislorical dala on KOA grade level	4 features	Mean KL grade over the 6 first visits
		Variance of KL grade over the 6 first visits
		Standard variation of KL grade over the 6 first visits
		KL grade at visit 6
Swartz features	6 features x 6 visits = 36 features	V06DAYLtMinS: Daily minutes of light activity (counts 100-573)
		V06DAYMVBoutMinS: Daily bout minutes of moderate/vigorous physical activity (574+)
		V06DAYMVMinS: Daily minutes of moderate/vigorous activity (counts 574+)
		V06DAYModMinS: Daily minutes of moderate activity (counts 574-4944)
		V06DAYVBoutMinS: Daily bout minutes of vigorous activity (counts 4945+)
		V06DAYVigMinS: Daily minutes of vigorous activity (counts 4945+)
Troiano features	6 features x 6 visits = 36 features	VO6DAYLtMinT: Daily minutes of light activity (counts 100-2019)
		V06DAYMVBoutMinT: Daily bout minutes of moderate/vigorous activity (counts 2020+)
		V06DAYMVMinT: Daily minutes of moderate/vigorous activity (counts 2020+)
		V06DAYModMinT: Daily minutes of moderate activity (counts 2020-5998)
		V06DAYVBoutMinT: Daily bout minutes of vigorous activity (counts 5999+)
		V06DAYVigMinT: Daily minutes of vigorous activity (counts 5999+)
Freedson features	6 features x 6 visits = 36 features	V06DAYLtMinF: Daily minutes of light activity (counts 100-1951)
		V06DAYMVBoutMinF: Daily bout minutes of moderate/ vigorous activity (counts 1952+)
		V06DAYMVMinF: Daily minutes of moderate/vigorous activity (counts 1952+)
		V06DAYModMinF: Daily minutes of moderate activity (counts 1952-5724)
		V06DAYVBoutMinF: Daily bout minutes of vigorous activity (counts 5725+)
		V06DAYVigMinF: Day minutes of vigorous activity (counts 5725+)

stopped attending the visits that were scheduled every 6 months. In this paper, we formulated a smaller in size dataset including data from those subjects that had participated in all the visits until month 48 (visit 6) and had also participated in the physical activity tests. Patients who did not follow up after the 6th visit were also dropped since they did not allow us to assess whether they progressed or not. The reason that we chose the 6th visit to be the cutoff for our study, is because of the fact that the availability of accelerometer data in the OAI database is limited between the 6th and 8th visit. This resulted in the dataset with a total number of 1120 samples.

Data Sampling. Stratified sampling was adopted to address the class imbalance problem in our generated dataset with the majority of the data being in the class of patients whose KL grade was unchanged. Specifically, the small class was oversampled to match the size of the big one. Before that, the dataset was split 70%–30% for the training and testing sets, respectively. The aforementioned stratified sampling mechanism was applied only in the train set, whereas the testing set was left unchanged. After this process, the training set had 719 observations and the testing set 224 observations.

Feature Selection. To identify the optimal set of features and extract valuable information with respect to features' discriminative capacity, data was organised into five different (5) feature subsets. The accelerometer data provided by the database has features from Swartz, Troiano and Freedson. These are cutting points, proposed by these 3 researchers [6, 14, 15] respectively. The five different (5) feature subsets as given below:

1. Past progress on KL grade, Swartz features, Troiano features and Freedson features.
2. Past progress on KOA features and Swartz features.
3. Past progress on KOA features and Troiano features.
4. Past progress on KOA features and Freedson features.
5. Swartz features and Troiano features and Freedson features.

Data Normalisation. Feature scaling is one of the most important steps during the preprocessing phase before creating the ML models. In our paper, each feature was normalised with respect to its standard deviation.

Data Labeling. Patients, whose KL grade in any of his/her legs progressed after visit 6, were grouped together forming the first class(Class 1). Class 2 was generated by grouping together all these subjects whose KL grade remained unchanged in both legs after visit 6.

3.2   Learning with Different ML Models

In this section we give some brief description of the ML algorithms that we used in our experimental comparative analysis.

Decision Tree. Decision Tree [12] is a supervised learning method used mostly for classification that learns from data with a set of if-then-else decision rule statements. A decision tree is a tree structure that breaks down a data set into smaller subsets producing decision nodes and leaf nodes. Each of these nodes has two or more branches, while every leaf node is either a classification or a decision.

k Nearest Neighbors. k Nearest Neighbors (kNN) is a non-parametric, lazy learning algorithm. Its purpose is to use a database, whose data points are separated into classes, in order to predict the classification of a new sample point. The kNN algorithm assumes that similar things exist in close distance, which means that they are near to each other. In a more elaborate way, kNN, for every data point in the dataset, calculates the distance between every query and the chosen data point; it sorts in an ascending order, and picks the first k entries. More information you can find here [5].

Support Vector Machines. Support Vector Machines (SVM) algorithm finds a separating line (or hyperplane) between data points that belong to two classes. It actually takes the data as input, and outputs a line that separates those classes. SVM finds the data points that are the closest to the line from the classes (these data points are called support vectors). Then, it computes the distance between the line and these two points, with a final goal to maximize the between-class distance. Kernels are applied to project data into higher dimensional spaces in order to transform the linear separations into powerful non-linear ones. You can find more information here [4].

Random Forest. Random Forest is simply a model made up of many decision trees or of any other week learner [1]. Its name is given due to: (i) the randomness in the sampling of training data points when building trees, and the (ii) random subsets of features considered when splitting nodes. During training, every tree in the forest is learning from a random sample of data. These random samples are taken with replacement, which means that each tree uses every sample of data multiple times. In this way, every decision tree in the forest has high variance, but overall the random forest will have lower variance. The decisions during testing are the average of predictions of every decision tree in the random forest.

Balanced Random Forest. A random forest induces every tree from a bootstrap sample of the training data. When data is imbalanced, the algorithm may perform very poorly in the prediction of the small class. The Balanced Random forest (BRF) solves this problem by inducing ensemble trees on balanced down-sampled data [2]. So at each iteration of the algorithm, it takes a bootstrap sample from the small class and then draws in random the same number of observations (with replacement), from the big class. It creates then a classification tree from the data with maximum height and without pruning. The tree is created with the CART algorithm and at every node a set of randomly selected variables is used. The process is repeated for a preferred number of iterations.

RUSBoostClassifier. This method is a random under-sampling integrating in the learning of an AdaBoost classifier [13]. AdaBoost actually improves the performance of any weak classifier (with the boosting method), taking into consideration the fact that the classifier results in better performance than random guessing. Unlike more complex algorithms for data sampling, RUSBoostClassifier does not remove observations from the train set with an "intelligent" method. It removes observations from the big class randomly until it achieves a given distribution.

3.3 Validation

A 70%–30% random data split was applied to generate the training and testing subsets, respectively. Learning of the ML was performed on the stratified version of the training sets and the final performance was estimated on the testing sets.

4 Results and Discussion

This section presents the classification results of the aforementioned ML algorithms on different feature subsets along with information for the selected hyperparameters per method.

4.1 Results on the Entire Feature Set

The discrimination capabilities of the proposed ML models on the full feature space were initially investigated including features from the KL grade history of the subjects as well as from all three feature categories that refer to physical activity (Swarz, Troiano and Freedson). Decision Trees (DT) achieved the best accuracy (74.1%) accomplishing moderate class accuracies (0.87% and 0.30% for classes 1 and 2, respectively). Random Forest (RF) and Support Vector Machines (SVM) accomplished accuracies at the same level (approximately 73%), whereas a prediction accuracy of 71.4% was achieved by Balanced Random Forest (BRF). RUSBoostClassifier (RBC) achieved a prediction accuracy of 67.4%. kNN was the worst performing model with a much lower prediction score 65.1% (Table 2).

Table 2. Performance of different ML models on the full feature space

Method	Selected hyperparameters	Overall accuracy
DT	max_features:auto, min_samples_split:2	74.1%
kNN	algorithm:auto, n_jobs:-1, n_neighbors:6, weights:distance	65.1%
RF	criterion:gini, min_samples_split:3, n_estimators:30,n_jobs:-1	73%
SVM	kernel = poly,degree = 5,C = 10000	73%
BRF	n_estimators = 100, criterion = gini	71.4%
RBC	n_estimators = 50, learning_rate = 1	67.4%

4.2   Results on Swartz Features

In this subsection, we present the results of the proposed ML models using only the Swartz features combined with the previous KL grades (prior to visit 6). The highest overall accuracy of 74.5% was achieved for the DT, RF and SVM models. The following remarks could be extracted from Table 3: (i) Swartz features are the most informative ones for the prediction of KOA, (ii) there is some redundancy in the information contained on the three feature sets (Swartz, Troiano and Freedson) and (iii) kNN was again the worst performing classifier achieving the same accuracy (67.4%) with BRF and RBC.

Table 3. Performance of different ML models using Swartz features combined with the KL grade history

Method	Selected hyperparameters	Overall accuracy
DT	max_features:log2, min_samples_split:2	74.5%
kNN	algorithm:auto, n_jobs:-1, n_neighbors:5, weights:distance	67.4%
RF	criterion:gini,min_samples_split:3, n_estimators:30, n_jobs:-1	74.5%
SVM	kernel = rbf,C = 10000	74.5%
BRF	n_estimators = 100, criterion = gini	67.4%
RBC	n_estimators = 50, learning_rate = 1	67.4%

4.3   Results on Troiano Features

Repeating the same analysis on the Troiano features combined with previous KL grades from visits 1 to 6 gives similar performances with the maximum of 74.5% to be achieved by RF and SVM. The second highest accuracy was received for DT (73.6%), whereas lower accuracies were obtained by the rest of the models. These results indicate the discrimination capacity of Troiano features similarly to the performance obtained with the Swartz features (Table 4).

Table 4. Performance of different ML models using Troiano features combined with the KL grade history

Method	Selected hyperparameters	Overall accuracy
DT	max_features:log2, min_samples_split:2	73.6%
kNN	algorithm:auto, n_jobs:-1, n_neighbors:6, weights:distance	66%
RF	criterion:gini, min_samples_split:3, n_estimators:20, n_jobs:-1	74.5%
SVM	kernel = rbf, C = 10000	74.5%
BRF	n_estimators = 100, criterion = gini	72.3%
RBC	n_estimators = 50, learning_rate = 1	61.1%

4.4   Results on Freedson Features

Slightly lower accuracies were achieved by the ML models trained on the Freedson features and the KL grade history. Again, DT, RF and SVM were the best classifiers (73.6%) with BRF being relatively less effective with an accuracy of 72.3%. kNN achieved a 66.5% and the worst performance was obtained by RBC (61.1%) (Table 5).

Table 5. Performance of different ML models using Freedson features combined with the KL grade history

Method	Selected hyperparameters	Overall accuracy
DT	max_features:log2, min_samples_split:2	73.6%
kNN	algorithm:auto, n_jobs:-1, n_neighbors:6, weights:distance	66.5%
RF	criterion:gini, min_samples_split:3, n_estimators:20,n_jobs:-1	73.6%
SVM	kernel = rbf, C = 10000	73.6%
BRF	n_estimators = 100, criterion = gini	72.3%
RBC	n_estimators = 50, learning_rate = 1	61.1%

4.5   Results on the Combination of Swartz, Freedson and Troiano Features

The last part of our experimentation involved training of the six classifiers on a combination of Swartz, Troiano and Freedson features without considering the KL grade progression before the sixth visit. Much lower accuracies were achieved for all the ML models highlighting the importance of the KL grade history as a risk factor on the final prediction outcome. The maximum accuracy (67.8%) was obtained by DT, RF and SVM had similar performances (67.4%), whereas RBC was the worst performing model with a 62% classification accuracy (Table 6).

Table 6. Performance of different ML models on the full feature space except to the KL grade history

Method	Selected hyperparameters	Overall accuracy
DT	max_features:auto, min_samples_split:2	67.8%
kNN	algorithm:auto, n_jobs:-1, n_neighbors:5, weights:distance	62.5%
RF	criterion:gini, min_samples_split:3, n_estimators:20,n_jobs:-1	67.4%
SVM	kernel = rbf, C = 1000	67.4%
BRF	n_estimators = 100, criterion = gini	65.6%
RBC	n_estimators = 50, learning_rate = 1	62%

5   Conclusions

The proposed methodology in this paper shows potential to predict, with relatively high accuracy, the KOA progression based on physical activity data measured by an accelerometer bracelet along with historical patient data with respect to KL grade. The best performances up to 74.5% were achieved with the second (Swartz) and third (Troiano) feature combinations. Adding Freedson features on the feature subsets led to small reductions in the prediction performance, whereas the KL grade history was proved to be a useful risk factor that contributed positively. The inclusion of more features from other heterogeneous sources (clinical data, demographics, nutrition etc) is considered as our future work towards the development of more robust predictive models.

Acknowledgments. This work has received funding from the European Community's H2020 Programme, under grant agreement Nr. 777159 (OACTIVE).

References

1. Breiman, L.: Random forests. Mach. Learn. **45**(1), 5–32 (2001). https://doi.org/10.1023/A:1010933404324
2. Chen, C., Breiman, L.: Using Random Forest to Learn Imbalanced Data. University of California, Berkeley (2004)
3. Christodoulou, E., Moustakidis, S., Papandrianos, N., Tsaopoulos, D., Papageorgiou, E.: Exploring deep learning capabilities in knee osteoarthritis case study for classification. In: 2019 10th International Conference on Information, Intelligence, Systems and Applications (IISA), pp. 1–6. IEEE (2019)
4. Cortes, C., Vapnik, V.: Support-vector networks. Mach. Learn. **20**(3), 273–297 (1995). https://doi.org/10.1023/A:1022627411411
5. Cover, T., Hart, P.: Nearest neighbor pattern classification. IEEE Trans. Inf. Theor. **13**(1), 21–27 (2006). https://doi.org/10.1109/TIT.1967.1053964
6. Freedson, P., Melanson, E., Sirard, J.: Calibration of the computer science and applications, inc. accelerometer. Med. Sci. Sports Exercise **30**(5), 777–781 (1998). https://doi.org/10.1097/00005768-199805000-00021
7. Lazzarini, N., et al.: A machine learning approach for the identification of new biomarkers for knee osteoarthritis development in overweight and obese women. Osteoarthritis Cartilage **25** (2017). https://doi.org/10.1016/j.joca.2017.09.001
8. Long, M., Papi, E., Duffell, L., McGregor, A.: Predicting knee osteoarthritis risk in injured populations. Clin. Biomech. **47**, 87–95 (2017). https://doi.org/10.1016/j.clinbiomech.2017.06.001
9. Moustakidis, S., Christodoulou, E., Papageorgiou, E., Kokkotis, C., Papandrianos, N., Tsaopoulos, D.: Application of machine intelligence for osteoarthritis classification: a classical implementation and a quantum perspective. Quantum Mach. Intell. 1–14 (2019)
10. Moustakidis, S., Theocharis, J., Giakas, G.: A fuzzy decision tree-based svm classifier for assessing osteoarthritis severity using ground reaction force measurements. Med. Eng. Phys. **32**(10), 1145–1160 (2010)
11. Phinyomark, A., Osis, S., Hettinga, B., Kobsar, D., Ferber, R.: Gender differences in gait kinematics for patients with knee osteoarthritis. BMC Musculoskeletal Disorders **17** (2016). https://doi.org/10.1186/s12891-016-1013-z

12. Quinlan, J.R.: Induction of decision trees. Mach. Learn. **1**(1), 81–106 (1986). https://doi.org/10.1023/A:1022643204877
13. Seiffert, C., Khoshgoftaar, T.M., Van Hulse, J., Napolitano, A.: Rusboost: a hybrid approach to alleviating class imbalance. Trans. Sys. Man Cyber. Part A **40**(1), 185–197 (2010). https://doi.org/10.1109/TSMCA.2009.2029559
14. Swartz, A.M., Strath, S.J., BASSETT, D.R., O'BRIEN, W.L., King, G.A., Ainsworth, B.E.: Estimation of energy expenditure using CSA accelerometers at hip and wrist sites. Med. Sci. Sports Exercise **32**(9), S450–S456 (2000)
15. Troiano, R.P., Berrigan, D., Dodd, K.W., Masse, L.C., Tilert, T., McDowell, M.: Physical activity in the United States measured by accelerometer. Med. Sci. Sports Exercise **40**(1), 181–188 (2008)
16. Vendittoli, P.A., Ouakrim, Y., Fuentes, A., Mitiche, A., Hagemeister, N., de Guise, J.: Mechanical biomarkers of medial compartment knee osteoarthritis diagnosis and severity grading: discovery phase. J. Biomech. **52** (2016). https://doi.org/10.1016/j.jbiomech.2016.12.022
17. Yoo, T., Kim, D.W., Choi, S., Oh, E., Park, J.S.: Simple scoring system and artificial neural network for knee osteoarthritis risk prediction: a cross-sectional study. PloS One **11**, e0148724 (2016). https://doi.org/10.1371/journal.pone.0148724

QPTAS for the CVRP with a Moderate Number of Routes in a Metric Space of Any Fixed Doubling Dimension

Michael Khachay[1,2,3(✉)] and Yuri Ogorodnikov[1,2(✉)]

[1] Krasovsky Institute of Mathematics and Mechanics, Ekaterinburg, Russia
{mkhachay,yogorodnikov}@imm.uran.ru
[2] Ural Federal University, Ekaterinburg, Russia
[3] Omsk State Technical University, Omsk, Russia

Abstract. The Capacitated Vehicle Routing Problem (CVRP) is the well-known combinatorial optimization problem having a host of valuable practical applications in operations research. The CVRP is strongly NP-hard both in its general case and even in very specific settings (e.g., on the Euclidean plane). The problem is APX-complete for an arbitrary metric and admits Quasi-Polynomial Time Approximation Scheme (QPTAS) in the Euclidean space of any fixed dimension (and even PTAS, under additional constraints). In this paper, we significantly extend the class of metric settings of the CVRP that can be approximated efficiently. We show that the metric CVRP admits QPTAS any time, when it is formulated in a metric space of a fixed doubling dimension $d > 1$ and is restricted to have an optimal solution of at most polylog n routes.

1 Introduction

The Capacitated Vehicle Routing Problem (CVRP) is the well-known combinatorial optimization problem having a lot of valuable practical applications in operations research. The problem was introduced by Dantzig and Ramser in their seminal paper [8] as a mathematical model for routing the fleet of gasoline trucks servicing a network of gas stations from a bulk terminal.

Since then, the field of the algorithmic design for the CVRP is developed in a number of research directions as follows. The first direction is based on a reduction of the problem in question to some appropriate mixed integer program and finding an optimal solution of this program using some of the well-known branch-and-price methods [25]. Recently, a significant success was achieved in development such algorithms and computational hardware [11,21]. Unfortunately, due to strongly NP-hardness of the CVRP, instances of this problem that are managed to be solved efficiently within this approach still remain quit modest.

Another direction is closely related to involving a wide range of heuristic algorithms and meta-heuristics including the local search [2], VNS [22], Tabu search [23], evolutionary and bioinspired methods [19], and their combinations [7, 18].

I. S. Kotsireas and P. M. Pardalos (Eds.): LION 14 2020, LNCS 12096, pp. 27–32, 2020.
https://doi.org/10.1007/978-3-030-53552-0_4

These algorithms often demonstrate an amazing performance finding close-to-optimal or even exact solutions to really huge instances of the CVRP coming from practice. Unfortunately, an absence of any theoretical guarantees implies additional computational expenses related to numerical evaluation of their accuracy and possible tuning during the transition to any novel class of instances. In addition, there are known cases when such a tuning is impossible at all, e.g. for the security reasons.

The third research direction is related to the design of approximation algorithms with theoretic performance guarantees and dates back to seminal papers of Haimovich and Rinnooy Kan [10], and Arora [3]. It is known that the CVRP is strongly NP-hard even on the Euclidean plane [20]. The problem is hardly approximable in general case, APX-complete for an arbitrary metric [4] and admits Quasi-Polynomial Time Approximation Schemes (QPTAS) in finite-dimensional Euclidean spaces [9]. For the planar CVRP with restricted capacity growth, there are known several Polynomial Time Approximation Schemes (PTAS), among them the PTAS proposed in [1] is the most general. The approach introduced in [10] is managed to extend to a number of modifications of the planar CVRP including the CVRP formulated in the Euclidean space of any fixed dimension [15,17], the case of multiple depots [12,16], the CVRP with Time Windows [13], and non-unit customer demand [14].

Thus, until now, the class of instances of the metric problems approximable by PTAS or QPTAS was exhausted by the Euclidean settings of the problem except maybe some special cases investigated in [6] Meanwhile, in recent papers by Talwar [24] and Bartal et al. [5] such a class for the closely related Traveling Salesman Problem (TSP) was substantially extended to include the instances of the problem in a metric space of an arbitrary fixed doubling dimension.

In this paper, we propose the first QPTAS for the CVRP formulated in such a space. Our contribution is as follows.

Theorem 1. *For the CVRP in a metric space of an arbitrary doubling dimension $d > 1$, an $(1 + O(\varepsilon))$-approximate solution can be found by a randomized approximation algorithm within time* $\operatorname{poly} n \cdot \left(m^2 n\right)^{m^2 \cdot \operatorname{polylog} n}$, *where* $m = O\left(\left(\frac{d(\log n - \log \varepsilon)}{\varepsilon}\right)^d\right)$. *The algorithm can be derandomized efficiently.*

The rest of the paper is structured as follows. In Sect. 2, we recall the statement of the CVRP. Then, in Sect. 3 we propose a short overview of the proposed approximation scheme. Finally, at Conclusion, we summarize the results obtained and overview some possible directions for the future work.

2   Problem Statement

In the classic Capacitated Vehicle Routing Problem (CVRP), we are given by a set of *customers* $X = \{x_1, \ldots, x_n\}$ having the same unit demand, which should be serviced by a *vehicle* located at some dedicated point y that is called *depot*. All vehicles have the same *capacity* q and visit the customers by cyclic routes, each of them departs from and arrives to the depot y. The goal is to provide a collection of the capacitated routes visiting each customer once and minimizing the total transportation costs.

Let $V = X \cup y$. An instance of the CVRP is specified by a complete undirected edge-weighted graph $G = (V, E, w)$ and an integer $q \geq 3$. The symmetric weighting function $w : E \to R_+$, to any edge $\{u, v\} \in E$, assigns the direct transportation cost $w(u, v)$. A simple cycle $\pi = y, x_{i_1}, x_{i_2}, \ldots, x_{i_s}, y$ in the graph G is referred to a *feasible route*, if it satisfies the capacity constraint, i.e. visits at most q customers. For the route π, its cost $w(\pi) = w(y, x_{i_1}) + w(x_{i_1}, x_{i_2}) + \cdots + w(x_{i_{s-1}}, x_{i_s}) + w(x_{i_s}, y)$. The goal is to find a family of feasible routes $\Pi = \{\pi_1, \ldots, \pi_k\}$ of the least total transportation cost that covers the total customer demand.

In this paper, we consider a restriction of the metric CVRP with the following additional constraints:

(i) for some $d > 1$, the weighting function w is a metric of doubling dimension d, i.e. for an arbitrary $v \in V$ and $R > 0$, there exist nodes $v_1, \ldots, v_M \in V$, such that the metric ball $B(v, R) \subseteq \bigcup_{j=1}^{M} B(v_j, R/2)$ and $M \leq 2^d$.

(ii) the problem is supposed to have an optimal solution, whose number of routes does not exceed polylog n.

3 Approximation Scheme: An Overview

The main idea of our approximation scheme extends the well-known Arora's PTAS for the Euclidean TSP and its generalization proposed in [5] to the TSP in a metric space of any fixed doubling dimension. The scheme consists of the following stages.

Accuracy-Driven Rounding. At this stage, given by $\varepsilon > 0$, to the initial instance, we assign a *rounded* one, such that each s $(1 + \varepsilon)$-approximate solution of the latter instance can be transformed in polynomial time to the appropriate $(1 + O(\varepsilon))$-approximate solution of the former one.

Without loss of generality, we assume that the diameter Δ of the set V is equal to n/ε (since otherwise we can rescale the initial metric by the factor $\frac{n}{\Delta\varepsilon}$). Then, we *round* each customer $x \in X$ to the nearest node $\xi \in X'$, where X' is some metric 1-net of the set X. Finally, we consider an auxiliary instance of the CVRP, specified by the set X' and inheriting all other parameters (y, q, and w) from the initial one. As a result, in the obtained rounded instance, each 'customer' ξ is counted with a multiplicity equal to the number of $x \in X$ assigned to it and, for any distinct 'customers' ξ_1 and ξ_2, $w(\xi_1, \xi_2) > 1$.

Randomized Hierarchical Clustering. Following to [5], we fix a number $s \geq 6$ and put $L = \lceil \log_s(n/\varepsilon) \rceil$. For any level $l = 0, \ldots, L + 1$, we construct an s^{L-l}-net N_l of the set $V' = X' \cup \{y\}$. Without loss of generality, we assume that N_0 is a singleton, $N_L = N_{L+1} = V'$ and $N_l \subset N_{l+1}$ for any l. We proceed with hierarchical clustering of the set V' by induction on l. For $l = 0$, we construct the only cluster $C_1^0 = V'$. Further, let $C_1^{l-1}, \ldots, C_K^{l-1}$ be the partition constructed at level $l - 1$. To proceed at level l, we partition each cluster C_j^{l-1} separately. To make such a partition, we take point by point from the net $\tilde{N}_l$ in a random order σ and, to each net point $\nu_{\sigma(i)}$, we assign a random radius $\eta \in [s^{L-l}, 2s^{L-l})$ from the uniform distribution. Then, the i-th subcluster of the cluster C_j^{l-1} is

$$C_{ji}^l = B(\nu_{\sigma(i)}, \eta) \cap C_j^{l-1} \setminus \bigcup_{t<i} C_{jt}^l.$$

By construction, all clusters at level $L + 1$ are singletons.

Following to [5], our scheme deals with approximate solutions of some special kind, which are referred to as *net-respecting and light*. To define this concept, we choose the number M as some degree of s, such that $M/s < d \cdot L/\varepsilon \leq M$. For any cluster C_j^l, each points from the s^{L-l}/M-net is called *portals*. As it follows from the well-known Packing Lemma (see, e.g. [24]), the number m of portals located in each cluster at an arbitrary level $l > 0$ does not exceed $(8 \cdot M)^d = O\left(\left(\frac{d(\log n - \log \varepsilon)}{\varepsilon}\right)^d\right)$. A route is called *net-respecting* if, for any its edge $\{u, v\}$ of length λ, both points u and v belong to the net N_l, where $s^{L-l} \leq \varepsilon\lambda < s^{L-l+1}$. Further, for some $r > 0$, a net-respecting route is called r-*light*, if it crosses the border of any cluster C_j^l (of any level $l > 0$) at most r times.

As it follows from the Structure Theorem [24], with high probability, for $r = m$, there exists an approximate solution of the CVRP, consisting of net-respecting r-light routes, whose total transportation cost is at most $(1+\varepsilon) \cdot \text{OPT}$. Therefore, to approximate the initial instance within the given accuracy, we can restrict ourselves on such solutions.

Dynamic Programming. For a given randomized clustering, we find the minimum-cost approximate solution consisting of net-respecting r-light routes using the dynamic program as follows. Entries of the DP table are defined by *configurations* that are assigned to each cluster C_j^l. For any cluster C_j^l, an associated configuration $\mathfrak{C}$ is a list of at most polylog n tuples (p_1, p_2, q_j, dep_j), each of them specifies a route segment entering and leaving this cluster at the portals p_1 and p_2 respectively, visiting q_j customers exactly and passing through the depot y or not depending on dep_j.

The table entries are computed bottom-up. Level $L + 1$ is the base case. Each configuration at this level can be computed trivially. Then, let C^l be some cluster at level l, $l = 0, \ldots, L$. To compute any configuration $\mathfrak{C}$ for the cluster C^l, we enumerate all combinations of the feasible configurations $\mathfrak{C}_1, \ldots, \mathfrak{C}_K$ associated with subclusters $C_1^{l+1}, C_2^{l+1}, \ldots, C_K^{l+1}$, $K = 2^{O(d)}$ to find such a combination that is compatible with the configuration $\mathfrak{C}$ and induces the set of route segments crossing the cluster C^l (maybe augmented by some routes contained in this cluster completely) of the minimum total cost. The required solution is obtained by minimization on the set of feasible configurations for the unique cluster at level 0.

Following to the approach proposed in [24], we can show that our algorithm admits an efficient derandomization.

4   Conclusion

In this paper we announce an approximation scheme for the CVRP in the metric space of an arbitrary doubling dimension $d > 1$. Our algorithm is a QPTAS, if the problem has an optimal solution, whose number of routes does not exceed polylog n. It is easy to verify that this condition holds, for instance, when $q = \Omega(n/\text{polylog } n)$. We postpone the proof of Theorem 1 to the forthcoming paper.

Although, to the best of our knowledge, the proposed algorithm appears to be the first approximation scheme for the metric CVRP for the spaces of fixed doubling dimension, the question: *'Can the QPTAS proposed by A.Das and C.Mathieu [9] for the Euclidean CVRP be extended to metric spaces of a fixed doubling dimension without any restriction on the capacity growth?'* still remains open. We'll try to bridge this gap in the future work.

References

1. Adamaszek, A., Czumaj, A., Lingas, A.: PTAS for k-tour cover problem on the plane for moderately large values of k. Int. J. Found. Comput. Sci. **21**(06), 893–904 (2010). https://doi.org/10.1142/S0129054110007623
2. Arnold, F., Sörensen, K.: Knowledge-guided local search for the vehicle routing problem. Comput. Oper. Res. **105**, 32–46 (2019). https://doi.org/10.1016/j.cor.2019.01.002
3. Arora, S.: Polynomial time approximation schemes for euclidean traveling salesman and other geometric problems. J. ACM **45**, 753–782 (1998)
4. Asano, T., Katoh, N., Tamaki, H., Tokuyama, T.: Covering points in the plane by k-tours: towards a polynomial time approximation scheme for general k. In: Proceedings of the Twenty-Ninth Annual ACM Symposium on Theory of Computing, STOC 1997, pp. 275–283. ACM, New York (1997). https://doi.org/10.1145/258533.258602
5. Bartal, Y., Gottlieb, L.A., Krauthgamer, R.: The traveling salesman problem: low-dimensionality implies a polynomial time approximation scheme. SIAM J. Comput. **45**(4), 1563–1581 (2016). https://doi.org/10.1137/130913328
6. Becker, A., Klein, P.N., Schild, A.: A PTAS for bounded-capacity vehicle routing in planar graphs. In: Friggstad, Z., Sack, J.-R., Salavatipour, M.R. (eds.) WADS 2019. LNCS, vol. 11646, pp. 99–111. Springer, Cham (2019). https://doi.org/10.1007/978-3-030-24766-9_8
7. Chen, J., Gui, P., Ding, T., Zhou, Y.: Optimization of transportation routing problem for fresh food by improved ant colony algorithm based on Tabu search. Sustainability **11** (2019). https://doi.org/10.3390/su11236584
8. Dantzig, G., Ramser, J.: The truck dispatching problem. Manag. Sci. **6**, 80–91 (1959)
9. Das, A., Mathieu, C.: A quasipolynomial time approximation scheme for euclidean capacitated vehicle routing. Algorithmica **73**(1), 115–142 (2014). https://doi.org/10.1007/s00453-014-9906-4
10. Haimovich, M., Rinnooy Kan, A.H.G.: Bounds and heuristics for capacitated routing problems. Math. Oper. Res. **10**(4), 527–542 (1985). https://doi.org/10.1287/moor.10.4.527

11. Hokama, P., Miyazawa, F.K., Xavier, E.C.: A branch-and-cut approach for the vehicle routing problem with loading constraints. Expert Syst. Appl. **47**, 1–13 (2016). https://doi.org/10.1016/j.eswa.2015.10.013
12. Khachai, M.Y., Dubinin, R.D.: Approximability of the vehicle routing problem in finite-dimensional euclidean spaces. Proc. Steklov Inst. Math. **297**(1), 117–128 (2017). https://doi.org/10.1134/S0081543817050133
13. Khachay, M., Ogorodnikov, Y.: Efficient PTAS for the euclidean CVRP with time windows. In: van der Aalst, W.M.P., et al. (eds.) AIST 2018. LNCS, vol. 11179, pp. 318–328. Springer, Cham (2018). https://doi.org/10.1007/978-3-030-11027-7_30
14. Khachay, M., Ogorodnikov, Y.: Approximation scheme for the capacitated vehicle routing problem with time windows and non-uniform demand. In: Khachay, M., Kochetov, Y., Pardalos, P. (eds.) MOTOR 2019. LNCS, vol. 11548, pp. 309–327. Springer, Cham (2019). https://doi.org/10.1007/978-3-030-22629-9_22
15. Khachay, M., Dubinin, R.: PTAS for the euclidean capacitated vehicle routing problem in R^d. In: Kochetov, Y., Khachay, M., Beresnev, V., Nurminski, E., Pardalos, P. (eds.) DOOR 2016. LNCS, vol. 9869, pp. 193–205. Springer, Cham (2016). https://doi.org/10.1007/978-3-319-44914-2_16
16. Khachay, M., Ogorodnikov, Y.: Towards an efficient approximability for the euclidean capacitated vehicle routing problem with time windows and multiple depots. IFAC PapersOnline **52**(13), 2644–2649 (2019). https://doi.org/10.1016/j.ifacol.2019.11.606
17. Khachay, M., Zaytseva, H.: Polynomial time approximation scheme for single-depot euclidean capacitated vehicle routing problem. In: Lu, Z., Kim, D., Wu, W., Li, W., Du, D.-Z. (eds.) COCOA 2015. LNCS, vol. 9486, pp. 178–190. Springer, Cham (2015). https://doi.org/10.1007/978-3-319-26626-8_14
18. Nalepa, J., Blocho, M.: Adaptive memetic algorithm for minimizing distance in the vehicle routing problem with time windows. Soft. Comput. **20**(6), 2309–2327 (2016). https://doi.org/10.1007/s00500-015-1642-4
19. Necula, R., Breaban, M., Raschip, M.: Tackling dynamic vehicle routing problem with time windows by means of ant colony system. In: 2017 IEEE Congress on Evolutionary Computation (CEC), pp. 2480–2487 (2017). https://doi.org/10.1109/CEC.2017.7969606
20. Papadimitriou, C.: Euclidean TSP is NP-complete. Theor. Comput. Sci. **4**, 237–244 (1977)
21. Pessoa, A.A., Sadykov, R., Uchoa, E.: Enhanced branch-cut-and-price algorithm for heterogeneous fleet vehicle routing problems. Eur. J. Oper. Res. **270**(2), 530–543 (2018). https://doi.org/10.1016/j.ejor.2018.04.009
22. Polat, O.: A parallel variable neighborhood search for the vehicle routing problem with divisible deliveries and pickups. Comput. Oper. Res. **85**, 71–86 (2017). https://doi.org/10.1016/j.cor.2017.03.009
23. Qiu, M., Fu, Z., Eglese, R., Tang, Q.: A tabu search algorithm for the vehicle routing problem with discrete split deliveries and pickups. Comput. Oper. Res. **100**, 102–116 (2018). https://doi.org/10.1016/j.cor.2018.07.021
24. Talwar, K.: Bypassing the embedding: algorithms for low dimensional metrics. In: Proceedings of the Thirty-Sixth Annual ACM Symposium on Theory of Computing, STOC 2004, pp. 281–290. Association for Computing Machinery, New York (2004). https://doi.org/10.1145/1007352.1007399
25. Toth, P., Vigo, D.: Vehicle Routing: Problems, Methods, and Applications. MOS-SIAM Series on Optimization, 2nd edn. SIAM, Philadelphia (2014)

Early Detection of Eating Disorders Through Machine Learning Techniques

Annabella Astorino[1]([✉])(iD), Rosa Berti[2], Alessandro Astorino[3],
Vincent Bitonti[3], Manuel De Marco[3], Valentina Feraco[3], Alexei Palumbo[3],
Francesco Porti[3], and Ilario Zannino[3]

[1] ICAR-CNR, 87036 Rende, Italy
`annabella.astorino@icar.cnr.it`
[2] IIS-Liceo Scientifico, 87055 San Giovanni in Fiore, Italy
`rosa.berti@istruzione.it`
[3] IIS-Liceo Scientifico – Classe V B - 2019/20, 87055 San Giovanni in Fiore, Italy

Abstract. In this work we analyze the relationships between nutrition, health status and well-being of the individual in evolutionary age, not only in consideration of the high prevalence of excess weight and the early appearance of metabolic pathologies, but also due to the significant presence of Eating Disorders (EDs). EDs, in fact, continue to be under-diagnosed by pediatric professionals and many adolescents go untreated, do not recover or reach only partial recovery.

We have observed the situation of young people at an Italian High School regarding EDs by carrying out a statistical survey on the students in relation to dietary habits, attitudes towards food and physical activity.

Finally, the collected data have been analyzed through statistical and machine learning techniques.

Keywords: Eating disorders detection · Psychometric test · Information extraction · SVM

1 Introduction

According to American Psychiatric Association [1], Eating Disorders (EDs) are *"illnesses in which the people experience severe disturbances in their eating behaviors and related thoughts and emotions"*. The relative medical complications are widespread and really serious because can affect every organ system. Then EDs are psychological problems, but if not properly treated they can lead to hospital admissions and in extreme cases to death. These disorders affect several million people at any given time and the most common EDs include anorexia nervosa, bulimia nervosa and binge eating disorder.

Anorexia nervosa is an eating disorder characterized by food restriction, fear of gaining weight and an anomaly in perceiving one's weight. It represents the

© Springer Nature Switzerland AG 2020
I. S. Kotsireas and P. M. Pardalos (Eds.): LION 14 2020, LNCS 12096, pp. 33–39, 2020.
https://doi.org/10.1007/978-3-030-53552-0_5

psychiatric illness with the highest mortality rate and, in female adolescents, it is the second leading cause of death after road accidents.

Bulimia nervosa is a mental disorder characterized by excessive and constant preoccupation with weight and shape, so the person starts to follow a strict diet, but presenting binges and self-induced vomiting. By vomiting or thought other compensation methods, bulimics believe they can reach their ideal form and at the same time be able to satisfy their need for food with binges.

People with anorexia nervosa and bulimia nervosa tend to be perfectionists with low self-esteem and are extremely critical of themselves and their bodies. They usually feel fat and see themselves as overweight, sometimes even despite malnutrition. An intense fear of gaining weight and of being fat may become all-pervasive. In early stages of these disorders, patients often deny that they have a problem [18].

People with binge eating disorder (BED) lose control over their eating. The BED is similar to bulimia with regard to the compulsion to binge eating, but differs from it because the periods of binge-eating are not followed by purging, excessive exercise, or fasting. As a result, people with BED often are overweight or obese, they live incredible feelings of intense guilt, shame and hate themselves after the binges.

Eating disorders can affect people of all ages, racial backgrounds, body weights and genders. They frequently appear during the teen years or young adulthood but may also develop during childhood or later in life [4]. These disorders affect both genders, although rates among women are higher than among men. Like women who have eating disorders, men also have a distorted sense of body image [12,17].

In many cases, eating disorders occur together with other psychiatric disorders like anxiety, panic, obsessive compulsive disorder and alcohol and drug abuse problems. There is no single cause of an eating disorder, although neurobiological and genetic predispositions are emerging as important. Without treatment of both the emotional and physical symptoms of these disorders, malnutrition, heart problems and other potentially fatal conditions can result. However, with proper medical care, people with EDs can resume suitable eating habits, and return to better emotional and psychological health. Then the early detection [11] of EDs together with an equally early intervention can be crucial for good prognosis: primary care providers can be key players in treatment success. It is therefore very important to identify good diagnostic criteria [16].

Tacking into account these considerations, in this work we analyze the relationships between nutrition, health status and well-being of the individual in evolutionary age with the following objectives:

1. to evaluate the prevalence of the risk condition for EDs distinguishing between male and female subjects;
2. to identify possible psychological elements most associated with the EDs condition, including distortion of body image;
3. to evaluate the relationship between the risk condition for EDs and some strategies for changing one's body.

In particular, we have observed the situation of young people at "I.I.S. – Liceo Scientifico – San Giovanni in Fiore (CS), Italy" High School regarding EDs, by carrying out a statistical survey on the students in relation to dietary habits, attitudes towards food and physical activity. Finally, the collected data have been analyzed through statistical and machine learning techniques [15].

The rest of the paper is organized as follows. In Sect. 2 we present our statistical survey, whereas in Sect. 3 we explain the numerical experiments by providing and analyzing the related results. Some conclusions are drawn in Sect. 4.

2 Statistical Survey

2.1 Participants and Procedure

In the school year 2018/19, we promoted a statistical survey on the students of "I.I.S. – Liceo Scientifico – San Giovanni in Fiore (CS), Italy" in relation to dietary habits, attitudes towards food and physical activity.

They were asked to complete, during school hours, an anonymous web-based questionnaire. More precisely, for the survey we have used tests of socio-demographic type, on physical activity and of psychometric type.

Data management was carried out in compliance with the Italian privacy law, in fact informed consent was signed by all participants and by at least one of the parents.

2.2 Socio-Demographic Form

We have elaborated a socio-demographic test to obtain information on students and their families of origin. It consists of 29 questions, which concern age, height, weight, weight problems and previous diets of the students, and composition, level of education, type of work, dietary habits of their families of origin. The purpose of the questionnaire is to provide information on the socio-demographic characteristics of the selected cohort.

2.3 Physical Activity Form

This form has been elaborated by using the questionnaire IPAQ (International Physical Activity Questionnaire) [10]. The questions refer to the activity carried out in the last seven days at school, to move from one place to another and during free time. The purpose of the questionnaire is to provide information on the amount and types of physical activities that people perform as part of daily life.

2.4 Psychometric Forms

Psychometric tests represent a systematic procedure for bringing out particular responses by subjects when a set of stimuli (questions, problems, tasks) is presented to them. These responses can be evaluated and quantitatively interpreted based on specific criteria or performance standards.

In our survey on EDs we have considered the following questionnaires:

- SCOFF (Sick, Control, One, Fat, Food) [13];
- BUTa (Body Uneasiness Test)[5,8];
- EAT6 (Eating Attitude Test 26) [14].

SCOFF (Sick, Control, One, Fat, Food)
The SCOFF questionnaire is a brief tool designed to early detect eating disorders and aid their treatment. It consists of 5 items to which you can answer "yes" or "no"; one point is given for every "yes"; a score greater or equal than 2 indicates a likely case of anorexia nervosa or bulimia. This test is designed to raise suspicion of a likely case rather than to diagnose.

BUTa (Body Uneasiness Test)
It is a test built for the psychometric evaluation of abnormal body image attitudes and eating disorders. It consists of 34 clinical items for measuring weight phobia, body image concerns, avoidance, compulsive self-monitoring, detachment and estrangement feelings towards one's own body (depersonalization). The average rating of all 34 items represents the GSI (Global Severity Index) related to own body image: the gravity is expressed on a scale from 0 to 5, where 0 corresponds to the absence of problems in that sector and 5 to the maximum gravity.

EAT6 (Eating Attitude Test 26)
It is a tool that measures pathological eating behaviors and the psychological attitude towards food and weight, typical of anorexia nervosa. The test consists of 26 items with multiple choice answers: Always; Very often; Often; Sometimes; Rarely; Never.

For all items, except for 25-th ("Have the impulse to vomit after meals?"), each of the answers receives the following value: Always $= 3$; Usually $= 2$; Often $= 1$; Sometimes $= 0$; Rarely $= 0$; Never $= 0$.

For 25-th item, the responses receive these values: Always $= 0$; Usually $= 0$; Often $= 0$; Sometimes $= 1$; Rarely $= 2$; Never $= 3$.

The scores of each item are added together. The subject obtaining a score over to 20 is considered "at risk" for eating disorders.

3   Numerical Experiments

The statistical survey has involved all students, from classes I to classes V, of the "I.I.S. – Liceo Scientifico – San Giovanni in Fiore (CS), Italy", for a total number, N, equal to 192, of which 97 males and 95 females. The average age of the sample has been 17 years.

By psychometric forms, we have obtained the following results:

- 44% of the sample tested positive for SCOFF,
- 12% of the sample tested positive for EAT26,
- 17% of the sample tested positive for BUTa.

As regards the first objective of the present study, i.e. to evaluate the prevalence of the risk condition for eating behavior disorders (EDs) distinguishing between male and female subjects, in the sample under examination ($N = 192$) 16 people (8% of the sample) obtained a pathological score on the SCOFF, EAT26 and BUTa scales: 8% of the sample is therefore at risk for EDs and, of these 16 people, only one is a male subject.

As for the second objective, the results of our analysis show that the subjects at risk for EDs come from families with a medium-high education level. In fact, of the 16 people at risk for EDs, 15 ones have a mother with a diploma (56%) or a university degree (38%) and 11 ones have a father with a diploma (33%), a university degree (33%) or a master degree (7%).

Finally, from the analysis of the data, some psychological characteristics of the subjects, such as perfectionism, have been also important for the risk of EDs: indeed, of the 16 people at risk for eating disorders, 12 ones were "very active" (75%), according to the IPAQ scale.

These considerations have been confirmed by the results obtained by addressing the problem of EDs detection through machine learning techniques. In particular, we have analyzed the problem of discriminating between the students "at risk" for EDs (positive for all the three used psychometric tests) and those "not at risk" (negative for at least one of the psychometric tests) by selecting only those parameters concerning sex, age and physical activity carried out by the subjects under observation, and educational level and type of employment of their parents, for a total of 11 features, including the class label (1 for students at risk and -1 for those not at risk). In this parameter space, we have used, as supervised classification technique, the SVM approach [7], more precisely the Libsvm code [6] under Weka [9], an open source Java based platform containing various machine learning algorithms.

In our experiments, in order to work with a balanced dataset, we have considered three possible configurations for it. Each configuration consists of 36 samples: all the 16 students at risk for EDs and 20 random students not at risk for EDs. Moreover, for each dataset configuration, we have performed a leave-one-out cross-validation. The results are listed in Table 1, where we report the average of the following standard quantities: Testing Correctness, True Positive Rate, False Positive Rate, Precision and F-Score.

Table 1. Classifier: LibSVM – C 100.0/polynomial kernel of degree 2

Dataset configuration	Testing correctness	TP rate	FP rate	Precision	F-Score	Class
1	86.11%	0.81	0.10	0.87	0.84	1
		0.90	0.19	0.86	0.88	−1
2	91.67%	1.00	0.15	0.84	0.91	1
		0.85	0.00	1.00	0.92	−1
3	86.11%	0.93	0.20	0.79	0.86	1
		0.80	0.06	0.94	0.87	−1

These results appear interesting, in fact they show that it is possible with good accuracy to classify subjects at risk for EDs, on which it is necessary to pay more attention and a more in-depth study, only by analyzing a few parameters of socio-demographic and psychological type.

4   Conclusions

In this paper we have presented an application of a supervised classification approach by analyzing the relationships between nutrition, health status and well-being of the individual in evolutionary age with the principal objective of identifying possible psychological elements most associated with the EDs condition. We have obtained promising results, both in terms of classification accuracy and sensitivity. After this first analysis, the next step of our research will be to expand the sample and to use other classification methods [2,3] in addition to SVM with the aim of the early detection of EDs signs on social media, so largely used by young people.

References

1. American Psychiatric Association: Diagnostic and Statistical Manual of Mental Disorders: DSM-5, 5th edn, Washington, DC (2013)
2. Astorino, A., Fuduli, A., Gorgone, E.: Non-smoothness in classification problems. Optim. Methods Softw. **23**(5), 675–688 (2008)
3. Astorino, A., Gaudioso, M., Seeger, A.: Conic separation of finite sets I. The homogeneous case. J. Convex Anal. **21**(1), 1–28 (2014)
4. Campbell, K., Peebles, R.: Eating disorders in children and adolescents: state of the art review. Pediatrics **134**(3), 582–592 (2014)
5. Carta, I., Zappa, L., Garghentini, G., Caslini, M.: Body image: a preliminary study of the administration of the body uneasiness test (BUT) to investigate specific features of eating disorders, anxiety, depression, and obesity. Ital. J. Psychopathol. **14**(1), 23–28 (2008)
6. Chang, C.C., Lin, C.J.: LIBSVM: a library for support vector machines. ACM Trans. Intell. Syst. Technol. **2**, 27:1–27:27 (2011)
7. Cristianini, N., Shawe-Taylor, J.: An Introduction to Support Vector Machines and Other Kernel-Based Learning Methods. Cambridge University Press, Cambridge (2000)
8. Cuzzolaro, M., Vetrone, G., Marano, G., Garfinkel, P.: The body uneasiness test (BUT): development and validation of a new body image assessment scale. Eat. Weight Disord. **11**(1), 1–13 (2006). https://doi.org/10.1007/BF03327738
9. Hall, M., Frank, E., Holmes, G., Pfahringer, B., Reutemann, P., Witten, I.H.: The WEKA data mining software: an update. SIGKDD Explor. **11**(1), 10–18 (2009)
10. Lee, P., Macfarlane, D., Lam, T., Stewart, S.: Validity of the international physical activity questionnaire short form (IPAQ-SF): a systematic review. Int. J. Behav. Nutr. Phys. Act. **8** (2011). https://doi.org/10.1186/1479-5868-8-115
11. Losada, D.E., Crestani, F., Parapar, J.: Early detection of risks on the internet: an exploratory campaign. In: Azzopardi, L., Stein, B., Fuhr, N., Mayr, P., Hauff, C., Hiemstra, D. (eds.) ECIR 2019. LNCS, vol. 11438, pp. 259–266. Springer, Cham (2019). https://doi.org/10.1007/978-3-030-15719-7_35

12. Moretti, P., Fontana, F., Eusebi, P., La Ferla, T.: Males eating disorders: importance of body image distortion. Ital. J. Psychopathol. **16**(2), 150–156 (2010)
13. Morgan, J., Reid, F., Lacey, J.: The SCOFF questionnaire: assessment of a new screening tool for eating disorders. Br. Med. J. **319**(7223), 1467–1468 (1999)
14. Olmsted, M., Bohr, Y., Garfinkel, P.: The eating attitudes test: psychometric features and clinical correlates. Psychol. Med. **12**(4), 871–878 (1982)
15. Paul, S., Kalyani, J., Basu, T.: Early detection of signs of anorexia and depression over social media using effective machine learning frameworks. In: CEUR Workshop Proceedings, vol. 2125 (2018)
16. Ramírez-Cifuentes, D., Mayans, M., Freire, A.: Early risk detection of anorexia on social media. In: Bodrunova, S.S. (ed.) INSCI 2018. LNCS, vol. 11193, pp. 3–14. Springer, Cham (2018). https://doi.org/10.1007/978-3-030-01437-7_1
17. Wyssen, A., Bryjova, J., Meyer, A., Munsch, S.: A model of disturbed eating behavior in men: the role of body dissatisfaction, emotion dysregulation and cognitive distortions. Psychiatry Res. **246**, 9–15 (2016)
18. Zamariola, G., Cardini, F., Mian, E., Serino, A., Tsakiris, M.: Can you feel the body that you see? On the relationship between interoceptive accuracy and body image. Body Image **20**, 130–136 (2017)

On Finding Minimum Cardinality Subset of Vectors with a Constraint on the Sum of Squared Euclidean Pairwise Distances

Anton V. Eremeev[1]([⊠]) [iD], Mikhail Y. Kovalyov[3] [iD], and Artem V. Pyatkin[1,2] [iD]

[1] Sobolev Institute of Mathematics, Novosibirsk, Russia
`eremeev@ofim.oscsbras.ru`
[2] Novosibirsk State University, Novosibirsk, Russia
[3] United Institute of Informatics Problems, Minsk, Belarus

Abstract. In this paper, we consider the problem of finding a minimum cardinality subset of vectors, given a constraint on the sum of squared Euclidean distances between all vectors of the chosen subset. This problem is closely related to the well-known Maximum Diversity Problem. The main difference consists in swapping the constraint with the optimization criterion. We prove that the problem is NP-hard in the strong sense. An exact algorithm for solving this problem is proposed. The algorithm has a pseudo-polynomial time complexity in the special case of the problem, where the dimension of the space is bounded from above by a constant and the input data are integer.

Keywords: Euclidean space · Subset of points · NP-hardness · Integer instance · Exact algorithm · Pseudo-polynomial time

1 Introduction

In this paper, we study a discrete extremal problem of searching a subset of vectors with minimum cardinality, given a constraint on the sum of squared Euclidean distances between all vectors of the chosen subset. We abbreviate this problem as MCSED and formulate it as follows.

Given: a set $\mathcal{Y} = \{y_1, \ldots, y_N\}$ of vectors from $\mathbb{R}^k$ and positive number a.
Find: a subset $\mathcal{C}^* \subseteq \mathcal{Y}$ of minimum cardinality such that

$$\sum_{y \in \mathcal{C}^*} \sum_{z \in \mathcal{C}^*} ||z - y||^2 \geq a, \tag{1}$$

where $|| \cdot ||$ is the Euclidean norm. In what follows, we denote $h(C) := \sum_{y \in C} \sum_{z \in C} ||z - y||^2$ for any subset $C \subseteq \mathcal{Y}$.

If one interprets the diversity of a finite set in $\mathbb{R}^k$ as the sum of squared Euclidean distances between all points of the set, then the MCSED problem has a clear interpretation in terms of computational geometry. This problem asks

I. S. Kotsireas and P. M. Pardalos (Eds.): LION 14 2020, LNCS 12096, pp. 40–45, 2020.
https://doi.org/10.1007/978-3-030-53552-0_6

for a minimum cardinality subset in the given finite set, such that the diversity of the chosen subset is at least a.

If the given vectors of the Euclidean space correspond to people so that the coordinates of vectors are equal to some characteristics of these people, then the MCSED problem may be treated as a problem of finding a sufficiently diverse group of people of minimum size.

MCSED problem is closely related to the Maximum Diversity Problem (see e.g. [2,3]). The main difference consists in swapping the constraint with the optimization criterion.

Our goals in this paper are to develop an exact algorithm for the special case of the MCSED problem where all vectors have integer coordinates, and to analyse the complexity of this problem.

2 Problem Complexity

It is known [8] that the classic NP-complete Independent Set problem [4] remains NP-complete for regular graphs:

Independent Set in a Regular Graph. Given a regular graph G of degree d and a positive integer m, find whether this graph contains a vertex subset of cardinality m such that every two vertices of this subset are not connected by an edge.

The following proof of intractability of the MCSED problem is based on a polynomial-time reduction of the Independent Set problem to the decision version of MCSED problem:

Given: A set $\mathcal{Y} = \{y_1, \ldots, y_N\}$ of Euclidean points from $\mathbb{R}^k$ and two positive integers M and K. *Question*: Is there a nonempty subset $\mathcal{C} \subseteq \mathcal{Y}$ such that

$$|\mathcal{C}| \leq M, \text{ and } h(\mathcal{C}) \geq K?$$

Theorem 1. *The MCSED problem is NP-hard in the strong sense.*

Proof. Suppose that an instance of the Independent Set problem is given by an integer m and a regular graph G of degree d with N vertices and $q = dN/2$ edges.

Construct the following instance of MCSED problem in the decision form. Put $k = q$ and assign to every vertex of the graph G a k-dimensional vector y whose i-th coordinate is 1 if the edge i is incident with this vertex, and is 0 otherwise. Then, for a pair of vectors y and z from $Y = \{y_1, \ldots, y_N\}$, clearly, $\|y - z\|^2 = 2d - 2$, if the vertices of G corresponding to y and z are adjacent, and $\|y - z\|^2 = 2d$, otherwise. Let $I(C)$ denote the set of vertices in G, that correspond to a subset of vectors C. Note that $I(C)$ is an independent set if $h(C) = 2d|C|(|C| - 1)$, and otherwise it holds that $h(C) < 2d|C|(|C| - 1)$. Besides that, obviously, $|I(C)| = |C|$. Therefore, the decision form of the MCSED problem with $M = m$ and $K = 2dm(m - 1)$ has a positive answer iff G contains an independent set of size exactly M.   □

The problem MCSED is NP-hard in the ordinary sense even for the three-dimensional case, which follows from the proof of Theorem 8 in [1], provided that this proof is modified by making all numbers integer or rational. The square root function, which is used in this proof, would have to be replaced by an appropriate approximation. The pseudo-polynomial algorithm in Sect. 4 below demonstrates that this problem in any fixed dimension cannot be NP-hard *in the strong sense*, if $P \neq NP$.

A feasible solution x for a minimization problem is called an $(1 + \varepsilon)$-*approximate* solution if for some given $\varepsilon > 0$ it satisfies the inequality $f(x) \leq (1 + \varepsilon)f^*$, where f^* is the objective function optimal value. An algorithm is called an $(1 + \varepsilon)$-approximation *algorithm* if in polynomial time it outputs an $(1 + \varepsilon)$-approximate solution for every solvable problem instance. A family of $(1 + \varepsilon)$-approximation algorithms parameterized by $\varepsilon > 0$, such that the time complexity of any of these algorithms is polynomially bounded by $1/\varepsilon$ and the problem instance length in binary encoding is called *a fully polynomial-time approximation scheme (FPTAS)*.

The objective function of the MCSED problem is integer valued. The objective function optimal value is polynomially bounded in binary incoding length. Therefore, see e.g. corollary of Theorem 6.8 in [4], the MCSED problem does not admit an FPTAS, unless $P = NP$. Note that the proof of Theorem 6.8 in [4] may be easily modified to show that any NP-hard (in the ordinary sense) optimization problem with a polynomially-bounded integer-valued objective function does not admit an FPTAS, unless $P = NP$. Therefore, the above-mentioned result from [1] implies that the there is no FPTAS for the MCSED problem even for any constant dimensionality $k \geq 3$, if $P \neq NP$.

3   Alternative Problem Formulation

Let $\bar{z}(C) := \frac{1}{|C|} \sum_{z \in C} z$ denote the centroid of a set $C \subseteq \mathcal{Y}$. The following "folklore" result is well-known.

Lemma 1. *For an arbitrary point $x \in \mathbb{R}^k$ and a finite set $C \subset \mathbb{R}^k$, it holds that*

$$\sum_{z \in C} ||z - x||^2 = \sum_{z \in C} ||z - \bar{z}(C)||^2 + |C| \cdot ||x - \bar{z}(C)||^2. \tag{2}$$

This observation turned to be useful in obtaining a number of results for similar problems [5–7]. The following equality (3) related to the function $h(C)$ follows from Lemma 1 by summation over $x \in C$.

$$f(C) := \sum_{z \in C} ||z - \bar{z}(C)||^2 = \frac{1}{2|C|} \sum_{z \in C} \sum_{y \in C} ||z - y||^2 = \frac{1}{2|C|} h(C). \tag{3}$$

The next lemma provides an equivalent formulation of the MCSED problem. Here we use the integer programming formulation, with one integer variable μ

and N binary variables $x_1, \ldots, x_N$, which give a natural representation of a subset $C \subseteq Y$, assuming $x_j = 1$ if point $y_j \in C$ and $x_j = 0$ otherwise, $j = 1, \ldots, N$:

$$\min \mu, \tag{4}$$

subject to

$$\sum_{j=1}^{N} x_j = \mu, \tag{5}$$

$$2\mu \sum_{j=1}^{N} \sum_{r=1}^{k} x_j \left(a_{rj} - \frac{\sum_{i=1}^{N} a_{ri} x_i}{\mu} \right)^2 \geq a, \tag{6}$$

$$\mu \in \mathbb{Z}, \ x_j \in \{0, 1\}, \ j = 1, \ldots, N. \tag{7}$$

Let $x := (x_1, \ldots, x_N)$. In what follows, for any $m \in \{1, \ldots, N\}$ we denote

$$f_N(m, x) := \max \sum_{j=1}^{N} \sum_{r=1}^{k} x_j \left(a_{rj} - \frac{\sum_{i=1}^{N} a_{ri} x_i}{m} \right)^2.$$

Lemma 2. [3] *If* $\sum_{j=1}^{N} x_j = m$ *then*

$$f_N(m, x) = \sum_{j=1}^{N} \frac{x_j}{m} \sum_{r=1}^{k} \left((m-1) a_{rj}^2 - 2 a_{rj} \sum_{i=1}^{j-1} a_{ri} x_i \right). \tag{8}$$

4   A Pseudo-Polynomial Time Algorithm for Bounded Dimension of Space

In this section, we show that in the case of a fixed space dimension k and integer coordinates of vectors from Y, the MCSED problem can be solved in a pseudo-polynomial time using the same approach as proposed in [3].

Consider the integer programming formulation (4)–(7) and introduce the following functions that evaluate partial sums in (8):

$$s_j(m, x_1, \ldots, x_j) = \frac{x_j}{m} \sum_{r=1}^{k} \left((m-1) a_{rj}^2 - 2 a_{rj} \sum_{i=1}^{j-1} a_{ri} x_i \right), \ j = 1, \ldots, N.$$

Then we have $f_N(m, x) = \sum_{j=1}^{N} s_j(m, x_1, \ldots, x_j)$.

Let $F_j(m, A_1, \ldots, A_k)$ be the maximum diversity for partial solutions $(x_1, \ldots, x_j)$ such that exactly $m \leq N$ components among $x_1, \ldots, x_j$ are equal to 1 and $\sum_{i=1}^{j-1} a_{ri} x_i = A_r$ for every $r = 1, \ldots, k$. Formally,

$$F_j(m, A_1, \ldots, A_k) = \max \sum_{i=1}^{j} s_i(x_1, \ldots, x_i), \tag{9}$$

subject to

$$\sum_{i=1}^{j} x_j = m, \tag{10}$$

$$\sum_{i=1}^{j-1} a_{ri} x_i = A_r, \quad r = 1, \ldots, k, \tag{11}$$

$$x_i \in \{0, 1\}, i = 1, \ldots, j. \tag{12}$$

In the case of integer inputs $y_j \in \mathbb{Z}^k$, $j = 1, \ldots, N$, the partial sums $\sum_{i=1}^{j-1} a_{ri} x_i$ take only integer values from $[-B, B]$ (recall that B denotes the maximum absolute coordinate value in the input set) and we have the Bellman Equation:

$$F_j(m, A_1, \ldots, A_k) =$$

$$\begin{cases} F_{j-1}(m, A_1, \ldots, A_k), & \text{if } x_j = 0, \\[2mm] \max_{x_j \in \{0,1\}} \begin{aligned} &F_{j-1}(m-1, A_1 - a_{1j}, \ldots, A_k - a_{kj}) \\ &+ \frac{1}{m} \sum_{r=1}^{k} \left((m-1) a_{rj}^2 - 2 a_{rj}(A_r - a_{rj}) \right), \end{aligned} & \text{otherwise.} \end{cases}$$

Our exact algorithm for instances with integer coordinates of the input points works as follows. Put $A := \max_{1 \le r \le k} \sum_{j=1}^{N} a_{rj} \le BN$. First, compute recursively the set of values $F_j(m, A_1, \ldots, A_k)$ for all $j = 1, \ldots, N$, $m = 1, \ldots, N$, and $A_1 = -A, \ldots, A$, where they are defined (otherwise assume $F_j(m, A_1, \ldots, A_k) = -\infty$). Note that the set of binary vectors $x_1, \ldots, x_j$ corresponding to each $F_j(m, A_1, \ldots, A_k)$ can be easily back-tracked. Then, compute

$$\min \left\{ m : \sum_{j=1}^{N} x_j = m, \ \sum_{j=1}^{N} s_j(m, x_1, \ldots, x_j) \ge \frac{a}{2m}, \ x \in \{0, 1\}^N \right\}$$

$$= \min \left\{ m : \max_{(A_1, \ldots, A_k) \in [-A, A]^k} F_N(m, A_1, \ldots, A_k) \ge \frac{a}{2m} \right\}, \tag{13}$$

backtrack to find the vector x^*, corresponding to the minimum in (13) and output a subset $C = \cup_{j : x_j^* = 1} \{y_j\}$ as a solution to the problem. This algorithm is called Algorithm DP in what follows. Lemma 3 below establishes a relation between optimal and algorithmic solutions.

Lemma 3. *Suppose that components of all points in Y are integers from the interval $[-B, B]$. Then the Algorithm DP finds an optimal solution to the MCSED problem.*

The following theorem establishes the time complexity of the Algorithm DP.

Theorem 2. *If components of all points in Y have integer values in the interval $[-B, B]$, then the running time of Algorithm DP is $\mathcal{O}(N^2(2BN+1)^k)$.*

Algorithm DP is pseudo-polynomial for a fixed space dimension k since the time complexity of this algorithm is $\mathcal{O}(N^2(BN)^k)$ that is polynomially bounded in terms of problem dimension N and the value of B.

5   Conclusions

The problem of finding a minimum cardinality subset of vectors, given a constraint on the sum of squared Euclidean distances between all vectors of the chosen subset is considered for the first time. It is shown that this problem is NP-hard in the strong sense and an exact dynamic programming algorithm for solving this problem is proposed. We prove a pseudo-polynomial time complexity bound for this algorithm in the special case, where the dimension of the space is bounded from above by a constant and the input data are integer. The computational complexity of the studied problem in one- and two-dimensional cases remains an open question.

Acknowledgements. The authors are grateful to Alexander Kel'manov for suggesting the research direction and to Yulia Kovalenko for her helpful comments. The study presented in Sect. 2 was supported by the RFBR grant 19-01-00308, the study presented Sect. 3 was supported by the Russian Ministry of Science and Education under the 5-100 Excellence Programme, the study presented Sect. 4 was supported by the Russian Academy of Sciences (the Program of basic research), project 0314-2019-0019.

References

1. Cevallos, A., Eisenbrand, F., Morell, S.: Diversity maximization in doubling metrics. In: 29th International Symposium on Algorithms and Computation, ISAAC 2018, pp. 33:1–33:12. Schloss Dagstuhl-Leibniz-Zentrum fuer Informatik, Dagstuhl, Germany (2018). Article No. 33
2. Cevallos, A., Eisenbrand, F., Zenklusen, R.: Max-sum diversity via convex programming. In: 32nd Annual Symposium on Computational Geometry (SoCG). LIPIcs, vol. 51, pp. 26:1–26:14. Schloss Dagstuhl-Leibniz-Zentrum fuer Informatik, Dagstuhl, Germany (2016). https://doi.org/10.4230/LIPIcs.SoCG.2016.26
3. Eremeev, A., Kel'manov, A., Kovalyov, M., Pyatkin, A.: Maximum diversity problem with squared Euclidean distance. In: Khachay, M., Kochetov, Y., Pardalos, P. (eds.) Mathematical Optimization Theory and Operations Research, MOTOR 2019. LNCS, vol. 11548, pp. 541–551. Springer, Cham (2019)
4. Garey, M.R., Johnson, D.S.: Computers and Intractability. A Guide to the Theory of NP-Completeness. W.H. Freeman and Company, San Francisco (1979)
5. Kel'manov, A.V., Romanchenko, S.M.: Pseudopolynomial algorithms for certain computationally hard vector subset and cluster analysis problems. Autom. Remote Control **73**(2), 349–354 (2012)
6. Kel'manov, A.V., Romanchenko, S.M.: An FPTAS for a vector subset search problem. J. Appl. Ind. Math. **8**(3), 329–336 (2014). https://doi.org/10.1134/S1990478914030041
7. Kel'manov, A.V., Motkova, A.V., Shenmaier, V.V.: An Approximation Scheme for a Weighted Two-Cluster Partition Problem. LNCS, vol. 10716, pp. 323–333 (2018)
8. Papadimitriou, C.H.: Computational Complexity. Addison-Wesley, New York (1994)

Practical Approximation Algorithms for Stabbing Special Families of Line Segments with Equal Disks

Konstantin Kobylkin[1,2]($\boxtimes$)(iD) and Irina Dryakhlova[2](iD)

[1] Krasovsky Institute of Mathematics and Mechanics, Ural Branch of RAS,
Sophya Kovalevskaya str. 16, 620108 Ekaterinburg, Russia
[2] Ural Federal University, Mira str. 19, 620002 Ekaterinburg, Russia
kobylkinks@gmail.com, iadryahlova@gmail.com

Abstract. An NP-hard problem is considered of stabbing a given set of n straight line segments on the plane with the least size subset of disks of fixed radii $r > 0$, where the set of segments forms a straight line drawing $G = (V, E)$ of a planar graph without proper edge crossings. Two $O\left(n^{3/2}\log^2 n\right)$-expected time algorithms are proposed for this stabbing problem considered on sets of segments, forming edge sets of special plane graphs, which are of interest in network applications. Namely, a 12-approximate algorithm is given for the problem, considered on edge sets of relative neighborhood graphs and 14-approximate algorithm is designed for edge sets of Gabriel graphs. The paper extends recent work where $O(n^2)$-time approximation algorithms are proposed with the same constant approximation factors for the problem on those two classes of sets of segments.

Keywords: Operations research · Computational geometry · Approximation algorithms · Straight line segment · Hippodrome

1 Introduction

Facility location represents an important class of practical problems from operations research, which can adequately be modeled by combinatorial optimization problems. A geometric modelling approach turns out to be successful for a variety of facility location problems where objects of interest, say, customers, roads, markets or inventories are geographically distributed. Under this approach an optimal placement of facilities (e.g of inventories, markets, petrol or charging stations) nearby objects of interest is to be found. Here objects of interest are modelled by simple geometric structures, which can be e.g. points, straight line segments or rectangles on the plane whereas locations of facilities are given by

This work was supported by Russian Foundation for Basic Research, project 19-07-01243.

I. S. Kotsireas and P. M. Pardalos (Eds.): LION 14 2020, LNCS 12096, pp. 46–51, 2020.
https://doi.org/10.1007/978-3-030-53552-0_7

translates of simple objects like unit disks, axis-parallel squares or rectangles. In its simplest form optimization is done over placement of facilities to achieve the minimum total distance from the placed facilities to the objects of interest. Alternatively, it might be aimed at minimizing total number of the placed facilities while serving needs of all objects of interest.

This alternative type of problems is of the form of a simple to formulate problem from computational geometry: given a set $\mathcal{K}$ of geometric objects on the plane, the smallest cardinality set $\mathcal{C}$ of objects is to be found on the plane, chosen from a class $\mathcal{F}$ of simply shaped objects, such that each object from $\mathcal{K}$ is intersected by an object from $\mathcal{C}$ in some prescribed way. In this paper, subquadratic time small constant factor approximation algorithms are designed for the following problem in which $\mathcal{F}$ is a set of radius r disks and $\mathcal{K}$ coincides with a finite set E of straight line segments on the plane.

INTERSECTING PLANE GRAPH WITH DISKS (IPGD): given a straight line drawing (or a plane graph) $G = (V, E)$ of an arbitrary simple planar graph without proper edge crossings and a constant $r > 0$, find the smallest cardinality set $\mathcal{C}$ of disks of radius r such that $e \cap \bigcup_{C \in \mathcal{C}} C \neq \varnothing$ for each edge $e \in E$. Here each isolated vertex $v \in V$ is treated as a zero-length segment $e_v \in E$. Moreover, the vertex set V is assumed to be in general position, i.e. no triple of points of V lies on any straight line.

Below the term "plane graph" is used to denote any straight line embedding of a planar graph whose (straight line) edges intersect at most at their endpoints.

The IPGD problem finds its applications in sensor network deployment and facility location. Suppose one needs to locate petrol or charging stations nearby all roads of a given road network. Geometrically, the network roads can be modeled by piecewise linear arcs on the plane. One can split these arcs into chains of elementary straight line segments such that any two of the resulting elementary segments intersect at most at their endpoints. To cover the road network with facility stations to some extent, it might be reasonable to place the minimum number of stations such that each piece of every road (represented by an elementary segment) is within a given distance from some of the placed stations. This modeling approach leads to a geometric combinatorial optimization model, which coincides with the IPGD problem.

The IPGD problem generalizes a classical NP-hard unit disk covering problem. In the unit disk covering problem one needs to cover a given finite point set E on the plane with the least cardinality set $\mathcal{C}$ of unit disks. In the IPGD problem setting E generally contains non-zero length segments instead of points.

The IPGD problem has close connections with the classical geometric HITTING SET problem on the plane. To describe a HITTING SET formulation of the IPGD problem, some notation is given below. Suppose $N_r(e) = \{x \in \mathbb{R}^2 : d(x, e) \leq r\}$, $\mathcal{N}_r(E) = \{N_r(e) : e \in E\}$ and $d(x, e)$ is Euclidean distance between a point $x \in \mathbb{R}^2$ and a segment $e \in E$; for a zero-length segment $x \in \mathbb{R}^2$ $N_r(x)$ denotes a radius r disk centered at x. Each object from $\mathcal{N}_r(E)$ is a Euclidean r-neighborhood of some segment of E also called r-*hippodrome* or r-*offset* in the literature [2].

The IPGD problem can equivalently be formulated as follows: given a set $\mathcal{N}_r(E)$ of r-hippodromes on the plane whose underlying straight line segments form an edge set of some plane graph $G = (V, E)$, find the minimum cardinality point set C such that $C \cap N \neq \varnothing$ for every $N \in \mathcal{N}_r(E)$. In fact, C represents a set of centers of radius r disks, forming a solution to the IPGD problem. In the sequel, a set $C_0 \subset \mathbb{R}^2$ is called a *piercing set* for $\mathcal{N}_r(E)$ when $C_0 \cap N \neq \varnothing$ for all $N \in \mathcal{N}_r(E)$.

1.1 Related Work and Our Results

As far as we know, settings close to the IPGD problem are originally considered in [2]. Motivated by applications from sensor monitoring for urban road networks, they explore the case in which $\mathcal{F}$ contains equal disks and E consists of n (generally properly overlapping) axis-parallel segments, giving 8-approximation $O(n \log n)$-time algorithm. Their algorithms can easily be extended to the case of sets E of straight line segments with bounded number of distinct orientations. A polynomial time approximation scheme (PTAS) is also proposed [10] for more general version of the IPGD problem in which disks of $\mathcal{C}$ are chosen from some prescribed finite set $\mathcal{H}$ of generally non-equal disks.

When pairs of segments of E are allowed to intersect properly and segments of E are admitted to have arbitrarily large number of distinct orientations, it is difficult to achieve a constant factor approximation at least by using known approaches. It is due to the non-constant lower bound obtained in [1] on integrality gap of a problem, which is close to the IPGD problem for $r = 0$.

In [6] constant factor approximation algorithms are first proposed for the IPGD problem. Namely, a 100-approximate $O(n^4 \log n)$-time algorithm is given for the problem in its general setting where E is formed by an edge set of an arbitrary plane graph. Moreover, due to applications, 68- and 54-approximate algorithms are given in [7] for special cases, where E is an edge set of a generalized outerplane graph and a Delaunay triangulation respectively as well as a 23-approximation algorithm is proposed under the assumption that all pairs of non-overlapping segments from E are at the distance more than r from each other.

Let us give some definitions. Let V be a finite point set in general position on the plane. Assuming that no 4 points of V lie on any circle, a plane graph $G = (V, E)$ is called a *Gabriel* graph [9] when $[u, v] \in E$ iff intersection of V is empty with interior of the disk with diameter $[u, v]$, where $[u, v]$ denotes a straight line segment with endpoints u and v. Under the same assumption a plane graph $G = (V, E)$ is called a *relative neighborhood* graph [3] when $[u, v] \in E$ iff $\max\{d(u, w), d(v, w)\} \geq d(u, v)$ for any $w \in V \backslash \{u, v\}$. Both types of plane graphs defined above appear in a variety of network applications. They represent convenient network topologies, simplifying routing and control in geographical (e.g. wireless) networks. They can also be applied when approximating complex networks.

In [8] faster $O(n^2)$-time 10-, 12- and 14-approximate algorithms are designed for the NP-hard ([5]) IPGD problem when E is being an edge set of a minimum Euclidean spanning tree, a relative neighborhood graph and a Gabriel graph respectively. This short paper extends this latter work by presenting much faster $O\left(n^{3/2}\log^2 n\right)$-expected time 12- and 14- approximation algorithms for the IPGD problem for classes of relative neighborhood graphs and Gabriel graphs respectively. Obtained gain in time performance of the resulting approximation algorithms can be useful for facility location problems on large networks.

2   Basic Algorithm

The following algorithm lies at the core of our $O(1)$-approximation algorithms. It operates on two concepts whose definitions are given below.

Definition 1. *A subset $\mathcal{I} \subseteq \mathcal{N}_r(E)$ is called a maximal (with respect to inclusion) independent set in $\mathcal{N}_r(E)$, if $I \cap I' = \varnothing$ for any $I, I' \in \mathcal{I}$, and for any $N \in \mathcal{N}_r(E)$ there is some $I \in \mathcal{I}$ with $N \cap I \neq \varnothing$.*

Definition 2. *Let $G = (V, E)$ be a plane graph and $f > 0$ be some (r-independent) absolute constant. An edge $e \in E$ is called f-coverable with respect to E, if for any constant $\rho > 0$ one can construct at most f-point piercing set $U(\rho, e, E) \subset \mathbb{R}^2$ for $\mathcal{N}_{\rho,e}(E) = \{N \in \mathcal{N}_\rho(E) : N \cap N_\rho(e) \neq \varnothing\}$ in polynomial time with respect to $|\mathcal{N}_{\rho,e}(E)|$.*

Let $G = (V, E)$ be either a Gabriel or a relative neighborhood graph and $r > 0$ be a constant, forming an input of the IPGD problem. Work of our algorithm below can be split into two phases. During its first phase it performs a pass through E to iteratively grow its subset E' by adding segments from E into E' such that $\mathcal{N}_r(E')$ finally becomes a maximal independent set in $\mathcal{N}_r(E)$. Within this phase it applies a special (randomized) geometric data structure [4], implementing a Voronoi diagram for straight line segments of E', assuming Euclidean distance between segments. This data structure allows to

1. return a segment $e' \in E'$ such that $N_r(e) \cap N_r(e') \neq \varnothing$ for a given straight line segment $e \in E \backslash E'$, or report that $N_r(e) \cap N_r(e') = \varnothing$ for all $e' \in E'$ in $O(\log^2 |E'|)$ expected time;
2. insert new segments of $E \backslash E'$ into E'.

Then, during the second phase, another pass is performed over the built set E' to construct piercing sets for subsets of the form $\mathcal{N}_{r,e}(E)$; each subset is defined by a segment $e \in E'$. Merging those piercing sets together into a point set $C \subset \mathbb{R}^2$, the algorithm yields a set $\mathcal{C} = \{N_r(c) : c \in C\}$ as an approximate solution to the IPGD problem instance, defined by G and r.

The algorithm implementation is based on a pseudo-code below. It contains a constant parameter $f > 0$, which is specific to the class of plane graphs from which G is chosen. More precisely, $f = 12$ for the class of relative neighborhood graphs and $f = 14$ for the class of Gabriel graphs.

MODIFIED COVERING OF LINE SEGMENTS WITH EQUAL DISKS.

Input: a constant $r > 0$ and a plane graph $G = (V, E)$;
Output: an f-approximate solution $\mathcal{C}$ of radius r disks for the IPGD problem instance, defined by G and r.

---

1. $E' := \varnothing$, $E_0 :=$ randomly shuffled E and $C := \varnothing$;
2. while $E_0 \neq \varnothing$, repeat steps 3–5: // *first phase: a single pass through* E
3. choose $e^* \in E_0$ and compute an edge $e' \in E'$ such that $N_r(e^*) \cap N_r(e') \neq \varnothing$ (if it exists) and set $flag := True$; otherwise, if either $N_r(e^*) \cap N_r(e') = \varnothing$ for all $e' \in E'$ or $E' = \varnothing$, set $flag := False$;
4. if $flag = True$, set $E_{e'} := E_{e'} \cup \{e^*\}$; otherwise, when $flag = False$, insert e^* into E' and set $E_{e^*} := \varnothing$;
5. set $E_0 := E_0 \backslash \{e^*\}$;
6. for each $e' \in E'$ repeat steps 7–8: // *second phase: a single pass through* E'
7. construct a piercing set $U(e')$ of at most f points for $\mathcal{N}_r(E_{e'})$, applying some auxiliary procedure;
8. set $C := C \cup U(e')$;
9. return $\mathcal{C} := \{N_r(c) : c \in C\}$ as an f-approximate solution.

At the basic algorithm step 7 within its second phase, to implement an auxiliary procedure of seeking a piercing set $U(e')$ for $\mathcal{N}_r(E_{e'})$, two special procedures are used, which are designed in [8]. Their performance is reported in two lemmas below (see lemmas 1 and 5 as well as implementations of those procedures in [8]).

Lemma 1. *Any edge $e \in E$ is 12-coverable of an arbitrary subgraph $G = (V, E)$ of a relative neighborhood graph. More precisely, for any $\rho > 0$ the respective piercing set $U(\rho, e, E)$ for $\mathcal{N}_{\rho, e}(E)$ can be found in $O(1)$ time.*

Lemma 2. *Any edge $e \in E$ is 14-coverable of an arbitrary subgraph $G = (V, E)$ of a Gabriel graph. Namely, for any $\rho > 0$ the respective piercing set $U(\rho, e, E)$ for $\mathcal{N}_{\rho, e}(E)$ can be found in $O(1)$ time.*

In [8] an analogous algorithm (to the basic algorithm above) is described, working in $O(n\text{OPT})$ time, where OPT is the problem optimum. Namely, it does $O(\text{OPT})$ "heavy" steps such that each step performs a single pass through a subset of E, which can be $\Omega(n)$-sized, thus, taking $O(n)$ time. In distinction to this latter algorithm, work of our algorithm is organized in the different way. It avoids doing those $O(\text{OPT})$ "heavy" steps and performs $O(n)$ "lighter" steps instead at its first phase, each of which mostly consists in querying and updating a nearest neighbor data structure, built on the $O(\text{OPT})$-sized subset of E. Its expected query times are polylogarithmic with respect to OPT.

3 Our Results

Applying our basic algorithm and $O(1)$-time auxiliary procedures from [8] at its step 7, which are titled PARTIAL r-DISK COVER SEARCH FOR $2r$-HIPPODROMES ON RNG EDGES and PARTIAL r-DISK COVER SEARCH FOR $2r$-HIPPODROMES ON GG EDGES, the following results can be obtained.

Theorem 1. *There is an $O\left(n^{3/2}\log^2 \text{OPT}\right)$-expected time f-approximate algorithm for the IPGD problem in the graph class $\mathcal{G}$, where*

1. *$f = 12$ and $\mathcal{G}$ is the class of subgraphs of relative neighborhood graphs;*
2. *$f = 14$ and $\mathcal{G}$ is the class of subgraphs of Gabriel graphs.*

References

1. Alon, N.: A non-linear lower bound for planar epsilon-nets. Discrete Comput. Geom. **47**(2), 235–244 (2011). https://doi.org/10.1007/s00454-010-9323-7
2. Dash, D., Bishnu, A., Gupta, A., Nandy, S.: Approximation algorithms for deployment of sensors for line segment coverage in wireless sensor networks. Wirel. Netw. **19**(5), 857–870 (2012). https://doi.org/10.1007/s11276-012-0506-4
3. Jaromczyk, J., Toussaint, G.: Relative neighborhood graphs and their relatives. Proc. IEEE **80**(9), 1502–1517 (1992). https://doi.org/10.1109/5.163414
4. Karavelas, M.I.: A robust and efficient implementation for the segment Voronoi diagram. In: Proceedings of the 1st International Symposium on Voronoi Diagrams in Science and Engineering, Tokyo, pp. 51–62 (2004)
5. Kobylkin, K.: Stabbing line segments with disks: complexity and approximation algorithms. In: van der Aalst, W., et al. (eds.) International Conference on Analysis of Images, Social Networks and Texts, AIST 2017. LNCS, vol. 10716, pp. 356–367. Springer, Heidelberg (2017). https://doi.org/10.1007/978-3-319-73013-4_33
6. Kobylkin, K.: Constant factor approximation for intersecting line segments with disks. In: Battiti, R., Brunato, M., Kotsireas, I., Pardalos, P. (eds.) Learning and Intelligent Optimization, LION 2018. LNCS, vol. 11353, pp. 447–454. Springer, Heidelberg (2018). https://doi.org/10.1007/978-3-030-05348-2_39
7. Kobylkin, K.: Efficient constant factor approximation algorithms for stabbing line segments with equal disks. CoRR abs/1803.08341 (2018). 31 p. https://arxiv.org/pdf/1803.08341.pdf
8. Kobylkin, K., Dryakhlova, I.: Approximation algorithms for piercing special families of hippodromes: an extended abstract. In: Khachay, M., Kochetov, Y., Pardalos, P. (eds.) Mathematical Optimization Theory and Operations Research, MOTOR 2019. LNCS, vol. 11548, pp. 565-580. Springer, Cham (2019). https://doi.org/10.1007/978-3-030-22629-9_40
9. Matula, D., Sokal, R.: Properties of Gabriel graphs relevant to geographic variation research and the clustering of points in the plane. Geogr. Anal. **12**(3), 205–222 (1980). https://doi.org/10.1111/j.1538-4632.1980.tb00031.x
10. Madireddy, R., Mudgal, A.: Stabbing line segments with disks and related problems. In: Proceedings of the 28th Canadian Conference on Computational Geometry, pp. 201–207. Simon Fraser University, Vancouver, Canada (2016)

A Class of Linear Programs Solvable by Coordinate-Wise Minimization

Tomáš Dlask$^{(\boxtimes)}$ and Tomáš Werner

Faculty of Electrical Engineering, Czech Technical University in Prague,
Prague, Czech Republic
dlaskto2@fel.cvut.cz

Abstract. Coordinate-wise minimization is a simple popular method for large-scale optimization. Unfortunately, for general (non-differentiable and/or constrained) convex problems it may not find global minima. We present a class of linear programs that coordinate-wise minimization solves exactly. We show that dual LP relaxations of several well-known combinatorial optimization problems are in this class and the method finds a global minimum with sufficient accuracy in reasonable runtimes. Moreover, for extensions of these problems that no longer are in this class the method yields reasonably good suboptima. Though the presented LP relaxations can be solved by more efficient methods (such as max-flow), our results are theoretically non-trivial and can lead to new large-scale optimization algorithms in the future.

Keywords: Coordinate-wise minimization · Linear programming · LP relaxation

1 Introduction

Coordinate-wise minimization, or *coordinate descent*, is an iterative optimization method, which in every iteration optimizes only over a single chosen variable while keeping the remaining variables fixed. Due its simplicity, this method is popular among practitioners in large-scale optimization in areas such as machine learning or computer vision, see e.g. [32]. A natural extension of the method is *block-coordinate minimization*, where every iteration minimizes the objective over a block of variables. In this paper, we focus on coordinate minimization with exact updates, where in each iteration a global minimum over the chosen variable is found, applied to convex optimization problems.

For general convex optimization problems, the method need not converge and/or its fixed points need not be global minima. A simple example is the unconstrained minimization of the function $f(x, y) = \max\{x - 2y, y - 2x\}$, which is unbounded but any point with $x = y$ is a coordinate-wise local minimum. Despite this drawback, (block-)coordinate minimization can be very successful for some large-scale convex non-differentiable problems. The prominent example is the class of *convergent message passing* methods for solving dual

© Springer Nature Switzerland AG 2020
I. S. Kotsireas and P. M. Pardalos (Eds.): LION 14 2020, LNCS 12096, pp. 52–67, 2020.
https://doi.org/10.1007/978-3-030-53552-0_8

linear programming (LP) relaxation of maximum a posteriori (MAP) inference in graphical models, which can be seen as various forms of (block-)coordinate descent applied to various forms of the dual. In the typical case, the dual LP relaxation boils down to the unconstrained minimization of a convex piece-wise affine (hence non-differentiable) function. These methods include max-sum diffusion [21,26,29], TRW-S [18], MPLP [12], and SRMP [19]. They do not guarantee global optimality but for large sparse instances from computer vision the achieved coordinate-wise local optima are very good and TRW-S is significantly faster than competing methods [16,27], including popular first-order primal-dual methods such as ADMM [5] or [8].

This is a motivation to look for other classes of convex optimization problems for which (block-)coordinate descent would work well or, alternatively, to extend convergent message passing methods to a wider class of convex problems than the dual LP relaxation of MAP inference. A step in this direction is the work [31], where it was observed that if the minimizer of the problem over the current variable block is not unique, one should choose a minimizer that lies in the *relative interior* of the set of block-optimizers. It is shown that any update satisfying this rule is, in a precise sense, not worse than any other exact update. Message-passing methods such as max-sum diffusion and TRW-S satisfy this rule. If max-sum diffusion is modified to violate the relative interior rule, it can quickly get stuck in a very poor coordinate-wise local minimum.

To be precise, suppose we minimize a convex function $f \colon X \to \mathbb{R}$ on a closed convex set $X \subseteq \mathbb{R}^n$. We assume that f is bounded from below on X. For brevity of formulation, we rephrase this as the minimization of the extended-valued function $\bar{f} \colon \mathbb{R}^n \to \mathbb{R} \cup \{\infty\}$ such that $\bar{f}(x) = f(x)$ for $x \in X$ and $\bar{f}(x) = \infty$ for $x \notin X$. One iteration of coordinate minimization with the relative interior rule [31] chooses a variable index $i \in [n] = \{1, \ldots, n\}$ and replaces an estimate $x^k = (x_1^k, \ldots, x_n^k) \in X$ with a new estimate $x^{k+1} = (x_1^{k+1}, \ldots, x_n^{k+1}) \in X$ such that[1]

$$x_i^{k+1} \in \operatorname{ri} \operatorname*{argmin}_{y \in \mathbb{R}} \bar{f}(x_1^k, \ldots, x_{i-1}^k, y, x_{i+1}^k, \ldots, x_n^k),$$

$$x_j^{k+1} = x_j^k \quad \forall j \neq i,$$

where $\operatorname{ri} Y$ denotes the relative interior of a convex set Y. As this is a univariate convex problem, the set $Y = \operatorname{argmin}_{y \in \mathbb{R}} \bar{f}(x_1^k, \ldots, x_{i-1}^k, y, x_{i+1}^k, \ldots, x_n^k)$ is either a singleton or an interval. In the latter case, the relative interior rule requires that we choose x_i^{k+1} from the interior of this interval. A point $x = (x_1, \ldots, x_n) \in X$ that satisfies

$$x_i \in \operatorname{ri} \operatorname*{argmin}_{y \in \mathbb{R}} \bar{f}(x_1, \ldots, x_{i-1}, y, x_{i+1}, \ldots, x_n)$$

for all $i \in [n]$ is called a (coordinate-wise) *interior local minimum* of function f on set X.

---

[1] In [31], the iteration is formulated in a more abstract (coordinate-free) notation. Since we focus only on coordinate-wise minimization here, we use a more concrete notation.

Some classes of convex problems are solved by coordinate-wise minimization exactly. E.g., for unconstrained minimization of a differentiable convex function, it is easy to see that any fixed point of the method is a global minimum; moreover, it has been proved that if the function has unique univariate minima, then any limit point is a global minimum [4, §2.7]. The same properties hold for convex functions whose non-differentiable part is separable [28]. Note that these classical results need not assume the relative interior rule [31].

Therefore, it is natural to ask if the relative interior rule can widen the class of convex optimization problems that are exactly solved by coordinate-wise minimization. Leaving convergence aside[2], more precisely we can ask for which problems interior local minima are global minima. A succinct characterization of this class is currently out of reach. Two subclasses of this class are known [18,26,29]: the dual LP relaxation of MAP inference with pairwise potential functions and two labels, or with submodular potential functions.

In this paper, we restrict ourselves to linear programs (where f is linear and X is a convex polyhedron) and present a new class of linear programs with this property. We show that dual LP relaxations of a number of combinatorial optimization problems belong to this class and coordinate-wise minimization converges in reasonable time on large practical instances. Unfortunately, the practical impact of this result is limited because there exist more efficient algorithms for solving these LP relaxations, such as reduction to max-flow. It is open whether there exist some useful classes of convex problems that are exactly solvable by (block-)coordinate descent but not solvable by more efficient methods. There is a possibility that our result and the proof technique will pave the way to such results.

2   Reformulations of Problems

Before presenting our main result, we make an important remark: while a convex optimization problem can be reformulated in many ways to an 'equivalent' problem which has the same global minima, not all of these transformations are equivalent with respect to coordinate-wise minimization, in particular, not all preserve interior local minima.

Example 1. One example is dualization. If coordinate-wise minimization achieves good local (or even global) minima on a convex problem, it can get stuck in very poor local minima if applied to its dual. Indeed, trying to apply (block-) coordinate minimization to the *primal* LP relaxation of MAP inference (linear optimization over the local marginal polytope) has been futile so far.

Example 2. Consider the linear program $\min\{x_1 + x_2 \mid x_1, x_2 \geq 0\}$, which has one interior local minimum with respect to individual coordinates that also corresponds to the unique global optimum. But if one adds a redundant constraint,

---

[2] We do not discuss convergence in this paper and assume that the method converges to an interior local minimum. This is supported by experiments, e.g., max-sum diffusion and TRW-S have this property. More on convergence can be found in [31].

namely $x_1 = x_2$, then any feasible point will become an interior local minimum w.r.t. individual coordinates, because the redundant constraint blocks changing the variable x_i without changing x_{3-i} for both $i \in \{1, 2\}$.

Example 3. Consider the linear program

$$\min \sum_{j=1}^{m} z_j \tag{1a}$$

$$z_j \geq a_{ij}^T x + b_{ij} \qquad\qquad \forall i \in [n], j \in [m] \tag{1b}$$

$$z \in \mathbb{R}^m, x \in \mathbb{R}^p \tag{1c}$$

which can be also formulated as

$$\min \sum_{j=1}^{m} \max_{i=1}^{n}(a_{ij}^T x + b_{ij}) \tag{2a}$$

$$x \in \mathbb{R}^p. \tag{2b}$$

Optimizing over the individual variables by coordinate-wise minimization in (1) does not yield the same interior local optima as in (2). For instance, assume that $m = 3$, $n = p = 1$ and the problem (2) is given as

$$\min \left(\max\{x, 0\} + \max\{-x, -1\} + \max\{-x, -2\}\right), \tag{3}$$

where $x \in \mathbb{R}$. Then, when optimizing directly in form (3), one can see that all the interior local optima are global optimizers.

However, when one introduces the variables $z \in \mathbb{R}^3$ and applies coordinate-wise minimization on the corresponding problem (1), then there are interior local optima that are not global optimizers, for example $x = z_1 = z_2 = z_3 = 0$, which is an interior local optimum, but is not a global optimum.

On the other hand, optimizing over blocks of variables $\{z_1, \ldots, z_m, x_i\}$ for each $i \in [p]$ in case (1) is equivalent to optimization over individual x_i in formulation (2).

3   Main Result

The optimization problem with which we are going to deal is in its most general form defined as

$$\min \left(\sum_{i=1}^{m} \max\{w_i - \varphi_i, 0\} + a^T \varphi + b^T \lambda + \sum_{j=1}^{p} \max\{v_j + A_{:j}^T \varphi + B_{:j}^T \lambda, 0\}\right) \tag{4a}$$

$$\underline{\varphi}_i \leq \varphi_i \leq \overline{\varphi}_i \ \forall i \in [m] \tag{4b}$$

$$\underline{\lambda}_i \leq \lambda_i \leq \overline{\lambda}_i \ \forall i \in [n], \tag{4c}$$

where $A \in \mathbb{R}^{m \times p}, B \in \mathbb{R}^{n \times p}, a \in \mathbb{R}^m, b \in \mathbb{R}^n, w \in \mathbb{R}^m, v \in \mathbb{R}^p, \underline{\varphi} \in (\mathbb{R} \cup \{-\infty\})^m, \overline{\varphi} \in (\mathbb{R} \cup \{\infty\})^m, \underline{\lambda} \in (\mathbb{R} \cup \{-\infty\})^n, \overline{\lambda} \in (\mathbb{R} \cup \{\infty\})^n$ (assuming $\underline{\varphi} < \overline{\varphi}$ and $\underline{\lambda} < \overline{\lambda}$). We optimize over variables $\varphi \in \mathbb{R}^m$ and $\lambda \in \mathbb{R}^n$. $A_{:j}$ and $A_{i:}$ denotes the j-th column and i-th row of A, respectively.

Applying coordinate-wise minimization with relative-interior rule on the problem (4) corresponds to cyclic updates of variables, where each update corresponds to finding the region of optima of a convex piecewise-affine function of one variable on an interval. If the set of optimizers is a singleton, then the update is straightforward. If the set of optimizers is a bounded interval $[a, b]$, the variable is assigned the middle value from this interval, i.e. $(a + b)/2$. If the set of optima is unbounded, i.e. $[a, \infty)$, then we set the variable to the value $a + \Delta$, where $\Delta > 0$ is a fixed constant. In case of $(-\infty, a]$, the variable is updated to $a - \Delta$. The details for the update in this setting are in Appendix A in [10].

Theorem 1. *Any interior local optimum of (4) w.r.t. individual coordinates is its global optimum if*

- *matrices A, B contain only values from the set $\{-1, 0, 1\}$ and contain at most two non-zero elements per row*
- *vector a contains only elements from the set $(-\infty, -2] \cup \{-1, 0, 1, 2\} \cup [3, \infty)$*
- *vector b contains only elements from the set $(-\infty, -2] \cup \{-1, 0, 1\} \cup [2, \infty)$.*

In order to prove Theorem 1, we formulate problem (4) as a linear program by introducing additional variables $\alpha \in \mathbb{R}^m$ and $\beta \in \mathbb{R}^p$ and construct its dual. The proof of optimality is then obtained (see Theorem 2) by constructing a dual feasible solution that satisfies complementary slackness.

The primal linear program (with corresponding dual variables and constraints on the same lines) reads

$$\min \sum_{i \in [m]} \alpha_i + \sum_{i \in [p]} \beta_i + a^T \varphi + b^T \lambda \qquad \max f(z, y, s, r, q, x) \tag{5a}$$

$$\beta_j - A_{:j}^T \varphi - B_{:j}^T \lambda \geq v_j \qquad\qquad x_j \geq 0 \quad \forall j \in [p] \tag{5b}$$

$$\alpha_i + \varphi_i \geq w_i \qquad\qquad s_i \geq 0 \quad \forall i \in [m] \tag{5c}$$

$$\varphi_i \geq \underline{\varphi}_i \qquad\qquad y_i \geq 0 \quad \forall i \in [m] \tag{5d}$$

$$\varphi_i \leq \overline{\varphi}_i \qquad\qquad z_i \leq 0 \quad \forall i \in [m] \tag{5e}$$

$$\lambda_i \geq \underline{\lambda}_i \qquad\qquad q_i \geq 0 \quad \forall i \in [n] \tag{5f}$$

$$\lambda_i \leq \overline{\lambda}_i \qquad\qquad r_i \leq 0 \quad \forall i \in [n] \tag{5g}$$

$$\varphi_i \in \mathbb{R} \qquad s_i + z_i + y_i - A_{i:}^T x = a_i \quad \forall i \in [m] \tag{5h}$$

$$\lambda_i \in \mathbb{R} \qquad r_i + q_i - B_{i:}^T x = b_i \quad \forall i \in [n] \tag{5i}$$

$$\beta_j \geq 0 \qquad\qquad x_j \leq 1 \quad \forall j \in [p] \tag{5j}$$

$$\alpha_i \geq 0 \qquad\qquad s_i \leq 1 \quad \forall i \in [m], \tag{5k}$$

where the dual criterion is

$$f(z, y, s, r, q, x) = \overline{\varphi}^T z + \underline{\varphi}^T y + w^T s + \overline{\lambda}^T r + \underline{\lambda}^T q + v^T x \tag{6}$$

and clearly, at optimum of the primal, we have

$$\alpha_i = \max\{w_i - \varphi_i, 0\} \qquad\qquad \forall i \in [m] \qquad (7a)$$

$$\beta_j = \max\{v_j + A_{:j}^T \varphi + B_{:j}^T \lambda, 0\} \qquad\qquad \forall j \in [p]. \qquad (7b)$$

The variables α, β were eliminated from the primal formulation (5) to obtain (4) due to similar reasoning as in Example 3. We also remark that setting $\overline{\varphi}_i = \infty$ (resp. $\underline{\varphi}_i = -\infty$, $\overline{\lambda}_i = \infty$, $\underline{\lambda}_i = -\infty$) results in $z_i = 0$ (resp. $y_i = 0$, $r_i = 0$, $q_i = 0$).

Even though the primal-dual pair (5) might seem overcomplicated, such general description is in fact necessary because as described in Sect. 2, equivalent reformulations may not preserve the structure of interior local minima and we would like to describe as general class, where optimality is guaranteed, as possible.

Example 4. To give the reader better insight into the problems (5), we present a simplification based on omitting the matrix A (i.e. $m = 0$) and setting $\underline{\lambda} = 0$, $\overline{\lambda} = \infty$, which results in $r_i = 0$ and variables q_i become slack variables in (5i). The primal-dual pair in this case then simplifies to

$$\min \sum_{i \in [p]} \beta_i + b^T \lambda \qquad\qquad \max v^T x \qquad\qquad\qquad (8a)$$

$$\beta_j - B_{:j}^T \lambda \geq v_j \qquad\qquad x_j \geq 0 \qquad\qquad \forall j \in [p] \qquad (8b)$$

$$\beta_j \geq 0 \qquad\qquad x_j \leq 1 \qquad\qquad \forall j \in [p] \qquad (8c)$$

$$\lambda_i \geq 0 \qquad\qquad -B_{i:}^T x \leq b_i \qquad\qquad \forall i \in [n]. \qquad (8d)$$

Theorem 2. *For a problem (4) satisfying conditions of Theorem 1 and a given interior local minimum* (φ, λ), *the values*[3]

$$x_j = \begin{cases} 0 & \text{if } A_{:j}^T \varphi + B_{:j}^T \lambda + v_j < 0 \\ \frac{1}{2} & \text{if } A_{:j}^T \varphi + B_{:j}^T \lambda + v_j = 0 \\ 1 & \text{if } A_{:j}^T \varphi + B_{:j}^T \lambda + v_j > 0 \end{cases} \quad s_i = \begin{cases} 1 & \text{if } w_i > \varphi_i \\ 0 & \text{if } w_i < \varphi_i \\ h_{[0,1]}(a_i + A_{i:}^T x) & \text{if } w_i = \varphi_i \end{cases}$$

$$r_i = \begin{cases} 0 & \text{if } \lambda_i < \overline{\lambda}_i \\ h_{\mathbb{R}_0^-}(b_i + B_{i:}^T x) & \text{if } \lambda_i = \overline{\lambda}_c \end{cases} \quad z_i = \begin{cases} 0 & \text{if } \varphi_i < \overline{\varphi}_i \\ h_{\mathbb{R}_0^-}(a_i + A_{i:}^T x - s_i) & \text{if } \varphi_i = \overline{\varphi}_i \end{cases}$$

$$q_i = \begin{cases} 0 & \text{if } \lambda_i > \underline{\lambda}_i \\ h_{\mathbb{R}_0^+}(b_i + B_{i:}^T x) & \text{if } \lambda_i = \underline{\lambda}_i \end{cases} \quad y_i = \begin{cases} 0 & \text{if } \varphi_i > \underline{\varphi}_i \\ h_{\mathbb{R}_0^+}(a_i + A_{i:}^T x - s_i) & \text{if } \varphi_i = \underline{\varphi}_i \end{cases}$$

are feasible for the dual (5) and satisfy complementary slackness with primal (5), where the remaining variables of the primal are given by (7).

---

[3] We define $h_{[x,y]}(z) = \min\{y, \max\{z, x\}\}$ to be the projection of $z \in \mathbb{R}$ onto the interval $[x, y] \subseteq \mathbb{R}$. The projection onto unbounded intervals $(-\infty, 0]$ and $[0, \infty)$ is defined similarly and is denoted by $h_{\mathbb{R}_0^-}$ and $h_{\mathbb{R}_0^+}$ for brevity.

It can be immediately seen that all the constraints of dual (5) are satisfied except for (5h) and (5i), which require a more involved analysis. The complete proof of Theorem 2 is technical (based on verifying many different cases) and given in Appendix B in [10].

4   Applications

Here we show that several LP relaxations of combinatorial problems correspond to the form (4) or to the dual (5) and discuss which additional constraints correspond to the assumptions of Theorem 1.

4.1   Weighted Partial Max-SAT

In weighted partial Max-SAT, one is given two sets of clauses, soft and hard. Each soft clause is assigned a positive weight. The task is to find values of binary variables $x_i \in \{0, 1\}$, $i \in [p]$ such that all the hard clauses are satisfied and the sum of weights of the satisfied soft clauses is maximized.

We organize the m soft clauses into a matrix $S \in \{-1, 0, 1\}^{m \times p}$ defined as

$$
S_{ci} = \begin{cases} 1 & \text{if literal } x_i \text{ is present in soft clause } c \\ -1 & \text{if literal } \neg x_i \text{ is present in soft clause } c \\ 0 & \text{otherwise} \end{cases}
$$

In addition, we denote $n_c^S = \sum_i \llbracket S_{ci} < 0 \rrbracket$ to be the number of negated variables in clause c. These numbers are stacked in a vector $n^S \in \mathbb{Z}^m$. The h hard clauses are organized in a matrix $H \in \{-1, 0, 1\}^{h \times p}$ and a vector $n^H \in \mathbb{Z}^h$ in the same manner.

The LP relaxation of this problem reads

$$\max \sum_{c \in [m]} w_c s_c \tag{10a}$$

$$s_c \leq S_{c:}^T x + n_c^S \qquad \forall c \in [m] \tag{10b}$$

$$H_{c:}^T x + n_c^H \geq 1 \qquad \forall c \in [h] \tag{10c}$$

$$x_i \in [0, 1] \qquad \forall i \in [p] \tag{10d}$$

$$s_c \in [0, 1] \qquad \forall c \in [m], \tag{10e}$$

where $w_c \in \mathbb{R}_0^+$ are the weights of the soft clauses $c \in [m]$. This is a sub-class of the dual (5), where $A = S$, $B = -H$, $a = n^S$, $b = 1 - n^H$, $\underline{\varphi} = 0$ ($y \geq 0$ are therefore slack variables for the dual constraint (5h) that correspond to (10b)), $\overline{\varphi} = \infty$ (therefore $z = 0$), $\underline{\lambda} = -\infty$ (therefore $q = 0$), $\overline{\lambda} = 0$ ($r \leq 0$ are slack variables for the dual constraint (5i) that correspond to (10c)), $v = 0$.

Formulation (10) satisfies the conditions of Theorem 1 if each of the clauses has length at most 2. In other words, optimality is guaranteed for weighted partial Max-2SAT.

Also notice that if we omitted the soft clauses (10b) and instead set $v = -1$, we would obtain an instance of Min-Ones SAT, which could be generalized to weighted Min-Ones SAT. This relaxation would still satisfy the requirements of Theorem 1 if all the present hard clauses have length at most 2.

Results. We tested the method on 800 smallest[4] instances that appeared in Max-SAT Evaluations [2] in years 2017 [1] and 2018 [3]. The results on the instances are divided into groups in Table 1 based on the minimal and maximal length of present clauses. We have also tested this approach on 60 instances of weighted Max-2SAT from Ke Xu [33]. The highest number of logical variables in an instance was 19034 and the highest overall number of clauses in an instance was 31450. It was important to separate the instances without unit clauses (i.e. clauses of length 1), because in such cases the LP relaxation (10) has a trivial optimal solution with $x_i = \frac{1}{2}$ for all $i \in V$.

Coordinate-wise minimization was stopped when the criterion did not improve by at least $\epsilon = 10^{-7}$ after a whole cycle of updates for all variables. We report the quality of the solution as the median and mean relative difference between the optimal criterion and the criterion reached by coordinate-wise minimization before termination.

Table 1 reports not only instances of weighted partial Max-2SAT but also instances with longer clauses, where optimality is no longer guaranteed. Nevertheless, the relative differences on instances with longer clauses still seem not too large and could be usable as bounds in a branch-and-bound scheme.

Table 1. Experimental comparison of coordinate-wise minimization and exact solutions for LP relaxation on instances from [2] (first 4 rows) and [33] (last row).

Instance Group Specification			Results	
Min CL	Max CL	#inst.	Mean RD	Median RD
≥ 2	any	91	0	0
1	2	123	$1.44 \cdot 10^{-9}$	$1.09 \cdot 10^{-11}$
1	3	99	$6.98 \cdot 10^{-3}$	$1.90 \cdot 10^{-7}$
1	≥ 4	487	$1.26 \cdot 10^{-2}$	$2.97 \cdot 10^{-3}$
1	2	60	$1.59 \cdot 10^{-9}$	$5.34 \cdot 10^{-10}$

---

[4] Smallest in the sense of the file size. All instances could not have been evaluated due to their size and lengthy evaluation.

4.2 Weighted Vertex Cover

Dual (8) also subsumes[5] the LP relaxation of weighted vertex cover, which reads

$$\min \left\{ \sum_{i \in V} v_i x_i \,\middle|\, x_i + x_j \geq 1 \;\forall\{i,j\} \in E, x_i \in [0,1] \;\forall i \in V \right\} \tag{11}$$

where V is the set of nodes and E is the set of edges of an undirected graph. This problem also satisfies the conditions of Theorem 1 and therefore the corresponding primal (4) will have no non-optimal interior local minima.

On the other hand, notice that formulation (11), which corresponds to dual (5) can have non-optimal interior local minima even with respect to all subsets of variables of size $|V| - 1$, an example is given in Appendix C in [10].

We reported the experiments on weighted vertex cover in [30] where the optimality was not proven yet. In addition, the update designed in [30] *ad hoc* becomes just a special case of our general update here.

4.3 Minimum st-Cut, Maximum Flow

Recall from [11] the usual formulation of max-flow problem between nodes $s \in V$ and $t \in V$ on a directed graph with vertex set V, edge set E and positive edge weights $w_{ij} \in \mathbb{R}_0^+$ for each $(i,j) \in E$, which reads

$$\max \sum_{(s,i) \in E} f_{si} \tag{12a}$$

$$0 \leq f_{ij} \leq w_{ij} \qquad\qquad \forall(i,j) \in E \tag{12b}$$

$$\sum_{(u,i) \in E} f_{ui} - \sum_{(j,u) \in E} f_{ju} = 0 \qquad\qquad \forall u \in V - \{s,t\}. \tag{12c}$$

Assume that there is no edge (s,t), there are no ingoing edges to s and no outgoing edges from t, then any feasible value of f in (12) is an interior local optimum w.r.t. individual coordinates by the same reasoning as in Example 2 due to the flow conservation constraint (12c), which limits each individual variable to a single value. We are going to propose a formulation which has no non-globally optimal interior local optima.

---

[5] It is only necessary to transform minimization to maximization of negated objective in (11).

The dual problem to (12) is the minimum st-cut problem, which can be formulated as

$$\max \sum_{(i,j)\in E} w_{ij}y_{ij} \tag{13a}$$

$$y_{ij} \leq 1 - x_i + x_j \qquad \forall (i,j) \in E, i \neq s, j \neq t \tag{13b}$$

$$y_{sj} \leq x_j \qquad \forall (s,j) \in E \tag{13c}$$

$$y_{it} \leq 1 - x_i \qquad \forall (i,t) \in E \tag{13d}$$

$$y_{ij} \in [0,1] \qquad \forall (i,j) \in E, \tag{13e}$$

$$x_i \in [0,1] \qquad \forall i \in V - \{s,t\}, \tag{13f}$$

where $y_{ij} = 0$ if edge (i,j) is in the cut and $y_{ij} = 1$ if edge (i,j) is not in the cut. The cut should separate s and t, so the set of nodes connected to s after the cut will be denoted by S and $T = V - S$ is the set of nodes connected to t. Using this notation, $x_i = [\![i \in S]\!]$. Formulation (13) is different from the usual formulation by replacing the variables y_{ij} by $1 - y_{ij}$, therefore we also maximize the weight of the not cut edges instead of minimizing the weight of the cut edges, therefore if the optimal value of (13) is O, then the value of the minimum st-cut equals $\sum_{(i,j)\in E} w_{ij} - O$.

Formulation (13) is subsumed by the dual (5) by setting $\underline{\varphi} = 0$, $\overline{\varphi} = \infty$ and omitting the B matrix. Also notice that each y_{ij} variable occurs in at most one constraint. The problem (13) therefore satisfies the conditions of Theorem 1 and the corresponding primal (4) is a formulation of the maximum flow problem, in which one can search for the maximum flow by coordinate-wise minimization. The corresponding formulation (4) reads

$$\min \Big(\sum_{(i,j)\in E} \max\{w_{ij} - \varphi_{ij}, 0\} + \sum_{(i,j)\in E, i \neq s} \varphi_{ij} +$$

$$+ \sum_{i\in V-\{s,t\}} \max\Big\{ \sum_{(j,i)\in E} \varphi_{ji} - \sum_{(i,j)\in E} \varphi_{ij}, 0 \Big\} \Big) \tag{14a}$$

$$\varphi_{ij} \geq 0 \quad \forall (i,j) \in E. \tag{14b}$$

Results. We have tested our formulation for coordinate-wise minimization on max-flow instances[6] from computer vision. We report the same statistics as with Max-SAT in Table 2, the instances corresponded to stereo problems, multiview reconstruction instances and shape fitting problems.

For multiview reconstruction and shape fitting, we were able to run our algorithm only on small instances, which have approximately between $8 \cdot 10^5$ and $1.2 \cdot 10^6$ nodes and between $5 \cdot 10^6$ and $6 \cdot 10^6$ edges. On these instances, the algorithm terminated with the reported precision in 13 to 34 min on a laptop.

---

[6] Available at https://vision.cs.uwaterloo.ca/data/maxflow.

Table 2. Experimental comparison of coordinate-wise minimization on max-flow instances, the references are the original sources of the data and/or to the authors that reformulated these problems as maximum flow. The first 6 rows correspond to stereo problems, the 2 following rows are multiview reconstruction instances, the last row is a shape fitting problem.

Instance group or instance		Results	
Name	#inst.	Mean RD	Median RD
BVZ-tsukuba [7]	16	$6.03 \cdot 10^{-10}$	$1.17 \cdot 10^{-11}$
BVZ-sawtooth [7,25]	20	$9.83 \cdot 10^{-11}$	$6.11 \cdot 10^{-12}$
BVZ-venus [7,25]	22	$3.40 \cdot 10^{-11}$	$2.11 \cdot 10^{-12}$
KZ2-tsukuba [20]	16	$2.69 \cdot 10^{-10}$	$1.77 \cdot 10^{-10}$
KZ2-sawtooth [20,25]	20	$4.08 \cdot 10^{-9}$	$1.56 \cdot 10^{-10}$
KZ2-venus [20,25]	22	$5.21 \cdot 10^{-9}$	$1.74 \cdot 10^{-10}$
BL06-camel-sml [23]	1	$1.21 \cdot 10^{-11}$	
BL06-gargoyle-sml [6]	1	$6.29 \cdot 10^{-12}$	
LB07-bunny-sml [22]	1	$1.33 \cdot 10^{-10}$	

4.4   MAP Inference with Potts Potentials

Coordinate-wise minimization for the dual LP relaxation of MAP inference was intensively studied, see e.g. the review [29]. One of the formulations is

$$\min \sum_{i\in V} \max_{k\in K} \theta_i^\delta(k) + \sum_{\{i,j\}\in E} \max_{k,l\in K} \theta_{ij}^\delta(k,l) \tag{15a}$$

$$\delta_{ij}(k) \in \mathbb{R} \ \forall \{i,j\} \in E, k \in K, \tag{15b}$$

where K is the set of labels, V is the set of nodes and E is the set of unoriented edges and

$$\theta_i^\delta(k) = \theta_i(k) - \sum_{j\in N_i} \delta_{ij}(k) \tag{16a}$$

$$\theta_{ij}^\delta(k,l) = \theta_{ij}(k,l) + \delta_{ij}(k) + \delta_{ji}(l) \tag{16b}$$

are equivalent transformations of the potentials. Notice that there are $2 \cdot |E| \cdot |K|$ variables, i.e. two for each direction of an edge. In [24], it is mentioned that in case of Potts interactions, which are given as $\theta_{ij}(k,l) = -[\![k \neq l]\!]$, one can add constraints

$$\delta_{ij}(k) + \delta_{ji}(k) = 0 \qquad\qquad \forall \{i,j\} \in E, k \in K \tag{17a}$$

$$-\tfrac{1}{2} \leq \delta_{ij}(k) \leq \tfrac{1}{2} \qquad\qquad \forall \{i,j\} \in E, k \in K \tag{17b}$$

to (15) without changing the optimal objective. One can therefore use constraint (17a) to reduce the overall amount of variables by defining

$$\lambda_{ij}(k) = -\delta_{ij}(k) = \delta_{ji}(k) \tag{18}$$

subject to $\frac{1}{2} \le \lambda_{ij}(k) \le \frac{1}{2}$. The decision of whether $\delta_{ij}(k)$ or $\delta_{ji}(k)$ should have the inverted sign depends on the chosen orientation E' of the originally undirected edges E and is arbitrary. Also, given values δ satisfying (17), it holds for any edge $\{i,j\} \in E$ and pair of labels $k, l \in K$ that $\max\limits_{k,l \in K} \theta^\delta_{ij}(k, l) = 0$, which can be seen from the properties of the Potts interactions.

Therefore, one can reformulate (15) into

$$\min \sum_{i \in V} \max_{k \in K} \theta^\lambda_i(k) \tag{19a}$$

$$-\tfrac{1}{2} \le \lambda_{ij}(k) \le \tfrac{1}{2} \quad \forall (i,j) \in E', k \in K, \tag{19b}$$

where the equivalent transformation in λ variables is given by

$$\theta^\lambda_i(k) = \theta_i(k) + \sum_{(i,j) \in E'} \lambda_{ij}(k) - \sum_{(j,i) \in E'} \lambda_{ji}(k) \tag{20}$$

and we optimize over $|E'| \cdot |K|$ variables λ, the graph (V, E') is the same as graph (V, E) except that each edge becomes oriented (in arbitrary direction). The way of obtaining an optimal solution to (15) from an optimal solution of (19) is given by (18) and depends on the chosen orientation of the edges in E'. Also observe that $\theta^\delta_i(k) = \theta^\lambda_i(k)$ for any node $i \in V$ and label $k \in K$ and therefore the optimal values will be equal. This reformulation therefore maps global optima of (19) to global optima of (15). However, it does not map interior local minima of (19) to interior local minima of (15) when $|K| \ge 3$, an example of such case is shown in Appendix D in [10].

In problems with two labels ($K = \{1, 2\}$), problem (19) is subsumed by (4) and satisfies the conditions imposed by Theorem 1 because one can rewrite the criterion by observing that

$$\max_{k \in \{1,2\}} \theta^\lambda_i(k) = \max\{\theta^\lambda_i(1) - \theta^\lambda_i(2), 0\} + \theta^\lambda_i(2) \tag{21}$$

and each $\lambda_{ij}(k)$ is present only in $\theta^\lambda_i(k)$ and $\theta^\lambda_j(k)$. Thus, $\lambda_{ij}(k)$ will have non-zero coefficient in the matrix B only on columns i and j. The coefficients of the variables in the criterion are only $\{-1, 0, 1\}$ and the other conditions are straightforward.

We reported the experiments on the Potts problem in [30] where the optimality was not proven yet. In addition, the update designed in [30] *ad hoc* becomes just a special case of our general update here.

4.5   Binarized Monotone Linear Programs

In [13], integer linear programs with at most two variables per constraint were discussed. It was also allowed to have 3 variables in some constraints if one of the variables occurred only in this constraint and in the objective function. Although the objective function in [13] was allowed to be more general, we will restrict ourselves to linear criterion function. It was also shown that such problems can be transformed into binarized monotone constraints over binary variables by introducing additional variables whose amount is defined by the bounds of the original variables, such optimization problem reads

$$\min \ w^T x + e^T z \tag{22a}$$
$$Ax - Iz \leq 0 \tag{22b}$$
$$Cx \leq 0 \tag{22c}$$
$$x \in \{0,1\}^{n_1} \tag{22d}$$
$$z \in \{0,1\}^{n_2}, \tag{22e}$$

where A, C contain exactly one -1 per row and exactly one 1 per row and all other entries are zero, I is the identity matrix. We refer the reader to [13] for details, where it is also explained that the LP relaxation of (22) can be solved by min-st-cut on an associated graph. We can notice that the LP relaxation of (22) is subsumed by the dual (5), because one can change the minimization into maximization by changing the signs in w, e. Also, the relaxation satisfies the conditions given by Theorem 1.

In the paper [13], there are listed many problems which are transformable to (22) and are also directly (without any complicated transformation) subsumed by the dual (5) and satisfy Theorem 1, for example, minimizing the sum of weighted completion times of precedence-constrained jobs (ISLO formulation in [9]), generalized independent set (forest harvesting problem in [14]), generalized vertex cover [15], clique problem [15], Min-SAT (introduced in [17], LP formulation in [13]).

For each of these problems, it is easy to verify the conditions of Theorem 1, because they contain at most two variables per constraint and if a constraint contains a third variable, then it is the only occurrence of this variable and the coefficients of the variables in the constraints are from the set $\{-1, 0, 1\}$.

The transformation presented in [13] can be applied to partial Max-SAT and vertex cover to obtain a problem in the form (22) and solve its LP relaxation. But this step is unnecessary when applying the presented coordinate-wise minimization approach.

5 Concluding Remarks

We have presented a new class of linear programs that are exactly solved by coordinate-wise minimization. We have shown that dual LP relaxations of several well-known combinatorial optimization problems (partial Max-2SAT, vertex cover, minimum st-cut, MAP inference with Potts potentials and two labels, and other problems) belong, possibly after a reformulation, to this class. We have shown experimentally (in this paper and in [30]) that the resulting methods are reasonably efficient for large-scale instances of these problems. When the assumptions of Theorem 1 are relaxed (e.g., general Max-SAT instead of Max-2SAT, or the Potts problem with any number of labels), the method experimentally still provides good local (though not global in general) minima.

We must admit, though, that the practical impact of Theorem 1 is limited because the presented dual LP relaxations satisfying its assumptions can be efficiently solved also by other approaches. Thus, max-flow/min-st-cut can be solved (besides well-known combinatorial algorithms such as Ford-Fulkerson) by message-passing methods such as TRW-S. Similarly, the Potts problem with two labels is tractable and can be reduced to max-flow. In general, all considered LP relaxations can be reduced to max-flow, as noted in Sect. 4.5. Note, however, that this does not make our result trivial because (as noted in Sect. 2) equivalent reformulations of problems may not preserve interior local minima and thus message-passing methods are not equivalent in any obvious way to our method.

It is open whether there are practically interesting classes of linear programs that are solved exactly (or at least with constant approximation ratio) by (block-) coordinate minimization and are not solvable by known combinatorial algorithms such as max-flow. Another interesting question is which reformulations in general preserve interior local minima and which do not.

Our approach can pave the way to new efficient large-scale optimization methods in the future. Certain features of our results give us hope here. For instance, our approach has an important novel feature over message-passing methods: it applies to a *constrained* convex problem (the box constraints (4b) and (4c)). This can open the way to a new class of applications. Furthermore, updates along large variable blocks (which we have not explored) can speed algorithms considerably, e.g., TRW-S uses updates along subtrees of a graphical model, while max-sum diffusion uses updates along single variables.

Acknowledgments. This work has been supported by the OP VVV project CZ.02.1.01/0.0/0.0/16_019/0000765, the Grant Agency of the Czech Technical University in Prague (grant SGS19/170/OHK3/3T/13) and the Czech Science Foundation (grant 19-09967S).

References

1. Ansotegui, C., Bacchus, F., Järvisalo, M., Martins, R., et al.: MaxSAT Evaluation 2017 (2017). https://helda.helsinki.fi/bitstream/handle/10138/233112/mse17proc.pdf
2. Bacchus, F., Järvisalo, M., Martins, R.: MaxSAT evaluation 2018: new developments and detailed results. J. Satisf. Boolean Model. Comput. **11**(1), 99–131 (2019). instances available at https://maxsat-evaluations.github.io/
3. Bacchus, F., Järvisalo, M.J., Martins, R., et al.: MaxSAT evaluation 2018 (2018). https://helda.helsinki.fi/bitstream/handle/10138/237139/mse18_proceedings.pdf
4. Bertsekas, D.P.: Nonlinear Programming, 2nd edn. Athena Scientific, Belmont (1999)
5. Boyd, S., Parikh, N., Chu, E., Peleato, B., Eckstein, J.: Distributed optimization and statistical learning via the alternating direction method of multipliers. Found. Trends Mach. Learn. **3**(1), 1–122 (2011)

6. Boykov, Y., Lempitsky, V.S.: From photohulls to photoflux optimization. In: Proceedings of the British Machine Conference, vol. 3, p. 27. Citeseer (2006)
7. Boykov, Y., Veksler, O., Zabih, R.: Markov random fields with efficient approximations. In: Proceedings of 1998 IEEE Computer Society Conference on Computer Vision and Pattern Recognition (Cat. No. 98CB36231), pp. 648–655. IEEE (1998)
8. Chambolle, A., Pock, T.: A first-order primal-dual algorithm for convex problems with applications to imaging. J. Math. Imaging Vis. **40**(1), 120–145 (2011). https://doi.org/10.1007/s10851-010-0251-1
9. Chudak, F.A., Hochbaum, D.S.: A half-integral linear programming relaxation for scheduling precedence-constrained jobs on a single machine. Oper. Res. Lett. **25**(5), 199–204 (1999)
10. Dlask, T., Werner, T.: A class of linear programs solvable by coordinate-wise minimization. arXiv.org (2020)
11. Fulkerson, D., Ford, L.: Flows in Networks. Princeton University Press, Princeton (1962)
12. Globerson, A., Jaakkola, T.: Fixing max-product: Convergent message passing algorithms for MAP LP-relaxations. In: Neural Information Processing Systems, pp. 553–560 (2008)
13. Hochbaum, D.S.: Solving integer programs over monotone inequalities in three variables: A framework for half integrality and good approximations. Eur. J. Oper. Res. **140**(2), 291–321 (2002)
14. Hochbaum, D.S., Pathria, A.: Forest harvesting and minimum cuts: a new approach to handling spatial constraints. For. Sci. **43**(4), 544–554 (1997)
15. Hochbaum, D.S., Pathria, A.: Approximating a generalization of MAX 2SAT and MIN 2SAT. Discret. Appl. Math. **107**(1–3), 41–59 (2000)
16. Kappes, J.H., et al.: A comparative study of modern inference techniques for structured discrete energy minimization problems. Int. J. Comput. Vis. **115**(2), 155–184 (2015). https://doi.org/10.1007/s11263-015-0809-x
17. Kohli, R., Krishnamurti, R., Mirchandani, P.: The minimum satisfiability problem. SIAM J. Discret. Math. **7**(2), 275–283 (1994)
18. Kolmogorov, V.: Convergent tree-reweighted message passing for energy minimization. IEEE Trans. Pattern Anal. Mach. Intell. **28**(10), 1568–1583 (2006)
19. Kolmogorov, V.: A new look at reweighted message passing. IEEE Trans. Pattern Anal. Mach. Intell. **37**(5), 919–930 (2015)
20. Kolmogorov, V., Zabih, R.: Computing visual correspondence with occlusions via graph cuts, Technical report. Cornell University (2001)
21. Kovalevsky, V.A., Koval, V.K.: A diffusion algorithm for decreasing the energy of the max-sum labeling problem. Glushkov Institute of Cybernetics. Kiev, USSR (approx. 1975, unpublished)
22. Lempitsky, V., Boykov, Y.: Global optimization for shape fitting. In: 2007 IEEE Conference on Computer Vision and Pattern Recognition, pp. 1–8. IEEE (2007)
23. Lempitsky, V., Boykov, Y., Ivanov, D.: Oriented visibility for multiview reconstruction. In: Leonardis, A., Bischof, H., Pinz, A. (eds.) ECCV 2006. LNCS, vol. 3953, pp. 226–238. Springer, Heidelberg (2006). https://doi.org/10.1007/11744078_18
24. Průša, D., Werner, T.: LP relaxation of the Potts labeling problem is as hard as any linear program. IEEE Trans. Pattern Anal. Mach. Intell. **39**(7), 1469–1475 (2017)
25. Scharstein, D., Szeliski, R.: A taxonomy and evaluation of dense two-frame stereo correspondence algorithms. Int. J. Comput. Vis. **47**(1–3), 7–42 (2002). https://doi.org/10.1023/A:1014573219977

26. Schlesinger, M.I., Antoniuk, K.: Diffusion algorithms and structural recognition optimization problems. Cybern. Syst. Anal. **47**, 175–192 (2011). https://doi.org/10.1007/s10559-011-9300-z
27. Szeliski, R., et al.: A comparative study of energy minimization methods for Markov random fields with smoothness-based priors. IEEE Trans. Pattern Anal. Mach. Intell. **30**(6), 1068–1080 (2008)
28. Tseng, P.: Convergence of a block coordinate descent method for nondifferentiable minimization. J. Optim. Theor. Appl. **109**(3), 475–494 (2001). https://doi.org/10.1023/A:1017501703105
29. Werner, T.: A linear programming approach to max-sum problem: a review. IEEE Trans. Pattern Anal. Mach. Intell. **29**(7), 1165–1179 (2007)
30. Werner, T., Průša, D., Dlask, T.: Relative interior rule in block-coordinate descent. In: Accepted to IEEE/CVF Conference on Computer Vision and Pattern Recognition (2020)
31. Werner, T., Průša, D.: Relative interior rule in block-coordinate minimization (2019). arXiv:1910.09488
32. Wright, S.J.: Coordinate descent algorithms. Math. Program. **151**(1), 3–34 (2015). https://doi.org/10.1007/s10107-015-0892-3
33. Xu, K., Li, W.: Many hard examples in exact phase transitions with application to generating hard satisfiable instances. arXiv.org (2003). Instances available at http://sites.nlsde.buaa.edu.cn/~kexu/benchmarks/max-sat-benchmarks.htm

Travel Times Equilibration Procedure for Route-Flow Traffic Assignment Problem

Alexander Krylatov[1,2(✉)] [ID] and Anastasiya Raevskaya[1] [ID]

[1] Saint Petersburg State University, Saint Petersburg, Russia
{a.krylatov,a.raevskaya}@spbu.ru
[2] Institute of Transport Problems RAS, Saint Petersburg, Russia

Abstract. The present paper is devoted to travel time equilibration procedure for solving traffic assignment problem with respect to route flows. We prove that equilibrium route-flow assignment can be easily obtained by implementing equilibration procedure together with simple rules for sequential extension and/or by reducing the set of active routes. Accurate route-flow solution for the small Sioux Falls network is found via the developed algorithm and demonstrates some important points related with visualization issues, decision making support and scenario analysis in the sphere of transportation planning.

Keywords: Route-flow traffic assignment · User equilibrium · Equilibration procedure

1 Introduction

The traffic assignment problem (TAP) is an optimization problem with non-linear objective function and linear constraints, which allows one to find traffic assignment in a road network by given travel demand values. The solution of TAP is proved to satisfy so called user equilibrium (UE) behavioural principle, formulated by J. G. Wardrop as follows: "*The journey times in all routes actually used are equal and less than those that would be experienced by a single vehicle on any unused route*" [15]. Actually, the first mathematical formulation of TAP was given by Beckmann *et al.* [2]. In this paper we consider an important case of a link-route TAP formulation, available under arc-additive travel time functions. The link-route formulation, first and foremost, allows to establish clear relationships between the link-flow and route-flow assignment patterns. In Sect. 3 we assume that the travel time functions are both separable and non-decreasing and prove that equilibrium route-flow assignment can be easily obtained by implementing simple rules for sequential extension and/or by reducing the set of active

The work was jointly supported by a grant from the Russian Science Foundation (No. 19-71-10012 Multi-agent systems development for automatic remote control of traffic flows in congested urban road networks).

I. S. Kotsireas and P. M. Pardalos (Eds.): LION 14 2020, LNCS 12096, pp. 68–79, 2020.
https://doi.org/10.1007/978-3-030-53552-0_9

routes. Pseudocode for an algorithm based on travel times equilibration procedure is given in Sect. 4. Section 5 demonstrates certain decision making aspects which appear when TAP is solved with respect to route flows under link-route formulation. Conclusion is given in Sect. 6.

2   Traffic Assignment Problem

Let us consider an urban road network presented by a directed graph $G = (E, V)$, where V represents a set of intersections, while $E \subseteq V \times V$ represents a set of available roads between the adjacent intersections. Define $W \subseteq V \times V$ as the ordered set of pairs of nodes with non-zero travel demand $F^w > 0$, $w \in W$. W is usually called as the set of origin-destination pairs (OD-pairs), $|W| = m$. Any set of sequentially linked edges initiating in the origin node of OD-pair w and terminating in the destination node of the OD-pair w we call *route* between the OD-pair w, $w \in W$. The ordered set of all possible routes between nodes of the OD-pair w we denote as R^w, $w \in W$, and ordered set of all possible routes between all OD-pairs we denote as R, $R = \cup_{w \in W} R^w$. Demand $F^w > 0$ seeks to be assigned between the available routes $R^w : \sum_{r \in R^w} f_r^w = F^w$, where f_r^w is a variable corresponding to a traffic flow through route $r \in R^w$ between nodes of OD-pair $w \in W$. Introduce the vector of demand $F = (F^1, \ldots, F^m)^{\mathrm{T}}$ and the vector f associated with the route-flow traffic assignment pattern, such that $f = f_R = (\ldots, f_r^w, \ldots)^{\mathrm{T}}$ is actually a vector of $\{f_r^w\}_{r \in R^w}^{w \in W}$.

Let us introduce differentiable strictly increasing functions on the set of real numbers $t_e(\cdot)$, $e \in E$. We suppose that $t_e(\cdot)$, $e \in E$, are non-negative and their first derivatives are strictly positive on the set of real numbers. By x_e we denote traffic flow on the edge e, while x is an appropriate vector of link-flows, $x = (\ldots, x_e, \ldots)^{\mathrm{T}}$, $e \in E$. Defined functions $t_e(x_e)$ are used to describe travel time on edges e, $e \in E$, and they are commonly called link *delay*, *cost* or *performance* functions. In this section we assume that the travel time function of the route $r \in R^w$ between OD-pair $w \in W$ is the sum of travel delays on all edges belonging to this route. Thus, we define travel time through the route $r \in R^w$ between OD-pair $w \in W$ as the following separable function

$$t_r^w(f) = \sum_{e \in E} t_e(x_e)\delta_{e,r}^w \quad \forall r \in R^w, w \in W, \tag{1}$$

where, by definition,

$$\delta_{e,r}^w = \begin{cases} 1, & \text{if edge } e \text{ belongs to the route } r \in R^w, \\ 0, & \text{otherwise.} \end{cases} \quad \forall e \in E, w \in W,$$

while, naturally,

$$x_e = \sum_{w \in W} \sum_{r \in R^w} f_r^w \delta_{e,r}^w \quad \forall e \in E, \tag{2}$$

i.e. traffic flow on the edge is the sum of traffic flows through all routes which include this edge or in a matrix form

$$x = \Delta_R f_R, \tag{3}$$

with a *link-route incidence matrix* Δ_R (generally, non-square) defined for the set of routes R, whereas Δ_R^T is a *route-link incidence matrix* defined for the set of routes R.

The *equilibrium traffic assignment problem* under separable travel time functions has the following form of optimization program:

$$\min_f \sum_{e \in E} \int_0^{x_e} t_e(u)du, \tag{4}$$

subject to

$$\sum_{r \in R^w} f_r^w = F^w, \quad \forall w \in W, \tag{5}$$

$$f_r^w \geq 0 \quad \forall r \in R^w, w \in W, \tag{6}$$

where, by definition,

$$x_e = \sum_{w \in W} \sum_{r \in R^w} f_r^w \delta_{e,r}^w \quad \forall e \in E. \tag{7}$$

A link-flow assignment pattern x and corresponding route-flow assignment pattern f, satisfying (4)–(7), are proved to reflect *user equilibrium* traffic assignment or such an assignment that

$$t_r^w(f) = \sum_{e \in E} t_e(x_e)\delta_{e,r}^w \begin{cases} = t^w, \text{ if } f_r^w > 0, \\ \geq t^w, \text{ if } f_r^w = 0, \end{cases} \quad \forall r \in R^w, w \in W, \tag{8}$$

where t^w is called an *equilibrium travel time* or travel time on actually used routes between OD-pair w, $w \in W$. Herewith, in case x is searched as a result of solving (4)–(7), then one deals with *link-flow equilibrium traffic assignment problem*. Otherwise, if f is searched, then *route-flow equilibrium traffic assignment problem* is under consideration.

Such formulation is highly fruitful: despite UE principle was formulated in terms of routes, the efficient approach for coping with TAP was firstly obtained in terms of links (arcs) independently by [10] and [12]. Although they just implemented a quadratic programming algorithm ([5]) to estimate UE arc flows, LeBlanc and Nguyen contributed significantly in the research area: when solving TAP, they offered to deal with arcs composing the most appropriate (shortest) paths only without storing all of the decision variables. Currently, storing data on the variables concerning the restricted set of routes and links, has become the most common (and even central) practice among algorithm developers in the field of equilibrium traffic flow assignment.

3   Descent Direction for Equilibrium Route-Flow Traffic Assignment Problem

Equilibration of travel times on any given set of routes decreases the goal function (4). This fact is proved by virtue of the following statements.

Theorem 1 ([3]). *If, for a flow pattern satisfying* (5), *there exist for some* $w \in W$, *paths* $p, q \in R^w$ *such that* $f^p > 0$, $f^q > 0$ *and*

$$t_p^w(f) > t_q^w(f),$$

then $\sum_{e \in E} \int_0^{x_e} t_e(u)du$ *is lowered by transferring flow from* p *to* q.

In its turn, this theorem allows us to prove easily the following corollary.

Corollary 1. *Let us consider a feasible route-flow pattern* $\bar{f}$. *If we reassign flows on routes from a set* $\hat{R} \subset R$ *only in such a way that equilibrates travel times on these routes in order to obtain the route-flow pattern* $\hat{f}$, *then the following inequality holds*

$$\sum_{e \in E} \int_0^{\hat{x}_e} t_e(u)du < \sum_{e \in E} \int_0^{\bar{x}_e} t_e(u)du,$$

where $\bar{x}$ *and* $\hat{x}$ *are corresponding link-flow patterns* $\bar{x} = \Delta_R \bar{f}$ *and* $\hat{x} = \Delta_R \hat{f}$.

Proof. It follows directly from the Theorem of [3].

Therefore, the iterative extension of *the set of active routes* (routes with non-zero flows) and equilibration of travel times on these routes can be treated as the descent direction for global optimization of the problem (4)–(7), since, according to Corollary 1, it decreases the goal function value. However, a route-flow assignment pattern obtained at a certain step of such iterative process can possess negative components.

Theorem 2. *Let us assume that the route-flow pattern* $\hat{f} \in \mathfrak{F}_{\hat{R}}$ *equilibrates travel times on the given set* $\hat{R} \subseteq R$, *and there exists* $\tilde{p} \in \hat{R}^v$ *for some* $v \in W$ *such that* $\hat{f}_{\tilde{p}}^v < 0$. *If the travel time functions* $\{t_r^w(f)\}_{r \in \hat{R}}^{w \in W}$ *are non-decreasing, then equilibration of travel times on routes of the set* $\hat{R} \backslash \tilde{p}$ *will decrease equilibrium travel time.*

Proof. Let us consider a route-flow assignment pattern $\hat{f} \in \mathfrak{F}_{\hat{R}}$. We believe that $\hat{f}$ equilibrates travel times on routes of some given set $\hat{R} \subseteq R$:

$$t_r^w(\hat{f}) = t_p^w(\hat{f}) \quad \forall r, p \in \hat{R}^w, w \in W$$
$$\hat{f}_p^w = 0 \quad \forall p \in R^w \backslash \hat{R}^w, w \in W,$$

and there exists $\tilde{p} \in \hat{R}^v$ for some $v \in W$ such that $\hat{f}_{\tilde{p}}^v < 0$. Hence it means that for this $v \in W$

$$t_r^v(\hat{f}) = t_{\tilde{p}}^v(\hat{f}) \quad \forall r \in \hat{R}^v,$$
$$\hat{f}_{\tilde{p}}^v < 0.$$

At the same time

$$\sum_{r \in \hat{R}^v} \hat{f}_r^v = \sum_{r \in \hat{R}^v \backslash \tilde{p}} \hat{f}_r^v + \hat{f}_{\tilde{p}}^v = F^v$$

or

$$\sum_{r \in \hat{R}^v \setminus \tilde{p}} \hat{f}_r^v = F^v - \hat{f}_{\tilde{p}}^v,$$

thus, since $\hat{f}_{\tilde{p}}^v < 0$, then

$$\sum_{r \in \hat{R}^v \setminus \tilde{p}} \hat{f}_r^v > F^v.$$

Since functions $\{t_e(x)\}_{e \in E}$ are non-decreasing, then there exists a unique equilibrium link-flow assignment pattern $\bar{x}$ for the set of active routes $\hat{R} \setminus \tilde{p}$ and, consequently there exists at least one corresponding route-flow pattern $\bar{f}$ which equilibrates travel times on routes of set $\hat{R} \setminus \tilde{p}$. Herewith

$$\sum_{r \in \hat{R}^v \setminus \tilde{p}} \bar{f}_r^v = F^v,$$

then

$$\sum_{r \in \hat{R}^v \setminus \tilde{p}} \bar{f}_r^v < \sum_{r \in \hat{R}^v \setminus \tilde{p}} \hat{f}_r^v$$

and

$$\sum_{w \in W} \sum_{r \in \hat{R}^w \setminus \tilde{p}} \bar{f}_r^w < \sum_{w \in W} \sum_{r \in \hat{R}^w \setminus \tilde{p}} \hat{f}_r^w.$$

Therefore, since functions $t_r^w(f)$, $r \in \hat{R}$, $w \in W$, are non-decreasing then

$$t_r^w(\bar{f}) \le t_r^w(\hat{f}), \quad \forall r \in \hat{R}^w \setminus \tilde{p}, w \in W.$$

Theorem is proved.

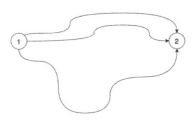

Fig. 1. Single-commodity network

Therefore, a negative component in route-flow assignment pattern, equilibrating travel times, actually indicates that the corresponding route is too "long". Indeed, consider an example network presented by a digraph with two nodes (Fig. 1). Let us assume that two short routes and one long route are available for travelling of 40 vehicles from node 1 to node 2. Travel time functions are presented by $t_1(f_1) = 12 + (f_1/10)^4$, $t_1(f_1) = 10 + (f_2/15)^4$ and

$t_3(f_3) = 30 + (f_3/12)^4$. If we equilibrate travel times on the set of all three routes then $t_1(f_1) = t_2(f_2) = t_3(f_3) = 31.5$ while $f_1 = 21$, $f_2 = 32.3$ and $f_3 = -13.3$. However, if we exclude the long route and equilibrate travel times on the set of two short routes, then $t_1(f_1) = t_2(f_2) = 17.4$, while $f_1 = 15.3$ and $f_2 = 24.7$. Hence, equilibrium travel time on two short routes is less then free travel time on the long one. Thus, negative component in a route-flow pattern which equilibrates travel times on available set of routes indicates that for a given travel demand the corresponding route with negative flow is actually excessive.

Corollary 2. *Minimal equilibrium travel times are reached on the set of actually used routes.*

Proof. Let us consider an arbitrary set of routes $\bar{R}$ and some route $p \notin \bar{R}$. If $\bar{f}$ and $\tilde{f}$ are route-flow patterns, equilibrating travel times on routes $\bar{R}$ and $\bar{R} \cup p$ correspondingly, which do not include negative components, then $t_r^w(\tilde{f}) \leq t_r^w(\bar{f})$ for any $r \in \bar{R}^w$, $w \in W$, since functions $\{t_r^w(f)\}_{r \in R^w}^{w \in W}$ are increasing. Therefore, extension of the set of routes decreases equilibrium travel time if route-flow pattern, equilibrating travel times on routes from extended set, has no negative components.

From the other hand, according to Theorem 2, if $\bar{f}$ and $\tilde{f}$ are route-flow patterns, equilibrating travel times of routes $\bar{R}$ and $\bar{R} \cup p$ correspondingly, in such way that $\bar{f}$ has no negative components while $\tilde{f}$ has a negative component, then $t_r^w(\tilde{f}) \geq t_r^w(\bar{f})$ for any $r \in \bar{R}^w$, $w \in W$.

Therefore, minimal travel times will be reached on such set of routes $\hat{R}$, whose corresponding route-flow pattern $\hat{f}$ has no negative components and equilibrium travel time on any route from $\hat{R}$ is less or equal to zero-route-flow travel time on any route $p \notin \hat{R}$. Consequently, by definition, $\hat{R}$ is the set of actually used routes.

Corollary is proved.

Thus, according to Theorem 2, the exclusion routes with negative flows from the set of active routes and equilibration of travel times on remaining routes can be also treated as descent direction for global optimization of the problem (4)–(7) since it minimizes equilibrium travel time and this, according to Corollary 2, leads to the equilibrium route-flow traffic assignment pattern.

Therefore, two following actions are proved to reflect the descent direction for solving the equilibrium route-flow traffic assignment problem with respect to the set of actually used routes:

- iterative extension of the set of active routes and equilibration of travel times on these routes;
- exclusion routes with negative flows from the set of active routes.

4   Algorithm

The statements that have been proved above allow us to give the general description of an algorithm (Algorithm 1) employing successive travel times equilibration for solving the route-flow traffic assignment problem (under $\mathfrak{C}$ we understand the travel times equilibration procedure, e.g. [7–9]). Since the set of all feasible routes between any OD-pair in an actual urban road network is certainly redundant, it is absolutely typical for the up-to-date algorithms to operate with restricted sets of feasible routes when solving the route-flow equilibrium traffic assignment problem.

Algorithm 1. Solving RF-TAP with respect to the set of actually used routes

1: $f \leftarrow \mathbb{O}$ & $\hat{R} \leftarrow \emptyset$
2: **for** $w = 1, \ldots, |W|$ **do**
3:     $p \leftarrow$ shortest path between OD-pair w
4:     $\hat{R} \leftarrow \hat{R} \cup p$
5: **end for**
6: $f \leftarrow \mathfrak{E}(f, \hat{R})$
7: **while** $\exists \, p \in R^w \backslash \hat{R}^w$, $w \in W : t_p^w(f) < t_r^w(f) \; \forall \, r \in \hat{R}^w$ **do**
8:     **for** $w = 1, \ldots, |W|$ **do**
9:         $p \leftarrow$ shortest path between OD-pair w on G congested by f
10:         $\hat{R} \leftarrow \hat{R} \cup p$
11:     **end for**
12:     $f \leftarrow \mathfrak{E}(f, \hat{R})$
13:     **for** $w = 1, \ldots, |W|$ **do**
14:         **if** $\exists \, r \in \hat{R}^w : f_r^w < 0$ **then**
15:             $\hat{R} \leftarrow \hat{R} \backslash r$
16:         **end if**
17:     **end for**
18:     $f \leftarrow \mathfrak{E}(f, \hat{R})$
19: **end while**
20: **return** f & $\hat{R}$

Theorem 3. *Algorithm 1 converges to an equilibrium route-flow traffic assignment pattern $\hat{f}$ with the set of actually used routes $\hat{R}$.*

Proof. According to Corollary 1, the iterative extension of the set of active routes and equilibration of travel times on these routes can be treated as the descent direction for global optimization of the problem (4)–(7). Thus, steps 8–12 of Algorithm 1 decrease the goal function value of the problem (4)–(7). On the other hand, according to Theorem 2, the exclusion routes with negative flows from the set of active routes minimizes the equilibrium travel time. Thus, according to Corollary 2, steps 13–18 leads to the equilibrium route-flow traffic assignment pattern as well.

Theorem is proved.

5   Numerical Result

Let us consider the Sioux Falls road network the complete input data and the best link-flow solution of which are available at:

https://github.com/bstabler/TransportationNetworks/tree/master/SiouxFalls.
We apply Algorithm 1 to solve TAP on the Sioux Falls network in order to obtain
a route-flow solution. Link delays are "traditionally" modeled by the following
BPR function:

$$t_e(x_e) = t_e^0 \left[1 + 0.15 \left(\frac{x_e}{c_e} \right)^4 \right] \quad \forall e \in E.$$

Notice, lines 7–19 in Algorithm 1 represent so-called *main iteration* of the
method. At this stage the algorithm checks if there are feasible routes to be
supplemented into the set of routes with non-zero flows. Zero main iteration
constructs the initial feasible route-flow pattern. Within coping with the present
case we constructed the initial pattern by disaggregating the master problem into
individual OD-pairs and solving TAP sequentially for each OD-pair. Herewith,
when solving TAP for any individual OD-pair we believed that the network is
congested by route flows obtained as solutions for already considered OD-pairs.
Eventually, Algorithm 1 with operator $\mathfrak{K}$ took seven main iterations to get highly
accurate solution as given in Table 1.

Table 1. Main iterations

Main iteration	0	1	2	3	4	5
Goal function value	47.738	43.558	42.753	42.427	42.325	42.317

Main iteration	6	7
Goal function value	42.31335287107793	42.31335287107442

Let us mention that the set of routes with non-zero flows stopped changing at
the sixth main iteration and the seventh main iteration was devoted to achieving
the higher accuracy level of the route-flow solution under already obtained set of
actually used routes. Moreover, it was 88% of computing time that was spent on
revealing the set of actually used routes, while improving the precision level of
the equilibrium route-flow assignment took the remaining 12%. Herewith, when
equilibrating travel times by the operator $\mathfrak{K}$ we used 10^{-12} level of precision
which means that we believed that travel times were equal, if they were equal
up to twelfth digit after the decimal separator. All the output data concerning
the obtained route-flow solution such as the set of actually used routes, route
flows etc. one can find at:
http://www.apmath.spbu.ru/ru/staff/krylatov/files/RFsolutionSiouxFalls.csv.
Let us emphasize that for any OD-pair obtained the actually used routes are independent and, consequently, the obtained equilibrium route-flow traffic assignment pattern is unique for this set of routes. Let us pay attention to visualization
ability of the route-flow TAP solution and its assistance for decision-making in
transportation planning.

First of all, the route-flow TAP solution allows one to make a matrix which
gives amount of actually used routes between each OD-pair. In fact, such a

Table 2. Amounts of actually used routes between OD-pairs in the Sioux Falls network

	1	2	3	4	5	6	7	8	9	10	11	12	13	14	15	16	17	18	19	20	21	22	23	24
1	–	1	1	1	1	1	1	1	1	1	1	1	1	1	1	2	1	1	1	1	1	1	1	1
2	1	–	1	1	1	1	1	1	1	1	1	1	1	1	1	1	1	–	1	1	–	1	–	–
3	1	1	–	1	1	1	1	2	1	1	1	1	1	1	1	1	1	–	–	–	–	1	1	–
4	1	1	1	–	1	1	1	1	1	1	1	1	1	1	1	1	1	1	1	1	1	1	1	1
5	1	1	1	1	–	1	1	1	1	1	1	1	1	1	1	1	1	–	1	1	1	1	1	–
6	1	1	1	1	1	–	1	1	1	1	1	1	1	1	1	1	1	1	1	1	1	1	1	1
7	1	1	1	1	1	1	–	1	1	1	1	1	2	1	1	1	1	1	1	1	1	1	1	1
8	1	1	1	1	1	1	1	–	1	1	1	1	3	1	1	1	1	1	1	1	1	1	1	1
9	1	2	1	1	1	1	1	1	–	1	1	1	1	1	1	3	1	1	1	1	1	1	2	1
10	1	1	1	1	1	1	1	1	1	–	1	2	1	1	1	1	1	1	2	1	1	2	2	2
11	1	1	1	1	1	1	1	1	1	1	–	1	1	1	2	1	1	1	1	1	1	1	1	1
12	1	1	1	1	1	1	1	1	2	1	1	–	1	1	1	2	2	1	1	1	1	1	1	1
13	1	1	1	1	1	1	2	1	1	1	1	1	–	1	2	1	1	1	3	1	1	1	1	1
14	1	1	1	1	1	1	1	1	1	1	1	1	2	–	1	1	1	1	1	1	1	2	1	1
15	1	1	1	1	1	1	1	1	1	1	2	1	3	1	–	1	1	1	1	1	1	1	1	1
16	1	1	1	1	2	1	1	1	3	1	1	1	1	1	1	–	1	1	2	1	1	1	1	1
17	1	1	1	1	1	1	1	1	1	1	1	1	1	1	1	1	–	1	1	1	2	1	1	2
18	1	–	–	1	–	1	1	1	1	1	1	1	1	1	1	1	1	–	1	1	1	1	1	–
19	1	1	–	1	1	1	1	1	1	1	1	1	2	1	1	1	1	1	–	1	2	1	1	1
20	1	1	–	1	1	1	1	1	1	2	1	1	1	1	1	1	1	1	1	–	1	1	1	1
21	1	–	–	1	1	1	1	1	1	1	1	1	1	1	1	1	2	1	2	1	–	1	1	1
22	1	1	1	1	1	1	1	1	1	1	2	1	2	1	1	1	1	1	1	1	1	–	1	1
23	1	–	1	1	1	1	1	1	1	2	1	1	1	1	1	1	1	1	1	1	1	1	–	1
24	1	–	–	1	–	1	1	1	1	2	1	1	1	1	1	1	1	–	1	1	1	1	1	–

matrix reflects how the road network is used by drivers, i.e. which OD-pairs are the most or the least loaded within available road infrastructure capability and overall travel demand. Moreover, changing of elements values in the matrix with amounts of actually used routes during successive periods of time indirectly demonstrates dynamics of road network operation. Thus, in case of the Sioux Falls road network we obtained the following amounts of actually used routes between OD-pairs (Table 2). Vast majority of travel demand values between given OD-pairs are satisfied by the single route. Indeed, among 528 OD-pairs there are only 36 OD-pairs with more than one route. Herewith, only five OD-pairs require three routes each, while the rest 31 OD-pairs need two routes each. A single actually used route between an OD-pair means that corresponding travel demand is met by the shortest path in the congested urban area, i.e. no driver deviates from this path. On the contrary, several actually used routes between an OD-pair mean that corresponding travel demand is met by the shortest path

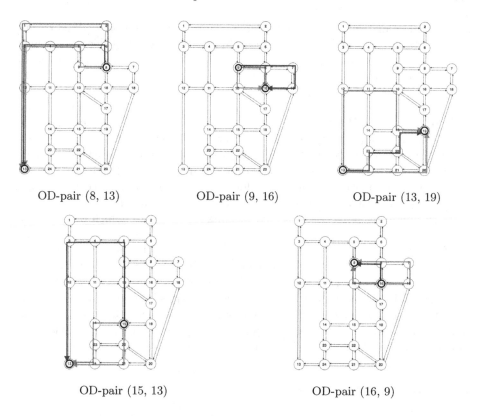

OD-pair (8, 13)   OD-pair (9, 16)   OD-pair (13, 19)

OD-pair (15, 13)   OD-pair (16, 9)

Fig. 2. Sioux Falls road network

in the congested urban area together with some other paths, i.e. drivers get to choose between certain alternatives.

Secondly, only the route-flow solution is actually able to show how the network users behave. Moreover, a decision maker or a traffic engineer can observe users behavior for each OD-pair separately when route-flow solution is obtained, that is undoubtedly very convenient. Meanwhile, user equilibrium is primarily a behavioral principle and so the behavioral strategies of users are expected to be obtained. Thus, a route-flow solution represents the average individual user strategies which is also very convenient from perspectives of decision making in network management. Indeed, consider, for instance, OD-pairs (8, 13), (9, 16), (13, 19), (15, 13) and (16, 9) with 3 actually used routes between every of them. In other words, due to clear visualization of actually used routes, a decision maker is able to see how the drivers behave under given traffic conditions and which routes appeared to be the most appropriate between an OD-pair in the congested road network.

Thirdly, topology of actually used routes can prompt feasible ways for improving road network operation. Thus, the actual routes from node 8 to node 13 have allocation that forms the bottleneck within traffic assignment of just a single OD-pair. Hence, the road interchange in the node 3 or roadbed extension on arcs (3, 12) and (12, 13) are quite feasible improving decisions which follow directly from given visualization. In turn, the actual routes from node 13 to node 19 allocate well without creating problem points in the road network. Meanwhile, the average travel time from node 15 to node 13 could be decreased by extension of roadbed on arcs (10, 11) and (11, 12). Finally, no improvements seem to be available for OD-pairs (9, 16) and (16, 9), since trips between nodes 9 and 16 are made in "dense" transit topology which follows from too close proximity of these nodes and high travel demand values. Nevertheless, let us accentuate that described decisions certainly cannot be considered as a solution of any kind in a mathematical sense (for example, solution of a corresponding network design problem). However, the decision maker can simulate changes in network operation based on obtained insights and evaluate which ones lead to better results.

Therefore, software on decision making support in traffic management or network design can be easily supplemented by two compact and clear tools. The first one is a matrix of amounts of actually used routes between OD-pairs (like in Table 2), while the second one is visualization of routes between any chosen OD-pair (like on Fig. 2). Hence, once usage of a network has been assessed by virtue of the matrix of amounts of actually used routes, a decision maker can visualize in details the most significant OD-pairs from perspectives of used routes.

6   Conclusion

An important case of a link-route TAP formulation, available under arc-additive travel time functions, is considered in this paper. We proved that equilibrium route-flow assignment can be easily obtained by implementing equilibration procedure together with simple rules for sequential extension and/or by reducing the set of active routes. Accurate route-flow solution for the small Sioux Falls network was given. The solution was found via the developed algorithm in order to demonstrate some important points related with visualization issues, decision making support and scenario analysis in the sphere of transportation planning.

A comprehensive empirical study on comparing the traffic assignment algorithms from these above mentioned groups has been recently made by [14]. They concluded that when choosing a suitable algorithm for solving TAP, the level of accuracy shall be one of the most important factors to be taken into account. Thus, the highest level of precision was demonstrated by bush-based approaches like the algorithm of [4] and the algorithm of [1]. Path-based methods as well as various advanced link-based algorithms were categorized as ones which achieve medium precision level in a reasonable amount of time. Low precision level was demonstrated by simple link-based algorithms employing the basic computational technique of Frank-Wolfe. However, while drawbacks and limits of link-based approaches applying the Frank-Wolfe algorithm for solving TAP had already been widely investigated by the end of the 20th century (see, for instance, [11], [13]) and no any other breakthroughs in this regard have been obtained in recent decades, the newest results on path-based approaches seem to be encouraging. Indeed, according to the computational study of [6], the algorithms based on path equilibration techniques are able to demonstrate state-of-the-art level of accuracy in a reasonable amount of time.

References

1. Bar-Gera, H.: Traffic assignment by paired alternative segments. Transp. Res. Part B **44**(8–9), 1022–46 (2010)
2. Beckmann, M., McGuire, C., Winsten, C.: Studies in the Economics of Transportation. Yale University Press, New Haven (1956)
3. Devarajan, S.: A note on network equilibrium and noncooperative games. Transp. Res. Part B **15**(6), 421–426 (1981)
4. Dial, R.: A path-based user-equilibrium traffic assignment algorithm that obviates path storage and enumeration. Transp. Res. Part B **40**, 917–936 (2006)
5. Frank, M., Wolfe, P.: An algorithm for quadratic programming. Naval Res. Logist. Q. **3**, 95–110 (1956)
6. Galligari, A., Sciandrone, M.: A computational study of path-based mathods for optimal traffic assignment with both inelastic and elastic demand. Comput. Oper. Res. **103**, 158–166 (2019)
7. Krylatov, A.Y.: Network flow assignment as a fixed point problem. J. Appl. Ind. Math. **10**(2), 243–256 (2016). https://doi.org/10.1134/S1990478916020095
8. Krylatov, A.: Reduction of a minimization problem for a convex separable function with linear constraints to a fixed point problem. J. Appl. Ind. Math. **12**(1), 98–111 (2018)
9. Krylatov, A., Shirokolobova, A.: Projection approach versus gradient descent for network's flows assignment problem. Lect. Notes Comput. Sci. **10556**, 345–350 (2017)
10. LeBlanc, L.: Mathematical programming algorithms for large scale network equilibrium and network design problem. Northwestern University, Department of Industrial Engineering and Management Science (1973)
11. Lupi, M.: Convergence of the Frank-Wolfe algorithm in transportation networks. Civ. Eng. Syst. **19**, 7–15 (1985)
12. Nguyen, S.: A mathematical programming approach to equilibrium methods of traffic assignment with fixed demands. Publication 138, Départment d'Informatique et de Recherche Opérationelle, Université de Montréal, Montréal (1973)
13. Patriksson, M.: The Traffic Assignment Problem: Models and Methods. VSP, Utrecht (1994)
14. Perederieieva, O., Ehrgott, M., Raith, A., Wang, J.: A framework for and empirical study of algorithms for traffic assignment. Comput. Oper. Res. **54**, 90–107 (2015)
15. Wardrop, J.G.: Some theoretical aspects of road traffic research. Proc. Inst. Civ. Eng. **2**, 325–378 (1952)

confStream: Automated Algorithm Selection and Configuration of Stream Clustering Algorithms

Matthias Carnein[1]([⊠]), Heike Trautmann[1], Albert Bifet[2],
and Bernhard Pfahringer[2]

[1] University of Münster, Münster, Germany
{carnein,trautmann}@wi.uni-muenster.de
[2] University of Waikato, Hamilton, New Zealand
{abifet,bernhard}@waikato.ac.nz

Abstract. Machine learning has become one of the most important tools in data analysis. However, selecting the most appropriate machine learning algorithm and tuning its hyperparameters to their optimal values remains a difficult task. This is even more difficult for streaming applications where automated approaches are often not available to help during algorithm selection and configuration. This paper proposes the first approach for automated algorithm selection and configuration of stream clustering algorithms. We train an ensemble of different stream clustering algorithms and configurations in parallel and use the best performing configuration to obtain a clustering solution. By drawing new configurations from better performing ones, we are able to improve the ensemble performance over time. In large experiments on real and artificial data we show how our ensemble approach can improve upon default configurations and can also compete with a-posteriori algorithm configuration. Our approach is considerably faster than a-posteriori approaches and applicable in real-time. In addition, it is not limited to stream clustering and can be generalised to all streaming applications, including stream classification and regression.

Keywords: Stream clustering · Data streams · Automated Machine Learning · Algorithm configuration · Algorithm selection

1 Introduction

Over the past decades, machine learning has revolutionised many of its application areas. However, due to the abundance of machine learning algorithms and application scenarios, it is often necessary to select an algorithm which is most suited for a given problem. In addition, machine learning algorithms tend to be very sensitive to their configuration and it is important to tune hyperparameters to their optimal values, which can be difficult even for experienced users. A first approach that tries to alleviate the user from this is Automated

© Springer Nature Switzerland AG 2020
I. S. Kotsireas and P. M. Pardalos (Eds.): LION 14 2020, LNCS 12096, pp. 80–95, 2020.
https://doi.org/10.1007/978-3-030-53552-0_10

Machine Learning [18]. It attempts to make design decisions such as the selection and configuration of machine learning algorithms automatically. Unfortunately, many of these automated approaches require multiple passes over the data and cannot adapt to changes over time. This makes them infeasible for any online or streaming application. However, many of today's data sources are data streams due to the widespread usage of sensors, the internet-of-things and social media. In this paper, we address the problem of automated algorithm selection and configuration of stream clustering algorithms which aim to maintain clusters over time in a stream of observations.

By using an ensemble of different algorithms and configurations, we are able to adapt the optimal algorithm and its hyperparameter settings over time. Our approach is not limited to stream clustering but can be applied to all streaming scenarios, including stream classification and regression. In our experiments, we use several state-of-the-art stream clustering algorithms. On multiple real and artificial data streams, we show that our ensemble-approach always performs better than the default configuration. Even compared to offline and a-posteriori configuration approaches it produces competitive results, while being much faster and applicable in real-time.

2   Background

2.1   Stream Clustering

Clustering is a popular tool for pattern recognition and is often used in marketing or network analysis. However, a major drawback of traditional clustering is that it requires a fixed data set. Whenever new data becomes available, the entire analysis needs to be repeated. This is time consuming, undesirable and often infeasible when working with data streams. An approach to solve this is stream clustering which is able to cluster a continuous and possibly infinite stream of observations. In stream clustering, relevant information is usually extracted into so called *micro-clusters* before discarding an observation. The micro-clusters are then "reclustered" into the final *macro-clusters* upon the user's request. A survey of stream clustering algorithms is available in [10].

Unfortunately, (stream) clustering algorithms usually require many hyperparameters to be set a-priori [10]. For example, typical implementations of `DenStream` [6] have 8 hyperparameters [5] ranging from distance and weight thresholds to window sizes. In practice, these settings are often difficult and unintuitive to choose. With streaming data, the hyperparameters also need to be adapted over time as the data changes. In this paper, we aim to automatically select the best algorithm and its optimal configuration over time.

2.2   Automated Machine Learning

Automated Machine Learning (AutoML) attempts to make the design decision in machine learning automatically [18]. For example, it tries tune the hyperparameters of algorithms automatically or select the most appropriate algorithms. Popular approaches in AutoML are `irace` [20], `SMAC` [16] or `ParamILS` [17].

`irace` for example can be used for automated algorithm configuration. It uses a racing procedure where configurations that perform statistically worse are removed after every race. New parameter configurations are drawn according to probability distributions and the sampling is biased towards better performing configurations. Unfortunately, the racing procedure makes it difficult to apply `irace` for streaming applications. However, we can draw some inspiration from the parameter sampling for our streaming case.

First ideas for Automated Algorithm Selection in the streaming scenario can be found in the stream *classification* [21–24], stream *regression* [25] and *online* learning [13,14] literature. First attempts used an ensemble approach where a meta-classifier periodically predicts the most suitable algorithm based on the stream's characteristics [23]. Similarly, the `BLAST` algorithm [22,24] also uses an ensemble of algorithms, but simply selects the best algorithm of the last window as the *active* classifier. In the following, we use the same idea as an inspiration to select and configure stream *clustering* algorithms.

In a first proof-of-concept [11], we already propose an ensemble approach for automated algorithm *configuration* of stream *clustering* algorithms, called `confStream`. For this, we use an ensemble of different configurations. Periodically, the clustering quality for every configuration is evaluated. Based on the observed performance, a regression model is trained to predict the performance of unknown configurations. Subsequently, a well performing configuration is sampled from the ensemble and used to create a new configuration from it. If its predicted performance is good enough, it replaces one of the configurations in the ensemble.

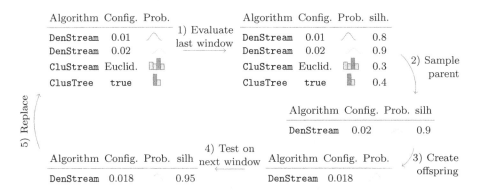

Fig. 1. First, the performance of all algorithms and configurations in the ensemble is evaluated. Afterwards, one algorithm is sampled to create an offspring and tested on the next window. If its performance is high enough, it is used to replace one of the algorithms in the ensemble. For brevity, only one hyperparameter per algorithm is shown.

3   Automated Algorithm Selection and Configuration for Stream Clustering

In this section, we extend our ensemble idea for the confStream algorithm. In our initial proposal, we only optimised one numeric parameter for a given algorithm. Here, we extend upon this and also include the algorithm *selection* problem. In particular, we treat the algorithm selection and algorithm configuration problem as one large optimisation task. In addition, we show how to optimise multiple parameters per algorithm which can be of different types such as numerical, categorical, integer, binary or ordinal. Finally, we also improve the selection process of new configurations and extensively test and compare the algorithm.

Our main idea is summarised in Fig. 1. Our algorithm uses a given starting configuration, i.e. a list of algorithms, their initial configurations and the corresponding parameter ranges. For example, this can be the default configuration of all available algorithms.

To apply our ensemble strategy, we process the stream in windows of size h. We use the observations in a window in order to train the algorithms in the current ensemble. After every window, we evaluate the clustering quality for every configuration (Step 1). In our experiments, we used the Silhouette Width as a measure of cluster quality due to its popularity but other quality metrics are equally applicable. For every observation i in the window, the Silhouette Width uses the average similarity to its own cluster $a(i)$ and compares it to the average similarity to its closest cluster $b(i)$:

$$s(i) = \frac{b(i) - a(i)}{\max\{a(i), b(i)\}}. \tag{1}$$

The Silhouette Width is usually averaged over all observations to obtain a single index. The algorithm with the highest cluster quality becomes the active clusterer or *incumbent* and provides the current solution of confStream.

To obtain new configurations, a configuration is sampled from the ensemble and serves as a parent (Step 2). The sampling is performed proportionally to the performance of the algorithms such that better performing configurations have a higher probability to be selected. In general, this can be implemented using a simple Roulette Wheel Selection. Note, however, that the Silhouette Width has a range of $[-1, 1]$. Since negative values cannot be used in Roulette Wheel Selection, we shift the values by $|\min x| - \min x$, where x are the Silhouette Widths of all configurations. This would turn a sequence $\{-0.5, 0.8, 1\}$ into $\{0.5, 1.8, 2\}$. Also note that we do not necessarily choose the best performing algorithm in order to increase the diversity in the ensemble.

The selected algorithm and configuration is then used as a parent in order to derive a new configuration from it (Step 3). For this, we use some of the ideas of irace [20]. Specifically, every parameter of every configuration has an associated probability distribution. For numerical parameters, every parameter maintains a truncated normal distribution $\mathcal{N}(\mu, \sigma)$ with expectation μ and standard deviation σ. The expectation of the truncated normal distribution is placed

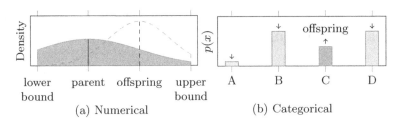

Fig. 2. For numerical parameters, a truncated normal distribution is placed at the parent to sample a new configuration. For categorical parameters, a new configuration is drawn according to a list of probabilities. Both approaches increase the probability for the offspring to favour promising solutions over time.

at the position of the parent to sample a new configuration. Subsequently, the standard deviation of the child is reduced to increase exploitation of this area (Fig. 2a). In our case, we use an exponential decay of the standard deviation according to a fading factor λ:

$$\sigma_{t+1} = \sigma_t \cdot 2^{-\lambda} \tag{2}$$

The idea is to reduce the standard deviation in exponentially smaller steps in order to narrow down a promising parameter region. Initially, the standard deviation is set to half the parameter range. For integer parameters, the same sampling strategy can be used, where the result is rounded to the nearest integer. Similarly, for ordinal parameters (e.g. strong, medium, weak) we can use the integer sampling strategy, where the resulting integer is used as the index in the list of possible outcomes.

For categorical parameters, a list of probabilities for every outcome is maintained. Starting with equal probabilities, the new configuration is sampled according to the list. Subsequently, the probability of the winning category is increased to facilitate exploitation (Fig. 2b). In our case, we increase the probability by the same amount that we reduce the standard deviation in the numerical case:

$$p_{t+1} = p_t \cdot \left(2 - 2^{-\lambda}\right). \tag{3}$$

Note that in both cases we use a factor of one as the baseline and either increase or decrease the factor by $1 - 2^{-\lambda}$ to decrease the standard deviation or increase the probability. To obtain probabilities, all values are then scaled to sum to one again.

Since the streams' distribution can change over time, it is not necessarily beneficial to exploit promising regions further. Instead, it is also necessary to explore new regions of the search space. For this, for a fraction of the ensemble, we reset the standard deviation or the probability vector to its initial value with probability p to explore new regions.

Note that adapting parameter values to create the child is not always easy. While some parameters, such as distance thresholds, can be easily changed "on-the-fly", i.e. while the algorithm is running, some parameters are much harder to change. For example, if the parameter influences a tree-height as in `ClusTree` [19], changing the parameter is often not possible due to the implications for the underlying data structure. For all cases where we cannot change the parameter, we instead initialise a new algorithm instance based on the new configuration. However, to keep as much information as possible, we then train the new algorithm with the micro-clusters of the parent configuration. For this, the centres of the micro-clusters are used as virtual points to train the child algorithm. While this will not reproduce the exact same micro-clusters, it passes on some information about the current clusters.

In our initial proposal [11], we trained a regression model [15] to predict the performance of the new configurations. However, we noticed that the regression model often favoured algorithms with many valid configurations (such as `BICO` [12]) whereas it disfavoured algorithms where some configurations perform exceptionally well but many fail. To prevent this, we eliminated the predictor and introduce a "test ensemble" instead where new configurations are evaluated on the actual stream before deciding whether to incorporate them into the ensemble (Step 4). After every window, we fill the test ensemble with a number of new configurations. The algorithms of the test ensemble are compared to algorithms in the active ensemble when clustering the next window. If a configuration in the test-ensemble outperforms a configuration in the active ensemble (on data both have never seen before), it is considered promising and replaces a configuration in the active ensemble. The decision to use or discard a new configuration is therefore lagged by one window.

Again, we sample the configuration that is replaced proportionally to its fitness, removing less fit solutions with higher probability. However, we never remove the best configuration of an algorithm and always keep at least one configuration per algorithm. This prevents that an algorithm is removed entirely and cannot be used for the generation of new configurations. For larger ensembles, it can also be beneficial to keep the default configurations in the ensemble as a fall-back. Note that we use an ensemble of fixed size here and initially fill the ensemble with new configurations to its maximum capacity.

4   Experiments

4.1   Experimental Setup

To evaluate our algorithm, we implemented it in Java[1] as a clustering algorithm for the `MOA` framework [4,5]. For our experiments, we aim to select the optimal algorithm and its configuration among the available stream clustering algorithms in `MOA` for a specific stream. For a fair comparison, we restrict ourselves to all clustering algorithms which expose the micro-clusters. Specifically, we optimise `DenStream` [6], `ClusTree` [19], `CluStream` [2] and `BICO` [12] which are all state-of-the-art stream clustering algorithms [10]. For every algorithm,

---

[1] Implementation available at: https://www.matthias-carnein.de/confStream.

we optimise all parameters that influence the clustering result in their full value range (Table 1). Unbounded value ranges are artificially capped using appropriate maximum values.

Table 1. Overview of optimised algorithms and parameters. Parameter names as used in MOA [5].

Algorithm	Configuration	Type	Range	default
DenStream	e	Numeric	$[0, 1]$	0.02
	b	Numeric	$[0, 1]$	0.2
	m	Integer	$\{0 \ldots 10000\}$	1
	o	Integer	$\{2 \ldots 20\}$	2
	l	Numeric	$[0, 1]$	0.25
ClusTree	H	Integer	$\{1 \ldots 20\}$	8
	B	Boolean	$\{1, 0\}$	0
CluStream	k	Integer	$\{2 \ldots 20\}$	5
	m	Integer	$\{1 \ldots 10000\}$	100
	t	Integer	$\{1 \ldots 10\}$	2
BICO	k	Integer	$\{2 \ldots 20\}$	5
	n	Integer	$\{1 \ldots 2000\}$	1000
	p	Integer	$\{1 \ldots 20\}$	10

To determine whether confStream is able to improve configurations over time, we first compare its performance to MOA's [4,5] default configurations. For this, we initialise our ensemble with the same default configurations and compare the results. Afterwards, we use irace [20] and optimise the parameter's a-posteriori to find the best overall configuration. We then compare whether confStream's adaptive approach is able to compete with the optimal result. For confStream, we set the ensemble size $e_{\text{size}} = 20$, fading factor $\lambda = 0.05$, reset probability $p = 0.01$ and evaluate the solutions every $h = 1000$ observations using the Silhouette Width. After each window, we create $e_{\text{test}} = 10$ new configurations.

We evaluate our algorithm on four data sets. Specifically, we use a Random Radial Basis Function (RBF) stream [4], the sensor stream[2], the power-supply stream[3] and the covertype data set[4]. All data sets are popular choices within the stream clustering and classification literature and for our analysis we use all numeric parameters of the streams. Note that the covertype data set is a static data set, which we turn into a data stream by processing observations one by one. This is a common strategy in stream clustering due to the limited

---

[2] http://db.csail.mit.edu/labdata/labdata.html.
[3] http://www.cse.fau.edu/~xqzhu/stream.html.
[4] http://archive.ics.uci.edu/ml/datasets/Covertype.

number of openly available data streams [1, 7, 8, 12]. An overview of all data sets is given in Table 2. To avoid differences in scale, we standardise the data sets by subtracting the mean and dividing by the standard deviation per feature. In real world scenarios, the values for normalisation can often be updated incrementally [3, 9]. Our goal was to include data streams with diverse characteristics. For this, we included both real and artificial data streams which all include different forms of concept drift, i.e. a shift of the underlying distribution. We also included a static data set which does not include any temporal changes. In addition, our data streams have between 2 and 10 dimensions and some have more than 2 million observations.

Table 2. Overview of the four data streams used in our experiments. All data streams are popular in the stream clustering literature [7].

Data set	n	d	Type	Drift
Random RBF	2,000,000	2	Artificial	✓
sensor	2,219,803	4	Real	✓
powersupply	29,928	2	Real	✓
covertype	581,012	10	Real	–

4.2   Results

Comparison to Default Configuration. For a start, we compare our conf-Stream algorithm with MOA's default algorithm configurations. For this, conf-Stream is initialised with the same default configurations (Table 1) but optimises the parameters. Figure 3 shows the Silhouette Width for every window of our test data streams. The boxplots on the right summarise the distribution of the Silhouette Width values along the stream. It is obvious that our ensemble approach produces a considerably better result throughout the entirety of the stream. For example, for the Random RBF stream, confStream yields a median Silhouette Width of 0.86 while the other algorithms perform much worse with median Silhouette Widths between 0.54 and 0.65. Similar results can also be observed for the remaining data streams. This is particularly visible for the covertype and powersupply data streams where most algorithms produce much worse results by default.

To evaluate how confStream optimises the algorithms in the ensemble over time, we analyse the best configuration for every clustering algorithm within the ensemble (Fig. 4a). We can see that throughout most segments of the stream, a configuration of ClusTree is the incumbent, i.e. the best configuration. BICO on the other hand tends to have the worst performance within the ensemble. This also shows in the ensemble composition (Fig. 4b). The ensemble is quickly filled with more configurations of ClusTree as it shows the best performance. Even

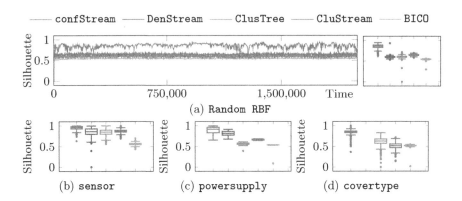

Fig. 3. Development of Silhouette Width for all data streams.

though the other algorithms have less relevance, we can see that their share in the ensemble increases whenever their performance improves. This is particularly visible for the periodic peaks of DenStream and also shows for CluStream as it improves after around 1.5 million observations.

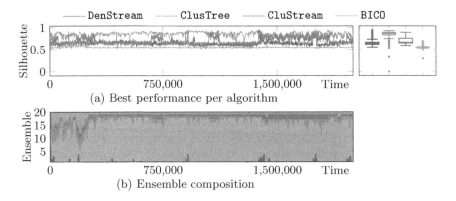

Fig. 4. Ensemble performance and composition for the Random RBF stream.

We can also see how confStream quickly adapts the parameter values depending on their performance. Figure 5 shows the current best algorithm and its parameter values for the first 100,000 observations of the Random RBF stream on a logarithmic scale. We can see that the algorithm initially switches between configurations of DenStream and CluStream, before eventually settling on ClusTree as the incumbent. After about 30.000 observations CluStream briefly becomes the incumbent and its parameters are improved before the algorithm goes back to ClusTree.

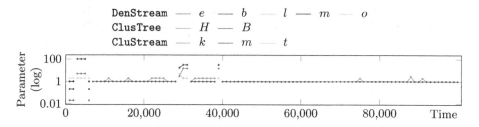

Fig. 5. Best algorithm and configuration over time for the first 100, 000 observations of the Random RBF stream. The currently best algorithm with its best configuration is shown on a logarithmic scale. The parameters are color coded based on the algorithm and the parameter names of MOA [4,5] are used.

Evaluation of Combined Algorithm Selection and Hyperparameter Configuration. In our experiments, we treated both algorithm selection and configuration as one large optimisation problem. This is also known as the Combined Algorithm Selection and Hyperparameter Configuration (CASH) [18] problem. In order to evaluate how this affects confStream's performance, we also compare it to individually configured stream clustering algorithms. For this, we use the default configuration per algorithm as separate starting configurations for confStream and optimise the algorithms separately using the same ensemble size and settings. Figure 6 compares the combined optimisation and individual optimisation. The results show that confStream's solution to the CASH problem is similar to the individual optimisation. This is a surprising result, given the tremendously increased search space of the combined optimisation problem. Generally, we would expect the individual optimisation to perform better. The fact that the CASH problem yields similar results shows that the algorithm can handle the increased search space. This shows that it is possible to perform algorithm selection and configuration simultaneously without sacrificing the clustering quality.

It is also interesting to compare the individual optimisation (Fig. 6) to the algorithms' performance of the combined optimisation problem (Fig. 4a). We can see, that ClusTree and BICO are almost as well configured in the combined optimisation as in the individual optimisation. DenStream on the other hand yields better results when optimised individually. This is likely due to the fact that the algorithm has many parameters and large parameter ranges which causes a large search space. Its weak performance, however, leads the ensemble to remove most instances of DenStream which yields too few instances to explore the large search space.

Comparison to Offline Optimisation. The above results show that confStream can considerably improve upon the default configurations with little additional knowledge. However, we should also compare our algorithm to optimised configurations using an a-posteriori scenario, i.e. given the knowledge of

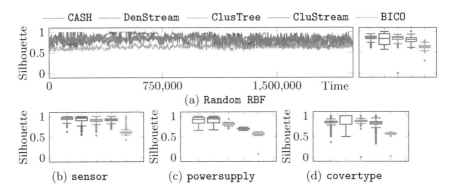

(a) Random RBF

(b) sensor   (c) powersupply   (d) covertype

Fig. 6. Combined algorithm selection and hyperparameter optimisation (CASH) compared to only hyperparameter optimisation for all data streams.

Table 3. Overview of optimal configurations as identified by `irace` (rounded).

Algorithm	Configuration	RBF	sensor	power	cover
DenStream	e	0.08	0.80	0.35	0.55
	b	0.32	0.26	0.02	0.61
	m	2913.1	9085.1	4027.1	282.4
	o	16.49	7.08	7.54	3.37
	l	0.10	0.07	0.88	0.11
ClusTree	H	8	3	1	1
	B	false	true	true	false
CluStream	k	5	8	5	3
	m	100	98	200	4
	t	2	2	2	2
BICO	k	2	6	14	16
	n	36	1880	53	637
	p	7	9	3	2

the total stream. In [7], we already used `irace` [20] to find configurations for some of the algorithms and data streams by optimising for the optimal adjusted Rand index and in [8] when optimising for the Sum of Squares (SSQ). In addition, we now also use `irace` and optimise for the Silhouette Width directly. This gives us three configurations for most algorithms, which have been optimised a-posteriori. We initialise `irace` with the default algorithm configuration and allow up to 150 evaluations for every data stream. Note that this would be equivalent to running `confStream` with an ensemble size of 150. In contrast, we only use 20 in our experiments, which gives `irace` a clear advantage. The configurations which produced the highest median Silhouette Width across the entire stream are listed in Table 3.

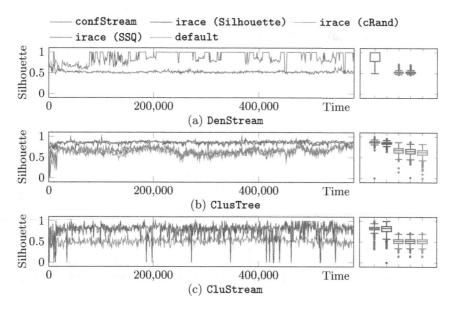

Fig. 7. Comparing performance of `confStream`, the optimal configuration found by `irace` optimised for the adjusted Rand Index (cRand) [7], the `SSQ` [8] and Silhouette Width as well as the default configuration for the `covertype` data stream.

Figure 7 compares the performance of `confStream` (without algorithm selection) with its default and the three optimised configurations for the `covertype` data set. For example, for the `ClusTree` algorithm (Fig. 7b), the default configuration yields the worst quality with a median Silhouette Widh of 0.63. As expected, all `irace` configurations perform better and the one optimised for the Silhouette Width also yields a higher Silhouette of 0.85. However, throughout the vast majority of the stream, `confStream` yields even higher quality than the a-posteriori solutions with a median Silhouette Width of 0.87. This is because `confStream` can adapt the configuration over time which allows it to adapt to changes in the stream. The results for `CluStream` are overall similar (Fig. 7c) and for `DenStream`, the default configuration and one of the optimised configurations did not produce a valid solution at all (Fig. 7a). `confstream` on the other hand produced good solutions throughout most of the stream.

Note that we highlight the results for the `covertype` data set here because we have the most complete set of configurations for this scenario. The same analysis for the other data streams is summarised in Fig. 8. For brevity, we only report the boxplots of the performance. Note that configurations of `irace` optimised for the adjusted Rand Index and SSQ are only available for some combinations. For all cases where they are available, `confStream` yields considerably better Silhouette Width. In comparison to `irace` optimised for the Silhouette Width, the quality of `confStream` is often similar. For example, for the `powersupply` data stream, the results of `confStream` is very slightly better than the a-posteriori

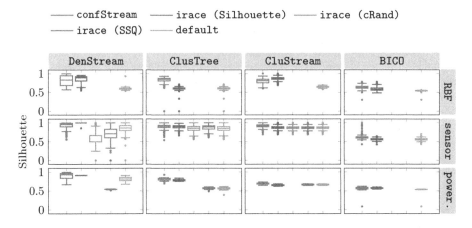

Fig. 8. Comparing performance of confStream, the optimal configuration found by irace optimised for the adjusted Rand Index (cRand) [7], the SSQ [8] and the Silhouette Width as well as the default configuration for the Random RBF, sensor and powersupply data streams.

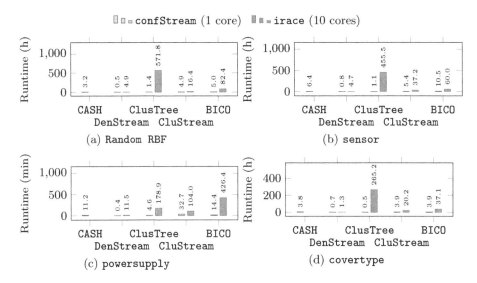

Fig. 9. Runtime of confStream and irace in hours or minutes. irace was paralellised with ten cores while confStream only used a single core. Experiments performed on an Intel Xeon CPU E5-2630 v4 with 2.20 GHz.

approach. Overall this shows, that confStream produces similar results in quality than a-posteriori optimisation with vastly less computational resources.

Throughout our experiments, we observed that the (online) confStream algorithm only requires a fraction of the number of evaluations compared to the (offline) irace approach. This also shows in the runtime of the algorithms as highlighted in Fig. 9. In our experiments, we parallelised the racing procedure of `irace` on ten cores. `confStream`, on the other hand, was run on a single core. Despite this considerable disadvantage, `confStream` is much faster for every single algorithm and data stream. This particularly shows for the `ClusTree` algorithm. For example, on the `Random RBF` data stream, `irace` required more than 23 days to optimise the parameters while `confStream` finished within less than one and a half hours. These results would further improve when parallelising the training of the ensemble for `confStream`. In addition, `confStream` is also able to solve the CASH problem. The runtime of the CASH problem mostly depends on the ensemble composition. As such, the runtime usually lies between the runtime of the fastest and slowest approach. In our experiments, the runtime of the CASH problem was often similar to the fastest algorithm in `irace`.

5 Conclusion

In this paper we proposed the first approach for automated algorithm selection and hyperparameter configuration of stream clustering algorithms. Our approach allows to apply stream clustering without expert knowledge and significantly facilitates the application of stream clustering in practice. In our approach, we train an ensemble of different algorithms and configurations in parallel to identify superior solutions. By drawing new configurations from the ensemble, we are able to improve solutions along the stream. Over time, sampling is biased towards more promising solutions. Our experiments on multiple state-of-the art algorithms, their hyperparameters as well as popular and diverse data streams have shown consistently good performances. The algorithm was able to quickly improve upon its initial configuration and to find valid configurations where the default configurations failed. We also compared the performance to `irace`, where we optimised the configurations a-posteriori. Even in this comparison, `confStream` found competitive solutions, despite working online and with far fewer iterations. While training multiple algorithms in parallel is slower, `confStream` was fast enough to work in real-time since the algorithms can be trained in parallel. In particular, it is much faster than a-posteriori approaches which are usually infeasible for data streams.

In future work we will evaluate our approach on a larger number of data streams and algorithms. In addition, we plan to apply our approach to other streaming applications such as stream classification or stream regression. Furthermore, we would like to revisit the idea of predicting the performance of new configurations. In a-posteriori approaches this has shown to produce good results [16] and we believe that this is also possible in the streaming scenario.

References

1. Ackermann, M.R., Märtens, M., Raupach, C., Swierkot, K., Lammersen, C., Sohler, C.: StreamKM++: a clustering algorithm for data streams. J. Exp. Algorithmics **17**, 2.4:2.1–2.4:2.30 (2012)

2. Aggarwal, C.C., Han, J., Wang, J., Yu, P.S.: A framework for clustering evolving data streams. In: Proceedings of the 29th International Conference on Very Large Data Bases (VLDB 2003), vol. 29, pp. 81–92 (2003)
3. Aggarwal, C.C., Han, J., Wang, J., Yu, P.S.: A framework for projected clustering of high dimensional data streams. In: Proceedings of the 30th International Conference on Very Large Data Bases (VLDB 2004), vol. 30, pp. 852–863 (2004)
4. Bifet, A., Gavalda, R., Holmes, G., Pfahringer, B.: Machine Learning for Data Streams with Practical Examples in MOA. MIT Press, Cambridge (2018)
5. Bifet, A., Holmes, G., Kirkby, R., Pfahringer, B.: MOA: massive online analysis. J. Mach. Learn. Res. **11**, 1601–1604 (2010)
6. Cao, F., Ester, M., Qian, W., Zhou, A.: Density-based clustering over an evolving data stream with noise. In: Proceedings of the Conference on Data Mining (SIAM 2006), pp. 328–339 (2006)
7. Carnein, M., Assenmacher, D., Trautmann, H.: An empirical comparison of stream clustering algorithms. In: Proceedings of the ACM International Conference on Computing Frontiers (CF 2017), pp. 361–365. ACM (2017)
8. Carnein, M., Trautmann, H.: Evostream – evolutionary stream clustering utilizing idle times. Big Data Res. **14**, 101–111 (2018)
9. Carnein, M., Trautmann, H.: Customer segmentation based on transactional data using stream clustering. In: Yang, Q., Zhou, Z.-H., Gong, Z., Zhang, M.-L., Huang, S.-J. (eds.) PAKDD 2019. LNCS (LNAI), vol. 11439, pp. 280–292. Springer, Cham (2019). https://doi.org/10.1007/978-3-030-16148-4_22
10. Carnein, M., Trautmann, H.: Optimizing data stream representation: an extensive survey on stream clustering algorithms. Bus. Inf. Syst. Eng. (BISE) **61**, 277–297 (2019). https://doi.org/10.1007/s12599-019-00576-5
11. Carnein, M., Trautmann, H., Bifet, A., Pfahringer, B.: Towards automated configuration of stream clustering algorithms. In: Cellier, P., Driessens, K. (eds.) ECML PKDD 2019. CCIS, vol. 1167, pp. 137–143. Springer, Cham (2020). https://doi.org/10.1007/978-3-030-43823-4_12
12. Fichtenberger, H., Gillé, M., Schmidt, M., Schwiegelshohn, C., Sohler, C.: BICO: BIRCH meets coresets for k-means clustering. In: Bodlaender, H.L., Italiano, G.F. (eds.) ESA 2013. LNCS, vol. 8125, pp. 481–492. Springer, Heidelberg (2013). https://doi.org/10.1007/978-3-642-40450-4_41
13. Fitzgerald, T., Malitsky, Y., O'Sullivan, B., Tierney, K.: ReACT: real-time algorithm configuration through tournaments. In: Edelkamp, S., Barták, R. (eds.) Proceedings of the Seventh Annual Symposium on Combinatorial Search (SOCS 2014) (2014)
14. Fitzgerald, T., O'Sullivan, B., Malitsky, Y., Tierney, K.: Online search algorithm configuration. In: Brodley, C.E., Stone, P. (eds.) Proceedings of the Twenty-Eighth AAAI Conference on Artificial Intelligence, pp. 3104–3105. AAAI Press (2014)
15. Gomes, H.M., Barddal, J.P., Ferreira, L.E.B., Bifet, A.: Adaptive random forests for data stream regression. In: Proceedings of the 26th European Symposium on Artificial Neural Networks (ESANN 2018) (2018)
16. Hutter, F., Hoos, H.H., Leyton-Brown, K.: Sequential model-based optimization for general algorithm configuration. In: Coello, C.A.C. (ed.) LION 2011. LNCS, vol. 6683, pp. 507–523. Springer, Heidelberg (2011). https://doi.org/10.1007/978-3-642-25566-3_40
17. Hutter, F., Hoos, H.H., Leyton-Brown, K., Stützle, T.: ParamILS: an automatic algorithm configuration framework. J. Artif. Intell. Res. **36**, 267–306 (2009)

18. Hutter, F., Kotthoff, L., Vanschoren, J. (eds.): Automated Machine Learning: Methods, Systems, Challenges. Springer, Cham (2019). https://doi.org/10.1007/978-3-030-05318-5
19. Kranen, P., Assent, I., Baldauf, C., Seidl, T.: Self-adaptive anytime stream clustering. In: Proceedings of the 9th IEEE International Conference on Data Mining (ICDM 2009), pp. 249–258, December 2009
20. López-Ibáñez, M., Dubois-Lacoste, J., Pérez Cáceres, L., Stützle, T., Birattari, M.: The irace package: Iterated racing for automatic algorithm configuration. Oper. Res. Perspect. **3**, 43–58 (2016)
21. Minku, L.L.: A novel online supervised hyperparameter tuning procedure applied to cross-company software effort estimation. Empir. Softw. Eng. **24**(5), 3153–3204 (2019). https://doi.org/10.1007/s10664-019-09686-w
22. van Rijn, J.N., Holmes, G., Pfahringer, B., Vanschoren, J.: Having a blast: meta-learning and heterogeneous ensembles for data streams. In: Proceedings of the 2015 IEEE International Conference on Data Mining (ICDM 2015), pp. 1003–1008, November 2015
23. van Rijn, J.N., Holmes, G., Pfahringer, B., Vanschoren, J.: Algorithm selection on data streams. In: Džeroski, S., Panov, P., Kocev, D., Todorovski, L. (eds.) DS 2014. LNCS (LNAI), vol. 8777, pp. 325–336. Springer, Cham (2014). https://doi.org/10.1007/978-3-319-11812-3_28
24. van Rijn, J.N., Holmes, G., Pfahringer, B., Vanschoren, J.: The online performance estimation framework: heterogeneous ensemble learning for data streams. Mach. Learn. **107**(1), 149–176 (2018). https://doi.org/10.1007/s10994-017-5686-9
25. Veloso, B., Gama, J., Malheiro, B.: Self hyper-parameter tuning for data streams. In: Soldatova, L., Vanschoren, J., Papadopoulos, G., Ceci, M. (eds.) DS 2018. LNCS (LNAI), vol. 11198, pp. 241–255. Springer, Cham (2018). https://doi.org/10.1007/978-3-030-01771-2_16

Randomized Algorithms for Some Sequence Clustering Problems

Sergey Khamidullin[1] , Vladimir Khandeev[1,2](✉) , and Anna Panasenko[1,2]

[1] Sobolev Institute of Mathematics, 4 Koptyug Avenue, 630090 Novosibirsk, Russia
{kham,khandeev,a.v.panasenko}@math.nsc.ru
[2] Novosibirsk State University, 2 Pirogova Street, 630090 Novosibirsk, Russia

Abstract. We consider two problems of clustering a finite sequence of points in Euclidean space. In the first problem, we need to find a cluster minimizing intracluster sum of squared distances from cluster elements to its centroid. In the second problem, we need to partition a sequence into two clusters minimizing cardinality-weighted intracluster sums of squared distances from clusters elements to their centers; the center of the first cluster is its centroid, while the center of the second one is the origin. Moreover, in the first problem, the difference between any two subsequent indices of cluster elements is bounded above and below by some constants. In the second problem, the same constraint is imposed on the cluster with unknown centroid. We present randomized algorithms for both problems and find the conditions under which these algorithms are polynomial and asymptotically exact.

Keywords: Clustering · Euclidean space · Minimum sum-of-squares · NP-hard problem · Randomized algorithm · Asymptotic accuracy

1 Introduction

The subject of this study are two strongly NP-hard problems of clustering a finite sequence of points in Euclidean space. Our goal is to construct a randomized algorithm for the problems. The research is motivated by the fact that the considered problems are related to mathematical time series analysis problems, approximation and discrete optimization problems, and also by their importance for applications such as signals analysis and recognition, remote object monitoring, etc. (see the next section and the papers therein).

The paper has the following structure. In Sect. 2, formulation of the problems is given. In the same Section, the known results are listed. The next Section contains the auxiliary problem and the algorithm for solving it, which are needed to construct our proposed algorithms. In Sect. 4, the randomized algorithms for the considered problems are presented.

© Springer Nature Switzerland AG 2020
I. S. Kotsireas and P. M. Pardalos (Eds.): LION 14 2020, LNCS 12096, pp. 96–101, 2020.
https://doi.org/10.1007/978-3-030-53552-0_11

2 Problems Formulation, Related Problems, and Known Results

We consider the following two problems.

Problem 1. *Given* a sequence $\mathcal{Y} = (y_1, \ldots, y_N)$ of points in $\mathbb{R}^d$ and positive integers $T_{\min}$, $T_{\max}$ and $M > 1$. *Find* a subset $\mathcal{M} = \{n_1, \ldots, n_M\} \subseteq \mathcal{N} = \{1, \ldots, N\}$ of the index set of $\mathcal{Y}$ such that

$$F_1(\mathcal{M}) = \sum_{j \in \mathcal{M}} \|y_j - \overline{y}(\mathcal{M})\|^2 \longrightarrow \min ,$$

where $\overline{y}(\mathcal{M}) = \frac{1}{|\mathcal{M}|} \sum_{i \in \mathcal{M}} y_i$ is the centroid of $\{y_j \,|\, j \in \mathcal{M}\}$, under the constraints

$$T_{\min} \leq n_m - n_{m-1} \leq T_{\max} \leq N, \quad m = 2, \ldots, M , \tag{1}$$

on the elements of the set $(n_1, \ldots, n_M)$.

Problem 2. *Given* a sequence $\mathcal{Y} = (y_1, \ldots, y_N)$ of points in $\mathbb{R}^d$ and positive integers $T_{\min}$, $T_{\max}$, and $M > 1$. *Find* a subset $\mathcal{M} = \{n_1, \ldots, n_M\} \subseteq \mathcal{N} = \{1, \ldots, N\}$ of the index set of $\mathcal{Y}$ such that

$$F_2(\mathcal{M}) = |\mathcal{M}| \sum_{j \in \mathcal{M}} \|y_j - \overline{y}(\mathcal{M})\|^2 + |\mathcal{N} \setminus \mathcal{M}| \sum_{i \in \mathcal{N} \setminus \mathcal{M}} \|y_i\|^2 \longrightarrow \min ,$$

where $\overline{y}(\mathcal{M}) = \frac{1}{|\mathcal{M}|} \sum_{i \in \mathcal{M}} y_i$ is the centroid of $\{y_j \,|\, j \in \mathcal{M}\}$, under the constraints (1) on the elements of the set $(n_1, \ldots, n_M)$.

Problem 1 is induced by the following applied problem. Given a sequence $\mathcal{Y}$ of N time-ordered measurements of d numerical characteristics of some object. M of these measurements correspond to a repeating (identical) state of the object. There is an error in each given measurement result. The correspondence of the measurement results to the states of the object is unknown. However, it is known that the time interval between two consecutive identical states is bound from above and below by the specified constants $T_{\min}$ and $T_{\max}$. It is required to find a subsequence of numbers corresponding to the measurements of the repeated state of the object.

In the special case when $T_{\min} = 1$ and $T_{\max} = N$, Problem 1 is equivalent to the well-known M-variance problem (see, e.g., [1]). A list of known results for M-variance problem can be found in [2].

When $T_{\min}$ and $T_{\max}$ are parameters, Problem 1 is strongly NP-hard for any $T_{\min} < T_{\max}$ [3]. When $T_{\min} = T_{\max}$, it is solvable in polynomial time.

In [4], a 2-approximation algorithm with $\mathcal{O}(N^2(MN + d))$ running time is proposed.

An exact algorithm for the case of integer inputs was substantiated in [5]. When the space dimension is fixed, the algorithm is pseudopolynomial and runs in $\mathcal{O}(N^3(MD)^d)$ time.

In [6], an FPTAS was presented for the case of Problem 1 when the space dimension is fixed. Given relative error ε, this algorithm finds a $(1 + \varepsilon)$-approximate solution to the problem in $\mathcal{O}(MN^3(1/\varepsilon)^{q/2})$ time.

Problem 2 simulates the following applied problem. As in Problem 1, we have a sequence $\mathcal{Y}$ of N time-ordered measurement results for d characteristics of some object. This object can be in two different states (active and passive, for example). Each measurement has an error and the correspondence between the elements of the input sequence and the states is unknown. One know that the object was in the active state exactly M times (or the probability of the active state is $\frac{M}{N}$) and the time interval between every two consecutive active states is bounded from below and above by some constants $T_{\min}$ and $T_{\max}$. It is required to find 2-partition of the input sequence and evaluate the object characteristics.

If $T_{\min} = 1$ and $T_{\max} = N$, Problem 2 is equivalent to *Cardinality-weighted variance-based 2-clustering with given center* problem. One can easily find a list of known results for this special case in [8].

Cardinality-weighted variance-based 2-clustering with given center problem is related but not equivalent to the well-known *Min-sum all-pairs 2-clustering* problem (see, e.g., [9,10]). Many algorithmic results are known for this closely related problem, but they are not directly applicable to *Cardinality-weighted variance-based 2-clustering with given center* problem.

Problem 2 is strongly NP-hard [11]. Only two algorithmic results have been proposed for this problem until now.

An exact pseudopolynomial algorithm was proposed in [11] for the case of integer instances and the fixed space dimension d. The running time of this algorithm is $\mathcal{O}(N(M(T_{\max}-T_{\min}+1)+d)(2MD+1)^d)$, where D is the maximum absolute value of coordinates of the input points.

In [12], a 2-approximation algorithm was presented. The running time of the algorithm is $\mathcal{O}(N^2(M(T_{\max} - T_{\min} + 1) + d))$.

The main results of this paper are randomized algorithms for Problems 1 and 2. These algorithms find $(1 + \varepsilon)$-approximate solution with probability not less than $1 - \gamma$ in $\mathcal{O}(dMN^2)$ time, for the given $\varepsilon > 0$, $\gamma \in (0,1)$ and under assumption $M \geq \beta N$ for $\beta \in (0,1)$. The conditions are found under which these algorithms are asymptotically exact (i.e. the algorithms find a $(1 + \varepsilon_N)$-approximate solutions with probability $1 - \gamma_N$, where $\varepsilon_N, \gamma_N \to 0$) and find the solutions in $\mathcal{O}(dMN^3)$ time.

3   Auxiliary Problem

To construct the algorithms for Problems 1 and 2, we need the following auxiliary problem.

Problem 3. Given a sequence $g(n)$, $n = 1, \ldots, N$, of real values, positive integers $T_{\min}$, $T_{\max}$ and $M > 1$. *Find* a subset $\mathcal{M} = \{n_1, \ldots, n_M\} \subseteq \mathcal{N}$ of indices of sequence elements such that

$$G(\mathcal{M}) = \sum_{i \in \mathcal{M}} g(i) \to \min ,$$

under constraints (1) on the elements of the tuple $(n_1, \ldots, n_M)$.

The following algorithm finds the solution of Problem 3.

Algorithm $\mathcal{A}$.

Input: a sequence $g(n)$, $n = 1, \ldots, N$, numbers $T_{\min}$, $T_{\max}$ and $M > 1$.

Step 1. Compute

$$
G_m(n) = \begin{cases} g(n), & \text{if } n \in \omega_1, \ m = 1 \, ; \\ g(n) + \max\limits_{j \in \gamma^-_{m-1}(n)} G_{m-1}(j), & \text{if } n \in \omega_m, \ m = 2, \ldots, M \, , \end{cases}
$$

where

$$
\omega_m = \big\{ n \,|\, 1 + (m-1)T_{\min} \le n \le N - (M-m)T_{\min} \big\}, m = 1, \ldots, M \, ,
$$

$$
\gamma^-_{m-1}(n) = \big\{ j \,|\, \max\{1 + (m-2)T_{\min}, n - T_{\max}\} \le j \le n - T_{\min} \big\},
$$
$$
n \in \omega_m, \ m = 2, \ldots, M \, .
$$

Step 2. Compute

$$
G^x_{\max} = \max\limits_{n \in \omega_M} G^x_M(n)
$$

and find the tuple $\mathcal{M} = (n_1, \ldots, n_M)$ by the formulae

$$
n^x_M = \arg \max\limits_{n \in \omega_M} G^x_M(n) \, ,
$$

$$
n^x_{m-1} = \arg \max\limits_{n \in \gamma^-_m(n^x_m)} G^x_m(n), \quad m = M, M-1, \ldots, 2 \, .
$$

Output: the tuple $\mathcal{M} = (n_1, \ldots, n_M)$.

Remark 1. It follows from [4,7] that Algorithm $\mathcal{A}$ finds the optimal solution of Problem 3 in $\mathcal{O}(NM(T_{\max} - T_{\min} + 1))$ time.

4   Randomized Algorithms

Below is a randomized algorithm for Problem 1.

Algorithm $\mathcal{A}_1$.

Input: a sequence $\mathcal{Y}$, positive integers $T_{\min}$, $T_{\max}$, M, a positive integer parameter k.

Step 1. Generate a multiset $\mathcal{T}$ of points by randomly and independently choosing k elements from $\mathcal{Y}$ with replacement.

Step 2. For every nonempty subset $\mathcal{H} \subseteq \mathcal{T}$ compute the centroid $\bar{y}(\mathcal{H})$ and find a solution $\mathcal{M} = \mathcal{M}(\mathcal{H})$ of Problem 3 for $g(n) = \|y_n - \bar{y}(\mathcal{H})\|^2$, $n = 1, \ldots, N$.

Step 3. From the family of solutions $\{\mathcal{M}(\mathcal{H}) \,|\, \mathcal{H} \subseteq \mathcal{T}\}$ found at Step 2, choose the set $\mathcal{M}_{\mathcal{A}_1} = \mathcal{M}(\mathcal{H})$ for which the value of $F_1(\mathcal{M}(\mathcal{H}))$ is minimal.

Output: the set $\mathcal{M}_{\mathcal{A}_1}$.

The next randomized algorithm allows one to find approximate solution of Problem 2.

Algorithm $\mathcal{A}_2$.

Input: a sequence $\mathcal{Y}$, positive integers $T_{\min}$, $T_{\max}$, M, a positive integer parameter k.

Step 1. Generate a multiset $\mathcal{T}$ of points by randomly and independently choosing k elements from $\mathcal{Y}$ with replacement.

Step 2. For every nonempty subset $\mathcal{H} \subseteq \mathcal{T}$ compute the centroid $\overline{y}(\mathcal{H})$ and find a solution $\mathcal{M} = \mathcal{M}(\mathcal{H})$ of Problem 3 for $g(n) = 2M\langle y_n, \overline{y}(\mathcal{H})\rangle - (2M - N)\|y_n\|^2 - M\|\overline{y}(\mathcal{H})\|^2$, $n = 1, \ldots, N$.

Step 3. From the family of solutions $\{\mathcal{M}(\mathcal{H}) \mid \mathcal{H} \subseteq \mathcal{T}\}$ found at Step 2, choose the set $\mathcal{M}_{\mathcal{A}_2} = \mathcal{M}(\mathcal{H})$ for which the value of $F_2(\mathcal{M}(\mathcal{H}))$ is minimal.

Output: the set $\mathcal{M}_{\mathcal{A}_2}$.

The following theorem describes the properties of algorithms $\mathcal{A}_1$ and $\mathcal{A}_2$.

Theorem 1. *Assume that in Problems 1 and 2, $M \geq \beta N$ for $\beta \in (0,1)$. Then, given $\varepsilon > 0$ and $\gamma \in (0,1)$, for a fixed parameter*

$$k = \max(\lceil \frac{2}{\beta}\lceil \frac{2}{\gamma\varepsilon}\rceil\rceil, \lceil \frac{8}{\beta}\ln\frac{2}{\gamma}\rceil)$$

algorithms $\mathcal{A}_1$ and $\mathcal{A}_2$ find $(1 + \varepsilon)$-approximate solutions of Problem 1 and 2 with probability $1 - \gamma$ in $\mathcal{O}(dMN^2)$ time.

Finally, in the next theorem, conditions are established under which algorithms $\mathcal{A}_1$ and $\mathcal{A}_2$ are polynomial and asymptotically exact.

Theorem 2. *Assume that in Problems 1 and 2, $M \geq \beta N$ for $\beta \in (0,1)$. Then, for fixed $k = \lceil \log_2 N \rceil$, algorithms $\mathcal{A}_1$ and $\mathcal{A}_2$ find $(1+\varepsilon_N)$-approximate solutions of Problem 1 and 2 with probability $1-\gamma_N$ in $\mathcal{O}(dMN^3)$ time, where $\varepsilon_N, \gamma_N \to 0$.*

The idea of proving Theorems 1 and 2 is to estimate the probability of events $F_i(\mathcal{M}_{\mathcal{A}_i}) \geq (1 + \frac{1}{\delta t})F_i(\mathcal{M}_i^*)$ in the case when the multiset $\mathcal{T}$ contains at least t elements of the optimal solution $\mathcal{M}_i^*$, where $\delta \in \mathbb{R}$, $t \in \mathbb{N}$, $i = 1, 2$. To do this, we use the Markov inequality. Then, using Chernov's inequality, we show that it is sufficient to put $\delta = \gamma/2$, $t = \lceil 2/(\gamma\varepsilon)\rceil$ in Theorem 1 and $\delta = (\log_2 N)^{-1/2}$, $t = \lceil kM/(2N)\rceil$ in Theorem 2.

5   Conclusion

In the present paper, we have proposed randomized algorithms for two sequence clustering problems. The algorithms find $(1 + \varepsilon)$-approximate solutions with probability not less than $1 - \gamma$ in $\mathcal{O}(dMN^2)$ time. Conditions are found under which the algorithms are polynomial and asymptotically exact.

In our opinion, the algorithms presented in this paper can be used to quickly obtain solutions to large-scale applied problems of signal analysis and recognition.

Acknowledgments. The study of Problem 1 was supported by the Russian Foundation for Basic Research, projects 19-01-00308 and 19-07-00397, by the Russian Academy of Science (the Program of basic research), project 0314-2019-0015, and by the Russian Ministry of Science and Education under the 5-100 Excellence Programme. The study of Problem 2 was supported by the Russian Foundation for Basic Research, project 19-31-90031.

References

1. Aggarwal, H., Imai, N., Katoh, N., Suri, S.: Finding K points with minimum diameter and related problems. J. Algorithms **12**(1), 38–56 (1991)
2. Kel'manov, A.V., Panasenko, A.V., Khandeev, V.I.: Randomized algorithms for some hard-to-solve problems of clustering a finite set of points in Euclidean space. Comput. Math. Math. Phys. **59**(5), 842–850 (2019). https://doi.org/10.1134/S0965542519050099
3. Kel'manov, A.V., Pyatkin, A.V.: On the complexity of some problems of choosing a subsequence of vectors. Zh. Vych. Mat. Mat. Fiz. (in Russian) **52**(12), 2284–2291 (2012)
4. Kel'manov, A.V., Romanchenko, S.M., Khamidullin, S.A.: Approximation algorithms for some intractable problems of choosing a vector subsequence. J. Appl. Indu. Math. **6**(4), 443–450 (2012)
5. Kel'manov, A.V., Romanchenko, S.M., Khamidullin, S.A.: Exact pseudopolynomial-time algorithms for some intractable problems of finding a subsequence of vectors. Zh. Vych. Mat. Mat. Fiz. (in Russian) **53**(1), 143–153 (2013)
6. Kel'manov, A.V., Romanchenko, S.M., Khamidullin, S.A.: An approximation scheme for the problem of finding a subsequence. Numer. Anal. Appl. **10**(4), 313–323 (2017). https://doi.org/10.1134/S1995423917040036
7. Kel'manov, A.V., Khamidullin, S.A.: Posterior detection of a given number of identical subsequences in a guasi-periodic sequence. Comput. Math. Math. Phys. **41**(5), 762–774 (2001)
8. Panasenko, A.: A PTAS for one cardinality-weighted 2-clustering problem. In: Khachay, M., Kochetov, Y., Pardalos, P. (eds.) MOTOR 2019. LNCS, vol. 11548, pp. 581–592. Springer, Cham (2019). https://doi.org/10.1007/978-3-030-22629-9_41
9. Sahni, S., Gonzalez, T.: P-complete approximation problems. J. ACM **23**, 555–566 (1976)
10. de la Vega, F., Karpinski, M., Kenyon, C., Rabani, Y.: Polynomial time approximation schemes for metric min-sum clustering. Electronic Colloquium on Computational Complexity (ECCC). Report No. 25 (2002)
11. Kel'manov, A., Khamidullin, S., Panasenko, A.: Exact algorithm for one cardinality-weighted 2-partitioning problem of a sequence. In: Matsatsinis, N.F., Marinakis, Y., Pardalos, P. (eds.) LION 2019. LNCS, vol. 11968, pp. 135–145. Springer, Cham (2020). https://doi.org/10.1007/978-3-030-38629-0_11
12. Kel'manov, A., Khamidullin, S., Panasenko, A.: 2-approximation polynomial-time algorithm for a cardinality-weighted 2-partitioning problem of a sequence. In: Sergeyev, Y.D., Kvasov, D.E. (eds.) NUMTA 2019. LNCS, vol. 11974, pp. 386–393. Springer, Cham (2020). https://doi.org/10.1007/978-3-030-40616-5_34

Hyper-parameterized Dialectic Search for Non-linear Box-Constrained Optimization with Heterogenous Variable Types

Meinolf Sellmann[1] and Kevin Tierney[2(✉)]

[1] General Electric, Niskayuna, NY, USA
meinolf@ge.com
[2] Bielefeld University, Bielefeld, Germany
kevin.tierney@uni-bielefeld.de

Abstract. We consider the dialectic search paradigm for box-constrained, non-linear optimization with heterogeneous variable types. In particular, we devise an implementation that can handle any computable objective function, including non-linear, non-convex, non-differentiable, non-continuous, non-separable and multi-modal functions. The variable types we consider are bounded continuous and integer, as well as categorical variables with explicitly enumerated domains. Extensive experimental results show the effectiveness of the new local search solver for these types of problems.

1 Introduction

Box-constrained optimization problems are ubiquitous. They appear when optimizing designs where objective functions estimate key performance characteristics of the design, when optimizing control parameters in simulated environments in simulation-based optimization, when optimizing function parameters to fit training samples in machine learning, when searching for new sample points over surrogate functions in Bayesian optimization, and when searching for optimal strategies in decision aiding. In many cases, these target functions are highly non-linear, non-convex, even non-differentiable, non-separable, or non-continuous. Moreover, frequently decision variables are discrete integer variables or even categorical variables that can only take specific values.

We tackle this problem using the dialectic search paradigm [8] in a hyper-parameterized setting [4]. The objective of this work is to show that a tuned, hyper-parameterized dialectic search program (that is implemented in C++ with less than 1,500 lines of code) can effectively tackle these problems. All that a user needs to provide is a short piece of code that computes the objective function given a variable assignment, as well as an incremental objective evaluation when changing two variable assignments from the last assignment. The latter can obviously be performed easily simply by reducing to the first evaluation function,

© Springer Nature Switzerland AG 2020
I. S. Kotsireas and P. M. Pardalos (Eds.): LION 14 2020, LNCS 12096, pp. 102–116, 2020.
https://doi.org/10.1007/978-3-030-53552-0_12

meaning our approach supports full black-box settings. However, in many cases, objectives can be evaluated more quickly when exploiting incrementality, giving the user the opportunity to gain efficiency in this way.

This paper makes the following contributions:

1. We extend the dialectic search paradigm to a general variable setting (continuous, integer and categorical).
2. We increase dialectic search's performance through a convex search procedure for its *synthesis* procedure, which can be seen as an alternative to path relinking.
3. We apply the hyper-parameterization paradigm from [4] to give this general-purpose tool the ability to be tuned for specific benchmarks and to adapt key search parameters dynamically during search.

In the following, we first review the concepts of dialectic search and hyper parameterization. Then, we apply these concepts to devise an algorithm implementation for solving box-constrained non-linear optimization problems with heterogeneous variable types. Following this detailed description, we then evaluate our implementation on various benchmarks to establish the efficacy of our new approach.

2   Dialectic Search

2.1   A Philosophical Framework as Metaheuristic

Dialectic search was first introduced in [8] as a metaheuristic for local search. The approach exists as a means to overcome the difficulties when applying a local search metaheuristic framework to a specific problem domain by enforcing a strict separation of intensification via greedy search and search space exploration via randomization. In [8], it is shown that dialectic search does not require any sophisticated tailoring for the concrete application domain at hand. At the same time, it significantly outperforms other metaheuristics, such as Tabu Search, Simulated Annealing, and Genetic Algorithms, on multiple benchmarks from constraint satisfaction and combinatorial and continuous optimization.

Dialectic search works as follows:

1. **Initialization:** We construct an assignment of variables to random values in their respective domains.
2. **Thesis:** We call the current assignment the "thesis" and greedily improve it until we find a local optimum. We can conduct a full (where we consider all potential moves before choosing the best) or a first-improvement (where we take the first possible move that improves the objective) greedy search. A search parameter determines the percentage of variables we consider in each greedy step.

3. **Antithesis:** We randomly modify the thesis by changing select variables' values, which can be seen as a type of perturbation like in Iterated Local Search [9]. This new assignment is called the "antithesis." A search parameter determines the percentage of variables to modify. Another search parameter determines the probability with which we greedily improve the antithesis. Again, we can conduct a full or a first-improvement greedy search. A separate search parameter determines how many variables to consider in each greedy step.

4. **Synthesis:** Next we search the space between thesis and antithesis by conducting a limited local search in the space that thesis and antithesis span, i.e., variables to whom both the thesis and antithesis assign the same value cannot be changed. The exact way how this nested local search is to be conducted is left open by dialectic search. For example, we could conduct a greedy path relinking [7] between the thesis and antithesis. Alternatively, we could start a more elaborate nested local search to find a favorable recombination of thesis and antithesis. In this paper, we will introduce yet another procedure, where we search over the space of convex combinations of thesis and antithesis. No matter how this step is implemented, we call the resulting assignment the "synthesis."

5. **Iteration and Restarts:** We next decide if we want to restart and return to Step 1. A search parameter determines the probability with which we restart. If we do not restart, we next test if the synthesis improves over the thesis. If so, the synthesis becomes the new thesis, and we continue with Step 2. If not, we continue with Step 3.

2.2   Hyper-parameterized Dialectic Search

While dialectic search was shown to give very good results even without sophisticated tuning, in [4] dialectic search was *hyper-parameterized* to tackle the MaxSAT problem. The core idea of hyper-parameterization is to enable a search heuristic like dialectic search to self-adapt search parameters *during* runtime based on runtime statistics. The parameters we adjust include the probability to restart, the size of the antithesis, the number of variables to open for the greedy heuristic, etc. Furthermore, the runtime statistics are, for example, the estimated number of local search moves until time limit, time since last overall improvement, time since last improvement in current restart, and the total time in the current restart. Hyper-parameterization has also been successfully applied to tabu search [3].

To hyper-parameterize dialectic search, [4] suggests to use a logistic regression function to transform normalized runtime statistics into probabilities (e.g., for restart probabilities) or percentages (e.g., for the antithesis size). These logistic regression functions in turn introduce new "hyper-parameters."

$$P_k = \frac{1}{1 + e^{\sum_i h_i^k s_i + h_0^k}} \tag{1}$$

In Eq. (1), the probability (for example for a restart in Step 5), or the percentage (for example to determine the percentage of variables to be modified in Step 3) P_k for the kth search parameter is derived from runtime statistics s. The key is that this transformation of current runtime statistics into search parameter values takes place *each time when the search parameter is needed*. For example, whenever we need to decide if we are going to restart the search in Step 5, we gather the current runtime statistics s and enter them into the logistic function using (static) hyper-parameters h^k to compute the probability of a restart. Then, we flip a coin, and with the computed probability we restart the search. So, to compute this probability/percentage, the current runtime statistics s are determined, the inner product with the search-parameter-specific hyper-parameters h^k is computed, the constant h_0^k added, and the result handed to the exponential function. This produces a value between minus infinity and infinity, which the logistic function transforms into a value between 0 and 1.

In contrast to regular parameterization, in which we would keep the restart probability and the percentage of variables we modify in the antithesis constant, hyper-parameters merely determine how the respective search parameters are derived from *current* runtime statistics. In [4], it is further suggested to use an instance-specific parameter tuner (e.g., [2] or [1]) to "learn" good settings for the hyper-parameters. For the MaxSAT problem, this resulted in a dialectic search portfolio that won four out of nine categories at the 2016 MaxSAT Evaluation [5].

3   Dialectic Search for Box-Constrained Non-Linear Optimization Problems with Continuous, Integer, and Categorical Variables

Having reviewed the general framework of hyper-parameterized dialectic search, we now introduce our new program for box-constrained non-linear optimization. In particular, our goal is to develop an implementation of dialectic search that works with any computable objective function and for mixtures of bounded continuous, integer, and explicitly enumerated categorical variables. In the following, we devise such a program that, in order to apply it to a new problem domain, only requires the implementation of an evaluation function of the objective to be minimized[1] for a given assignment of variables. For reasons of efficiency, we also require a second objective evaluation function that returns the objective value for the same assignment as was given for the last evaluation when two variables are changed to new values. This can be implemented easily by altering the last assignment and evaluating the objective from scratch using the first evaluation function. However, for many objectives it is possible to evaluate a new assignment incrementally and much faster than by re-evaluating the new assignment from scratch.

To instantiate dialectic search, we need to define each of the five steps (Initialization, Thesis, Antithesis, Synthesis, and Restarting) of dialectic search, as

---

[1] Note that the latter easily allows maximization as well, simply by having the function return the negative of the actual objective value.

well as determine how to hyper-parameterize the resulting algorithm. In the following, we describe in detail how each of these aspects is implemented.

3.1 Initialization

The dialectic search receives three parameters: 1. The number of variables n. 2. A vector of length n of a composite class that describes each variable's type, as well as its lower and upper bound or, in case of categorical variable, and explicit enumeration of the values that the variable can take. 3. A pointer to an evaluation class that has two evaluation functions, one that takes an entire assignment vector of length n as input and returns a rational number as output, the second taking two variable indices and two new values for the respective variables as input and returning a rational number as output. To initialize the local search we simply assign a random value within each variable's domain to the respective variable.

3.2 Thesis

The new assignment generated is labeled as the *thesis*. To greedily improve the thesis, we proceed as follows:

1. In each greedy step, we first select a random subset of variables, How many is determined by a search parameter.
2. Next we consider the variables in random order. For categorical variables, we consider each possible value and evaluate the objective if we change the respective variable to the new value. For ordinal and continuous variables, we conduct a pseudo-convex optimization as follows. First, we choose a number $\alpha \in [0,1]$ uniformly at random. Next we construct the interval $[x_1, x_4]$ where $x_1 = \max\{v - \alpha/2, L\}$ and $x_4 = \min\{v + \alpha/2, U\}]$, where v is the current value of the variable, and L, U are its lower and upper bound, respectively. We evaluate the objective when setting the variable to each end point of the interval, as well as at two intermediate points x_2 and x_3. The point x_3 is at $\frac{200}{\sqrt{(5)}+1}\% \simeq 61.8\%$ inside the interval and x_2 is at $100 - \frac{200}{\sqrt{(5)}+1}\% \simeq 38.2\%$ inside the interval. Note that for ordinal variables, we round to the nearest integer for evaluating the objective. However, the computation below continues using the actual fractional values.
   If the objective at x_1 is strictly less than at the other three points, or if the objective at x_2 is lower than at x_3, we reduce the interval to $[x_1, x_3]$. In this case, the point x_2 automatically finds itself at 61.8% of the new interval. Therefore, to continue, we only need to evaluate one new point at 38.2% of the new interval to iterate the search. On the other hand, if x_4 is strictly less than the other three points, or if x_3 results in a better objective than x_2, we continue the search in $[x_2, x_4]$. In this case, x_3 already finds itself at 38.2% of the new interval, so we only need to evaluate one more point at 61.8% to conduct the next iteration. The search stops when the interval length shrinks below a certain minimum length, which is set by a search parameter.

3. Having thus found an optimal or at least very favorable value for the respective variable in this way, we check if it improves the objective. We conduct a first-improvement greedy search. In this case, we next commit the variable to this value and continue with Step 1. If the best variable assignment found does not lead to an improvement of the objective, we continue with the next variable in the random selection, in random order.
4. The greedy search ends when no value can be found for any variable in the random selection that improves on the objective value.

3.3   Antithesis

To select an antithesis, we first choose a random subset of variables. How many variables is determined by a search parameter. Then, for each variable in the selection we choose a new value uniformly at random. Depending on another search parameter, we may conduct a greedy search to improve the antithesis (in the same manner as we just improved the thesis) before we move to the next step.

3.4   Synthesis

The purpose of this step is to opportunistically search the space between the thesis and antithesis. All variables to which the antithesis and thesis assign the same value are not changed. In a first step, we aim to recombine the variable settings between thesis and antithesis via path-relinking. We start at the thesis and consider all variables in turn to assess which variable, when set to the value given by the antithesis, would give the best objective value. This variable is then set to the antithesis value, regardless whether this leads to a worsening of the objective or not. We proceed in this way until all variables are set to the respective antithesis values. We next consider the assignment with the best objective value on this chain of assignments that "connect" thesis and antithesis. If the best assignment on this chain improves over the thesis, then this synthesis becomes the new thesis and we continue by jumping back to Step 2.

If the best recombination found does not improve over the current thesis, we next consider convex-combinations between thesis and anti-thesis. That is, we interpolate ordinal and continuous parameters between the respective values in the thesis and antithesis, and "round" categorical variables to the "nearest" thesis or antithesis value. For example, assume we have three parameters, the first is categorical (say it can take values red, blue and green), the second parameter is ordinal, and the third is continuous. Assume the thesis assignment is [green, -2, 0.2] and the antithesis is [red, 5, 0.7]. For the categorical parameter, we associate value 0 with the thesis value, and 1 with the antithesis value. Any interpolation value below 0.5 is then "rounded down" to the thesis value, all values greater or equal 0.5 get "rounded up" to the antithesis value. Similarly, for the ordinal parameters we round to the nearest integer. Then, the 0.6 interpolation between thesis and antithesis is, for example, [red, 2, 0.5]. Having thus defined how points on the "line" between thesis and antithesis map to assignments, we can conduct a pseudo-convex optimization, exactly as we did earlier in the greedy improvement of ordinal and continuous variables in Step 2. If this procedure results in an assignment that improves over the thesis, this synthesis becomes the new thesis. If not, then we keep the old thesis and consider a new randomized antithesis in Step 3.

3.5   Restarting

At the end of each synthesis step, we flip a coin to determine if we restart the search instead of attempting to further improve the current thesis assignment. The probability of a restart is again given by a search parameter. When a restart is triggered, we apply a randomized modification of the current thesis before we jump back to Step 2. This modification works exactly like the way how an antithesis is constructed, whereby we use a different search parameter to determine how much the new starting point will differ from the current thesis. Finally, when the time-limit of the search is exceeded, we return the best assignment ever encountered during the search.

3.6   Hyper-parameterization

The outline of our dialectic search instantiation shows that there are numerous search parameters influencing how the search progresses. For example, if we use a very high restart probability, we will primarily perform randomly restarted greedy searches. Alternatively, if we consider a very high percentage of variables to construct the antithesis, the bulk of the search effort will consist in synthesis steps (Fig. 1).

We list the search parameters within our implementation below:

g: The size of the greedy candidate set as percentage of all variables in the problem.

a_l, a_u: A lower and upper bound on the percentage of variables to be changed to construct an antithesis. The exact size of the change is then chosen uniformly at random in the interval given whenever a new antithesis is generated.

p_a: The probability of greedily improving the antithesis.

p_r: The probability of restarting the search.

r_l, r_u: A lower and upper bound on the percentage of variables to be changed to construct a new starting point when a restart is triggered. The exact size of the change is then chosen uniformly at random in the interval $[r_l, r_u]$.

As reviewed earlier, [4] proposes not to assign static values to these search parameters, but to allow them to dynamically adapt to the way in which the search progresses. Equation (1) governs how we derive probabilities and percentages from runtime statistics. Therefore, the only information needed at this point is which runtime statistics s we will track. In the following, we list the statistics we track during the search:

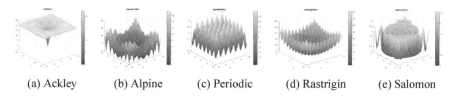

(a) Ackley   (b) Alpine   (c) Periodic   (d) Rastrigin   (e) Salomon

Fig. 1. Optimization functions over two continuous variables. Images from http:// benchmarkfcns.xyz/.

1. Time elapsed as percentage of total time before timeout.
2. Number of restarts conducted as a percentage of total restarts expected to be completed within the time limit.
3. Number of moves as a percentage of the total moves expected to be completed within the time limit.
4. Number of steps as a percentage of the total steps expected to be completed within the time limit.
5. Total number of improving syntheses found over the total number of dialectic moves expected to be completed within the time limit.
6. Number of moves in the current restart over the total number of dialectic moves expected to be completed within the time limit.
7. Number of moves since the current best known solution was found over the total number of dialectic moves expected to be completed within the time limit.
8. Number of moves since the last thesis update in the current restart over the total number of dialectic moves expected to be completed within the time limit.
9. Number of steps in the current restart over the total number of steps expected to be completed within the time limit.
10. Number of steps since the current best known solution was found over the total number of steps expected to be completed within the time limit.
11. Number of steps since the last thesis update in the current restart over the total number of steps expected to be completed within the time limit.

The resulting dialectic search with seven search parameters is now *hyper-configurable* with 84 (7 times 12) hyper-parameters. To tune these parameters for specific instance distributions, we employ the parameter tuner GGA++ [1].

4 Experimental Results

We now evaluate our approach on various benchmarks to assess its effectiveness. We consider the following function classes:

- **Ackley Instances:** This class has n continuous decision variables $X_i \in [-500, 500]$, m integer decision variables $Y_j \in [-500, 500]$, n dependent modeling variables $A_i = X_i - t_i$ for n given values $t_i \in [-500, 500]$ and m dependent modeling variables $B_j = Y_j - s_j$ for m given values $s_j \in [-500, 500]$. The objective is to minimize $b + f(n + m) -$

$f(n+m)b - \sum_i \frac{A_i^2}{10n} - \sum_j \frac{B_j^2}{10m} - b\sum_i \frac{\cos(2\pi A_i^2)}{2n} + \sum_j \frac{\cos(2\pi B_j^2)}{2m}$ for given parameters $f, b \in \mathbb{N}$. An instance to this problem is therefore fully described by the vector $(n, t_1, \ldots, t_n, m, s_1, \ldots, s_m, f, b)$.

- **Alpine Instances:** This class has n continuous decision variables $X_i \in [-10, 10]$ and n dependent modeling variables $A_i = X_i - t_i$ for n given values in $[-10, 10]$. The objective is to minimize $\sum_i |A_i \sin A_i + 0.1 A_i|$. An instance to this problem is therefore fully described by the vector $(n, t_1, \ldots, t_n)$.

- **Griewank Instances:** This class has n continuous decision variables $X_i \in [-600, 600]$ and n dependent modeling variables $A_i = X_i - t_i$ for n given values in $[-600, 600]$. The objective is to minimize $1 + \sum_i \frac{A_i^2}{4000} - \prod_i \cos(\frac{A_i}{\sqrt{i}})$. An instance to this problem is therefore fully described by the vector $(n, t_1, \ldots, t_n)$.

- **Periodic Instances:** This class has n continuous decision variables $X_i \in [-10, 10]$ and n dependent modeling variables $A_i = X_i - t_i$ for n given values in $[-10, 10]$. The objective is to minimize $1 + 10 \sum_i \sin^2(A_i) - e^{-\sum_i A_i^2}$. An instance to this problem is therefore fully described by the vector $(n, t_1, \ldots, t_n)$.

- **Rastrigin Instances:** This class has n continuous decision variables $X_i \in [-5.12, 5.12]$ and n dependent modeling variables $A_i = X_i - t_i$ for n given values in $[-5.12, 5.12]$. The objective is to minimize $10n + \sum_i A_i^2 - 10 \cos(2\pi A_i)$. An instance to this problem is therefore fully described by the vector $(n, t_1, \ldots, t_n)$.

- **Salomon Instances:** This class has n continuous decision variables $X_i \in [-100, 100]$ and $n + 1$ dependent modeling variables $A_i = X_i - t_i$ for n given values in $[-100, 100]$ and $Z = \sqrt{\sum_i A_i^2}$. The objective is to minimize $1 - \cos(2\pi Z) + 0.1Z$. An instance to this problem is therefore fully described by the vector $(n, t_1, \ldots, t_n)$.

- **Mixed Instances:** This class has n continuous decision variables $X_i \in [-1, 1]$ for $i \in \{0, \ldots, n - 1\}$, m integer decision variables $Y_j \in [-25, 5]$ for $j \in \{0, \ldots, m - 1\}$, and o categorical decision variables $Z_k \in \{2, 3, 4\}$ for $k \in \{0, \ldots, o - 1\}$. For each instance, we are given n continuous target values $t_i \in [0, 1]$, m integer target values $s_j \in [-25, 5]$, and o categorical targets $r_k \in \{2, 3, 4\}$. We use $n + m + o$ continuous modeling constants $a_l = \frac{t_{l\%n} t_{(l+1)\%n} + s_{(l-1)\%m} - s_{(l+1)\%m}}{r_{l\%o}}$, whereby $\%$ symbolizes the modulo operation. The objective is to

$$\text{Minimize} \sum_{l=0}^{n+m+o-1} \left(\frac{X_{l\%n} X_{(l+1)\%n} + Y_{(l-1)\%m} - Y_{(l+1)\%m}}{Z_{l\%o}} - a_l \right)^2.$$

An instance to this problem is therefore fully described by the vector

$$(n, t_0, \ldots, t_{n-1}, m, s_0, \ldots, s_{m-1}, o, r_0, \ldots, r_{o-1}).$$

Note that, in the above, we use modeling variables A, B to simplify the definition of these problems. In the actual implementation, we immediately substitute these variables with the respective expression over the decision variables in the objective function.

Table 1. Comparison of the new dialectic search approach (DS) with the dialectic search from [8] (DS-old) and simulated annealing approaches from [10] (SA-0.98 and SA-0.99)

Function	Dims	DS		DS-old		SA-0.98		SA-0.99	
		Value	Evals	Value	Evals	Value	Evals	Value	Evals
Alpine	20	10^{-3}	**21K**	10^{-3}	86K	10^{-3}	1M	10^{-3}	2M
	50	10^{-3}	**364K**	10^{-3}	458K	10^{-3}	2.9M	10^{-3}	5.8M
Rastrigin	20	10^{-3}	**18K**	10^{-3}	208K	24.4	3.4M	22.4	6.8M
	50	10^{-3}	**138K**	10^{-3}	818K	87.3	8.3M	86.8	9.9M

4.1 Minimizing the Number of Function Evaluations

As our approach works with any computable objective function, it is ideally suited for black-box optimization. The main objective is to minimize the number of black-box function evaluations that are needed to reach a certain approximation level of the global optimum. We use randomly generated Rastrigin and Alpine instances for this purpose.

In Table 1, we compare against the results published in [8]. There, a dialectic search (DS-old) algorithm was introduced for continuous optimization problems, and compared with a simulated annealing approach using two different cooling factors (SA-0.98 and SA-0.99) [10]. While the old dialectic search already gave very competitive results, we clearly see the benefits of the new approach, that has three main improvements over the old dialectic search approach: 1. The line search used as greedy step in DS-old is replaced with the pseudo-convex search approach we outlined in Step 2 in the Section "Thesis." 2. The new DS approach conducts interpolation searches on top of recombination searches to produce an improving synthesis, as outlined in Section "Synthesis." And 3. The new approach is hyper-parameterized, which not only allows the approach to adjust otherwise static parameters during the search, but also to tailor the search behavior for the respective benchmark – fully automatically, thanks to parameter tuners like GGA++ [1].

Overall, on average the new approach lowers the number of function evaluations by a factor 3. On 20-dimensional Rastrigin functions, the improvement is a whole order of magnitude, while for 50-dimensional Alpine instances we "only" save 20% of function evaluations, compared to the old DS approach. When compared to the better of the SA approaches, the hyper-reactive dialectic search approach only requires 3.5% of the function evaluations required by SA-0.98.

4.2 Optimizing Functions Within a Given Time Limit

We now change the setting to the more realistic scenario where we have been given a timelimit and have to find the best solution possible in the given time. We first examine the performance of DS versus a well-known continuous black-box optimizer, LSHADE [11], which we use thanks to its performance and its

Table 2. Comparison of DS to LSHADE on the Alpine and Rastrigin functions with 50 instances for each function/dimension pair.

Function	Dims	# ≤ 1e-3		Mean		Geo. Mean	
		LSHADE	DS	LSHADE	DS	LSHADE	DS
Alpine	20	49	50	0.021	0.001	0.001	0.001
	50	49	50	0.021	0.001	0.001	0.001
	200	4	50	3.500	0.001	1.652	0.001
	500	0	50	57.140	0.001	56.369	0.001
Rastrigin	20	33	50	0.836	0.001	0.013	0.001
	50	30	50	1.228	0.001	0.021	0.001
	200	0	50	416.282	0.001	410.934	0.001
	500	0	50	3010.779	0.001	2457.697	0.001

freely available source code. For this comparison, we again consider the Alpine and Rastrigin benchmarks. Later in this section, we will expand our analysis to the other benchmarks, including ones with discrete parameters, and include the optimizer LocalSolver [6] for comparison. In all experiments, we impose a 60 second CPU timelimit.

Comparison with LSHADE. Table 2 provides a comparison of dialectic search (DS) and LSHADE. We note that LSHADE is generally used on lower dimensional problems. For 20 and 50 dimensional instances, it performs reasonably well, even though it does not manage to find a solution below 1e-3 for all instances. However, many real-world optimization problems contain thousands of variables, thus we also examine LSHADE on larger instances with 200 and 500 decision variables. Here, LSHADE requires all 60 seconds of CPU time to find solutions far from the optimal. In contrast, DS requires on average 3.6 seconds for even the hardest benchmark in the table, the Rastrigin function with 500 dimensions. Given these encouraging results, we move on from conventional continuous, black-box approaches to compete on harder problems.

Comparison with LocalSolver. LocalSolver is a general purpose heuristic solver that supports essentially any white-box optimization problem. It offers an interface for problem modelling similar to those of mixed-integer programming solvers. We compare to LocalSolver[2] on randomly generated Ackley, Alpine, Griewank, Mixed, Periodic, Rastrigin, and Salomon functions.

In Tables 3 and 4 we present our results on the set of instances used to train the hyper-parameters of DS trained on each target problem individually, and

---

[2] We note that we are unable to tune LocalSolver's parameters with GGA due to LocalSolver's license restrictions, meaning our results should only be seen as a lower bound on performance.

Table 3. LocalSolver (LS) versus Dialectic Search (DS) on continuous and mixed continuous functions evaluated on our training set. DS's hyperparameters are trained specifically on the function it is being tested on, whereas DS-A is tuned on all functions.

	# ≤ 1e−3			Mean			Geo. Mean			Stdev.		
	LS	DS	DS-A	LS	DS	DS-A	LS	DS	DS-A	LS	DS	DS-A
Ackley	**97**	1	1	**0.001**	0.066	0.080	**0.001**	0.032	0.039	0.000	0.072	0.079
Alpine	41	**68**	**68**	0.002	0.003	**0.001**	0.002	**0.001**	**0.001**	0.001	0.009	0.001
Griewank	23	45	**85**	**0.002**	0.055	0.002	**0.001**	0.007	**0.001**	0.003	0.004	0.004
Mixed	7	7	7	6.823	**1.605**	1.611	2.108	**0.649**	0.650	6.167	1.176	1.185
Periodic	**4**	0	0	**0.934**	1.000	1.000	**0.699**	1.000	1.000	0.244	0.000	0.000
Rastrigin	45	**46**	41	73.712	**0.002**	**0.002**	0.160	**0.001**	0.002	143.144	0.001	0.001
Salomon	**84**	0	0	0.215	**0.100**	0.507	**0.003**	0.100	0.390	0.748	0.000	0.319
Ø Total	**43**	23.86	28.86	11.67	**0.40**	0.46	**0.0203**	0.0283	0.0296	59.829	0.744	0.747

Table 4. LocalSolver (LS) versus Dialectic Search (DS) on continuous and mixed continuous functions evaluated on our test set. DS's hyperparameters are trained specifically on the function it is being tested on, whereas DS-A is tuned on all functions.

	# ≤ 1e−3			Mean			Geo. Mean			Stdev.		
	LS	DS	DS-A	LS	DS	DS-A	LS	DS	DS-A	LS	DS	DS-A
Ackley	**99**	3	0	**0.001**	0.070	0.085	**0.001**	0.036	0.044	0.000	0.084	0.089
Alpine	39	**62**	61	0.002	0.004	0.003	0.002	0.002	**0.001**	0.001	0.009	0.008
Griewank	20	**37**	21	**0.003**	0.063	0.255	**0.002**	0.011	0.030	0.008	0.003	0.402
Mixed	**12**	10	10	7.177	1.680	**1.679**	1.340	0.591	**0.475**	6.696	1.300	1.300
Periodic	1	0	0	**0.964**	1.000	1.000	**0.875**	1.000	1.000	0.178	0.000	0.000
Rastrigin	48	**51**	49	92.320	**0.002**	**0.002**	0.167	**0.002**	**0.002**	155.001	0.001	0.001
Salomon	**92**	0	0	0.055	**0.100**	0.462	**0.001**	0.100	0.368	0.408	0.00	0.276
Ø Total	**44.43**	23.29	20.04	14.36	**0.26**	0.50	**0.0186**	0.0369	0.0464	66.764	0.790	0.784

DS-A, which is DS trained on a subset of training instances from all functions but Ackley. We also provide results on a test set of instances that were not used for the development or training of the new approach. For each benchmark function, we generate 200 instances, 100 for training and another 100 for testing. The dimensionality for each instance is chosen uniformly at random from the following ranges: $[250, 1000]$ (for both continuous and ordinal variables) for Ackley, $[1500, 5000]$ for Alpine, $[500, 5000]$ for Griewank, $[10, 50]$ for each variable type for Mixed, $[125, 1000]$ for Periodic, $[500, 5000]$ for Rastrigin and $[25, 200]$ for Salomon.

The tables give us multiple insights. First, we see that both DS (trained on each individual benchmark) and DS-A (trained on a mix of instances from all benchmarks, with the exception of Ackley instances) generalize well to the formerly unseen test instances. Second, we can see here that DS-A also does a reasonable job on the Ackley benchmark it was not trained for. In fact, it achieves almost the same performance as DS. Finally, we observe that there is

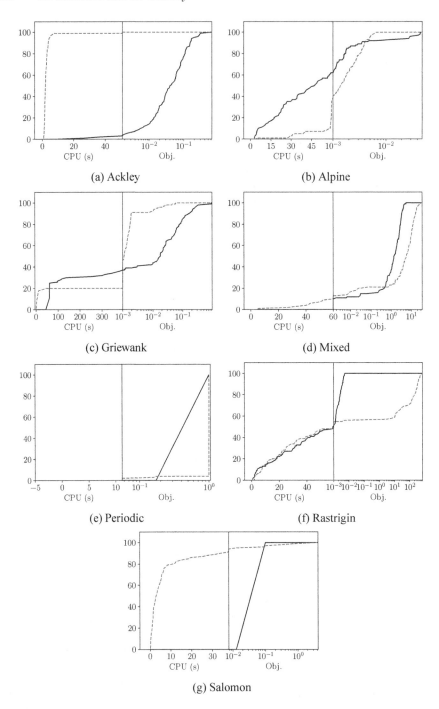

Fig. 2. Number of instances solved to 1e-3 in a given time limit (left) and the number of instances with an objective function better than the given value (right). LS is shown as a blue, dashed line and DS as a black, solid line.

a reasonable benefit of tuning DS for each individual benchmark. However, this is clearly not the main determining factor for DS' performance.

With respect to LocalSolver, we see that it solves more instances close to optimality than DS and DS-A. However, when looking at the mean gap over all instances, LS performs massively worse. We conclude that DS performs with less variance.

Figure 2 gives a detailed overview of how long it takes to solve instances on each function, and the quality of the solutions on those instances that aren't solved. Clearly, arguments can be made for both DS and LS depending on the function being solved and the amount of time available for solving. On Griewank and Alpine, DS is able to quickly get to 1e-3 on many instances, but has trouble on others, whereas the performance of LS remains mostly constant throughout all instances. In contrast, on the Rastrigin function, DS and LS solve roughly the same amount of instances, but the unsolved instances from DS are of significantly higher quality than those of LS.

5 Conclusion

We presented a novel dialectic search procedure for box-constrained, non-linear optimization problems with heterogeneous variable types. Our approach introduces a convex search procedure for *synthesizing* the thesis and antithesis of the search procedure, allowing it to highly effectively move through the search space. Moreover, we hyper-parameterized the resulting approach, allowing the meta-heuristics to adapt key search parameters during search based on runtime statistics characterizing the progress of the search. We compared our approach to three state-of-the-art procedures, the previous version of dialectic search, LSHADE and LocalSolver, and showed that the new dialectic search is able to compete, or even outperform these approaches, on occasion by very large margins.

Acknowledgements. The authors would like to thank the Paderborn Center for Parallel Computation (PC^2) for the use of the OCuLUS cluster.

References

1. Ansotegui, C., Malitsky, Y., Samulowitz, H., Sellmann, M., Tierney, K.: Model-based genetic algorithms for algorithm configuration. In: IJCAI, pp. 733–739 (2015)
2. Ansotegui, C., Sellmann, M., Tierney, K.: A gender-based genetic algorithm for the automatic configuration of algorithms. In: CP, pp. 142–157 (2009)
3. Ansótegui, C., Heymann, B., Pon, J., Sellmann, M., Tierney, K.: Hyper-reactive tabu search for MaxSAT. In: Battiti, R., Brunato, M., Kotsireas, I., Pardalos, P.M. (eds.) LION 12 2018. LNCS, vol. 11353, pp. 309–325. Springer, Cham (2019). https://doi.org/10.1007/978-3-030-05348-2_27
4. Ansótegui, C., Pon, J., Sellmann, M., Tierney, K.: Reactive dialectic search portfolios for MaxSAT. In: AAAI Conference on Artificial Intelligence (2017)

5. Argelich, J., Li, C., Manyà, F., Planes, J.: MaxSAT Evaluation (2016). www. maxsat.udl.cat
6. Benoist, T., Estellon, B., Gardi, F., Megel, R., Nouioua, K.: 4or. Localsolver 1. x: a black-box local-search solver for 0–1 $9(3)$, 299 (2011)
7. Glover, F., Laguna, M., Marti, R.: Fundamentals of scatter search and path relinking. Control Cybern. 39, 653–684 (2000)
8. Kadioglu, S., Sellmann, M.: Dialectic search. In: CP, pp. 486–500 (2009)
9. Lourenço, H., Martin, O., Stützle, T.: Iterated local search. In: Glover, F., Kochenberger, G.A. (eds.) Handbook of Metaheuristics, vol. 57, pp. 320–353. Springer, Boston (2003). https://doi.org/10.1007/0-306-48056-5_11
10. SIMANN: Fortran simulated annealing code (2004)
11. Tanabe, R., Fukunaga, A.S.: Improving the search performance of shade using linear population size reduction. In: 2014 IEEE Congress on Evolutionary Computation (CEC), pp. 1658–1665. IEEE (2014)

Least Squares K-SVCR Multi-class Classification

Hossein Moosaei[1,2]([✉]) [iD] and Milan Hladík[2] [iD]

[1] Department of Mathematics, Faculty of Science, University of Bojnord,
Bojnord, Iran
hmoosaei@gmail.com

[2] Department of Applied Mathematics, School of Computer Science,
Faculty of Mathematics and Physics, Charles University, Prague, Czech Republic
hladik@kam.mff.cuni.cz

Abstract. The support vector classification-regression machine for K-class classification (K-SVCR) is a novel multi-class classification method based on "1-versus-1-versus-rest" structure. In this paper, we propose a least squares version of K-SVCR named as LSK-SVCR. Similarly as the K-SVCR algorithm, this method assess all the training data into a "1-versus-1-versus-rest" structure, so that the algorithm generates ternary output $\{-1, 0, +1\}$. In LSK-SVCR, the solution of the primal problem is computed by solving only one system of linear equations instead of solving the dual problem, which is a convex quadratic programming problem in K-SVCR. Experimental results on several benchmark data set show that the LSK-SVCR has better performance in the aspects of predictive accuracy and learning speed.

Keywords: SVM · K-SVCR · Multi-class classification · Least squares

1 Introduction

Support vector machines (SVM) were proposed for binary classification problems by Vapnik and his colleagues [3,4]. The idea of this method is based on finding the maximum margin between two hyperplanes, which leads to solving a constraint convex quadratic programming problem (QPP).

Whereas there are many methods for binary classification [2,7–9,12], multi-class classification is often accrued in practical problems and real life [11]. Due to its wide range of applications, Angulo et al. [1] introduced a new method for multi-class classification based on "1-versus-1-versus-rest" structure with ternary output $\{-1, 0, +1\}$ for K-class classification. This method constructs $\frac{k(k-1)}{2}$ K-SVCR classifiers. It should be noted that, since all samples are given for construction of classifiers, the K-SVCR provides better performance than SVM methods for multi-class problems.

The authors were supported by the Czech Science Foundation Grant P403-18-04735S.

I. S. Kotsireas and P. M. Pardalos (Eds.): LION 14 2020, LNCS 12096, pp. 117–127, 2020.
https://doi.org/10.1007/978-3-030-53552-0_13

In this study, we propose a least squares version of K-SVCR, named as the least squares K-class support vector classification-regression machine (LSK-SVCR). In LSK-SVCR, we need to solve only one system of linear equations rather than solving a QPP in K-SVCR.

Numerical experiments on several benchmark data set indicate that the suggested LSK-SVCR has higher accuracy with lower computational time than K-SVCR.

Notations. Let $a = [a_i]$ be a vector in R^n. If f is a real valued function defined on the n-dimensional real space R^n, the gradient of f respect to x is denoted by $\frac{\partial f}{\partial x}$, which is a column vector in R^n. By A^T we mean the transpose of matrix A. For two vectors x and y in the n-dimensional real space, $x^T y$ denotes the scalar product. For $x \in R^n$, $\|x\|$ denotes 2-norm. A column vector of ones of arbitrary dimension is indicated by e. For $A \in R^{m \times n}$ and $B \in R^{n \times l}$, the kernel $k(A, B)$ is an arbitrary function which maps $R^{m \times n} \times R^{n \times l}$ into $R^{m \times l}$. In particular, if x and y are column vectors in R^n and $A \in R^{m \times n}$, then $k(x^T, y)$ is a real number, $k(x^T, A^T)$ is a row vector in R^m, and $k(A, A^T)$ is an $m \times m$ matrix. The identity $n \times n$ matrix is denoted by I_n, and $[A; B]$ stands for the matrix operation

$$[A;\, B] = \begin{bmatrix} A \\ B \end{bmatrix}.$$

The rest of this paper is organized as follows: Sect. 2 briefly describes SVM, and K-SVCR is then introduced in Sect. 3. Section 4 presents our LSK-SVCR in linear and non-linear cases as well as a classification decision rule. We analyse the computational complexity of the methods in Sect. 5. Section 6 presents experimental results on benchmark data set to show the efficiency of the proposed algorithm, and concluding remarks are given in Sect. 7.

2   Support Vector Machine for Classification

For a classification problem, a data set (x_i, y_i) is given for training with the input $x_i \in R^n$ and the corresponding target value or label $y_i = 1$ or -1, i.e.,

$$(x_1, y_1), \ldots, (x_m, y_m) \in R^n \times \{\pm 1\}. \tag{1}$$

The two parallel supporting hyperplanes are defined as follow:

$$w^T x - b = +1 \quad \text{and} \quad w^T x - b = -1.$$

In canonical form, the optimal hyperplanes are found by solving the following primal optimization problem [13]:

$$\min_{w,b,\xi} \frac{1}{2} w^T w + c e^T \xi$$
$$\text{subject to } \tilde{D}(Aw - eb) \geq e - \xi, \tag{2}$$
$$\xi \geq 0,$$

where the matrix $A \in R^{m \times n}$ records the whole data, the diagonal matrix $\tilde{D} \in R^{m \times m}$ (with ones or minus ones along its diagonal) is according to membership of each point in the classes $+1$ or -1 , $c > 0$ is the regularization parameter, and ξ is a slack variable. As for the primal problem, SVM solves its Lagrangian dual problem as follows:

$$\min_{\alpha} \sum_{i=1}^{m}\sum_{j=1}^{m} \alpha_i \alpha_j y_i y_j x_i x_j - \sum_{i=1}^{m} \alpha_i$$

$$\text{subject to } \sum_{i=1}^{m} y_i \alpha_i = 0, \tag{3}$$

$$0 \leq \alpha_i \leq c, \quad i = 1, \ldots, m,$$

where α_is are the Lagrange multipliers.

3   Support Vector Classification-Regression Machine for K-Class Classification

K-SVCR, which is a new method of multi-class classification with ternary output $\{-1, 0, +1\}$, has been proposed in [1]. This method introduces the support vector classification-regression machine for K-class classification. This new machine evaluates all the training data into a "1-versus-1-versus-rest" structure during the decomposing phase using a mixed classification and regression support vector machine (SVM). Figure 1 illustrates the K-SVCR method graphically.

Throughout this paper, we suppose without loss of generality that there are three classes $A_{m_1 \times n}$, $B_{m_2 \times n}$ and $C_{m_3 \times n}$ marked by class labels $+1$, -1, and 0, respectively. K-SVCR can be formulated as a convex quadratic programming problem as follows:

$$\min_{w,b,\zeta_1,\zeta_2,\phi,\phi^*} \frac{1}{2}\|w\|^2 + c_1(e_1^T \zeta_1 + e_2^T \zeta_2) + c_2 e_3^T(\phi + \phi^*) \tag{4}$$

$$\text{subject to } Aw + e_1 b \geq e_1 - \zeta_1,$$
$$- (Bw + e_2 b) \geq e_2 - \zeta_2,$$
$$- \delta e_3 - \phi^* \leq Cw + e_3 b \leq \delta e_3 + \phi,$$
$$\zeta_1, \zeta_2, \phi, \phi^* \geq 0.$$

Where $c_1 > 0$ and c_2 are the regularization parameters, ζ_1, ζ_2, ϕ and ϕ^* are positive slack variables, and e_1, e_2, and e_3 are vectors of ones with proper dimensions. To avoid overlapping, positive parameter δ must be lower than 1.

The dual formulation of (4) can be expressed as

$$\max_{\gamma} q^T \gamma - \frac{1}{2}\gamma^T H\gamma, \tag{5}$$

$$\text{subject to } 0 \leq \gamma \leq F,$$

where $Q = \begin{bmatrix} A^T & -B^T & C^T & -C^T \end{bmatrix}$, $H = Q^T Q$, $q = \begin{bmatrix} e_1^T & e_2^T & -\delta e_3^T & -\delta e_3^T \end{bmatrix}$, and $F = \begin{bmatrix} c_1 e_1; & c_1 e_2; & c_2 e_3; & c_2 e_3 \end{bmatrix}$. By solving this quadratic box constraint optimization problem, we can obtain the separating hyperplane $f(x) = w^T x + b$ and the decision function can be written as:

$$f(x_i) = \begin{cases} 1 & w^T x_i + b \geq \delta, \\ -1 & w^T x_i + b \leq -\delta, \\ 0 & \text{otherwise.} \end{cases}$$

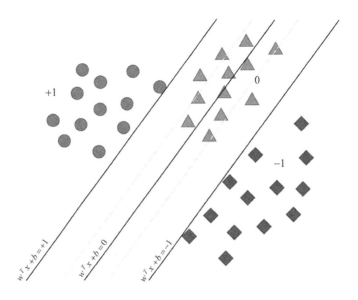

Fig. 1. Geometric representation of K-SVCR method

4   Least Square K-SVCR

In this section, we propose a least squares type of K-SVCR method called LSK-SVCR in both linear and nonlinear cases.

4.1   Linear Case

We modify the primal problem (4) of K-SVCR as (6), which uses the square of 2-norm of slack variables ζ_1, ζ_2, ϕ and ϕ^* instead of 1-norm slack variables in the objective function and uses equality constraint instead of inequality constraint in K-SVCR. Then, the following minimization problem can be considered:

$$\min_{w,b,\zeta_1,\zeta_2,\phi,\phi^*} \frac{1}{2}\|w\|^2 + c_1\left(\|\zeta_1\|^2 + \|\zeta_2\|^2\right) + c_2\|\phi\|^2 + c_3\|\phi^*\|^2 \qquad (6)$$

$$\text{subject to } e_1 - (Aw + e_1 b) = \zeta_1,$$
$$e_2 + (Bw + e_2 b) = \zeta_2,$$
$$Cw + e_3 b - \delta e_3 = \phi^*,$$
$$-Cw - e_3 b - \delta e_3 = \phi.$$

Where ζ_1, ζ_2, ϕ and ϕ^* are positive slack variables, and c_1, c_2 and c_3 are penalty parameters and positive parameter δ is a lower than 1.

Now by substituting the constraint into the objective function, we have the following unconstrained QPP

$$\min_{w,b} \frac{1}{2}\|w\|^2 + c_1\|e_1 - Aw - e_1 b\| + c_1\|e_2 + Bw + e_2 b\|$$
$$+ c_2\| - Cw - e_3 b - \delta e_3\|^2 + c_3\|Cw + e_3 b - \delta e_3\|. \qquad (7)$$

The objective function of problem (7) is convex, so for obtaining the optimal solution, we set the gradient of this function with respect to w and b to zero. Then we have

$$\frac{\partial f}{\partial w} = w + 2c_1(-A^T)(e_1 - Aw - e_1 b) + 2c_1 B^T(e_2 + Bw + e_2 b)$$
$$+ 2c_2(-C^T)(-Cw - e_3 b - \delta e_3) + 2c_3 C^T(Cw + e_3 b - \delta e_3) = 0,$$

$$\frac{\partial f}{\partial b} = 2c_1(-e_1^T)(e_1 - Aw - e_1 b) + 2c_1 e_2^T(e_2 + Bw + e_2 b)$$
$$+ 2c_2(-e_3^T)(-Cw - e_3 b - \delta e_3) + 2c_3 e_3^T(Cw + e_3 b - \delta e_3) = 0.$$

The above equation can be displayed in the matrix form as

$$2c_1 \begin{bmatrix} A^T A & A^T e_1 \\ e_1^T A & e_1^T e_1 \end{bmatrix} \begin{bmatrix} w \\ b \end{bmatrix} + 2c_1 \begin{bmatrix} B^T B & B^T e_2 \\ e_2^T B & e_2^T e_2 \end{bmatrix} \begin{bmatrix} w \\ b \end{bmatrix} + 2(c_2 + c_3) \begin{bmatrix} C^T C & C^T e_3 \\ e_3^T C & e_3^T e_3 \end{bmatrix} \begin{bmatrix} w \\ b \end{bmatrix}$$
$$+ \begin{bmatrix} 2c_1(-A^T)e_1 + 2c_1 B^T e_2 + 2c_2 C^T \delta e_3 + 2c_3 C^T(-\delta e_3) \\ 2c_1(-e_1^T e_1) + 2c_1 e_2^T e_2 + 2c_2 \delta e_3^T e_3 + 2c_3 e_3^T(-\delta e_3) \end{bmatrix} = 0.$$

Therefore w and b can be computed as follows:

$$\begin{bmatrix} w \\ b \end{bmatrix} = \begin{bmatrix} c_1(A^T A + B^T B) + (c_2 + c_3)C^T C & c_1(A^T e_1 + B^T e_2) + (c_2 + c_3)C^T e_3 \\ c_1(e_1^T A + e_2^T B) + (c_2 + c_3)e_3^T C & c_1(e_1^T e_1 + e_2^T e_2) + (c_2 + c_3)e_3^T e_3 \end{bmatrix}^{-1}$$
$$\begin{bmatrix} c_1(-A^T)e_1 + c_1 B^T e_2 + c_2 C^T \delta e_3 + c_3 C^T(-\delta e_3) \\ c_1(-e_1^T e_1) + c_1 e_2^T e_2 + c_2 \delta e_3^T e_3 + c_3 e_3^T(-\delta e_3) \end{bmatrix}.$$

We rewrite it as

$$\begin{bmatrix} w \\ b \end{bmatrix} = - \left[c_1 \begin{bmatrix} A^T \\ e_1^T \end{bmatrix} [A\ e_1] + c_1 \begin{bmatrix} B^T \\ e_2^T \end{bmatrix} [B\ e_2] + c_2 \begin{bmatrix} C^T \\ e_3^T \end{bmatrix} [C\ e_3] + c_3 \begin{bmatrix} C^T \\ e_3^T \end{bmatrix} [C\ e_3] \right]^{-1}$$
$$\left(-c_1 \begin{bmatrix} A^T e_1 \\ m_1 \end{bmatrix} + c_1 \begin{bmatrix} B^T e_2 \\ m_2 \end{bmatrix} + c_2 \delta \begin{bmatrix} C^T e_3 \\ m_3 \end{bmatrix} - c_3 \delta \begin{bmatrix} C^T e_3 \\ m_3 \end{bmatrix} \right).$$

Denote $E = [A \ e_1]$, $F = [B \ e_2]$, and $G = [C \ e_3]$, then we can obtain the separating hyperplane by solving a system of linear equations as follows:

$$\begin{bmatrix} w \\ b \end{bmatrix} = - \left[c_1 E^T E + c_1 F^T F + c_2 G^T G + c_3 G^T G \right]^{-1}$$
$$(-c_1 E^T e_1 + c_1 F^T e_2 + c_2 \delta G^T e_3 - c_3 \delta G^T e_3).$$

4.2   Nonlinear Case

In the real world problems, a linear kernel cannot always separate most of the classification tasks. To make the nonlinear types of problems separable, the samples are mapped to a higher dimensional feature space. Thus, in this subsection, we extend the linear case of LSK-SVCR to the nonlinear case, and we would like to find the following kernel surface:

$$k(x^T, D^T)w + eb = 0,$$

where $k(\cdot, \cdot)$ is an appropriate kernel function and $D = [A; B; C]$. After a careful selection of the kernel function, the primal problem of (4) becomes:

$$\min_{w,b,\zeta_1,\zeta_2,\phi,\phi^*} \frac{1}{2}\|w\|^2 + c_1(\|\zeta_1\|^2 + \|\zeta_2\|^2) + c_2\|\phi\|^2 + c_3\|\phi^*\|^2,$$
$$\text{subject to } e_1 - (k(A, D^T)w + e_1 b) = \zeta_1,$$
$$e_2 + (k(B, D^T)w + e_2 b) = \zeta_2,$$
$$k(C, D^T)w + e_3 b - \delta e_3 = \phi^*,$$
$$- k(C, D^T)w - e_3 b - \delta e_3 = \phi.$$

By substituting the constraint into the objective function, the problem takes the form

$$\min_{w,b} \frac{1}{2}\|w\|^2 + c_1\|e_1 - k(A, D^T)w - e_1 b\| + c_1\|e_2 + k(B, D^T)w + e_2 b\|$$
$$+ c_2\| - k(C, D^T)w - e_3 b - \delta e_3\|^2 + c_3\|k(C, D^T)w + e_3 b - \delta e_3\|.$$

Similarly to linear case, the solution of this convex optimization problem can be derived as follows:

$$\begin{bmatrix} w \\ b \end{bmatrix} = - \left[c_1 M^T M + c_1 N^T N + c_2 P^T P + c_3 P^T P \right]^{-1}$$
$$(-c_1 M^T e_1 + c_1 N^T e_2 + c_2 \delta P^T e_3 - c_3 \delta P^T e_3),$$

where $M = [k(A, D^T) \ e_1] \in R^{m_1 \times (m+1)}$, $N = [k(B, D^T) \ e_2] \in R^{m_2 \times (m+1)}$, $P = [k(C, D^T) \ e_3] \in R^{m_3 \times (m+1)}$, $D = [A; B; C]$ and $m = m_1 + m_2 + m_3$.

The solution to the nonlinear case requires the inversion of a matrix of size $(m+1) \times (m+1)$. In general, a matrix has a special form if the number of features (nF) is much less than the number of samples (nS), i.e., $nS \gg nF$, and in this case the inverse matrix can be inverted by inverting a smaller $nF \times nF$ matrix by using the Sherman–Morrison–Woodbury (SMW) [5] formula. Therefore, in this paper, to reduce the computational cost, the SMW formula is applied. More concretely, the SMW formula gives a convenient expression for the inverse matrix $A + UV^T$, where $A \in R^{n \times n}$ and $U, V \in R^{n \times K}$, as follows:

$$(A + UV^T)^{-1} = A^{-1} - A^{-1}U(I + V^T A^{-1}U)^{-1}V^T A^{-1}.$$

Herein, A and $I + V^T A^{-1}U$ are nonsingular matrices.

By using this formula, we can reduce the computational cost and rewrite the above formula for the hyperplane as follows:

$$\begin{bmatrix} w \\ b \end{bmatrix} = - \left(Z - ZM^T \left(\frac{1}{c_1}I_{m_1} + MZM^T \right)^{-1} MZ \right) \left(- c_1 M^T e_1 \right.$$

$$\left. + c_1 N^T e_2 + (c_2 - c_3)\delta P^T e_3 \right),$$

where $Z = \left(c_1 N^T N + (c_2 + c_3)P^T P \right)^{-1}$. When we apply SMW formula on Z again, then we have

$$Z = \frac{1}{c_2 + c_3} \left(Y - YN^T \left(\frac{c_2 + c_3}{c_1}I_{m_2} + NYN^T \right)^{-1} NY \right),$$

where $Y = (P^T P)^{-1}$. Due to possible ill-conditioning of $P^T P$, we use a regularization term αI, ($\alpha > 0$ and small enough). Then we have $Y = \frac{1}{\alpha}(I_{m_3} - P^T(\alpha I + PP^T)^{-1}P)$.

4.3   Decision Rule

The multi-class classification techniques evaluate all training points into the "1-versus-1-versus-rest" structure with ternary output $\{-1, 0, 1\}$. For a new testing point x_i, we predict its class label by the following decision functions:

For linear K-SVCR and LSK-SVCR :

$$f(x_i) = \begin{cases} +1, & x_i^T w + b \geq \delta, \\ -1, & x_i^T w + b \leq -\delta, \\ 0, & \text{otherwise.} \end{cases}$$

For nonlinear K-SVCR and LSK-SVCR :

$$f(x_i) = \begin{cases} +1, & k(x_i^T, D^T)w + eb \geq \delta, \\ -1, & k(x_i^T, D^T)w + eb \leq -\delta, \\ 0, & \text{otherwise.} \end{cases}$$

For k-class classification problem, the "1-versus-1-versus-rest" constructs $K(K - 1)/2$ classifiers in total, and for decision about final class label of testing sample x_i we get a total votes of each class. So the given testing sample will be assigned to the class label that gets the most votes (i.e. max-voting rule).

5 Computational Complexity

In this subsection, we discuss computational complexity of K-SVCR, and LSK-SVCR. In three-class classification problems, suppose the total size of each class is equal to $m/3$ (where $m = m_1 + m_2 + m_3$). Since samples in the third class are used twice in the constraints of K-SVCR problem, there are $4m/3$ inequality constraints in total. Therefore the computational complexity of K-SVCR is the complexity of solving one convex quadratic problem in dimension $n+1$ and with $4m/3$ constraints, where n is the dimension of the input space.

In our proposed methods for linear LSK-SVCR, we need to compute only one square system of linear equation of size $n + 1$.

In nonlinear LSK-SVCR, the inverse of a matrix of size $(m + 1) \times (m + 1)$ must be computed. The Sherman–Morrison–Woodbury (SMW) formula reduces the computational cost by finding the inverses of three matrices of smaller sizes $m_1 \times m_1$, $m_2 \times m_2$ and $m_3 \times m_3$.

6 Numerical Experiments

To assess performance of the proposed method, we apply LSK-SVCR on several UCI benchmark data sets [10] and compare our method with the K-SVCR. All experiments were carried out in Matlab 2019b on a PC with Intel core 2 Quad CPU (2.50 GHZ) and 8 GB RAM. For solving the dual problem of K-SVCR, we used "quadprog.m" function in Matlab. Also, we used 5-fold cross-validation to assess the performance of the algorithms in aspect of accuracy and training time. Note that in 5-fold cross-validation, the dataset is split randomly into five almost equal-size subsets, and one of them is reserved as a test set and the others play the role of a training set. This process is repeated five times, and the average accuracy of five testing results was used as the classification performance measure. Notice that the accuracy is defined as the number of correct predictions divided by the total number of predictions; to display it into a percentage we multiplied it by 100.

6.1 Parameter Selection

It must be noted that the performance of the algorithms depends on the choice of parameters. In the experiments, we opt for the Gaussian kernel function $k(x_i, x_j) = \exp(\frac{-\|x_i - x_j\|^2}{\gamma^2})$. The best parameters are then obtained by the grid search method [6,7].

In this paper, the optimal value for c_1, c_2, c_3, were selected from the set $\{2^i | i = -8, -7, \ldots, 7, 8\}$, the parameters of the Gaussian kernel γ were selected from the set $\{2^i | i = -6, -5, \cdots, 5, 6\}$, and parameter δ in K-SVCR and LSK-SVCR was chosen from set $\{0.1, 0.3, \ldots, 0.9\}$.

Table 1. The characterization of data sets.

Data set	Number of instances	Number of attributes	Number of classes
Iris	150	4	3
Balance	625	4	3
Soybean	47	35	4
Wine	178	13	3
Breast tissue	106	10	4
Hayes-Roth	132	5	3
Ecoli	327	7	5
Teaching	150	5	3
Thyroid	215	6	3

Table 2. Performance of K-SVCR and LSK-SVCR with Gaussian kernel.

Data set	K-SVCR Acc ± std	K-SVCR Time (s)	LSK-SVCR Acc ± std	LSK-SVCR Time (s)
Iris	96.54 ± 2.04	5.27	**98.67 ± 1.82**	**1.54**
Balance	94.21 ± 2.04	89.56	**94.89 ± 2.01**	**4.48**
Soybean	**100.00 ± 0.00**	1.65	**100.00 ± 0.00**	**1.59**
Wine	98.81 ± 2.51	7.06	**99.45 ± 1.24**	**0.79**
Breast tissue	**47.06 ± 9.73**	2.91	46.59 ± 15.39	**1.12**
Hayes-Roth	46.33 ± 12.86	12.70	**75.72 ± 8.81**	**0.58**
Ecoli	77.36 ± 4.28	21.23	**89.01 ± 5.89**	**5.57**
Teaching	63.68± 5.58	10.63	**70.19 ± 7.46**	**0.96**
Thyroid	83.25 ± 6.02	21.26	**93.49 ± 2.55**	**1.38**

6.2 UCI Data Sets

In this subsection, to compare the performance of K-SVCR with LSK-SVCR, we ran these algorithms on several benchmark data sets from UCI machine learning repository [10], which are described in Table 1.

To analyse the performance of the K-SVCR and LSK-SVCR algorithms, Table 2 shows a comparison of classification accuracy and computational time for K-SVCR and LSK-SVCR on nine benchmark datasets available at the UCI machine learning repository. This table indicates that for Iris dataset, the accuracy of LSK-SVCR (accuracy: 98.67, time: 0.03 s) was higher than K-SVCR (accuracy: 96.54, time: 3.51 s), so our proposed method was more accurate and faster than original K-SVCR. A similar discussion can be made for Balance, Soyabean, Wine, Brest Tissue, Hayes-Roth, Ecoli, Teaching, and Thyroid datasets.

The analysis of experimental results on nine UCI datasets revealed that the performance of LSK-SVCR was slightly better than the original K-SVCR. We should note that for Brest Tissue, although the K-SVCR is a little more accurate than LSK-SVCR, the LSK-SVCR is faster. Therefore, according to the experimental results in Table 2, LSK-SVCR not only yielded higher prediction accuracy but also had lower computational times.

7   Conclusion

The support vector classification-regression machine for K-class classification (K-SVCR) is a novel multi-class method. In this paper, we proposed a least squares version of K-SVCR named as LSK-SVCR for multi-class classification. Our proposed method leads to solving a simple system of linear equations instead of solving a hard QPP in K-SVCRKindly provide the page range for Ref. [2].. The K-SVCR and LSK-SVCR evaluates all training data into "1-versus-a-versus-rest" structure with ternary output $\{-1, 0, +1\}$.

The computational results performed on several UCI data set demonstrate that, compared to K-SVCR, the proposed LSK-SVCR has better efficiency in terms of accuracy and training time.

References

1. Angulo, C., Català, A.: K-SVCR. a multi-class support vector machine. In: López de Mántaras, R., Plaza, E. (eds.) ECML 2000. LNCS (LNAI), vol. 1810, pp. 31–38. Springer, Heidelberg (2000). https://doi.org/10.1007/3-540-45164-1_4
2. Bazikar, F., Ketabchi, S., Moosaei, H.: DC programming and DCA for parametric-margin ν-support vector machine. Appl. Intell. 1–12 (2020)
3. Boser, B.E., Guyon, I.M., Vapnik, V.N.: A training algorithm for optimal margin classifiers. In: Proceedings of the fifth annual workshop on Computational learning theory. COLT 1992, pp. 144–152, Association for Computing Machinery, New York (1992). https://doi.org/10.1145/130385.130401
4. Cortes, C., Vapnik, V.: Support-vector networks. Mach. Learn. 20(3), 273–297 (1995). https://doi.org/10.1007/BF00994018
5. Golub, G.H., Van Loan, C.F.: Matrix Computations. Johns Hopkins University Press, Baltimore (2012)
6. Hsu, C.W., Chang, C.C., Lin, C.J., et al.: A practical guide to support vector classification (2003)
7. Ketabchi, S., Moosaei, H., Razzaghi, M., Pardalos, P.M.: An improvement on parametric ν -support vector algorithm for classification. Ann. Oper. Res. 276(1–2), 155–168 (2019)
8. Kumar, M.A., Gopal, M.: Least squares twin support vector machines for pattern classification. Expert Syst. Appl. 36(4), 7535–7543 (2009). https://doi.org/10.1016/j.eswa.2008.09.066
9. Lee, Y.J., Mangasarian, O.: SSVM: a smooth support vector machine for classification. Comput. Optim. Appl. 20(1), 5–22 (2001). https://doi.org/10.1023/A:1011215321374

10. Lichman, M.: UCI machine learning repository (2013). http://archive.ics.uci.edu/ml
11. Tang, L., Tian, Y., Pardalos, P.M.: A novel perspective on multiclass classification: regular simplex support vector machine. Inf. Sci. **480**, 324–338 (2019)
12. Tang, L., Tian, Y., Yang, C., Pardalos, P.M.: Ramp-loss nonparallel support vector regression: robust, sparse and scalable approximation. Knowl.-Based Syst. **147**, 55–67 (2018)
13. Vapnik, V.N., Chervonenkis, A.J.: Theory of Pattern Recognition. Nauka (1974)

Active Learning Based Framework for Image Captioning Corpus Creation

Moustapha Cheikh[1,3(✉)] and Mounir Zrigui[2,3]

[1] Faculty of Economics Management, University of Sfax, Sfax, Tunisia
moustapha.ml.cheikh@gmail.com
[2] Faculty of sciences of Monastir, University of Monastir, Monastir, Tunisia
mounir.zrigui@fsm.rnu.tn
[3] Research Laboratory in Algebra, Numbers Theory and Intelligent Systems, University of Monastir, Monastir, Tunisia

Abstract. Image captioning aims at analyzing the content of an image in order to subsequently generate a textual description through verbally expressing the important aspects of it. In spite of the fact that the task of automatic image description is not bound to the English language, yet, the recent advances mostly focus on English descriptions. Collecting captions for images is an expensive process that requires time and labor cost. In this paper, we introduce a novel active learning framework with human in the loop for image captioning corpus creation, using a translated version of existing datasets. We implemented this framework to create a new dataset called ArabicFlickr1K. This dataset has 1095 images, each is associated with three to five descriptions. We also propose a neural network architecture to automatically generate Arabic captions for images. This architecture relies on an encoder-decoder framework. Our model scored 47% on BLUE-1.

Keywords: Image captioning · Computer vision · Natural language processing

1 Introduction

Overs the last few years has been renewed interest in tasks that require a combination of linguistic and visual information [1]. This interest is largely motivated by the amount of available data on the internet and the recent advances in computer vision.

Image captioning [10] has become a key task with the interest of both natural language processing (NLP) and computer vision communities. This task consists of analyzing the content of an image in order to subsequently generate a textual description of it by verbally expressing the important aspects of that image.

Image captioning may play an important role in many applications. For instance the generated captions can be used in text based information retrieval [12], video indexing [44] and several other NLP applications.

© Springer Nature Switzerland AG 2020
I. S. Kotsireas and P. M. Pardalos (Eds.): LION 14 2020, LNCS 12096, pp. 128–142, 2020.
https://doi.org/10.1007/978-3-030-53552-0_14

The description could be difficult because it could in principle, taking into consideration any visual aspect of the image, include the description of the objects and their properties, as well as the way in which people and objects of the image are interacting. Nevertheless, image captioning is a complex task since it requires not only a complete understanding of the image, but also a sophisticated generation of natural language.

A brief look at an image is enough for a human being to point out and describe an important amount of details about the visual scene. Our visual system can't recognize a lot of gray shares compared to hundreds of thousands of different color shades and intensities. The images considered in this work are color images, and this is due to the immense deal of information that can be found in color images.

In an image captioning system, we have as an input an RGB image I and we are required to generate a sequence of words $= (s_1, s_2, ..., s_N)$. The possible words $s_i \in V$ at time-step i are subsumed in a discrete set V of options. The number of possible options $|V|$ easily reaches several thousands. There are special tokens in the set of option V that mark any word that is not in the set, the start of a sequence and its end. In practice, those tokens are used to identify whether a word exists in the set of options V or it is either the start or the end of a sequence.

Given a training set $D = \{(I, S^*)\}$, which contains pairs (I, S^*) of image input I and corresponding ground-truth caption $S^* = (s_1^*, s_2^*, ..., s_N^*)$, consisting of words $s_i^* \in V, I \in \{1, 2, .., N\}$, we maximize, with respect to parameters W, a probability model $P_W(s_1, s_2, ..., s_N | I)$.

Collecting captions for images or videos [26] is an expensive process. This is not unique to image caption, however, it's much easier to create a corpus, compared to other tasks in NLP, that suffers from lack of resources [25].

The main contribution of this paper is the focus on solving the lack of resource for Arabic image captioning. Our contributions are as follows:

- We proposed a novel active learning framework with human in the loop for image captioning corpus creation, using a translated version of existing datasets.

  Human annotators help refine the translation of the automatic translation model and identify the correct one. As annotators label quality of the translations, our system ranks the rest of the translated sentences and propose new instances that have the highest probability of being correct for human verification. The idea behind this is to reduce the time that would be spent to find the correct translation in the translated version.
- We proposed a new dataset of Arabic image captions named ArabicFlickr1k. This dataset contains 1095 images, every image is associated with at least three captions.
- We introduced a deep learning model based on Encoder-Decoder architecture for Arabic image captioning.

The remainder of the paper is organized in sections as follows: Sect. 2 aims at presenting a detailed review of existing datasets and approaches for automatic image captioning in the literature. Section 3 is about providing a description of each component of the proposed active learning framework for image captioning corpus creation. In Sect. 4, we are describing end to end architecture for Arabic image captioning. Last but not least, the experimental evaluation and the results are provided. Finally, in Sect. 6 conclusion and future research directions are presented.

2 Literature Review

The availability of datasets, containing images mapped to their descriptions, has contributed to the advance of image captioning research. Image captioning model benefits from the quality and the size of this datasets. Serious progress has been made in the English language. However, other languages are behind, given the scarcity of image captions corpora [2]. The following datasets are the most commonly used of the literature.

2.1 Datsets

Flickr8k [15] is a collection of 8092 images taken from the Flickr website and made public by the University of Illinois. The images contain no person or famous place, so that the entire image can be described according to all the different objects of the image. Each image contains five different captions for reference with an average length of 11.8 words written by humans. They used Amazon Mechanical Turk crowdsourcing service to collect this descriptions. They asked people on the platform to describe the objects, scenes and activities in the images without providing them any information about the images. Only With the information that can be found in the image they were able to collect conceptual descriptions that describe the images.

Flickr30k [42] is an extension of Flickr8k. It contains 31,783 images of people involved in everyday activities and events from the Flickr website. Each image is associated with five descriptions in English, which were collected from Amazon Mechanical Turk. These descriptions are required to accurately describe the objects, scenes, and activities displayed in the image. The dataset contains 158,915 descriptions. Usually, 1000 images are selected as validation data, 1000 images as test data, and the remaining images are used as train data.

Microsoft COCO [23] is a large scale dataset, containing 123,287 images. Most images contain multiple objects and meaningful contextual information we encounter in everyday scenes, and each image is accompanied by five English descriptions annotated by humans. Microsoft COCO is widely used for various computer vision tasks.

STAIR Captions [40] is a Japanese version of Microsoft COCO, it consists of 820,310 Japanese captions for 164,062 images. The authors proposed a model combining the English and Japanese captions. The resulting bilingual model has better performance when compared with the monolingual model that uses only the Japanese caption corpus.

Multi30K [9] is the German version of Flickr30K. The authors extended the Flick30k dataset by collecting five descriptions in German for the 31014 images. They used the Crowdflower platform to hire 185 people for 31 days to describe each image. They collected five independent descriptions for each image. They also translated 31,000 descriptions (about 6200 images) of the English version, translated by professional translators without seeing the images.

Recently, a growing number of research focused on the task of associating images and sentences from both the computer vision and the NLP researchers. In The literature, there are two traditional well-studied directions. The first approaches are known as the language model-based approaches or generation based approaches. They start by converting images into words describing a fixed number of scenes, objects, their attributes and their spatial relations. After that, they formulate new coherent sentences from those words. The second approaches are known as retrieving based approaches. They produce the description by transferring existing description from other images. In the remaining of this section, we will see works done on those approaches and other approaches based on neural networks.

2.2 Generation Based Approaches

Generation based approaches differ in the way they represent images and the technique they use to generate the descriptions.

We mention in this category [22]. Their approach comprises a first step that uses Image Recognition models to extract visual information from the image [11]. They extract a fixed number of objects, including things like birds and cars, they also extract stuff like grass and water. For each object extracted from the image, they also extract their attributes like color. Finally, they extract the special relationships between those objects (near, under). This information is next used for composing sentences to describe the image. The generation step relies on Web-scale N-grams [5]. They did not take actions into consideration in the extraction step.

Another work that considers actions, by relying on an external corpus to predict the relationships between objects, is [39]. In which they fill in a sentence template by predicting the likely objects, verbs, scenes and prepositions that make up the core sentence structure. This is based on a Hidden Markov Model (HHM). They started by detecting objects and scenes from the image using the state of art image recognition models at that time. After this step, they used a language model trained on the English Gigaword corpus to predict the verbs given the objects detected in the image. Using the predicted actions, they estimated the probability that a preposition co-locates with a scene using an existing data. They used a HMM model to find the likely sentence structure given the predicted objects, verbs, scenes and propositions. The last step is the generation, using the results from the previous steps they fill in a fixed sentence template. They limited the number of objects, Verbs, Scenes and Prepositions to cover only what is commonly encountered in images. In addition, the sentence generated for each image is limited to at most two objects occurring in a unique scene.

In this context too, [20] generate descriptions from using pre-trained object detectors and a fixed template based method for description generation. Their system use object recognition models to detect objects (bird, car person, grass, trees). For each detected object they pass it to an attribute classifier and store the detected attributes. Same for every object and region, they predict preposi-tional relationships. They combine the output of the above in a CRF to produce input for language generation methods and generate the description using a fixed template.

[28] used computer vision models to predict the bounding boxes of objects in the image. For each detected object in the image, they extract attributes such as shape and texture. They also associate detected actions from the image to objects. Finally, they use preposition functions to predict a set of spatial relations that is held between each pair of objects based on their bounding box. A step before the description generation filters detected attributes that are unlikely and place objects into an ordered syntactic structure. Finally, they generate a large set of syntactically well-formed sentence fragments and then recombine these using a tree-substitution grammar.

2.3   Retrieval Based Approaches

Retrieval based approaches in the literature can be branded into two main cat-egories. The first one uses a visual space to extract similar image for a given query image. The other category combines textual and visual information in one space.

[21] work falls into this category. For a given test image, their system retrieves visually similar images from the training data. From those images, they extract segments of their corresponding descriptions that are potentially useful. Then they selectively use those text segments to produce a new description. In order to compose the description, they proposed a new stochastic composing algorithm. A downside of their system is that the produced sentences rely on how correctly the retrieved text segments can describe the given image.

A collection of one million images associated with visually relevant descrip-tions was introduced in [30]. These descriptions are written by people on the Flicker website. The authors also proposed two methods for automatic descrip-tions generation. The first method uses two global image descriptors to retrieve similar images. The second method integrates global descriptors and specific con-tent estimators. The specific content estimators extract objects, actions, staff, attributes and scenes from the image. Relaying on the large parallel corpus that they collected, they used both methods to produce relevant image descrip-tions. Since the descriptions associated with images were written by humans, this corpus enabled the proposed methods to yield descriptions that have a high linguistic quality.

The second category of retrieval-based approaches produces a co-embedding of images and descriptions in the same space. Among the works that have opted for this approach, we find [13]. Where they proposed Stacked Auxiliary Embedding (SAE); an approach based on weakly annotated images data. They were able to improve the performance of description retrieval using SAE, to transfer knowledge from a large-scale data of weakly annotated images. Even with large amount of dataset, retrieval-based approaches do not have the ability to generate new description for unseen image with new combination of objects.

2.4 Retrieval Based Approaches

Recently research in image automatic description has been limited by the existing techniques in image recognition systems and their efficacy. However, this systems begins to improve with the advances of neural network approaches [19].

[18] is the first to use only neural networks for automatic image description in the same period as [35] where they proposed a representation that map images and sentences into the same space using recursive neural networks. After that, they can map a given image into this space, rank all the sentences and chose the first one as description. However, unlike Socher, Kiros proposed a multi-layer perceptron (MLP) that uses a group of word representation vectors biased by features from the image. This means the image features condition the language model output. The image features are extracted from a convolutional neural network.

Advances in machine translation and computer vision enabled [37] to produce a new model based on deep neural network for image description. Their model consists of a convolutional neural network that represents the image in a context vector which then is passed to a language model based on LSTM. The joined model takes an image as input and is trained to maximize the probability of a description associated with a the given image. The model is fully trainable using Stochastic gradient descent and has the state of the art performance at that time on MS COCO. Similar work at the same period has been done by [17] where they used VGGNet [34] to represent the images and obtained the state of the art performance on Flickr8K, Flickr30K and MSCOCO.

Karpathy and Vinyals models provide the image features vector as an input only at the first step of the generation. In [8], the proposed model uses the image features vector at each time step. They represent the images using a pretrained CNN model (VGGNet [34] and CaffeNet [16]) on ImageNet. They explored three variants of their image description architecture, and evaluated the effect of depth in the LSTM language model. Their work also covered video description and Activity Recognition.

Work cited above that uses neural networks do not pay attention to a particular area or objects in the image when generating the description. The concept of attention was first introduced by [38] for image description. They proposed two variations of spatial attention and demonstrate that their models are able to focus on specific region in the image while generating the description. This can be used to gain insight on how their models work.

The success of spatial attention proposed by [38] was followed by the semantic attention in [41]. Spatial attention enables the generation component to focus on relevant places and regions in the image to compose more accurate image description. While, the semantic attention helps the generation step to incorporate semantically relevant concepts like actions and objects detected from the image. First, they map visual concepts (regions, objects, attributes, etc.) detected in the image to words. After that, they use a pretrained convolutional neural network to extract visual features. Then, the model learns through the semantic attention to selectively fuse this words with the visual features into the hidden states and outputs of recurrent neural networks that generates the image description. This words are represented with word embeddings, witch means that they can use external resource, not only for the image representation, but also for text representation.

While [41] extracted visual concepts and used them to help the visual attention, another text based type of attention was proposed by [29], in which they used the description associated with an image to guide the visual attention. During the training; they guide the visual attention with the description to help the model focus only on relevant visual objects in the image. This description is retrieved from visually similar images in the training dataset. They showed that this approach yield better performance on MS-coco at that time.

Neural networks based approaches before that [24] typically provide the language model with the image features at every step of the description generation. The authors argue that the language model does not need visual features to generate every word in the description. They introduced an adaptive attention encoder-decoder model. This model can automatically choose to ignore visual features when generating the next word of the description and to only use the language model. The adaptive attention decides when the language model should look at the image and also where it should look. This is done by a new extension of LSTM that relies on a new spacial attention that they introduced. They reported a significant performance over previous methods on MS COCO and Flickr30.

3   Corpus Creation Framework

There is a wide variety of resources on the internet like Facebook, Flickr and other websites from which we can collect images with captions. The only problem with those captions is that they do not describe the image specifically, but rather they give information about what cannot be seen in the image. In [15], the authors suggested that the description should focus on conceptual information that refer to objects, attributes, events and other literal content of the image.

While the task of automatic image description is not bound to the English language, yet, the recent advances have been mostly focusing on English descriptions. It is clear that the creation of resources like [9] costs tens of thousands of dollars and is a time-consuming task. However, The creation of new resources for Arabic image captioning will have a great impact on future research.

In the following subsections we explain in details the different components of our active learning based framework for image captioning corpus creation.

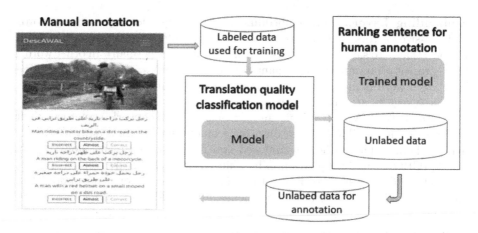

Fig. 1. Active learning based framework for image captioning corpus creation.

The first component is a manual annotation tool created with Django and Vue js to speed the process for annotators. This tool is used by Arabic native speakers to classify the quality of the translation presented for them. We use a translation quality classification model to rank and decide what images be passed to the manual verification.

3.1 Manual Annotation

This step consists of an interface that present an image with the associated descriptions to the annotators. Every description comes with an Arabic translation. The human annotators are instructed to verify the quality of the translation on a scale of incorrect, almost correct and incorrect. Almost correct is for caption that needs one or two word refinements. The next component is responsible for choosing which image to be present for annotation to minimize the human effort of finding good translations.

Initially, we picked a random set of descriptions and presented them to human annotators. We used their annotations as initial training data for the next component of the active learning framework.

3.2 Translation Quality Classification Model

The translation quality classification model is an essential component. We use a text classification [43] model. After training, we use this model to classify a small set of the unlabeled data and rank them by the model confidence. We then choose the first batch and pass it to human annotators. The idea is to get more correct translations in a batch compared with random selection. The model is a combination of different layers. The details of each layer are presented next.

Embedding Layer. Word embedding is a technique for representing words by fixed-size vectors, so that words that have a similar meaning also have close vectors (that is to say, vectors whose Euclidean distance is small). In other words, the representation implies the semantic meaning of words [3]. In the Embedding layer, each word is mapped to a dense vector of dimension d.

These vectors are initialized based on the word embeddings model that was proposed in [27]. With a simple and efficient neural network structure, their model made it possible to train on a huge amounts of textual data in a short period of time. The authors introduced two models called Continuous Bag of Words (CBOW) and Skip-Gram. The Skip-Gram architecture tries to predict the context of a given word. Both of these models have become very popular in recent years, showing several improvements in the field of NLP.

Bidirectional Gated Recurrent Units Layer. The units of this layer are composed of the GRU architecture proposed in [6]. GRUs are a more recent variation of LSTM networks. GRU first calculates an update gate based on the current input vector and the hidden state.

$$z^{(t)} = \sigma(W_z x^{(t)} + R_z h^{(t-1)} + b_z) \tag{1}$$

Then, it calculate the reset gate in a similar way but with different weights using a new memory content. If the reset gate is 0, then it skips the previous memory and stores only the new information. The final memory at the time step combines the current and previous time steps.

$$\bar{h}^{(t)} = \sigma(W_h x^{(t)} + r^t \odot R_h h^{(t-1)} + b_h) \tag{2}$$

$$h^{(t)} = z^{(t)} \odot h^{(t-1)} + (1 - z^{(t)}) \odot \bar{h}^{(t)} \tag{3}$$

Bidirectional GRUs process data in both directions with forward and backward hidden layers. Compared with the unidirectional, the number of parameters doubles. Bidirectional GRU returns a vector for each direction. The average of the outputs is taken, giving a vector with the dimension equals to the number of GRU units in the layer. It has been shown that GRUs work better than regular LSTMs and are faster thanks to a simpler architecture [7].

Attention Layer. Taking the representation sequence h, outputted by the BiGRU layer as input, the attention layer produces a new representation vector c with the dimension equal to time steps. This attention is proposed in [32].

$$c = \sum_{t=1}^{T} \alpha_t h_t \tag{4}$$

Where

$$e_t = a(h_t), \alpha_t = \frac{\exp(e_t)}{\sum_{k=1}^{T} \exp(e_k)}, \tag{5}$$

$$a\left(h_t\right) = \tanh\left(Wh_t + b\right) \tag{6}$$

W, b are learned with the model.

Output Layer. This layer takes as an input the output of the attention layer. The input is fed to a feed forward neural network, with output going through the Softmax function to give the predictions.

3.3   Ranking Sentence for Human Annotation

Initially, we choose random batch of descriptions for annotation, this is done in the first few batches. We use those descriptions as a training data for the translation quality classification model.

After the model is trained, we predict the classes of a random set of descriptions. The model outputs the probability of every class (Correct, Incorrect, and mostly correct) for a given translation instance. We then sort the images based on the number of correct translations and the degree of confidence given by the model. After that, we chose set from the top and send them to human for annotation. After this step, we retrain the model again and repeat.

4   Arabic Image Captioning Model

The encoder-decoder architecture was introduced for the first time in [36], Since then, it has become the standard Neural Machine Translation (NMT) approach. This architecture especially if given large amount of data, outperforms classical Machine translation (MT) methods [4].

Our model was inspired from this architecture. In image captioning, the core idea is to utilize a Convolutional Neural Network (CNN) as an encoder to extract visual features and a Recurrent Neural Networks (RNN) as a decoder to generate the caption.

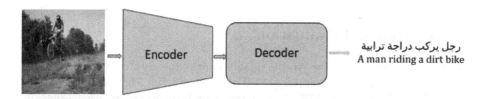

Fig. 2. Arabic image captioning system based on encoder-decoder architecture.

4.1 Encoder

For many years, training deep Neural network was difficult because of a problem known as vanishing gradient. The gradient of the loss function shrinks to zero when the chain rule is applied several times. This prevented the network weights from getting updated so the learning is not performed. ResNets [14] solve this problem where the gradient flow backwards using the skip connections.

To extract hierarchical visual information from the image, we used a pre-trained Resnet101 on Imagenet. In our model Fig. 2, the encoder is first applied to extract both global and regional visual information from the input image. Then we pass those features to the encoder to generate a description. The encoder can be fine-tuned during the training phase.

4.2 Decoder

The main idea behind the decoder is that of the conditional language model. A language model calculates the probability of a sentence by the following equation, where x_i is the next word and $x_1, x_2, ..., x_{i-1}$ represent the context.

$$P(X) = \prod_{i=1}^{n} P(x_i | x_1, x_2, ..., x_{i-1}) \tag{7}$$

To model the image caption generation problem, we use a conditional language model with the image I. s_j is the next word in the image description and $I, x_1, x_2, ..., x_{i-1}$ represent the context used to generate s_j.

$$P(S|I) = \prod_{j=1}^{n} P(s_j | I, s_1, s_2, ..., s_{j-1}) \tag{8}$$

We used an embedding layer, followed by an LSTM layer and a feed forward network layer. The model is trained end to end using cross-entropy loss. At each step, the decoder produces a probability distribution over possible next works. The embedding layer is initialized with pre-trained word2vec model on Arabic Wikipedia.

5 Experimental Evaluation

The proposed framework for Arabic images captioning corpus creation is based on the translation of existing dataset. We translated the Flick30K [42] dataset using Google Translation API. To evaluate this framework, we used Flick30K but the same steps are valid for a much bigger dataset like MS coco.

First, the annotators were given a set of 3430 descriptions. They were asked to classify them into three classes. Correct if the translation corresponding to the original English version is correct, almost correct if the translation needs one or two word editing and incorrect otherwise. On average, we found about 6% incorrect translation, 29% almost correct and the rest 65% is correct.

Then we used the result from the above step to train the translation quality classification model to classify the quality based on two classes (correct and incorrect). We could use the three classes, but we focused only on the correct translations. The embedding layer is initialized with word2vec weights trained on Arabic Wikipedia articles using Gensim [33].

We chose a random batch of 2000 images and classified their translated descriptions using the translation quality classification model and passed the top 885 images with the correct translations ranked by the model confidence to the annotators. We found 73% correct descriptions, that is an improvement on the random selection strategy. Finally, we ended up with 1095 images, each image has at least three correct descriptions validated by humans.

The caption model was implemented in Pytorch with the help of Scikit-learn and Tensorflow. All the experiments were done on an Ubuntu system. We used one NVIDIA 1080Ti and 32 GB RAM. We split our data set to 895 images for training, 100 for validation and 100 for the test. We applied some transformation to the images before feeding them to the encoder. All images are scaled to $3 * 224 * 224$ and normalized. We prepossessed all captions. We started by tokenizing and then removing words that occur less than two times and then added tokens to mark the start and the end of each caption. In the encoder layer we used an LSTM layer with 512 units. We used 300 for the embedding layer size.

All metrics use for language evaluation output a score indicating a similarity between the candidate sentence and the reference sentences. A popular metric used for automatic image captioning evaluation is BLEU.

مجموعة من الناس
في الخارج في مدينة مزدحمة
A group of people outside
in a crowded city

رجل يركب دراجة ترابية
A man riding a dirt bike

لاعب كرة قدم يرتدي قميصًا
أحمر اللون في الملعب
A footballer wearing a red
shirt on the pitch

امرأة في سترة حمراء تحمل
الزهور من باقة
A woman in a red jacket holds
a bouquet of flowers

Fig. 3. Arabic image descriptions generated using the proposed model with their translation in English.

BLEU (Bilingual evaluation understudy) [31] computes the geometric mean of n-gram precision scores multiplied by a brevity penalty in order to avoid overly short sentences. It is a metric that can be used to measure the quality of machine generated text in tasks like text summarization, Speech recognition and automatic image captioning. This metric was first introduced for machine translation as a reasonable correlation with human judgments of quality.

The caption model is trained end to end with the cross entropy loss. The performance of the proposed model on the test set gave a promising result of 47 for the BLEU-1, 24 for the BLEU-2, 20 for the BLEU-3 and 11 for the BLEU-4.

6 Conclusions

In this paper, we proposed a novel active learning based framework for Arabic image captioning corpus creation. This framework relies on the translations of existing datasets. We also proposed a new corpus for Arabic image captioning (ArabicFlickr1K). We did a detailed review of the literature and the existing resources. We introduced a deep learning model based on the Encoder-Decoder architecture for Arabic image captioning. Our model scored 47% on BLUE-1. Future research directions will go towards leveraging unsupervised data, using more complex language models in the Decoder and more supervised fine-tuning in the training phase.

References

1. Agrawal, A., et al.: VQA: visual question answering. Int. J. Comput. Vision **123**(1), 4–31 (2017)
2. Al-Muzaini, H.A., Al-Yahya, T.N., Benhidour, H.: Automatic Arabic image captioning using RNN-LSTM-based language model and CNN. Database **9**(6) (2018)
3. Ayadi, R., Maraoui, M., Zrigui, M.: LDA and LSI as a dimensionality reduction method in Arabic document classification. In: Dregvaite, G., Damasevicius, R. (eds.) ICIST 2015. CCIS, vol. 538, pp. 491–502. Springer, Cham (2015). https://doi.org/10.1007/978-3-319-24770-0_42
4. Bacha, K., Zrigui, M.: Machine translation system on the pair of Arabic/English. In: KEOD, pp. 347–351 (2012)
5. Brants, T.: Web 1t 5-gram version 1. http://www.ldc.upenn.edu/Catalog/CatalogEntry.jsp?catalogId=LDC2006T13 (2006)
6. Cho, K., et al.: Learning phrase representations using RNN encoder-decoder for statistical machine translation. arXiv preprint arXiv:1406.1078 (2014)
7. Chung, J., Gulcehre, C., Cho, K., Bengio, Y.: Empirical evaluation of gated recurrent neural networks on sequence modeling. arXiv preprint arXiv:1412.3555 (2014)
8. Donahue, J., et al.: Long-term recurrent convolutional networks for visual recognition and description. In: Proceedings of the IEEE Conference on Computer Vision and Pattern Recognition, pp. 2625–2634 (2015)
9. Elliott, D., Frank, S., Sima'an, K., Specia, L.: Multi30k: multilingual English-German image descriptions. arXiv preprint arXiv:1605.00459 (2016)
10. Farhani, N., Terbeh, N., Zrigui, M.: Image to text conversion: state of the art and extended work. In: 2017 IEEE/ACS 14th International Conference on Computer Systems and Applications (AICCSA), pp. 937–943. IEEE (2017)
11. Felzenszwalb, P.F., Girshick, R.B., McAllester, D., Ramanan, D.: Object detection with discriminatively trained part-based models. IEEE Trans. Pattern Anal. Mach. Intell. **32**(9), 1627–1645 (2010)

12. Filipe, J., Fred, A.L.N. (eds.): ICAART 2013 - Proceedings of the 5th International Conference on Agents and Artificial Intelligence, Barcelona, Spain, 15–18 February 2013, vol. 2. SciTePress (2013)

13. Gong, Y., Wang, L., Hodosh, M., Hockenmaier, J., Lazebnik, S.: Improving image-sentence embeddings using large weakly annotated photo collections. In: Fleet, D., Pajdla, T., Schiele, B., Tuytelaars, T. (eds.) ECCV 2014. LNCS, vol. 8692, pp. 529–545. Springer, Cham (2014). https://doi.org/10.1007/978-3-319-10593-2_35

14. He, K., Zhang, X., Ren, S., Sun, J.: Deep residual learning for image recognition. In: Proceedings of the IEEE Conference on Computer Vision and Pattern Recognition, pp. 770–778 (2016)

15. Hodosh, M., Young, P., Hockenmaier, J.: Framing image description as a ranking task: data, models and evaluation metrics. J. Artif. Intell. Res. **47**, 853–899 (2013)

16. Jia, Y., et al.: Caffe: convolutional architecture for fast feature embedding. In: Proceedings of the 22nd ACM International Conference on Multimedia, pp. 675–678. ACM (2014)

17. Karpathy, A., Fei-Fei, L.: Deep visual-semantic alignments for generating image descriptions. In: Proceedings of the IEEE Conference on Computer Vision and Pattern Recognition, pp. 3128–3137 (2015)

18. Kiros, R., Salakhutdinov, R., Zemel, R.: Multimodal neural language models. In: International Conference on Machine Learning, pp. 595–603 (2014)

19. Krizhevsky, A., Sutskever, I., Hinton, G.E.: ImageNet classification with deep convolutional neural networks. In: Advances in Neural Information Processing Systems, pp. 1097–1105 (2012)

20. Kulkarni, G., et al.: Baby talk: understanding and generating simple image descriptions. In: 2011 IEEE Conference on Computer Vision and Pattern Recognition, CVPR 2011, pp. 1601–1608 (2011)

21. Kuznetsova, P., Ordonez, V., Berg, T., Choi, Y.: TREETALK: composition and compression of trees for image descriptions. Trans. Assoc. Comput. Linguist. **2**(1), 351–362 (2014)

22. Li, S., Kulkarni, G., Berg, T.L., Berg, A.C., Choi, Y.: Composing simple image descriptions using web-scale N-grams. In: Proceedings of the Fifteenth Conference on Computational Natural Language Learning, pp. 220–228. Association for Computational Linguistics (2011)

23. Lin, T.-Y., et al.: Microsoft COCO: common objects in context. In: Fleet, D., Pajdla, T., Schiele, B., Tuytelaars, T. (eds.) ECCV 2014. LNCS, vol. 8693, pp. 740–755. Springer, Cham (2014). https://doi.org/10.1007/978-3-319-10602-1_48

24. Lu, J., Xiong, C., Parikh, D., Socher, R.: Knowing when to look: adaptive attention via a visual sentinel for image captioning. In: Proceedings of the IEEE Conference on Computer Vision and Pattern Recognition (CVPR), vol. 6, p. 2 (2017)

25. Mahmoud, A., Zrigui, M.: Artificial method for building monolingual plagiarized Arabic corpus. Computación Sistemas **22**(3), 767–776 (2018)

26. Mansouri, S., Charhad, M., Zrigui, M.: A heuristic approach to detect and localize text in Arabic news video. Computación Sistemas **22**(1), 75–82 (2018)

27. Mikolov, T., Sutskever, I., Chen, K., Corrado, G.S., Dean, J.: Distributed representations of words and phrases and their compositionality. In: Advances in Neural Information Processing Systems, pp. 3111–3119 (2013)

28. Mitchell, M., et al.: Midge: generating image descriptions from computer vision detections. In: Proceedings of the 13th Conference of the European Chapter of the Association for Computational Linguistics, pp. 747–756. Association for Computational Linguistics (2012)

29. Mun, J., Cho, M., Han, B.: Text-guided attention model for image captioning. In: AAAI, pp. 4233–4239 (2017)
30. Ordonez, V., Kulkarni, G., Berg, T.L.: Im2Text: describing images using 1 million captioned photographs. In: Advances in Neural Information Processing Systems, pp. 1143–1151 (2011)
31. Papineni, K., Roukos, S., Ward, T., Zhu, W.J.: BLEU: a method for automatic evaluation of machine translation. In: Proceedings of the 40th Annual Meeting on Association for computational Linguistics, pp. 311–318. Association for Computational Linguistics (2002)
32. Raffel, C., Ellis, D.P.: Feed-forward networks with attention can solve some long-term memory problems. arXiv preprint arXiv:1512.08756 (2015)
33. Řehůřek, R., Sojka, P.: Software framework for topic modelling with large corpora. In: Proceedings of the LREC 2010 Workshop on New Challenges for NLP Frameworks, pp. 45–50. ELRA, Valletta, Malta, May 2010. http://is.muni.cz/publication/884893/en
34. Simonyan, K., Zisserman, A.: Very deep convolutional networks for large-scale image recognition. arXiv preprint arXiv:1409.1556 (2014)
35. Socher, R., Karpathy, A., Le, Q.V., Manning, C.D., Ng, A.Y.: Grounded compositional semantics for finding and describing images with sentences. Trans. Assoc. Comput. Linguis. 2(1), 207–218 (2014)
36. Sutskever, I., Vinyals, O., Le, Q.V.: Sequence to sequence learning with neural networks. In: Advances in Neural Information Processing Systems, pp. 3104–3112 (2014)
37. Vinyals, O., Toshev, A., Bengio, S., Erhan, D.: Show and tell: a neural image caption generator. In: Proceedings of the IEEE Conference on Computer Vision and Pattern Recognition, pp. 3156–3164 (2015)
38. Xu, K., et al.: Show, attend and tell: neural image caption generation with visual attention. In: International Conference on Machine Learning, pp. 2048–2057 (2015)
39. Yang, Y., Teo, C.L., Daumé III, H., Aloimonos, Y.: Corpus-guided sentence generation of natural images. In: Proceedings of the Conference on Empirical Methods in Natural Language Processing, pp. 444–454. Association for Computational Linguistics (2011)
40. Yoshikawa, Y., Shigeto, Y., Takeuchi, A.: Stair captions: constructing a large-scale Japanese image caption dataset. arXiv preprint arXiv:1705.00823 (2017)
41. You, Q., Jin, H., Wang, Z., Fang, C., Luo, J.: Image captioning with semantic attention. In: Proceedings of the IEEE Conference on Computer Vision and Pattern Recognition, pp. 4651–4659 (2016)
42. Young, P., Lai, A., Hodosh, M., Hockenmaier, J.: From image descriptions to visual denotations: new similarity metrics for semantic inference over event descriptions. Trans. Assoc. Comput. Linguist. 2, 67–78 (2014)
43. Zrigui, M., Ayadi, R., Mars, M., Maraoui, M.: Arabic text classification framework based on latent dirichlet allocation. J. Comput. Inf. Technol. 20(2), 125–140 (2012)
44. Zrigui, M., Charhad, M., Zouaghi, A.: A framework of indexation and document video retrieval based on the conceptual graphs. J. Comput. Inf. Technol. 18(3), 245–256 (2010)

Reducing Space Search in Combinatorial Optimization Using Machine Learning Tools

Flavien Lucas[1,2(✉)], Romain Billot[2], Marc Sevaux[1], and Kenneth Sörensen[3]

[1] Université Bretagne Sud, Lab-STICC, UMR 6285, CNRS,
Lorient, France
{flavien.lucas,marc.sevaux}@univ-ubs.fr
[2] IMT Atlantique, Lab-STICC, UMR 6285, CNRS, Brest, France
{flavien.lucas,romain.billot}@imt-atlantique.fr
[3] Department of Engineering Management, ANT/OR Group,
University of Antwerp, Antwerp, Belgium
kenneth.sorensen@uantwerpen.be

Abstract. A new metaheuristic, called Feature-Guided MNS (FG-MNS) is proposed, combining well-known local search with simple machine learning techniques. In this metaheuristic, a solution is represented by features (mean depth of each route, standard deviation of the length of each route, etc.). The solver uses decision trees to define promising areas in the features space. The search is mainly focused on the promising areas, in order to minimize the exploration time, and to improve the quality of the found solutions. Additional neighborhoods, guided by the features are proposed.

Keywords: Metaheuristic · Machine learning · Data mining · Vehicle routing

1 Introduction

The VRP is one of the most studied problems in the optimization field. Even with thousands of papers, solving to optimality still takes too long, in particular for large problems with multiple constraints. As a consequence, metaheuristics are still largely used to solve instances of the VRP. In our work, we will mainly focus on local search based heuristics, like MNS (Multiple Neighborhood Search) and ALNS (Adaptive Large Neighborhood Search) [2].

With the fast improvement of machine learning tools a few years ago, more and more scientists develop ideas to improve optimization methods with machine learning techniques [1]. The method in this paper uses decision trees to guide the local search.

After introducing the core data and heuristic used, we will present the new meta-heuristic developed, its first results, and suggests ways to improve the method.

I. S. Kotsireas and P. M. Pardalos (Eds.): LION 14 2020, LNCS 12096, pp. 143–150, 2020.
https://doi.org/10.1007/978-3-030-53552-0_15

2   Which Features Define VRP Solutions?

In the literature, solutions are considered as sets of permutations, and the data used are patterns from the permutations [3,4]. The patterns are very easy to create and use. In [5], F. Arnold and K Sörensen create features specific to the VRP: average number of crosses, width and depth per route, etc. The features are then mainly used to evaluate, with a good accuracy, the quality of the solutions. The experiments show the high correlation between number of inter-route crosses, and the cost of the solution. As a consequence, the general knowledge provided by data inspired the authors to create a new heuristic, reducing first the number of inter-route crosses.

Due to the good accuracy of prediction in Arnold and Sörensen's work, we use most of the features described in their paper. To increase the prediction accuracy, we add some new features, presented in Table 1.

Table 1. List of features used

• Average width of each route	• Length of the longest interior edge of each route divided by the length of each route. (∗)
• Standard deviation of the width of each route. (∗)	• Mean length of the first and last edges of each route
• Average span of each route	• Demand of the first and last customer of each route. (∗)
• Standard deviation of the span of each route. (∗)	• Demand of the farthest customer of each route. (∗)
• Average depth of each route	• Standard deviation of the demand of the farthest customer of each route. (∗)
• Standard deviation of the depth of each route. (∗)	• Standard deviation of the length of each route. (∗)
• Distance of the first and last edges of each route, divided by the length of each route. (∗)	• Mean distance between each route
• Mean length of the longest edge of each route. (∗)	• Standard deviation of the number of customers
• Length of the longest edge of each route divided by the length of each route. (∗)	• Degree of neighborhood. (∗)

(∗) New or modified features

3   Metaheuristic Framework

The main contribution of this paper is the addition of machine learning techniques within a metaheuristic creating good solutions in very short time.

The solver is a two-phase metaheuristic. First, a set of solutions is generated, then, the solutions are improved by local search.

Phase 1: Generation of Solutions A GRASP, based on the Clarke and Wright constructive algorithm generated a set of k_1 solutions. Only the k_2 best solutions are chosen, the other are deleted.

Numbers k_1 and k_2 are parameters of the solver.

Phase 2: Local Search The core of the second phase is a Multiple Neighborhood Search: a set of neighborhoods is used. The neighborhoods are the following:

- Intra-route reallocation
- Intra-route Swap
- 2-opt
- Inter-route reallocation
- Inter-route path move
  A sequence of 2 consecutive customers are removed from one route and place in another route
- Inter-route swap
- Inter-route Path-exchange
  2 sequences of 2 consecutive customers in two different routes are exchanged.
- Split algorithm [6]
  All route are merged, and split by Bellman-Ford algorithm. This neighborhood is executed only for heterogeneous instances.

The sequence of execution of neighborhood is randomly chosen. Each solution is improved by different sequences of neighborhoods. For each neighborhood, the local search is based on Tabu-search, stop after 5 consecutive non-improving steps. The MNS stops when, for a solution found, no neighborhood is able to improve the solution Fig. 1.

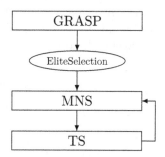

Fig. 1. MNS-TS

Performances

The metaheuristics can be executed on different types of instances. In this section, we will focus on 2 difficult types of instances: very large scale instances, and heterogeneous instances.

The first set of instances (available in [7]), created by F. Arnold and K. Sörensen [8] is a set of classical Capacitated VRP instances, with euclidean and symmetric distances. The specificity of theses instances, is the number of customers: from 3,000 to 30,000.

The authors used a new algorithm, called A-G-S, combining different local searches in reduced space. The paper described two versions of the A-G-S: a long version (resp. a short version), with a time limit equal to 5 min (resp. 3 min) per thousands of customers. Table 2 describes the performance of the MNS-TS and the A-G-S (short version). The best known solutions BKS are the best solutions found with the A-G-S (long version).

Table 2. Comparison of A-G-S and MNS-TS on XXL instances

Instances	Number of customers	A-G-S (short version)		MNS-TS	
		Gap to BKS (%)	CPU time (s)	Gap to BKS(%)	CPU time (s)
L1	3.000	1.36	180	1.67	1.91
L2	4.000	2.62	240	5.15	2.47
A1	6.000	0.22	360	1.67	7.20
A2	7.000	1.68	420	3.69	5.59
G1	10.000	0.35	600	1.44	12.46
G2	11.000	1.27	660	2.99	17.94
B1	15.000	0.77	900	2.05	24.75
B2	16.000	1.89	960	2.79	16.86
F1	20.000	0.42	1200	1.6	57.90
F2	30.000	2.04	1800	2.88	108.46
Mean		1.14	732	2.59	25.55

The MNS-TS has an higher GAP to the best known solution (+ 1.45% in average) than the faster A-G-S algorithm. However, the MNS-TS is, on average, 28 times faster than the A-G-S.

The second set is the Duhamel-Lacomme-Prodhon (DLP) set of 96 instances (available in [10]). Each instance contains 2 to 8 categories of vehicles, with their own capacity, fixed cost and consumption cost. Each category contains a limited number of vehicles. Table 3 summarizes the performance of the solver. Set A corresponds to instances solved by optimality in the literature (see [11]) and set B corresponds to the others instances.

Table 3. Performances on DLP instances

Set of instances	Number of customers	Number of vehicles	Mean GAP to BKS (%)	Mean CPU time (s)
Set A	20 to 186	3 to 8	6.56	2.61
Set B	77 to 256	2 to 8	7.78	5.45

As a conclusion, the metaheuristic presented above is able to found quite good solutions in very short time. The time save by the metaheuristic is used in a second phase, using the knowledge learned with the previous phase, to improve the quality of the solution founds.

4   Feature-Guided MNS: A New Metaheuristic

The Feature-Guided MNS (FG-MNS) is based on MNS-TS and divided in two steps. Firstly, the solver will collect data and learn the link between features and quality. Secondly, the solver will use the knowledge to guide the solver, to the optimal, using a score (describes in Sect. 4).

The main idea is represented on Fig. 2.

Let X the solution space and f the objective function to minimize (curve dashed in blue in Fig. 2). Let x_{opt} the optimal solution, and x^* the solution predicted as optimal, by the machine learning method. Let s the score function (red full curve in Fig. 2), representing the distance between the current solution and x^*. The main idea of the FG-MNS is to alternate between minimizing function f and function s.

Example. Solution x_0 is in a local minimum, meaning a local search focus on $f(x)$ can't improve x. However, a local search focus on $s(x)$ can reduce the distance, in the solution space, between the current solution and x_{opt}.

After a local search, the solution x^* is found. x^* is worth than x_0 ($f(x^*) > f(x_0)$), but near x_{opt}.

A local search centered in x^* focus on $f(x)$ is sufficient to find x_{opt}.

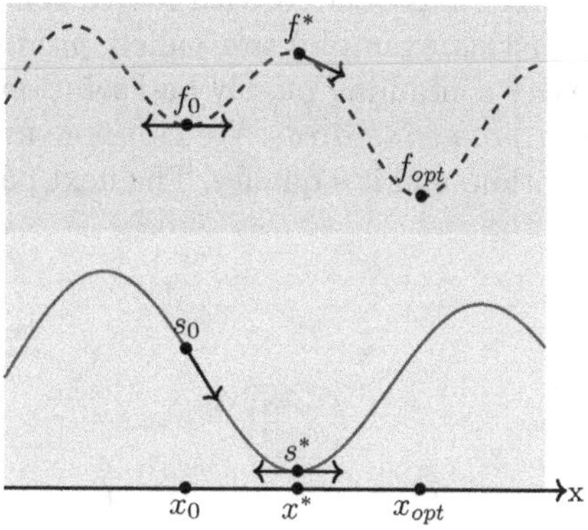

Fig. 2. Two scores for one instance (Color figure online)

Learning Phase

This part is initialize by the framework described in Sect. 3, in order to collect data. After collecting enough various solutions, the learning phase really began. The main goal, as seen in example above, is to localise the supposed optimum solution. One of the most appropriate machine learning methods for this purpose is the Decision Tree. This method cut the feature space in two subspaces, such that, each subspace is more homogeneous than the initial space. We repeat the cut for each subspace, until cutting will not increase the homogeneity of the subspace. Finally, the initial space is partition in subspaces, as homogeneous as possible.

Example. Let's define a solution as good when the GAP to BKS is $< 1\%$ and bad otherwise. Let F1 and F2 the 2 features used in this example. Let's use a decision to determined the feature spaces containing good solutions. This decision is showed in Fig. 3. The tree split initially the feature space in 2 subspaces:

- One subspace with solutions such that Feature 1 ≥ 0.37. This subspace contains 5032 good solutions and 300 bad solutions.
- One subspace with solutions such that Feature 1 < 0.37 This subspace contains 1661 good solutions and 6442 bad solutions.

Each subspace is split in 2 areas. Splitting each area don't increase the homogeneity of each leaf, the decision tree stop. At last, we can define 4 areas (see Fig. 4) corresponding to the 4 final leaves.

- Leave 1: $F1 \geq 0.37$ and $F2 < 786$. The associate area contains mostly good solutions.
- Leave 2: $F1 \geq 0.37$ and $F2 \geq 786$. The associate area contains mostly bad solutions.
- Leave 3: $F1 < 0.31$. The associate area contains mostly good solutions.
- Leave 4: $0.31 \leq F1 \leq 0.37$. The associate area contains mostly bad solutions.

Areas associated with leaves containing mostly good solutions (leaves 1 and 3 in this example) are called *promising areas*. Oppositely, areas associate to leaves containing mostly bad solutions (leaves 2 and 4 in this example) are called *non promising areas*. We can now understand the rule between the feature of a solution, and it's quality. The next part of the solver with exploit this knowledge.

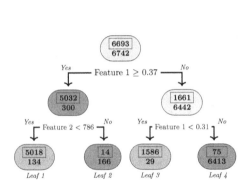

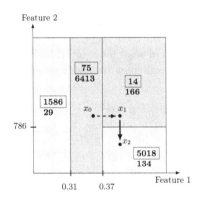

Fig. 3. Example of decision tree (Color figure online)

Fig. 4. Areas corresponding to the rules (Color figure online)

Numbers boxed in blue represents the number of good solutions, and the number in **black** *without box represents the number of bad solutions.*

Exploitation Phase

In this phase, the solver will use the rules generated by the decision tree, to be guided. The solver work differently, depending on the area of the current solution.

- Case 1: The current solution is in promising area.
  Most neighbors solutions are good, the cost can be reduced with local search, as described in Sect. 3.

– Case 2: The current solution is in non promising area.

  Most neighbors solutions have a high objective value. The solver will avoid this area bu guided this solution to a promising area. The solution will be modified by a feature-guided local search. More precisely, the FG-MNS tries to identify which feature need to be modified in priority. Once the solution is in a promising area, a local search is executed, as described in case 1.

Example. The initial solution x_0 (see Fig. 4) is in non promising area. The nearest promising area, is defined by $F1 \geq 0.37$ and $F2 < 786$. The features values of x_0 are respectively 0.36 and 788.

Thus, the two features need to be changed. In the decision tree (Fig. 3), the leaf associated to this promising area is first defined by F1 (higher in the tree) and then by F2 (lower in the tree). The solver will first try to correct the value of F1. A local search minimizing the distance between the new value of F1, and the interval of authorized value for F1 is applied. After a few moves, a new solution x_1 with a value of F1 higher than 0.37 is found. However, the value of F2 associated with x_1 is higher than 786. A second local search, minimizing now F2 is called.

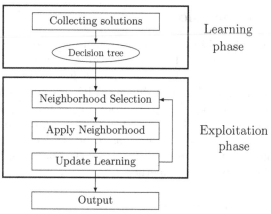

Fig. 5. Diagram of FG-MNS

When the solution is a local minimum, features of all solutions that have led to the local minimum are added to the data, and a new decision tree is built, using the old and new data. A new generated solution can be improved, using the updated decision rules.

The metaheuristic FG-MNS is summarized by Fig. 5.

5 Experiments

The quality of the decision tree is tested on a set of 3000 randomly generated instances. For each instance, thousands of solutions are created. In this section we will consider a solution good when the gap to the best known solution is lower than 1%. In a similar way, we will consider a solution bad when the gap to the best known solution is higher than 2%. The solutions with gap between 1% and 2% are not considered. The SMOTE algorithm [9] is used to balanced the number of good and bad solutions. Then, the solutions are split in 2 sets:

66% of the solutions are put in the training set. The 33% others formed the test set. Rules are created by a decision tree, with the solutions on the training set (Fig. 3).

The test set solutions are used to determine the performance of the decision tree. On average, 95% of good solutions are in promising areas. Furthermore, 60% of the solutions in promising areas are good solutions.

Using local search algorithms exploring only promising areas seems a good methodology. Indeed, it will reduce the search space, and therefore, the computation time. The highest proportion of good solutions appears to reduce the probability to be blocked in a bad local minimum.

6   Conclusion

The FG-MNS is a new type of metaheuristic, combining our knowledge in operation research and machine learning. This metaheuristic is a modification of the MNS-TS, a metaheuristic able to find good solutions in very short time, for different variations of difficult VRP instances. Adding decision tree allows the method to focus only on promising solutions, reducing the computation time, and increasing the probability to find a good solution. The learning part needs at least hundreds of solutions to be efficient. A preliminary decision tree, using solutions from similar instances would be a good way to improve the FG-MNS, as a pre-processing method.

References

1. Bengio, Y., Lodi, A., Prouvost, A.: Machine learning for combinatorial optimization: a methodological tour d'horizon. arXiv preprint arXiv:1811.06128 (2018)
2. Grangier, P., Gendreau, M., Leuédé, F., Rousseau, L.M.: An adaptive large neighborhood search for the two-echelon multiple-trip vehicle routing problem with satellite synchronization. Eur. J. Oper. Res. **254**(1), 80–91 (2016)
3. Arnold, F., Santana, Í., Sörensen, K., Vidal, T.: PILS: exploring high-order neighborhoods by pattern mining and injection. arXiv preprint arXiv:1912.11462 (2019)
4. Tarantilis, C.D., Kiranoudis, C.T.: BoneRoute: an adaptive memory-based method for effective fleet management. Ann. Oper. Res. **115**(1–4), 227–241 (2002)
5. Arnold, F., Sörensen, K.: What makes a VRP solution good? The generation of problem-specific knowledge for heuristics. Comput. Oper. Res. **106**, 280–288 (2019)
6. Duhamel, C., Lacomme, P., Prodhon, C.: A GRASPxELS with depth first search split procedure for the HVRP. Res. Rep. LIMOS/RR-10-08 (Inst. Sup er. Inform., Mod el. Appl., Aubiere, France (2010). http://www.isima.fr/~lacomme/doc/RR_HVRP1-4_V1.pdf (2010)
7. XXL instances. https://antor.uantwerpen.be/xxlrouting/
8. Arnold, F., Gendreau, M., Sörensen, K.: Efficiently solving very large-scale routing problems. Comput. Oper. Res. **107**, 32–42 (2019)
9. Chawla, N.V., Bowyer, K.W., Hall, L.O., et al.: SMOTE: synthetic minority oversampling technique. J. Artif. Intell. Res. **16**, 321–357 (2002)
10. DLP instances. http://fc.isima.fr/~lacomme/hvrp/hvrp.html
11. Sadykov, R., Uchoa, E., Pessoa, A.: A bucket graph based labeling algorithm with application to vehicle routing. Cadernos do LOGIS, Universidade Federal Fluminense, vol. 7 (2017)

Impact of the Discretization of VOCs for Cancer Prediction Using a Multi-Objective Algorithm

Sara Tari[1(✉)], Lucien Mousin[1,2], Julie Jacques[1,2], Marie-Eleonore Kessaci[1], and Laetitia Jourdan[1]

[1] University of Lille, CNRS, UMR 9189 CRIStAL, 59650 Villeneuve d'Ascq, France
{sara.tari,marie-eleonore.kessaci,laetitia.jourdan}@univ-lille.fr
[2] Lille Catholic University, Faculté de Gestion, Economie et Sciences, Lille, France
{lucien.mousin,julie.jacques}@univ-catholille.fr

Abstract. Volatile organic compounds (VOCs) are continuous medical data regularly studied to perform non-invasive diagnosis of diseases using machine learning tasks for example. The project PATHACOV aims to use VOCs in order to predict invasive diseases such as lung cancer. In this context, we propose to use a multi-objective modeling for the partial supervised classification problem and the MOCA-I algorithm specifically designed to solve these problems for discrete data, to perform the prediction. In this paper, we apply various discretization techniques on VOCs data, and we analyze their impact on the performance results of MOCA-I. The experiments show that the discretization of the VOCs strongly impacts the classification task and has to be carefully chosen according to the evaluation criterion.

Keywords: Supervised classification · Medical data · Multi-objective optimization

1 Introduction

Human bodies emit a wide range of volatile organic compounds (VOCs), some of which are odorous. The composition of VOCs produced by a given individual corresponds to a unique signature odor. Age, sex, diet are among many factors that can influence this unique fingerprint, as well as diseases. These modifications often result in smell changes and explain what allowed Hippocrates to report changes related to the presence of certain diseases in the smell of urine and sputum. Nowadays, the composition of VOCs produced by individuals is regularly studied as a non-invasive way to detect pathologies [5,7,8]. The project PATHACOV[1] aim at designing a classifier based on VOCs data in order to predict invasive diseases, with a major focus on lung cancer. Thus, we propose to

---

[1] This project is funded by the Interreg France-Wallonie-Vlaanderen program, with the support of the European Regional Development Fund see www.pathacov-project.com for more information.

© Springer Nature Switzerland AG 2020
I. S. Kotsireas and P. M. Pardalos (Eds.): LION 14 2020, LNCS 12096, pp. 151–157, 2020.
https://doi.org/10.1007/978-3-030-53552-0_16

use an approach based on the Pittsburgh representation and where the classification task is modeled as a multi-objective optimization problem. The medical datasets have specific characteristics; in particular, the number of attributes is significantly higher than the number of individuals, and the classes are regularly imbalanced. Most frequent disease like diabetes only occurs on less than 6% of the population. These characteristics strongly impact on the performance of classification techniques. Therefore, the algorithm MOCA-I (Multi-Objective Classification Algorithm for Imbalanced data) [3], designed for a multi-objective modeling and these types of characteristics, has been chosen to identify the relevant VOCs. However, MOCA-I requires discrete attributes, while VOCs are continuous data.

This paper presents our resolution approach for the detection of diseases using VOCs and an experimental study where various discretization techniques and their impact on the performance of MOCA-I to produce good models are analyzed. The experiments are conducted on three different medical datasets with VOCs.

The outline of the paper is as follows. Section 2 presents the proposed approach and various data discretization techniques. Section 3 describes the datasets and the experimental protocol before giving and analyzing the results. Finally, Sect. 4 provides a discussion about this study and points out future work.

2   Proposed Resolution Approach

Bronchopulmonary cancer is often discovered late. The objective of the PATHA-COV project is to detect it earlier by non-invasive means with a low-cost breath test, by measuring exhaled VOCs. For each individual, we can measure the VOCs produced and their quantities. They may vary significantly from an individual to another. Moreover, none of the individuals emit all the VOCs present in the dataset. This task can be seen as a supervised partial classification problem, where we want to identify which VOCs can predict Bronchopulmonary cancer.

2.1   Description

This problem can be modelized as a multi-objective optimization problem. Since the VOCs profile may vary from an individual to another, we opted for a Pittsburgh modelization, where each solution is a ruleset. Hence, several profiles can fit into several rules. Moreover, Pittsburgh is a white box modelization, which means it is compatible with November 2018 CCNE[2] (French National Consultative Ethics Committee)' recommendations about AI and health, suggesting to use AI approaches that the care team can criticize or challenge.

For this problem, three objectives are considered. The *sensitivity* – to maximize – will measure the ability of the model to detect a high proportion of

---

[2] https://www.ccne-ethique.fr/en/.

patients with the disease. The *confidence* – to maximize – will measure if the predicted patients are correctly identified. Moreover, *sensitivity* and *confidence* are two classical machine learning complementary metrics that are adapted to deal with imbalanced and medical data [6]. We also want to minimize the number of VOCs used in each model: this will generate models easier to understand.

We will use the MOCA-I (multi-objective classification algorithm for imbalanced data) algorithm, which implements the preceding modelization. It uses a multi-objective local search (MOLS) to tackle the resulting problem. MOCA-I was initially developed for handling discrete medical data. Thus, each VOC amount will be discretized, and the objective of this paper is to determine which is the impact of discretization on the cancer prediction. Since a classification task generates only one model and MOCA-I produces a Pareto set of equivalent solutions, the solution of best *G-mean* is selected among this set.

2.2 Data Discretization Techniques

In this work, we consider nine discretization techniques, that are briefly described in Table 1, following the taxonomy of [2].

Table 1. Description of discretization techniques.

Method	Static	Supervised	Separation	Global	Direct	Measure
10-bin	Yes	No	Yes	Yes	Yes	Bin.
1R	Yes	Yes	Yes	Yes	Yes	Bin.
CAIM	Yes	Yes	Yes	Yes	No	Stat.
Chi2	Yes	Yes	No	Yes	No	Stat.
ChiMerge	Yes	Yes	No	Yes	No	Stat.
Fayyad	Yes	Yes	Yes	No	No	Info.
FUSINTER	Yes	Yes	No	Yes	No	Info.
ID3	No	Yes	Yes	No	No	Info.
Zeta	Yes	Yes	Yes	Yes	Yes	Stat.

Following this taxonomy, a discretization technique can be *static* or *dynamic*, depending on when it is applied respectively before or during the learning algorithm. A *supervised* method takes into account the class to construct the intervals. For the *separation* approach, a single initial interval is produced and is then progressively split into several intervals. The opposite approach is *fusion*, where many intervals are produced and then merged. A *global* method may use the entirety of the available data for the discretization process, whereas a *local* one only uses a subset of the data. *Direct* approaches define a single interval at each iteration, while incremental approaches create many intervals at each step. The evaluation measure is used to select the best solution produced by the discretization technique.

In the following, we will test these techniques to discretize VOCs data in our resolution approach.

3   Experiments

This section presents the datasets and the experimental protocol of our approach. Then the results of these experiments are given and an analysis is drawn.

3.1   Datasets

In this study, we use three medical datasets with VOCs (see Table 2). The datasets T3 and T4 have been provided by our partners of the PATHACOV project and come from dialysis patients while P1 has been taken from the literature [4]. Note that T3 and T4 contain the VOCs of respectively 36 and 37 patients before and after dialysis, meaning that a given individual provides two samples (a positive one and a negative one) and that the extraction of biomarkers is probably easier to perform on these datasets.

Table 2. Description of real datasets resulting from patients samples.

Name	Diagnosis	#individuals	#positive	#attributes
T3	Dialysis	72	36	346
T4	Dialysis	74	37	341
P1	Prostate cancer	103	59	137

3.2   Experimental Protocol

The purpose of this work is to predict a class. Since we have only three datasets, we use a 5-fold cross-validation protocol to limit *overfitting* as follows. Each dataset is separated in five same-size folds, then four folds are combined into a training set, while the remaining one corresponds to the test set. This process is repeated for each fold's combinations and creates five training sets associated with 5 test sets. For each discretization method, we conduct 6 independent runs of MOCA-I on each training set, leading to 30 runs per dataset.

We used the software KEEL [1] to discretize the datasets. Note that in order to reduce the bias when assessing the efficiency of the discretization methods, we limit the risk to overfit the data by discretizing each training set independently.

MOCA-I parameters correspond to the default parameters proposed by [3]: initial population of 100 solutions, 10 rules maximum per ruleset, a maximal archive size of 500. At each iteration, the multi-objective local search under consideration selects one solution in the archive and explores the whole neighborhood of this solution. Note that, the non-dominated neighbors are considered, which explains the use of a bounded archive.

We compare the effect of the discretization methods according to four machine learning metrics: *sensitivity, specificity, geometric mean* (G-mean), and Matthew's correlation coefficient (MCC). MCC is comprised between -1 and 1, where 1 corresponds to the best performance and 0 to the theoretical performance of a random classifier. The other metrics' values are comprised between 0 and 1, where 1 corresponds to the highest performance and 0.5 to a performance that is not better than a random classifier.

3.3   Results

Table 3 presents the ranks of the nine discretization techniques according to the four considered measures (Sensitivity, Specificity, G-Mean and MCC) for each dataset. Bold types means that the discretization techniques are statistically equivalent according to the statistical test of Friedman.

Table 3. Ranking of the discretization methods in function of the average sensitivity (top-left), specificity (top-right), G-mean (bottom-left) and MCC (bottom-right).

Sensitivity			Specificity		
T3	T4	P1	T3	T4	P1
#1 Chi2	**#1 ID3**	**#1 Fayyad**	#1 Chi2	**#1 ID3**	#1 ID3
#2 10bin	**#2 CAIM**	#2 Fusinter	#2 ID3	**#2 10bin**	#2 Fusinter
#3 Zeta	**#3 1R**	#3 Chi2	#3 1R	#2 Fayyad	#3 1R
#3 Fusinter	**#4 10bin**	#4 CAIM	#3 Fusinter	#3 1R	#4 CAIM
#4 ID3	**#4 Fayyad**	#5 1R	#4 10bin	#4 Chi2	#4 Zeta
#4 ChiMerge	**#5 ChiMerge**	#6 ChiMerge	#5 CAIM	#5 CAIM	#5 10bin
#5 CAIM	**#5 Zeta**	#7 ID3	#6 ChiMerge	#6 Fusinter	#6 Chi2
#6 Fayyad	#6 Fusinter	#8 Zeta	#7 Zeta	#7 Zeta	#7 ChiMerge
#7 1R	#7 Chi2	#9 10bin	#8Fayyad ·	#8 ChiMerge	#8 Fayyad
G-mean			Matthew's Correlation Coefficient (MCC)		
T3	T4	P1	T3	T4	P1
#1 Chi2	**#1 ID3**	**#1 Fusinter**	**#1 Chi2**	**#1 ID3**	**#1 Fusinter**
#2 Fusinter	**#2 1R**	**#2 ID3**	#2 ID3	**#2 1R**	**#2 1D3**
#2 ID3	**#3 Fayyad**	#3 CAIM	#3 10bin	**#3 10bin**	#3 1R
#2 10bin	**#4 CAIM**	#4 1R	#4 Fusinter	#4 Fayyad	#3 CAIM
#3 CAIM	**#5 10bin**	#5 Chi2	#5 ChiMerge	#5 CAIM	#4 Chi2
#4 1R	#6 Zeta	#6 Zeta	#6 1R	#6 1R	#5 Zeta
#5 ChiMerge	#7 ChiMerge	#7 10bin	#6 CAIM	#7 ChiMerge	#7 10bin
#8 Zeta	#8 Fusinter	#8 ChiMerge	#7 Zeta	#8 Chi2	#7 ChiMerge
#9 Fayyad	#7 Chi2	#9 Fayyad	#8 Fayyad	#9 Fusinter	#8 Fayyad

The results are heterogeneous between the datasets, the discretization techniques, and the quality measures. For example, for the sensitivity, the best-ranked techniques Chi2 and Fayyad for the datasets T3 and P1 respectively are statistically different from the other techniques. In contrast, for dataset T4, seven of the nine techniques give equivalent results. For the specificity, numerous discretization techniques are equivalent for datasets T3 and T4, while only three techniques are equivalent for dataset P1. Besides, for dataset T3, Chi2 leads to the best average score for each metric, while ID3 leads to the most efficient rulesets for dataset T4. For dataset P1, Fusinter and ID3 lead to the best specificity, G-mean, and MCC while Fayyad gives the best sensitivity, and it is last ranked for the three other measures. This behavior is probably due to the presence of several zeros in the samples for each attribute that leads most VOCs to have a single interval $((-inf; +inf))$ after the application of Fayyad. ID3 is among the best techniques for seven of the twelve experiments.

4   Discussion

In this work, we observed the impact of different discretization methods on the models produced by MOCA-I. In particular, we focused on real health data, where a sample corresponds to quantities of VOCs emitted by individuals. The aim was to determine which discretization method is the most suited for this type of data. The results on our datasets highlight that the ID3 discretization method seems to be suited to the case of VOCs.

In the future, we will perform these experiments on other datasets containing VOCs, in particular, datasets with more individuals provided by the PATHA-COV project and imbalanced datasets. We also plan to study the impact of discretization methods with different parameters for MOCA-I, since their values may influence the quality of the resulting ruleset. In order to compare our approach to classical machine learning algorithms, we will study the impact of the discretization methods on their efficiency.

References

1. Alcalá-Fdez, J., et al.: Keel: a software tool to assess evolutionary algorithms for data mining problems. Soft. Comput. **13**(3), 307–318 (2009). https://doi.org/10. 1007/s00500-008-0323-y
2. Garcia, S., Luengo, J., Sáez, J.A., Lopez, V., Herrera, F.: A survey of discretization techniques: taxonomy and empirical analysis in supervised learning. IEEE Trans. Knowl. Data Eng. **25**(4), 734–750 (2012)
3. Jacques, J., Taillard, J., Delerue, D., Jourdan, L., Dhaenens, C.: The benefits of using multi-objectivization for mining Pittsburgh partial classification rules in imbalanced and discrete data. In: Proceedings of the 15th Annual Conference on Genetic and Evolutionary Computation, pp. 543–550. ACM (2013)
4. Khalid, T., et al.: Urinary volatile organic compounds for the detection of prostate cancer. PLoS ONE **10**(11), e0143283 (2015)

5. Leunis, N., et al.: Application of an electronic nose in the diagnosis of head and neck cancer. Laryngoscope **124**(6), 1377–1381 (2014)
6. Ohsaki, M., Abe, H., Tsumoto, S., Yokoi, H., Yamaguchi, T.: Evaluation of rule interestingness measures in medical knowledge discovery in databases. Artif. Intell. Med. **41**(3), 177–196 (2007)
7. Phillips, M., et al.: Volatile biomarkers of pulmonary tuberculosis in the breath. Tuberculosis **87**(1), 44–52 (2007)
8. Sakumura, Y., et al.: Diagnosis by volatile organic compounds in exhaled breath from lung cancer patients using support vector machine algorithm. Sensors **17**(2), 287 (2017)

AUGMECON2 Method for a Bi-objective U-Shaped Assembly Line Balancing Problem

Ömer Faruk Yılmaz[(✉)] [ID]

Department of Industrial Engineering, Karadeniz Technical University,
Trabzon, Turkey
omerfarukyilmaz@ktu.edu.tr

Abstract. This study explores the bi-objective U-shaped assembly line balancing problem (UALBP) by considering several scenarios constructed based on worker skill levels. Because the investigated problem has two objectives, namely the minimizing the number of stations and maximum workload imbalance, an improved ε-constrained based method is employed to find the Pareto-optimal solutions to the problem. This method is called to be the second version of the augmented ε-constrained (AUGMECON2) and it is highly effective to find the Pareto-optimal solutions within reasonable CPU time. In order to investigate the impact of workers' inherent on both of the objectives, a set of scenarios is considered. Each scenario is determined based on the nature of the worker pool in which workers are assigned to the stations. An optimization model is presented for the problem and it is solved with a real case study from the industry. The computational results indicate that the scenarios have a great impact on the workload imbalance objective. In particular, it is revealed that while the skill levels of workers increases, the workload imbalance decreases. However, the same impact is not observed for the number of stations.

Keywords: U-shaped assembly line balancing · Bi-objective optimization · AUGMECON2 method · Optimization model

1 Introduction

Today's competitive market environment pushes the manufacturing companies to take some necessary actions to become more robust. In this manner, the companies have converted their traditional systems into a more reliable system so as to satisfy customers' expectations within a shorter manufacturing lead time. Assembly lines are effective in a way that large units can be produced shorter lead time by applying one-piece flow principle. On one hand, utilizing traditional assembly lines provides many benefits, such as economy of scale and competitiveness. On the other hand, it leads to some disadvantages that can be eliminated, such as great shop-floor requirement, preventing group working and skill enhancement [1]. Besides, it cannot fully provide some other benefits, such as visibility, motivation, and communication. In order to utilize the advantages of assembly lines, traditional assembly lines have been converted to U-shaped lines following the lean manufacturing principles [2, 3]. After the U-shaped assembly line is designed, it needs to be balanced accordingly [4]. Assembly

© Springer Nature Switzerland AG 2020
I. S. Kotsireas and P. M. Pardalos (Eds.): LION 14 2020, LNCS 12096, pp. 158–167, 2020.
https://doi.org/10.1007/978-3-030-53552-0_17

line balancing requires assigning the tasks to the stations in such a way that all tasks are assigned under necessary constraints, such as the precedence relations and pre-determined cycle time. Besides, some assumptions can be considered while balancing the assembly line. Because U-shaped assembly lines are designed in a compact manner and the precedence constraints allow assigning a task with its predecessors and successors to the same station, it has several advantages compared to the traditional assembly lines [5]. The manual operations are carried out inside a U-shaped assembly line and workers' skills play a major role to enhance the system performance. That is why, it is plausible to determine some scenarios representing a real manufacturing environment properly.

In this study, a bi-objective U-shaped assembly line balancing problem (UALBP) is investigated with the heterogeneity inherent of workers. An optimization model is presented for the addressed problem. A set of scenarios is determined to investigate how skill sharing in the worker pool affects the system performance from two perspectives: (i) the number of stations and (ii) the workload imbalance. According to the computational results, when the number of high skill level workers increase in the worker pool, the system performance increase with respect to the workload imbalance. However, the same effect is not valid for the number of stations objective.

The rest of the paper is organized as follows. The studies regarding the multi-objective UALBP are reviewed in Sect. 2. The optimization model and the implemented method are presented in Sect. 3. The computational results with the real case study are given in Sect. 4. Concluding remarks are provided in Sect. 5.

2   Literature Review

In this section, the studies are reviewed regarding multi-objective U-shaped assembly line balancing problem (UALBP).

Hwang et al. [6] proposed a multi-objective genetic algorithm (moGA) for UALBP. A comparison was made between traditional and U-shaped lines. Sirovet-nukul and Chutima [7] considered three different objective functions, number of workers, deviation of operation times, and walking times. Non-dominated sorting genetic algorithm-II (NSGA-II) and COINcidence algorithms were employed to the problem. Sirovetnukul and Chutima [8] applied the Particle Swarm Optimization with Negative Knowledge (PSONK) algorithm to UALBP. The NSGA-II algorithm is used for comparison. Farkhondeh et al. [9] first used goal programming to solve the model, and then the efficiencies of the optimal solution were evaluated employing data envelopment analysis (DEA). Dong et al. [10] provided a multi-objective genetic algorithm for solving UALBP. Two different performance criteria were considered: (i) number of workstations and (ii) workload variation. Alavidoost et al. [11] considered four different objectives for multi-objective UALBP with fuzzy processing time. A fuzzy adaptive genetic algorithm was employed for the problem. Besides, Taguchi design was used for parameter analysis. Manavizadeh et al. [12] focused on both the scheduling and sequencing problems simultaneously. The objectives were the minimizing cycle time, wastages, and work overload. A novel heuristic algorithm was employed for the problems. Alavidoost et al. [13] proposed a two-stage interactive fuzzy programming approach for bi-objective ALBP in which triangular fuzzy numbers were employed. They stated that the proposed approach can be implemented in other multi-objective problems. Rabbani et al. [14] employed multi-objective evolutionary algorithms (MOEA) and particle swarm optimization (PSO) for the type II robotic mixed-model UALBP. Babazadeh et al. [15] developed an efficient multi-objective genetic algorithm for both fuzzy straight and U-shaped ALBP. The algorithm was compared to exact methods with small-sized instances. Zhang et al. [16] constructed an optimization model for U-shaped robotic ALBP. Pareto artificial bee colony algorithm was used for this NP-hard problem. Chutima and Suchanun [17] investigated parallel adjacent U-shaped ALBP. A hybrid algorithm combining MOEA and PSO was implemented to the problem. Babazadeh and Javadian [18] developed a novel meta-heuristic approach for bi-objective UALBP. A modified NSGA-II algorithm was implemented to the problem and it was compared to well-known algorithms in the existing literature.

The survey of previous studies reveals some interesting insights as follows. Several algorithms are applied to the UALBP problem. NSGA-II algorithm is used for comparison purposes. The fuzzy optimization technique is used to model uncertainty in ALBP. Several performance criteria are considered. From this review, it is observed that there is still a need to explore some interesting insights regarding UALBP with heterogeneity inherent of workers.

3 Optimization Model and AUGMECON2 Method

3.1 Optimization Model

In this section, an optimization model is constructed for the bi-objective U-shaped assembly line balancing problem by considering workers' skill levels. The model introduced by [19] is modified accordingly and presented. First, indices, parameters, variables are described. Afterwards, the optimization model is presented. The detailed explanations regarding the model are given after the equations.

Assumptions
- The operation times of tasks are known in advance.
- The workers can have different skill levels for the tasks.
- The tasks are available to be assigned to the stations.
- The precedence relations of tasks are known in advance.
- The demand is known in advance.
- There are enough workers in the worker pool.

Indices
i: Indices defined for tasks
j, k: Indices defined for stations
w: Indices defined for workers

Parameters

n: number of tasks

t_i: time to operate task i

m_{max}: number of stations allowed to be utilized

C: pre-determined cycle time according to customer demand

W_j: determined tasks for station j

$\|W_j\|$: number of tasks in W_j

$L(r, s)$: set defined for tasks preceding task s

p_{wi}: proficiency of worker w for task i

Variables

A_{jw}: if worker w is employed in station j, 1; otherwise, 0

x_{ij}: if task i is allocated to forward of station j, 1; otherwise, 0

y_{ij}: if task i is allocated to backward of station j, 1; otherwise, 0

k_{ij}: if task i is allocated to station k, 1; otherwise, 0

z_j: if station j is utilized, 1; otherwise 0

Objective function

$$\min f_1 = objective1 \tag{1}$$

$$\min f_2 = objective2 \tag{2}$$

$$\sum_{j=1}^{J} k_{ij} = 1 \qquad \forall i \tag{3}$$

$$x_{ij} + y_{ij} = k_{ij} \qquad \forall i, j \tag{4}$$

$$A_{jw} k_{ij} = s_{ijw} \qquad \forall i, j, w \tag{5}$$

$$\sum_{j=1}^{J} (m_{max} - j + 1)(x_{rj} - x_{sj}) \geq 0 \qquad \forall (r, s) \in L \tag{6}$$

$$\sum_{j=1}^{J} (m_{max} - j + 1)(y_{rj} - y_{sj}) \geq 0 \qquad \forall (s, r) \in L \tag{7}$$

$$\sum_{i \in W_j} k_{ij} - \|W_j\| z_j \leq 0 \qquad \forall j \tag{8}$$

$$\sum_{j=1}^{J} A_{jw} \leq 1 \qquad \forall w \tag{9}$$

$$\sum_{w=1}^{W} A_{jw} \leq 1 \qquad \forall j \tag{10}$$

$$z_j - \sum_{w=1}^{W} A_{jw} = 0 \qquad \forall j \tag{11}$$

$$A_{jw} + k_{ij} \geq 2s_{ijw} \qquad \forall i, j, w \tag{12}$$

$$A_{jw} + k_{ij} \leq 1 + s_{ijw} \qquad \forall i, j, w \tag{13}$$

$$\sum_i^I \sum_w^W t_i \times s_{ijw} \leq C \qquad \forall j \tag{14}$$

$$\sum_i^I \sum_w^W t_i \times s_{ijw} = C1_k \qquad \forall j \tag{15}$$

$$wd_{j,k} = C1_j - C1_k \qquad \forall j, k; j > k \tag{16}$$

$$objective2 \geq wd_{j,k} \qquad \forall j, k \tag{17}$$

$$objective1 = \sum_{j=1}^J z_j \tag{18}$$

$$x_{ij}, y_{ij}, Z_j, A_{jw}, s_{ijw}, yr_{ijw}, k_{ij} \in \{0, 1\} \tag{19}$$

As indicated earlier, there are two different objectives for the addressed ALBP problem. The first objective is minimizing the number of stations utilized in the U-shaped assembly line. With this objective, the operational cost is trying to be reduced as much as possible since each opened station leads to an operational cost. Equation (1) represents the first objective function. The second objective is related to the maximum workload imbalance between workers. Each station is considered to be occupied by only one worker and that is why computing the workload difference between stations is enough to compute those between stations. Equation (2) represents the second objective function. In U-shaped ALBP problem, two different precedence relationship diagrams are employed to consider forward and backward assignments of tasks to the stations simultaneously. Equations (3) states that a task can be assigned to only one station. Equation (4) is constructed to allow the assignment of task to both forward and backward of a station. Equation (5) implies that if a worker is assigned to a station and a task assigned to the same station, then a new auxiliary variable is defined for worker-task-station assignment. Equations (6) and (7) are employed to satisfy the precedence constraints for both precedence and phantom networks. Equation (8) obligates that even if one task is assigned to a station, then this station must be opened and other tasks can also be assigned to that station by satisfying other constraints. Equation (9) implies that a worker can be assigned only one station. Equation (10) states that a station is operated only one worker. It also means that multi-manned stations are not allowed. Equation (11) is used to compute the value of variable that represents whether the station is opened or not. Equations (12) and (13) are used to ensure the linearity of the proposed model. Equation (14) guarantees that the sum of task times, which are assigned to the same station, cannot exceed the pre-determined cycle time according to the demand of customers. Equation (15) is employed to compute the sum of task times assigned to a station. Equation (16) is used to compute the workload difference between stations. As indicated earlier, computing the workload difference between stations is the same to compute difference between stations since each station is operated by one worker. Equation (17) implies that maximum workload difference, which is the second objective function, is greater than or equal to all workload differences. Equation (18) is used to compute the total number of opened stations for the first objective. Equation (19) represents binary restrictions of the variables.

3.2   AUGMECON2 Method

In the existing academic literature, there are different types of ε-constrained based methods and these methods are widely employed to the multi-objective problems [20–22]. In this study, an improved ε-constrained method, namely second version of the augmented ε-constrained (AUGMECON2), is applied to the problem to find Pareto-optimal solutions to the problem. This method uses the procedure behind the lexico-graphic optimization. Because it is an effective method to find all solutions by applying payoff tables [23], the same procedure is also implemented to the problem.

In the following, the details of the method are presented with equations.
Minimize

$$minf_1(x) + eps \times \left(\frac{s_2}{f\,max_2 - f\,min_2}\right) \tag{20}$$

Subject to:

$$f_2(x) + s_2 = f\,min_2 + t \times (f\,max_2 - f\,min_2)/q_2 \tag{21}$$

$$x \in S \ and \ s_i \in R^+ \tag{22}$$

In these equations, f_1 and f_2 correspond to the number of opened stations and maximum workload imbalance between workers objectives, respectively. The slack variable s_2 is used to find Pareto-optimal solutions within acceptable CPU time by eliminating unnecessary search steps [22].

In Eq. (21), whilst t is used to count the interval, q_i represents the length of intervals. On the other hand, the *eps* in Eq. (20) represents a small number between 10^{-6} and 10^{-3}. In order to obtain the solutions, GAMS® 23.5/CPLEX 12.2 optimization solver is employed by changing the right-hand side value of Eq. (21) each time.

4   Computational Results

In this section, an industrial case study is presented with real data taken from a water-meter producer. In order to analyze the impact of worker inherent in detail, four different scenarios are considered. Each scenario is constructed based on the workers' skill levels.

Figure 1 gives the precedence diagram in which the numbers inside the circles represent tasks, while the others represent the processing times.

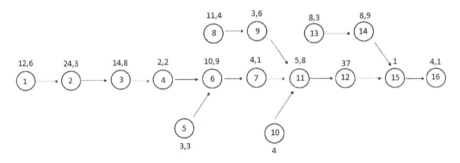

Fig. 1. Precedence diagram of tasks

In the following, the scenarios constructed for the problem are presented.

– Scenario 1: Each worker has medium skill levels in the worker pool.
– Scenario 2: The worker pool consists of workers who have high, medium, and low skill levels with the ratio by 33%, 33%, and 33%, respectively.
– Scenario 3: The worker pool consists of workers who have high and low skill levels with the ratio by 66% and 33%, respectively.
– Scenario 4: The worker pool consists of workers who have high and low skill levels with the ratio by 33% and 66%, respectively.

Table 1 presents the Pareto-optimal solutions for each scenario. These values also include payoff values. For instance, for scenario 1 and C = 80, the payoff values are (2–4; 79.2-15). Three different cycle time value is determined to analyze the results from a different perspective. From this table, it is observed that the number of stations objective does not change along with scenarios or cycle times. So as to further analyze the results Fig. 2 is constructed. When Fig. 2 is analyzed, it is observed that scenarios do not affect the first objective. In other words, workers' skill levels do not have a strong impact on the number of stations. On the other hand, the cycle time has an impact on the number of stations, which is observed when the cycle time increases to 80.

It is also interesting to state that while the cycle time increase, the second objective (maximum workload imbalance among workers) also increase. It means that the increase in cycle time negatively affects workload imbalance. Besides, scenario 2 does not affect the workload imbalance. That is to say, when there is equal number of workers from each skill set in the system, the performance does not change in terms of the workload imbalance. Furthermore, the workload imbalance is at the lowest level for scenario 3 in which 66% of the workers have a high skill level. The workload imbalance is at the highest level for scenario 4 in which 66% of the workers have a low skill level. Overall, it can be easily concluded that the workload imbalance is positively affected when the skill levels of workers increase in the system.

Table 1. Computational results (Pareto-optimal solutions including payoff values)

C=60			**C=70**			**C=80**		
1 scenario							o1	o2
	o1	o2		o1	o2	Min obj 1	2.0	79.2
Min obj 1	3.0	53.6	Min obj 1	3.0	67.5		3.0	63.8
Min obj 2	4.0	13.8	Min obj 2	4.0	15.0	Min obj 2	4.0	15.0
2 scenario						Min obj 1	2.0	79.2
Min obj 1	3.0	53.6	Min obj 1	3.0	67.5		3.0	63.8
Min obj 2	4.0	13.8	Min obj 2	4.0	15.0	Min obj 2	4.0	15.0
3 scenario						Min obj 1	2.0	71.4
Min obj 1	3.0	48.5	Min obj 1	3.0	55.7		3.0	59.6
Min obj 2	4.0	10.0	Min obj 2	4.0	12.5	Min obj 2	4.0	13.0
4 scenario						Min obj 1	2.0	84.6
Min obj 1	3.0	58.5	Min obj 1	3.0	74.0		3.0	66.5
Min obj 2	4.0	18.0	Min obj 2	4.0	19.5	Min obj 2	4.0	20.3

o1: objective1 (number of stations); o2: objective2 (workload imbalance)

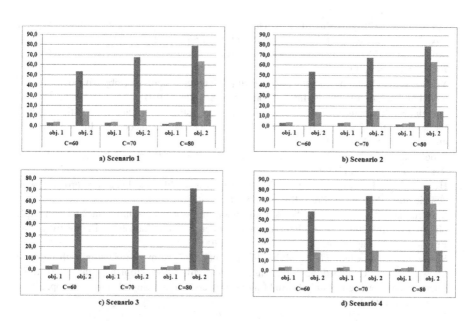

Fig. 2. Computational results with respect to scenarios

5 Concluding Remarks

This study deals with a bi-objective U-shaped assembly line balancing problem with heterogeneity inherent of workers. An optimization model is proposed for the problem by modifying a model from existing academic literature. A real case study is presented with real data from the water-meter producer for the problem. In order to analyze the impact of skill levels on the objectives, which are the number of opened stations and the maximum workload imbalance between workers, four different scenarios are constructed. AUGMECON2 method is employed to obtain Pareto-optimal solutions for the problem with respect to scenarios and three different cycle times.

According to the computational results, while the skill levels of workers increase the system performance also increases in terms of the workload imbalance. On the other hand, the skill levels do not have any impact on the number of opened stations in the system. It means that the operational cost of the system does not change with the skill of workers. However, the cycle time increase affects the number of opened stations, which is an expected conclusion.

For the future research directions, (i) uncertainties can be considered for the bi-objective problem with robust formulation, (ii) stochastic or fuzzy modeling techniques can be applied to the problem, last but not least (iii) other scenarios can be included to extend the sensitivity analysis.

References

1. Cevikcan, E., Durmusoglu, M.B., Unal, M.E.: A team-oriented design methodology for mixed model assembly systems. Comput. Ind. Eng. **56**, 576–599 (2009)
2. Miltenburg, J.: U-shaped production lines: a review of theory and practice. Int. J. Prod. Econ. **70**(3), 201–214 (2001)
3. Yılmaz, Ö.F.: Operational strategies for seru production system: a bi-objective optimisation model and solution methods. Int. J. Prod. Res. **58**, 1–25 (2019)
4. Satoglu, S.I., Durmusoglu, M.B., Ertay, T.: A mathematical model and a heuristic approach for design of the hybrid manufacturing systems to facilitate one-piece flow. Int. J. Prod. Res. **48**(17), 5195–5220 (2010)
5. Scholl, A., Klein, R.: ULINO: optimally balancing u-shaped JIT assembly lines. Int. J. Prod. Res. **37**(4), 721–736 (1999)
6. Hwang, R.K., Katayama, H., Gen, M.: U-shaped assembly line balancing problem with genetic algorithm. Int. J. Prod. Res. **46**(16), 4637–4649 (2008)
7. Sirovetnukul, R., Chutima, P.: Worker allocation in U-shaped assembly lines with multiple objectives. In: 2009 IEEE International Conference on Industrial Engineering and Engineering Management, pp. 105–109. IEEE (2009)
8. Sirovetnukul, R., Chutima, P.: Multi-objective particle swarm optimization with negative knowledge for U-shaped assembly line worker allocation problems. In: 2010 IEEE International Conference on Industrial Engineering and Engineering Management, pp. 2033–2038. IEEE (2010)

9. Farkhondeh, H., Hassanzadeh, R., Mahdavi, I., Mahdavi-Amiri, N.: A DEA approach for comparing solution efficiency in U-line balancing problem using goal programming. Int. J. Adv. Manuf. Technol. **61**(9–12), 1161–1172 (2012). https://doi.org/10.1007/s00170-011-3773-9

10. Dong, H., Cao, J.H., Zhao, W.L.: A genetic algorithm for the mixed-model U-line balancing problem. In: Advanced Materials Research, vol. 694, pp. 3391–3394. Trans Tech Publications Ltd. (2013)

11. Alavidoost, M.H., Tarimoradi, M., Zarandi, M.F.: Fuzzy adaptive genetic algorithm for multi-objective assembly line balancing problems. Appl. Soft Comput. **34**, 655–677 (2015)

12. Manavizadeh, N., Rabbani, M., Radmehr, F.: A new multi-objective approach in order to balancing and sequencing mixed model assembly line problem: a proposed heuristic algorithm. Int. J. Adv. Manuf. Technol. **79**(1-4), 415–425 (2015). https://doi.org/10.1007/s00170-015-6841-8

13. Alavidoost, M.H., Babazadeh, H., Sayyari, S.T.: An interactive fuzzy programming approach for bi-objective straight and U-shaped assembly line balancing problem. Appl. Soft Comput. **40**, 221–235 (2016)

14. Rabbani, M., Mousavi, Z., Farrokhi-Asl, H.: Multi-objective metaheuristics for solving a type II robotic mixed-model assembly line balancing problem. J. Ind. Prod. Eng. **33**(7), 472–484 (2016)

15. Babazadeh, H., Alavidoost, M.H., Zarandi, M.F., Sayyari, S.T.: An enhanced NSGA-II algorithm for fuzzy bi-objective assembly line balancing problems. Comput. Ind. Eng. **123**, 189–208 (2018)

16. Zhang, Z., Tang, Q., Li, Z., Zhang, L.: Modelling and optimisation of energy-efficient U-shaped robotic assembly line balancing problems. Int. J. Prod. Res. **57**(17), 5520–5537 (2019)

17. Chutima, P., Suchanun, T.: Productivity improvement with parallel adjacent U-shaped assembly lines. Adv. Prod. Eng. Manag. **14**(1), 51–64 (2019)

18. Babazadeh, H., Javadian, N.: A novel meta-heuristic approach to solve fuzzy multi-objective straight and U-shaped assembly line balancing problems. Soft. Comput. **23**(17), 8217–8245 (2018). https://doi.org/10.1007/s00500-018-3457-6

19. Oksuz, M.K., Buyukozkan, K., Satoglu, S.I.: U-shaped assembly line worker assignment and balancing problem: a mathematical model and two meta-heuristics. Comput. Ind. Eng. **112**, 246–263 (2017)

20. Hamacher, H.W., Pedersen, C.R., Ruzika, S.: Finding representative systems for discrete bicriterion optimization problems. Oper. Res. Lett. **35**(3), 336–344 (2007)

21. Mavrotas, G.: Effective implementation of the ε-constraint method in multi-objective mathematical programming problems. Appl. Math. Comput. **213**(2), 455–465 (2009)

22. Mavrotas, G., Florios, K.: An improved version of the augmented ε-constraint method (AUGMECON2) for finding the exact pareto set in multi-objective integer programming problems. Appl. Math. Comput. **219**(18), 9652–9669 (2013)

23. Bal, A., Satoglu, S.I.: A goal programming model for sustainable reverse logistics operations planning and an application. J. Clean. Prod. **201**, 1081–1091 (2018)

Two-Channel Conflict-Free Square Grid Aggregation

Adil Erzin[1,2] and Roman Plotnikov[1(✉)]

[1] Sobolev Institute of Mathematics, SB RAS, Novosibirsk 630090, Russia
{adilerzin,prv}@math.nsc.ru
[2] Saint Petersburg State University, St Petersburg 199034, Russia

Abstract. The conflict-free data aggregation problem in an arbitrary wireless network is NP-hard, both in the case of a limited number of frequencies (channels) and with an unlimited number of channels. However, on graphs with a particular structure, this problem sometimes becomes polynomially solvable. For example, when the network is a square grid (lattice), at each node of which there is a sensor, and the transmission range does not exceed 2, the problem is polynomially solvable. In this paper, we consider the problem of conflict-free data aggregation in a square grid, when network elements use two frequencies, and the transmission range is at least 2. It consists in finding an energy-efficient conflict-free (we will give later the definition of a conflict) schedule of minimum length for the transfer of aggregated data from all vertices of the lattice to the center node (base station).

We find polynomially solvable cases, and also develop an efficient algorithm that builds a schedule with a guaranteed accuracy estimate. For example, when the transmission range is 2, the algorithm constructs either an optimal schedule or a schedule whose length exceeds the optimal latency by no more than 1. For a transmission range more than 2, an estimate of the reduction in the length of the schedule is obtained compared to the case when only one frequency is used.

Keywords: Multichannel aggregation · Square grid · Conflict-free scheduling

1 Introduction

Data transmission in wireless networks, such as sensor networks, is carried out using radio communications. During *convergecasting*, each network element transmits a packet of *aggregated* data received from its children, as well as its data

The research is partly supported by the Russian Science Foundation (projects 18–71–00084, section 3 and 19–71–10012, sections 1, 2, 3.1, 4).
Submitted to Special Session 4 "Intractable Problems of Combinatorial Optimization, Computational Geometry, and Machine Learning: Algorithms and Theoretical Bounds".

I. S. Kotsireas and P. M. Pardalos (Eds.): LION 14 2020, LNCS 12096, pp. 168–183, 2020.
https://doi.org/10.1007/978-3-030-53552-0_18

to the parent vertex once during the entire aggregation session. This requirement is dictated by the extreme power consumption of the transmission process and entails the need to build a spanning *aggregation tree* (AT) with arcs directed to the sink, which is called the *base station* (BS). The faster the aggregated data reaches the BS, the better the schedule. In a TDMA scheduling, time is divided in equal-length slots under assumptions that each slot is long enough to send or receive one packet [4]. Minimizing time for the aggregated convergecast in this case is equivalent to minimizing the number of time slots required for all packets to reach the sink [17].

The solution of the problem includes two components: an AT, and a schedule, which assigns a transmitting time slot for each node so that every node transmits after all its children in the tree have, and potentially interfered links scheduled to send in different time slots. The last condition means that the TDMA schedule should be interference-free, i.e., no receiving node is within the interference range of the other transmitting node. There are two types of interferences or collisions in wireless networks: primary and secondary. A primary collision occurs when more than one node transmits to the same destination. In tree-based aggregation, it corresponds to the case when two or more children of the same parent node send their packets in the same time slot. A secondary collision occurs when a node overhears transmissions intended for another node. Links in the underlying communication graph cause such kind of collision, but not in the aggregation tree. If the links employ different frequencies (communication channels), then this type of conflict does not occur.

The conflict-free data aggregation problem was proven to be NP-hard [3] even if AT is known [5]. Therefore, almost all existing results in literature are polynomial algorithms for finding *approximate* solutions when the network elements use one channel [1,3,10–14,16–18] or several channels [2,9,15]. In [1] presented a novel cross-layer approach for reducing the latency in disseminating aggregated data to the BS over multi-frequency radio links. Their approach forms the aggregation tree to increase the simultaneity of transmissions and reduce buffering delay. Aggregation nodes picked, and time slots are allocated to the individual sensors so that the most number of ready nodes can transmit their data without delay. The use of different radio channels allows to avoid colliding transmissions. Their approach is validated through simulation and outperforms previously published schemes. [3] considered a min-length scheduling problem to aggregate data in the BS. The authors study the problem with an equal transmission range of all sensors. They assume that, in each time slot, data sent by a sensor reaches exactly all sensors within its transmission range, and a sensor receives data if it is the only data that reaches the sensor during this time slot. They first prove that the problem is NP-hard even when all sensors have deployed a grid, and data from all sensors are required to be aggregated in the BS. A $(\Delta - 1)$-approximation algorithm is designed, where $\Delta + 1$ equals the maximum number of sensors within the transmission range of any sensor. The authors also simulate the proposed algorithm and compare it with the existing algorithm. The obtained results show that their algorithm has much better performance in practice than the theoretically proved guarantee and outperforms other algorithms.

In [12], the authors investigate the question: "How fast can information be collected from a wireless sensor network organized as a tree?" To address this, they explore and evaluate several different techniques using realistic simulation models under the many-to-one communication paradigm known as a converge-casting. First, min-time scheduling on a single frequency channel considered. Next, they combine scheduling with transmission power control to mitigate the effects of interference and show that while power control helps in reducing the schedule length under a single frequency, scheduling transmissions using multiple frequencies is more efficient. The authors gave lower bounds on the schedule length without interference conflicts, and proposed algorithms that achieve these bounds. They also evaluate the performance of various channel assignment methods and find empirically that for moderate size networks of about 100 nodes, the use of multifrequency scheduling can suffice to eliminate most of the interference. Then, the data collection rate no longer remains limited by interference but by the topology of the routing tree. To this end, they construct degree-constrained spanning trees and capacitated minimal spanning trees and show significant improvement in scheduling performance over different deployment densities. Lastly, they evaluate the impact of different interference and channel models on the schedule length.

In [16], the authors consider the problem of aggregation convergecast scheduling. The solution to aggregation convergecast satisfies the aggregation process, expressed as precedence constraints, combined with the impact of the shared wireless medium, expressed as resource constraints. Both sets of constraints influence the routing and scheduling. They propose an aggregation tree construction suitable for aggregation convergecast that is a synthesis of a tree tailored to precedence constraints and another tree tailored to resource constraints. Additionally, they show that the scheduling component modeled as a mixed graph coloring problem. Specifically, the extended conflict graph introduced, and through it, a mapping from aggregation convergecast to mixed graphs described. Bounds for the graph coloring provided and a branch-and-bound strategy developed from which the authors derive numerical results that allow comparison against the current state-of-the-art heuristic.

[18] focuses on the latency of data aggregation. Since the problem is NP-hard, many approximate algorithms have proposed to address this issue. Using maximum independent set and first-fit algorithms, in this study a scheduling algorithm, Peony-tree-based Data Aggregation (PDA), designed which has a latency bound of $15R + \Delta - 15$, where R is the network radius (measured in hops) and Δ is the maximum node degree. They theoretically analyze the performance of PDA based on different network models and further evaluate it through extensive simulations. Both the analytical and simulation results demonstrate the advantages of PDA over the state-of-art algorithm, which has a latency bound of $23R + \Delta - 18$.

In [9], the authors focus on designing a multi-channel minimum latency aggregation scheduling protocol, named MC-MLAS, using a new joint approach for tree construction, channel assignment, and transmission scheduling. To the best knowledge of the authors, this is the first work in the literature that combines orthogonal channels and partially overlapping channels to consider the total latency involved in data aggregation. Extensive simulations verify the superiority of MC-MLAS in WSNs.

In [15], the authors consider a problem of minimum length scheduling for the conflict-free aggregation convergecast in wireless networks in a case when each element of a network uses its frequency channel. This problem is equivalent to the well-known NP-hard problem of telephone broadcasting since only the conflicts between the children of the same parent taken into account. They propose a new integer programming formulation and compare it with the known one by running the CPLEX software package. Based on the results of a numerical experiment, they concluded that their formulation is preferable in practice to solve the considered problem by CPLEX than the known one. The authors also propose a novel heuristic algorithm, which based on a genetic algorithm and a local search metaheuristic. The simulation results demonstrate the high quality of the proposed algorithm compared to the best-known approaches.

However, if the network has a regular structure, for example, it is a lattice, the problem is solved in polynomial time. Known that in a square lattice, in each node of which information is located, the process of single-channel data aggregation is simple [8]. Moreover, in some cases, for example, when the transmission range equals 1 [8] or 2 [6], one can build an optimal schedule. If the transmission range is greater than 2, then one can find a solution close to the optimal [5, 7].

1.1 Our Contribution

In this paper, for the first time, the problem of two-channel conflict-free aggregation in a square lattice is considered, when the transmission distance is not less than 2. If the transmission distance is 1, then the problem does not differ from single-channel aggregation and is solved completely [8]. We have developed and analyzed an efficient algorithm that builds either an optimal or near-optimal solution. We estimated the reduction in the length of the schedule compared with the case when one channel is used.

The rest of the paper is organized as follows. Section 2 contains the statement of the problem. The main results with the description and analysis of the algorithm make up the contents of Sect. 3. Section 4 concludes the paper.

2 Problem Formulation

We suppose that the network elements are positioned at the nodes of a square grid of size $(n + 1) \times (m + 1)$. For convenience, we will call the network elements sensors, vertices, or nodes equivalently. A sink node (or BS) is located at the point $(0, 0)$. At each time slot, any sensor except the sink node can either be idle, send the data to another sensor within its transmission range, or receive the data from another sensor within its transmission range. We assume that each sensor has the same transmission distance $d \geq 2$ in L_1 metric. A sink node can only receive the data at any time slot. Each data transmission is performed using one of the available frequency channels (for short, further, they are referred to as *channels*) and each sensor can use any channel for data transmission and receiving. Besides, we suppose that the following conditions met:

- each vertex sends a message only once during the aggregation session (except the sink which always can only receive messages);
- once a vertex sends a message, it can no longer be a destination of any transmission;
- if some vertex sends a data packet by the channel c, then during the same time slot none of the other vertices within a receiver's interference range can post a message by the channel c;
- a vertex cannot receive and transmit at the same time slot.

For simplicity, we assume that the interference range equals the transmission range. We have precisely two available channels for data transmission: 0 and 1. The problem consists in constructing the conflict-free min-length schedule of the data aggregation from all the vertices to the BS.

3   Building a Conflict-Free Schedule with Different Transmission Ranges

As mentioned before, the considered problem is NP-hard in the general case. Therefore, our goal is to find some individual cases and propose polynomial algorithms that construct either optimal solutions or solutions with guaranteed estimations on the schedule length. In this section, we offer such algorithms and estimates for the different values of the transmission range.

Let us introduce some notations that are used further.

Definition 1. *Let the* distance *from the vertex i to the sink be the minimum number of time slots that are necessary to transmit the data from i to the sink.*

Definition 2. *The* most remote vertex *(MRV) is such vertex, that the distance from it to the sink is maximum among all vertices.*

In square grid $(n+1) \times (m+1)$, the distance from the node (x, y) to the sink equals $\lceil (x+y)/d \rceil$, where d is transmission range and $\lceil a \rceil$ is a smallest integer not less than a. Obviously, the vertex (n, m) is MRV. There can be more than one MRV, and the distance depends of the remainder of division of n and m by d.

We will also refer to the distance from a vertex to the sink as its *remoteness*. Let D be the remoteness of MRV. The following two propositions are obvious:

Proposition 1. *The aggregation time cannot be less than D.*

Proposition 2. *If remoteness of at least two different vertices is R, then the aggregation time cannot be less than $R + 1$.*

The validity of the last statement follows from the property that any vertex during one time round can receive no more than one message.

For convenience, we call the *row* a set of sensors that are positioned at the points with the same ordinates. All the presented below algorithms consist of two stages: vertical and horizontal aggregation. During the vertical transmissions, all sensors except ones in the row 0 transmit the data downwards or upwards. At the end of the vertical stage, all data are aggregated in row 0. At the second stage, the sensors in the row 0 transmit the data horizontally until the sink receives all the data.

Further, we will consider separately two cases: when $d = 2$ and when $d > 2$.

3.1   The Transmission Range is 2

Suppose that $d = 2$. Let us prove the following

Lemma 1. *In the square grid $(n + 1) \times (m + 1)$, if the transmission range is 2 and the number of channels is 2, then the length of any feasible aggregation schedule is not less than $\lfloor n/2 \rfloor + \lfloor m/2 \rfloor + 1$, where $\lfloor a \rfloor$ is the largest integer not exceeding a.*

Proof. If at least one of two values, n and m, is odd, then the remoteness of (n, m) is $\lfloor n/2 \rfloor + \lfloor m/2 \rfloor + 1$, and, according to the Proposition 1, the aggregation time cannot be less than $\lfloor n/2 \rfloor + \lfloor m/2 \rfloor + 1$ in this case. If n and m are even then there are three MRV whose remoteness is $\lfloor n/2 \rfloor + \lfloor m/2 \rfloor$. Therefore, in this case, according to the Proposition 2, aggregation time cannot be less than $\lfloor n/2 \rfloor + \lfloor m/2 \rfloor + 1$.

Lemma 2. *In the square grid $(n+1) \times (m+1)$, if n and m are odd, the transmission range is 2 and the number of channels is 2, then the length of any feasible aggregation schedule is not less than $\lfloor n/2 \rfloor + \lfloor m/2 \rfloor + 2$.*

Proof. In this case, three MRV, $(n-1, m)$, $(n, m-1)$, and (n, m) are at the distance $\lfloor n/2 \rfloor + \lfloor m/2 \rfloor + 1$ from the sink. Therefore, according to the Proposition 2, aggregation time cannot be less than $\lfloor n/2 \rfloor + \lfloor m/2 \rfloor + 2$.

Let us describe the Algorithm 1 of aggregation scheduling in the square grid $(n+1) \times (m+1)$ when the transmission range is 2, and the number of channels is 2. The pseudo-code of the vertical aggregation stage is given in the Algorithm 1. The examples of the proposed vertical aggregation algorithm are illustrated in Fig. 1. As it follows from lines 3–8, during the first $\lceil m/2 \rceil - 1$ time slots, at each time slot, the highest two rows that did not transmit yet simultaneously transmit the data downwards at a distance 2. Subroutine $TransmitRow(j, k, ch, t)$ which is called in lines 5–6 assigns transmission of each sensor in j-th row vertically to the corresponding sensors in k-th row at the time slot t using the channel

ch. Here one row uses channel 0, another one - channel 1, and therefore there are no conflicts in such data transmission. After that, only one or two rows left that did not transmit (except the row 0, which does not transmit at this stage): one in a case when m is odd and two – if it is even. Then, according to lines 9–19, at each time slot, the highest row, that did not transmit yet, transmits the data downwards at a distance 1 using two channels. It is performed without conflicts because vertices with different parity of abscissa use different channels. If m is odd, then the data transmission that is described in lines 9–19 requires one time slot. Subroutine $TransmitSensor((i,j),(l,k),ch,t)$ which is called in line 19 assigns transmission of sensor (i,j) to sensor (l,k) at the time slot t using the channel ch. If m is even, it requires two time slots. Overall, the described vertical aggregation takes $\lfloor m/2 \rfloor + 1$ time slots.

Algorithm 1. Transmission range equals 2. Vertical aggregation

1: $t \leftarrow 0$;
2: $j \leftarrow m$;
3: **while** $j > 2$ **do**
4:     $t \leftarrow t + 1$;
5:     $TransmitRow(j, j - 2, 0, t)$;
6:     $TransmitRow(j - 1, j - 3, 1, t)$;
7:     $j \leftarrow j - 2$;
8: **end while**
9: **while** $j > 0$ **do**
10:     $t \leftarrow t + 1$;
11:     **for all** $i \in \{0, ..., n\}$ **do**
12:         $ch \leftarrow 0$;
13:         **if** i is odd **then**
14:             $ch \leftarrow 1$;
15:         **end if**
16:         $TransmitSensor((i,j),(i,j-1),ch,t)$;
17:         $i \leftarrow i - 1$;
18:     **end for**
19: **end while**

Remark 1. The time complexity of vertical aggregation is $O(m)$. During one time round, all the vertices of the two highest layers send. After they have transferred, it is enough to remember only the number of the highest layer, whose vertex has not transmitted yet. Therefore, space (additional memory) needed to implement vertical aggregation is $O(1)$.

The pseudo-code for the horizontal aggregation stage is presented in Algorithm 2. The examples that show how horizontal aggregation is performed are presented in Fig. 2. According to lines 3–8, during the first $\lceil n/2 \rceil - 1$ time slots, at each time slot, the most remote two vertices that did not transmit yet simultaneously transmit the data to the left at a distance 2 using different channels.

Algorithm 2. Transmission range equals 2. Horizontal aggregation

1: $t \leftarrow 0$;
2: $i \leftarrow n$;
3: **while** $i > 2$ **do**
4:    $t \leftarrow t + 1$;
5:    $TransmitSensor((i, 0), (i - 2, 0), 0, t)$;
6:    $TransmitSensor((i - 1, 0), (i - 3, 0), 1, t)$;
7:    $i \leftarrow i - 2$;
8: **end while**
9: **while** $i > 0$ **do**
10:    $t \leftarrow t + 1$;
11:    $TransmitSensor((i, 0), (i - 1, 0), 0, t)$;
12:    $i \leftarrow i - 1$;
13: **end while**

After that, as it follows from lines 9–13, at each time slot, the most remote vertex that did not transmit yet transmits the data to the left at a distance 1 until such vertex exists. The last subroutine requires one time slot if n is odd, and two time slots if it is even. Eventually, the horizontal aggregation takes $\lfloor n/2 \rfloor + 1$ time slots.

Remark 2. The time complexity of horizontal aggregation is $O(n)$, and the additional space required to implement the procedure is limited to $O(1)$. Consequently, the complexity of constructing a two-channel conflict-free aggregation schedule in the $(n + 1) \times (m + 1)$ grid with a transmission distance of 2 is $O(m + n)$, with additional memory space $O(1)$. Although each vertex of the grid must store its state, the total memory for storing incoming and current information is $O(mn)$.

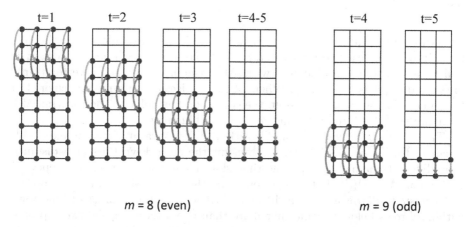

$m = 8$ (even) $m = 9$ (odd)

Fig. 1. Example of vertical data aggregation when transmission distance equals 2.

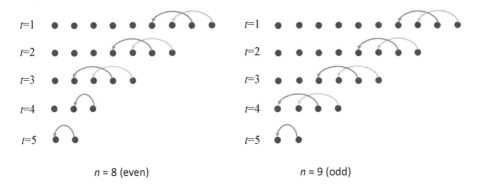

Fig. 2. Example of horizontal data aggregation when transmission distance equals 2.

As a result, the entire aggregation process takes $\lfloor m/2 \rfloor + \lfloor n/2 \rfloor + 2$ time slots. Due to the Lemma 2, a solution that is constructed by the proposed algorithm is *optimal* if n and m are odd. According to the Lemma 1, in all other cases, the length of a constructed schedule does not exceed the minimum schedule length by more than 1. As it is proved in [6], in a case when the only channel is used the minimum length of an aggregation schedule is $\lfloor n/2 \rfloor + \lfloor m/2 \rfloor + 3$. Therefore, the usage of two channels allows decreasing the length of a schedule by one.

3.2   The Transmission Range is at Least 3

In this section we propose the algorithm that constructs a schedule of two-channels aggregation on a square grid $(n + 1) \times (m + 1)$ when d is arbitrary not less than 3. Assume that $m = Md + r_v, r_v < d$ and $n = Nd + r_h, r_h < d$. As well as the algorithm described in previous section, this algorithm consists of two stages: vertical and horizontal aggregation. We will describe each stage separately.

Vertical Aggregation. For convenience purposes, let us colorize all the vertices. Initially, let each vertex $(i, j), i, j \leq m$ be colored in red if $m - j \equiv 0 \pmod{d}$ and let it be colored in blue otherwise. After the moment when a vertex transmits the data, it becomes grey. When all the vertices of the same row colored in the same color, we will also assign the corresponding color to the row. Initially there are $\lfloor m/d \rfloor + 1$ red rows and $m - \lfloor m/d \rfloor$ blue rows.

The pseudo-code has given in Algorithm 3. At first, as it stated in lines 2–14, a pair of top blue rows iteratively transmit the data downstairs at a distance d by different channels during one time slot. These transmissions are repeated until all of $(M - 1)(d - 1)$ top blue rows send their data. Meanwhile, the top red row transmits downwards at a distance d as soon as the number of blue rows within $2d$ rows below becomes not more than one, otherwise, the two top blue rows would transmit at the same time slot, and this would generate a conflict. After every $(M - 1)(d - 1)$ top blue rows transmitted, in lines 15–19, the top red rows sequentially transmit downwards at a distance d until the moment when only two red rows remain. Note that after that exactly $d + r_v + 1$ non-grey rows left. Then, in lines 20–29, the top r_v rows transmit downstairs during the next $\lceil r_v/2 \rceil$ time slots. After that, in lines 30–41, the rows between row 0 and row d transmit the data. Here the data is transmitted at less distance than d. It is easy to see that to transmit all the data of some row to another row at distance $d - k$ without conflicts by two channels $\lceil (k + 1)/2 \rceil$ time slots are required. Note that for this, we slightly changed a signature of the method $TransmitRow$ to denote such data transmission. Finally, only two non-grey rows remain: row 0 and row d. At the last time slot, row d transmits at distance d downwards. Overall, the described vertical aggregation uses

$$\lceil (M - 1)(d - 1)/2 \rceil + \lceil r_v/2 \rceil + 2 \sum_{i=2}^{\lfloor (d-1)/2 \rfloor + 1} \lceil i/2 \rceil + x_d \lceil (d/2 + 1)/2 \rceil + 2$$

time slots, where $x_d = 1$ if d is even and 0 otherwise. An example of vertical aggregation is presented in Fig. 3.

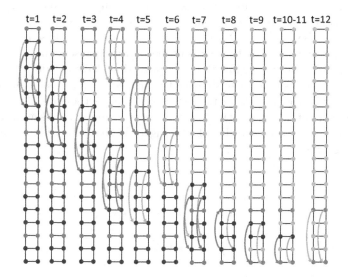

Fig. 3. Example of two-channel vertical data aggregation when $d = 4$. (Color figure online)

Horizontal Aggregation. The horizontal aggregation performed on row 0 of the grid when all the data aggregated to this row. At first, we assume that $n = Nd$ and propose the algorithm for this case. For simplicity, in this subsection we refer to sensor $(i, 0)$ as sensor i, $i = 0, \ldots, n$. Suppose that each sensor is colored in blue or red color: sensor i is colored in red if $i \equiv 0 \pmod{d}$, and it is colored in blue otherwise. As well as it was previously, the sensor becomes grey as soon as it transmits the data. With such colorizing, the blue sensors are divided by the red sensors into *groups* of $d - 1$ sensors. We classify these groups into 3 types: A, B, and C. The group $\{i, \dots, i + d - 2\}$ has type A (B or C) if $\lfloor (n - i)/d \rfloor \equiv 0$ (1 or 2) $(\bmod\,3)$. For example, the group $\{n - d + 1, \dots, n - 1\}$ has type A. The idea of the algorithm is that in each time slot, sensors of groups of the same type perform the same transmissions. For this reason, we enumerate the sensors of each group from 1 to $d - 1$.

The pseudo-code of the proposed procedure is given in Algorithm 4. Here we use the method *TransmitSensor* whose signature differs from the similar method used above. For instance, *TransmitSensor*$(1, 0, A, 0, t)$ means that by channel 0 at time slot t sensor $n - d + 1$ transmits the data to sensor $n - d$, sensor $n - 4d + 1$ transmits to sensor $n - 4d$, and so on. As noted in lines 6–16, during first $\lfloor (d - 2)/2 \rfloor$ time slots, only sensors in groups of types A and B transmit the data, while sensors in groups of type C remain in the idle state because transmission from any of them would lead to a conflict. After that, in each group of types A and B, only two blue sensors remain if d is odd and one – if it is even. The last blue sensors of groups A and B transmit the data in lines 18–30. And, when it is possible, some sensors of groups of type C transmit the data at the same time. In lines 33–39, the last blue sensors of groups C transmit the data during the next $\lfloor (d - 2)/2 \rfloor$ time slots. In total, all blue sensors transmit the data during the first $d - 1$ time slots of the aggregation session. In lines 40–44, red sensors sequentially transmit the data from right to left. Note that the first red sensor transmits simultaneously with the last data transmission in groups C. After that, during the next $N - 1$ time slots, other red sensors transmit the data. Overall, the horizontal aggregation takes $N + d - 2$ time slots. As an illustration, two examples are presented in Fig. 4 and Fig. 5. In first one d is odd (5) and in second it is even (6).

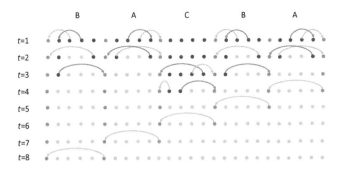

Fig. 4. Example of two-channel horizontal data aggregation when $d = 5$. (Color figure online)

It is easy to observe that if n is not a multiple of d, then the horizontal aggregation time increases by 1. Indeed, the sensors $1, \dots, n - d\lfloor n/d \rfloor - 1$ can

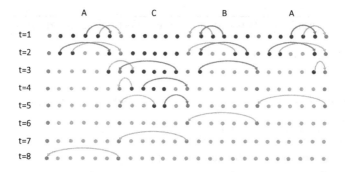

Fig. 5. Example of two-channel horizontal data aggregation when $d = 6$.

transmit the data during the first $d-1$ time slots in any case, and it is required a time slot more to aggregate the data from the red sensors. Eventually, horizontal aggregation takes $\lceil n/d \rceil + d - 2$ time slots.

In total the proposed procedure constructs a schedule of length

$$\lceil (\lfloor m/d \rfloor - 1)(d-1)/2 \rceil + \lceil n/d \rceil + \lceil r_v/2 \rceil + 2 \sum_{i=2}^{\lfloor (d-1)/2 \rfloor + 1} \lceil i/2 \rceil + x_d \lceil (d/2+1)/2 \rceil + d.$$

Remark 3. Note that the time complexity of the described algorithm has the same order as the length of the schedule $O(n + m)$, and the additional memory space required for its implementation is $O(1)$. However, the length of the input data determined by the size of the grid; therefore, when implementing the algorithm, $O(nm)$ memory space is used.

We compared the schedule length obtained by this algorithm with the best-known approach for the one-channel aggregation, which proposed in [7]. The results presented in Table 1. In most cases, the two-channel aggregation schedule constructed by the proposed algorithm has less length than the length of the one-channel aggregation schedule.

Table 1. Comparison of schedule length for aggregation by 1 and 2 channels

d	11×11		26×26		51×51		101×101	
	2 ch	1 ch	2 ch	1 ch	2 ch	1 ch	2 ch	1 ch
3	12	15	22	25	38	43	72	75
4	14	19	24	25	39	39	61	61
5	15	20	24	26	36	36	56	56
7	21	35	29	41	43	43	59	59
10	30	49	40	61	52	57	67	67

Algorithm 3. Arbitrary transmission range. Vertical aggregation

---

1: $t \leftarrow 0$;
2: **while** The number of blue rows exceeds $d - 1 + r_v$ **do**
3:    $t \leftarrow t + 1$;
4:    $j_0 \leftarrow$ the maximum number of red row;
5:    **if** There are not more than one blue row between j_0-th and $(j_0 - 2d)$-th rows
   then
6:       $TransmitRow(j_0, j_0 - d, 1, t)$;
7:    **end if**
8:    $j_1 \leftarrow$ the maximum number of blue row;
9:    $TransmitRow(j_1, j_1 - d, 0, t)$;
10:    $j_2 \leftarrow$ the maximum number of blue row less than j_1;
11:    **if** $j_2 > d + r_v$ **then**
12:       $TransmitRow(j_2, j_2 - d, 1, t)$;
13:    **else**
14:       $j \leftarrow$ the maximum number of red row;
15:       $TransmitRow(j, j - d, 1, t)$;
16:    **end if**
17: **end while**
18: **while** The number of red rows exceeds 2 **do**
19:    $t \leftarrow t + 1$;
20:    $j \leftarrow$ the maximum number of red row;
21:    $TransmitRow(j, j - d, 0, t)$;
22: **end while**
23: $j \leftarrow d + r_v$
24: **while** $j > d$ **do**
25:    $t \leftarrow t + 1$;
26:    $TransmitRow(j, j - d, 0, t)$;
27:    $j \leftarrow j - 1$
28:    **if** $j > d$ **then**
29:       $TransmitRow(j, j - d, 1, t)$;
30:       $j \leftarrow j - 1$
31:    **end if**
32: **end while**
33: **for all** $k = 1, \ldots \lfloor (d - 1)/2 \rfloor$ **do**
34:    $t_\delta \leftarrow \lceil (k + 1)/2 \rceil$;
35:    $TransmitRow(k, d, \{0, 1\}, [t + 1, t + t_\delta])$;
36:    $t \leftarrow t + t_\delta$;
37:    $TransmitRow(d - k, 0, \{0, 1\}, [t + 1, t + t_\delta])$;
38:    $t \leftarrow t + t_\delta$;
39: **end for**
40: **if** d is even **then**
41:    $t_\delta \leftarrow \lceil (d/2 + 1)/2 \rceil$;
42:    $TransmitRow(d/2, 0, \{0, 1\}, [t + 1, t + t_\delta])$;
43:    $t \leftarrow t + t_\delta$;
44: **end if**
45: $TransmitRow(d, 0, 0, [t + 1, t + t_\delta])$;
46: $t \leftarrow t + t_\delta$;

Algorithm 4. Arbitrary transmission range. Horizontal aggregation

---

1: $t \leftarrow 0$;
2: $s_{A0} \leftarrow d - 2$;
3: $s_{A1} \leftarrow d - 3$;
4: $s_{B0} \leftarrow 2$;
5: $s_{B1} \leftarrow 3$;
6: **for all** $i = 1, \ldots, \lfloor (d-2)/2 \rfloor$ **do**
7:     $t \leftarrow t + 1$;
8:     $TransmitSensor(s_{A0}, d, A, 0, t)$;
9:     $s_{A0} \leftarrow s_{A0} - 2$;
10:     $TransmitSensor(s_{A1}, d - 1, A, 1, t)$;
11:     $s_{A1} \leftarrow s_{A1} - 2$;
12:     $TransmitSensor(s_{B0}, 0, B, 0, t)$;
13:     $s_{B0} \leftarrow s_{B0} + 2$;
14:     $TransmitSensor(s_{B1}, 1, B, 1, t)$;
15:     $s_{B1} \leftarrow s_{B1} + 2$;
16: **end for**
17: $t \leftarrow t + 1$;
18: **if** d is odd **then**
19:     $TransmitSensor(1, d, A, 0, t)$;
20:     $TransmitSensor(d - 1, 0, A, 1, t)$;
21:     $TransmitSensor(d - 1, 0, B, 0, t)$;
22:     $t \leftarrow t + 1$;
23:     $TransmitSensor(1, d, B, 1, t)$;
24:     $TransmitSensor(d - 2, d, C, 0, t)$;
25:     $TransmitSensor(d - 1, 0, C, 1, t)$;
26: **else**
27:     $TransmitSensor(d-1, d+1, A, 0, t)$ except sensor $n-1$ that transmits to sensor n;
28:     $TransmitSensor(1, 0, B, 0, t)$;
29:     $TransmitSensor(d - 1, 0, C, 1, t)$;
30: **end if**
31: $s_{C0} \leftarrow 1$;
32: $s_{C1} \leftarrow 2$;
33: **for all** $i = 1, \ldots, \lfloor (d-2)/2 \rfloor$ **do**
34:     $t \leftarrow t + 1$;
35:     $TransmitSensor(s_{C0}, 0, C, 0, t)$;
36:     $s_{C0} \leftarrow s_{C0} + 2$;
37:     $TransmitSensor(s_{C1}, d, C, 1, t)$;
38:     $s_{C1} \leftarrow s_{C1} + 2$;
39: **end for**
40: $TransmitMostRemoteRedSensor(t)$;
41: **for all** $i = 1, \ldots, N - 1$ **do**
42:     $t \leftarrow t + 1$;
43:     $TransmitMostRemoteRedSensor(t)$;
44: **end for**

4   Conclusion

In this paper, we considered the problem of two-channel conflict-free dater aggregation in a square lattice and proposed an efficient algorithm that yields a better solution than the convergecasting using one frequency. However, with increasing transmission distance, the advantages of two-channel aggregation disappear. This drawback is associated with insufficient consideration of the specifics of two-channel aggregation. We plan to fix this shortcoming in the future.

References

1. Bagaa, M., et al.: Data aggregation scheduling algorithms in wireless sensor networks: solutions and challenges. IEE Commun. Surv. Tutor. **16**(3), 1339–1368 (2014)
2. Bagaa, M., Younis, M., Badache, N.: Efficient data aggregation scheduling in wireless sensor networks with multi-channel links. In: MSWiM 2013, Barcelona, Spain, 3–8 November 2013 (2014). https://doi.org/10.1145/2507924.2507995
3. Chen, X., Hu, X., Zhu, J.: Minimum data aggregation time problem in wireless sensor networks. In: Jia, X., Wu, J., He, Y. (eds.) MSN 2005. LNCS, vol. 3794, pp. 133–142. Springer, Heidelberg (2005). https://doi.org/10.1007/11599463_14
4. Demirkol, I., Ersoy, C., Alagoz, F.: MAC protocols for wireless sensor networks: a survey. IEEE Commun. Mag. **44**, 115–121 (2006)
5. Erzin, A., Pyatkin, A.: Convergecast scheduling problem in case of given aggregation tree. The complexity status and some special cases. In: 10th International Symposium on Communication Systems, Networks and Digital Signal Processing, Article 16, 6 p. IEEE-Xplore, Prague (2016)
6. Erzin, A.: Solution of the convergecast scheduling problem on a square unit grid when the transmission range is 2. In: Battiti, R., Kvasov, D.E., Sergeyev, Y.D. (eds.) LION 2017. LNCS, vol. 10556, pp. 50–63. Springer, Cham (2017). https://doi.org/10.1007/978-3-319-69404-7_4
7. Erzin, A., Plotnikov, R.: Conflict-free data aggregation on a square grid when transmission distance is not less than 3. In: Fernández Anta, A., Jurdzinski, T., Mosteiro, M.A., Zhang, Y. (eds.) ALGOSENSORS 2017. LNCS, vol. 10718, pp. 141–154. Springer, Cham (2017). https://doi.org/10.1007/978-3-319-72751-6_11
8. Gagnon, J., Narayanan, L.: Minimum latency aggregation scheduling in wireless sensor networks. In: Gao, J., Efrat, A., Fekete, S.P., Zhang, Y. (eds.) ALGOSENSORS 2014. LNCS, vol. 8847, pp. 152–168. Springer, Heidelberg (2015). https://doi.org/10.1007/978-3-662-46018-4_10
9. Ghods, F., et al.: MC-MLAS: multi-channel minimum latency aggregation scheduling in wireless sensor networks. Comput. Netw. **57**, 3812–3825 (2013)
10. Guo, L., Li, Y., Cai, Z.: Minimum-latency aggregation scheduling in wireless sensor network. J. Comb. Optim. **31**(1), 279–310 (2014). https://doi.org/10.1007/s10878-014-9748-7
11. Ha, N.P.K., Zalyubovskiy, V., Choo, H.: Delay-efficient data aggregation scheduling in duty-cycled wireless sensor networks. In Proceedings of ACM RACS, October 2012, pp. 203–208 (2012)
12. Incel, O.D., Ghosh, A., Krishnamachari, B., Chintalapudi, K.: Fast data collection in tree-based wireless sensor networks. IEEE Trans. Mob. Comput. **11**(1), 86–99 (2012)

13. Li, J., et al.: Approximate holistic aggregation in wireless sensor networks. ACM Trans. Sensor Netw. **13**(2), 1–24 (2017)
14. Malhotra, B., Nikolaidis, I., Nascimento, M.A.: Aggregation convergecast scheduling in wireless sensor networks. Wirel. Netw. **17**, 319–335 (2011). https://doi.org/10.1007/s11276-010-0282-y
15. Plotnikov, R., Erzin, A., Zalyubovskiy, V.: Convergecast with unbounded number of channels. In: MATEC Web of Conferences, vol. 125, p. 03001 (2017). https://doi.org/10.1051/matecconf/20171250
16. de Souza, E., Nikolaidis, I.: An exploration of aggregation convergecast scheduling. Ad Hoc Netw. **11**, 2391–2407 (2013)
17. Wan, P.-J. et al.: Minimum-latency aggregation scheduling in multihop wireless networks. In: Proceedings of ACM MOBIHOC, May 2009, pp. 185–194 (2009)
18. Wang, P., He, Y., Huang, L.: Near optimal scheduling of data aggregation in wireless sensor networks. Ad Hoc Netw. **11**, 1287–1296 (2013)

Online Stacking Using RL with Positional and Tactical Features

Martin Olsen[✉][iD]

Department of Business Development and Technology, Aarhus University,
Aarhus, Denmark
martino@btech.au.dk

Abstract. We study the scenario where some items are stored temporarily in stacks and where it is not allowed to put an item on top of another item leaving earlier. An arriving item is assigned to a stack based only on information on the arrival and departure times for the new item and items currently stored. The objective is to minimize the maximum number of stacks used over time. This problem is referred to as online stacking. We use Reinforcement Learning (RL) techniques to improve heuristics earlier presented in the literature. Using an analogy to chess, we look at positional and tactical features where the former give high priority to stacking configurations that are well suited to meet the challenges on a long-term basis and the latter focus on using few stacks on a short-term basis. We show how the RL approach finds the optimal mix of positional and tactical features to be used at different stages of the stacking process. We document quantitatively that positional features play a bigger role at stages of the stacking process with few items stored. We believe that the RL approach combining positional and tactical features can be used in many other online settings within operations research.

Keywords: Online algorithms · Reinforcement learning · Stacking

1 Introduction

The challenge of stacking items temporarily in a storage area in an optimal manner is a problem that has many applications within logistics. Some notable examples of items to consider are containers in a container terminal or on a container ship [2], trains at a train station [3,5], or steel bars [12].

In this paper, we focus on the online version of the problem where an arriving item is assigned to a stack with no information on future items to arrive. The objective is to minimize the maximum number of stacks in use over time with the constraint that we cannot put an arriving item on top of an item that has to leave the storage area before the arriving item. In other words, we do not allow *overstowage* using terminology from the shipping industry.

We consider stacking heuristics that are controlled by so-called *features*. A feature is a function assigning a value to every possible action at every state

I. S. Kotsireas and P. M. Pardalos (Eds.): LION 14 2020, LNCS 12096, pp. 184–194, 2020.
https://doi.org/10.1007/978-3-030-53552-0_19

of the stacking process. For a given state of the stacking process, the heuristics that we develop greedily pick an action that maximizes a feature – or a linear combination of features – for that state. We combine/mix heuristics controlled by different features by constructing hybrid heuristics that are controlled by linear combinations of the features of the component heuristics and we allow the weights (coefficients) of the linear combinations to depend on the number of stacks that are in use at the different stages of the stacking process. This makes it possible to let some features be active when only a few stacks are in use and let other features be active at busy periods for the storage area.

In chess, a common strategy is to try to establish a strong *position* of your pieces – for example a position where your pieces dominate the center of the chess board. At some point this might lead to a situation where you can take advantage of the strong position and switch to a more *tactical* type of play – for example going for a specific pawn of the opponent. This is clearly expressed by a famous quote of a former world chess champion:

"Tactics flow from a superior position". (Bobby Fischer)

Using an analogy from chess, we develop stacking heuristics that are guided by an optimal mix of *positional* and *tactical* features. Positional features steer the heuristics towards stacking configurations with a long-term positional advantage as opposed to tactical features that focus on using few stacks on a short-term basis. The intuition is that the positional features are active at the quiet stages of the process in such a way that the stacking configurations are well-formed at the entry points of busy stages where the tactical features take over.

1.1   Contribution

The contribution of the paper can be split into three parts. The first part is directly related to stacking and the two other parts are on a more generic and methodological level. We believe that the methodological parts of the contribution are interesting for a broader audience working with online algorithms.

We use simple Reinforcement Learning (RL) tools to improve natural heuristics earlier presented in the literature [2, 6, 7, 15]. The heuristics are trained using Markov Decision Processes as models for the stacking environment.

On the methodological level, we quantitatively justify the intuition expressed above. The numerical data from our experiments directly show that the positional features are more active for the improved heuristics when few stacks are used and that the tactical features to a certain extent take over when the number of stacks increase.

A second part of the contribution at the methodological level is that we demonstrate how RL can be used to find an optimal combination of heuristics for a given problem where the optimal combination might be different at the various stages when instances are processed. In other words, we present a technique for constructing a hybrid heuristic by forming an optimal dynamic combination of component heuristics. It is very important to note that the RL approach – by the nature of RL – adapts to the stochastic environment that generates the instances and that the approach is a generic approach not restricted to stacking.

1.2 Related Work

Demange et al. [5] develop lower and upper bounds for the competitive ratio for online stacking algorithms in the context of assigning trains to tracks at a train station and Demange and Olsen [4] present some improvements both for the offline and online case. Simple heuristics for online stacking are presented by Borgman et al. [2], Duinkerken et al. [6], Hamdi et al. [7], and Wang et al. [15]. More details on these heuristics will be treated later in a separate section in this paper. Olsen and Gross [10] have constructed a polynomial time algorithm for online stacking and shown that the competitive ratio for the algorithm converges to 1 in probability assuming that the arrival times and departure times for the items are picked uniformly at random.

The stacking problem is a hard problem to solve. The offline version where all information on future items is available is NP-hard for stack capacity $h \geq 6$ [3] and it is also NP-hard for unbounded stack capacity [1]. To the best of our knowledge, the computational complexity for the case $2 \leq h \leq 5$ is an open problem for the offline case. Tierney et al. [14] show that it is possible to decide in polynomial time whether the items can be stacked using a *fixed* number of bounded capacity stacks but the running time of their offline algorithm is very high even if the fixed number of stacks is small. For the online case, the competitive ratio is unbounded for *any* online algorithm for unbounded stack capacity as shown by Demange et al. [5].

The stacking problem also has applications within the steel industry as demonstrated by the work of Rei and Pedroso [12] and König et al. [9] and the shipping industry as shown in the PhD-thesis of Pacino [11] on container ship stowage. Finally, we mention that Hirashima et al. [8] use Q-learning for container transfer scheduling in a container terminal.

The preliminaries are presented in Sect. 2 and our RL approach is covered in Sect. 3 followed by Sect. 4 where we document our experiments and discuss the results.

2   Preliminaries

An item to be stored is represented by an interval $[x, y]$ where x is the arrival time and y is the departure time for the item. Two intervals $[x_1, y_1]$ and $[x_2, y_2]$ are said to *overlap* if we are not allowed to put the corresponding items in the same stack: $x_1 < x_2 < y_1 < y_2$ or $x_2 < x_1 < y_2 < y_1$.

In order to provide a clear and unambiguous definition, the stacking problem is formulated as a coloring problem where we add an additional natural constraint for the stacking height represented by the number h:

Definition 1. *The h-STACKING problem:*

- *Instance: A set of n intervals $\{I_1, I_2, \ldots, I_n\}$.*
- *Solution: A coloring of the intervals using a minimum number of colors such that the following two conditions are satisfied:*
  1. *Two overlapping intervals receive different colors.*
  2. *For any real number z and any color c there will be no more than h intervals with color c containing z as an interior point.*

A stacking instance consisting of 8 items/intervals with stacking height constraint $h = 2$ is shown in Fig. 1[1]. An optimal solution using 3 colors is shown as well.

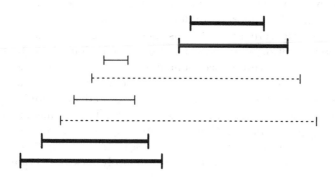

Fig. 1. A stacking instance with $n = 8$ and $h = 2$. An optimal solution uses 3 colors. There is more than one optimal solution.

For the online version of the problem, the intervals are colored in increasing order with respect to the arrival times using no information on future intervals.

2.1   Reinforcement Learning

We will put our problem into a Reinforcement Learning (RL) context in order to use simple RL tools to improve heuristics previously presented in the literature. The RL introduction and notation in this paper is based on the work by Sutton and Barto [13] and the reader is referred to this book for more details.

The online case is considered so the intervals are – as mentioned earlier – presented to a stacking agent in chronological order defined by the arrival times. If two items arrive at the same time, then the item with the latest departure will be presented first to allow the agent to put the items in the same stack if the agent decides to do so. The stacking environment is modelled by an episodic Markov Decision Process (MDP) as follows:

- A state s consists of an arriving item and a coloring of all items that have arrived earlier. The arriving item is not colored in the state s. There is also an initial state modelling the beginning of time when the storage area is empty and a terminal state that is reached when the last item has been processed.

---

[1] Figure from [10].

- An action a represents the color/stack that is assigned to the arriving item.
- The reward r for an action is 0 for all actions except for the final action where r is the number of colors used multiplied by -1 (we want to use a low number of colors). We do not use discounting ($\gamma = 1$).

The transition probabilities depend on the way the stacking instances are generated and we use a model-free approach that does not require information on these probabilities.

The agent uses a policy π_Θ defined by a parameter vector $\Theta \in \mathbb{R}^d$ to take decisions. As mentioned earlier, we consider policies that are based on features where we for every state s and action a has a feature vector $x(s, a) \in \mathbb{R}^d$ containing d features extracting information for taking the action a given state s. Every time we mention a feature, the feature is containing information on the coloring or stacking configuration appearing after taking an action a in a state s and the argument (s, a) to a feature is occasionally omitted to improve the readability. The objective is to find a value for Θ maximizing the expected reward for π_Θ.

We consider two types of policies, stochastic and deterministic, that are guided by linear combinations of features $\Theta x(s, a)^T$. The stochastic policies have probability $\pi_\Theta(a|s)$ of taking action a at state s where π_Θ is computed applying the softmax function on the vector with values $\Theta x(s, a)^T$, $a \in \mathcal{A}$, where $\mathcal{A}$ is the set of actions. The deterministic policies simply pick the action a with the highest value of $\Theta x(s, a)^T$ for any state s.

The overall strategy of our paper is to use RL tools to develop good stochastic policies that converge to deterministic policies improving heuristics earlier presented in the literature.

2.2 Stacking Heuristics

If we decide to put an arriving item on top of another item, it is intuitively appealing to try to improve the chance to do the same for the items to follow. It seems natural to improve this chance by placing the arriving item on an item leaving as early as possible (after the arriving item). Heuristics guided by this fundamental idea are presented several places in the literature [2,7,15] where the top priority is to assign an arriving item to a non-empty stack and the secondary priority is to place the new item on an item leaving early.

In terms of RL, these heuristics can be viewed as partly controlled by a feature, TIL (Top Interval Length), defined as the sum of the remaining time for the items that are situated at the top of a stack,

$$TIL = \sum_{\text{stacks}} (y_{top} - x_{new}),$$

where y_{top} is the departure time for the item on top after taking action a in state s and x_{new} is the arrival time for the new item. The other feature that controls these heuristics is the number of stacks, st, currently in use after taking action a in state s. It is very important to stress that the features are computed for the stacking configuration appearing if action a was taken for every possible action

$a \in \mathcal{A}$. To use the terminology and notation from above, these heuristics can be seen as deterministic policies using the feature vector $x = (TIL, (-st))$ and parameter vector $\Theta = (1, M)$ where M is a sufficiently big number.

Duinkerken et al. [6] propose heuristics controlled by a feature, RSC (Remaining Stack Capacity), that is similar to the TIL-feature with the exception that the heights of the stacks are taken into account. In our implementation, the remaining time for the top item is multiplied with the difference between the stack capacity h and the current height of a stack - before the sum is taken:

$$RSC = \sum_{stacks} (h - h_{stack})(y_{top} - x_{new}).$$

The RSC-heuristic can be viewed as a deterministic policy using $x = (RSC, (-st))$ and the same Θ as above. It should be noted that the TIL-heuristics and RSC-heuristics presented here seem to perform quite well.

3   The RL Approach

The underlying policies for the heuristics presented so far are defined by a linear combination of features:

$$\pi \sim w_{TIL} \cdot TIL + w_{RSC} \cdot RSC + w_{st} \cdot (-st), \tag{1}$$

where the w-weights are fixed constants. As the author of this papers sees it, the $(-st)$-feature is a tactical feature and the TIL- and RSC-features are positional features implying that the heuristics presented above assign a very high weight to the tactical feature at all stages of the stacking process.

Now imagine an item arriving to a storage area at a time with only a few stacks in use and imagine that the arriving item will only stay for a short time while all other items have to stay for a long time. What would be the most sensible action to take in this case? The best thing to do seems to be to ignore the new item and assign it to an empty stack. This would allow us to build up a positional advantage for other items arriving in the near future. To be more specific, we would increase the chance of assigning incoming items to stacks with items with similar departure times.

These considerations suggest that w_{st} should be smaller at quiet stages for the storage area and lead us to the following key question: Will our heuristics improve if the w-weights are allowed to vary as a function of the number of stacks in use?

To answer this question, we seek an optimal stochastic policy π_Θ defined by $\Theta x(s, a)^T$, $\Theta \in \mathbb{R}^9$, where the feature vector x consists of 9 component features as follows where the number of intervals, n, is used for rescaling:

$$x = \left(TIL, \left(\tfrac{st}{n}\right) TIL, \left(\tfrac{st}{n}\right)^2 TIL, RSC, \left(\tfrac{st}{n}\right) RSC, \left(\tfrac{st}{n}\right)^2 RSC, -st, \left(\tfrac{st}{n}\right)(-st), \left(\tfrac{st}{n}\right)^2 (-st)\right).$$

Let $\Theta = (\theta_1, \theta_2, \ldots, \theta_9)$. The right hand side of (1) now matches $\Theta x(s, a)^T$ with w_{TIL}, w_{RSC}, and w_{st} as quadratic polynomials in st for fixed n:

$$w_{TIL} = \theta_1 + \theta_2 \left(\frac{st}{n}\right) + \theta_3 \left(\frac{st}{n}\right)^2$$

$$w_{RSC} = \theta_4 + \theta_5 \left(\frac{st}{n}\right) + \theta_6 \left(\frac{st}{n}\right)^2$$

$$w_{st} = \theta_7 + \theta_8 \left(\frac{st}{n}\right) + \theta_9 \left(\frac{st}{n}\right)^2 .$$

In this way, we can directly see how the optimal mix between tactical and positional features vary at the different stages of the stacking process as the number of stacks in use, st, changes.

We use the REINFORCE algorithm that is a simple and well-known episodic Monte-Carlo Policy-Gradient algorithm to develop a stochastic policy π_Θ with a high expected reward. The TIL-heuristic is used as a baseline. We simulate the MDP by repeatedly generating random instances. When an instance is generated, it is processed using the current policy π_Θ (starting in the initial state) until the final item has been colored (ending at the terminal state). Every time we end up at the terminal state, we update Θ taking the observed reward and the values of the observed features into account. When we update a component θ_i of Θ, we change θ_i by $\Delta\theta_i$ where $\Delta\theta_i$ is proportional to the reward obtained and a number measuring how active the corresponding feature was when processing the instance – this number, which might be negative, measures the difference between the feature values observed and the expected values for that feature. After the update, we generate a new instance and process it by running the MDP from the initial state again. When many instances have been processed, θ_i will end up having a high value if the corresponding feature leads the heuristic in the right direction. The reader is referred to [13] for the details of the REINFORCE algorithm that are beyond the scope of this paper.

4   Experiments

We have carried out two experiments where the arrival times and departure times have been picked uniformly at random from the interval $[0, 1]$. For an interval representing an item, we simply pick two numbers and let the arrival time and departure time be the smallest and largest number, respectively. In order to allow the possibility to have two intervals with the same arrival time or departure time we have rounded the numbers to two decimal places. For the first experiment, we have been fixing the number of items to 30 and the height capacity of a stack to 4: $n = 30$ and $h = 4$. For the second experiment, we have been using $n = 100$ and $h = 5$.

The REINFORCE algorithm is converging slowly and the experiments were carried out on a simple MacBook Pro using Jupyter Notebooks/Python so the training took several days. We believe that the time for training could be significantly reduced using stronger hardware and more advanced RL algorithms.

4.1   Positional vs. Tactical Features

The stochastic policies developed by the REINFORCE algorithm for both of the experiments tell an interesting story about the optimal mix of positional and tactical features for the different stages at the stacking process. As explained earlier, our RL approach allows us to express the weights of the features in (1), w_{TIL}, w_{RSC}, and w_{st}, as quadratic polynomials of the number of stacks in use. Figure 2 displays a graph with the weights obtained for the experiment with $n = 100$ and $h = 5$.

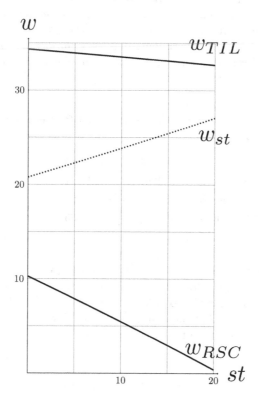

Fig. 2. The relationship between the numbers of stacks, st, in use at the storage area and the weights appearing in (1) for the policy developed for the experiment with $n = 100$ and $h = 5$. The positional features are displayed using solid curves and the tactical feature is using a dashed curve.

The coefficients for the quadratic terms are relatively small producing almost linear relationships. It is very clear that the positional features play a bigger role when only a few stacks are used in the storage area and that the tactical feature comes more into play when the storage area is busy. The RSC-feature is the positional feature with the most drastic weight change and this can maybe be explained by observing that the RSC-feature is a bit more strategic than the TIL-feature – at least as the authors sees it. The TIL-feature appears to be a good all-round feature to use at all stages. Similar results are obtained for the experiment with $n = 30$ and $h = 4$.

4.2   Performance of the New Heuristics

For both of the experiments, we have turned the stochastic policies produced by the RL approach into deterministic policies and compared the corresponding heuristics to a TIL-heuristic and RSC-heuristic implemented as described in Sect. 2.2. We have generated 1 million random instances for each of the experiments and computed the average number of colors/stacks used for the three types of heuristics for each experiment. It is very important to note that these instances have not been used for training when simulating the MDP. The results of the experiments are shown in Table 1.

Table 1. The average number of colors/stacks used for the heuristics for 1 million instances.

	$n = 30, h = 4$	$n = 100, h = 5$
TIL-heuristic	7.336	16.215
RSC-heuristic	7.351	16.228
Our heuristic	7.254	15.651

The TIL-heuristic performs marginally better than the RSC-heuristic while our new heuristic uses an average number of stacks that is 1.1% and 3.5% better than the TIL-heuristic for the experiments with $n = 30$ and $n = 100$, respectively. It should be noted that such improvements could imply significant savings in real world applications within logistics and that our new heuristic is very simple to implement.

5   Conclusion

By using RL techniques, we have developed simple and efficient heuristics for online stacking that perform better than heuristics earlier presented in the literature for the specific stochastic environment considered. The heuristics are trained by simulating the stacking environment to obtain an optimal mix of tactical and positional features.

The numbers from our experiments directly demonstrate that online algorithms can benefit from letting the balance between positional and tactical features vary as instances are processed. Our RL approach produces a hybrid heuristic combining the component heuristics in an optimal and dynamical way. By the nature of RL, our RL approach is adaptive to the stochastic environment generating instances and it is not restricted to stacking so it seems very interesting to investigate whether it can improve heuristics for other problems in other environments.

The performance of the online stacking heuristics can probably be improved using more advanced RL techniques like for example actor-critic methods involving neural networks. Our results also suggest that it maybe is possible to develop a powerful and simple universal online stacking heuristic based on linear functions for the weights for the features.

References

1. Avriel, M., Penn, M., Shpirer, N.: Container ship stowage problem: complexity and connection to the coloring of circle graphs. Discret. Appl. Math. **103**(1–3), 271–279 (2000). https://doi.org/10.1016/S0166-218X(99)00245-0
2. Borgman, B., van Asperen, E., Dekker, R.: Online rules for container stacking. OR Spectr. **32**(3), 687–716 (2010). https://doi.org/10.1007/s00291-010-0205-4
3. Cornelsen, S., Stefano, G.D.: Track assignment. J. Discret. Algorithms **5**(2), 250–261 (2007). https://doi.org/10.1016/j.jda.2006.05.001
4. Demange, M., Olsen, M.: A note on online colouring problems in overlap graphs and their complements. In: Rahman, M.S., Sung, W.-K., Uehara, R. (eds.) WALCOM 2018. LNCS, vol. 10755, pp. 144–155. Springer, Cham (2018). https://doi.org/10.1007/978-3-319-75172-6_13
5. Demange, M., Stefano, G.D., Leroy-Beaulieu, B.: On the online track assignment problem. Discret. Appl. Math. **160**(7–8), 1072–1093 (2012). https://doi.org/10.1016/j.dam.2012.01.002
6. Duinkerken, M.B., Evers, J.J.M., Ottjes, J.A.: A simulation model for integrating quay transport and stacking policies on automated container terminals. In: Proceedings of the 15th European Simulation Multiconference (ESM2001) (2001)
7. Hamdi, S.E., Mabrouk, A., Bourdeaud'Huy, T.: A heuristic for the container stacking problem in automated maritime ports. IFAC Proc. Vol. **45**(6), 357–363 (2012). https://doi.org/10.3182/20120523-3-RO-2023.00410
8. Hirashima, Y., Takeda, K., Harada, S., Deng, M., Inoue, A.: A Q-learning for group-based plan of container transfer scheduling. JSME Int. J. Ser. C **49**(2), 473–479 (2006). https://doi.org/10.1299/jsmec.49.473
9. König, F.G., Lübbecke, M., Möhring, R., Schäfer, G., Spenke, I.: Solutions to real-world instances of PSPACE-complete stacking. In: Arge, L., Hoffmann, M., Welzl, E. (eds.) ESA 2007. LNCS, vol. 4698, pp. 729–740. Springer, Heidelberg (2007). https://doi.org/10.1007/978-3-540-75520-3_64
10. Olsen, M., Gross, A.: Probabilistic analysis of online stacking algorithms. In: Corman, F., Voß, S., Negenborn, R.R. (eds.) ICCL 2015. LNCS, vol. 9335, pp. 358–369. Springer, Cham (2015). https://doi.org/10.1007/978-3-319-24264-4_25
11. Pacino, D., Jensen, R.M.: Fast generation of container vessel stowage plans: using mixed integer programming for optimal master planning and constraint based local search for slot planning. Ph.D. thesis, IT University of Copenhagen (2012)
12. Rei, R.J., Pedroso, J.P.: Tree search for the stacking problem. Ann. OR **203**(1), 371–388 (2013). https://doi.org/10.1007/s10479-012-1186-2
13. Sutton, R.S., Barto, A.G.: Reinforcement Learning: An Introduction, 2nd edn. The MIT Press, Cambridge (2018)

14. Tierney, K., Pacino, D., Jensen, R.M.: On the complexity of container stowage planning problems. Discret. Appl. Math. **169**, 225–230 (2014). https://doi.org/10.1016/j.dam.2014.01.005

15. Wang, N., Zhang, Z., Lim, A.: The stowage stack minimization problem with zero rehandle constraint. In: Ali, M., Pan, J.-S., Chen, S.-M., Horng, M.-F. (eds.) IEA/AIE 2014. LNCS (LNAI), vol. 8482, pp. 456–465. Springer, Cham (2014). https://doi.org/10.1007/978-3-319-07467-2_48

Power Consumption Estimation in Data Centers Using Machine Learning Techniques

Emmanouil Karantoumanis[ID] and Nikolaos Ploskas[✉][ID]

Department of Electrical and Computer Engineering,
University of Western Macedonia, 50100 Kozani, Greece
{dece00034,nploskas}@uowm.gr

Abstract. Large data centers consume large amounts of electricity. Estimating the energy consumption in a data center can be of great importance to data centers administrators in order to know the energy-consuming tasks and take actions for reducing the total energy consumption. Smart workflow mechanisms can be built to reduce the energy consumption of data centers significantly. In this paper, we are investigating the factors that affect the energy consumption of scientific applications in data centers. We also use eight machine learning methods to estimate the energy consumption of multi-threaded scientific applications. Extensive computational results on a computer with 20 cores show that the CPU usage is the most important parameter in the power consumed by an application. However, better results can be obtained when the CPU utilization is combined with other parameters. We generate various regression models that predict the energy consumption of an application with an average accuracy of 99%. Simpler models with one and two parameters can achieve comparable accuracy with more complex models. We also compare various machine learning methods for their ability to obtain accurate predictions using as few parameters as possible.

Keywords: Data centers · Energy consumption · Machine learning · Cloud computing

1 Introduction

Cloud computing has evolved in the last decade to become the technological backbone for most modern enterprises. With the increase of the data hosted in data centers and the applications ported in them, new larger data centers are needed to meet the demands of the users. However, data centers consume large amounts of electricity. For example, Google's data center used about 2.26 million MW hours of electricity to run its operations in 2010 [10]. Energy consumption increased by 90% from 2000 to 2005, but only by 4% from 2010 to 2014, and this is due to the optimization of energy consumption that most data centers apply [6]. In addition, the total carbon dioxide emissions of the Information and

© Springer Nature Switzerland AG 2020
I. S. Kotsireas and P. M. Pardalos (Eds.): LION 14 2020, LNCS 12096, pp. 195–200, 2020.
https://doi.org/10.1007/978-3-030-53552-0_20

Communication Technology (ICT) sector keep increasing. The carbon dioxide emissions of the ICT sector are equal to those of the aviation sector [1]. An average data center consumes energy equivalent to 25,000 households and the environmental impact of data centers was estimated at 116.2 million tons of carbon dioxide in 2006 [13,15]. Therefore, new solutions to minimize the energy consumption of data centers are of great importance.

Various works have studied the energy consumption problem in data centers. Some researchers focused on the workload and consumption prediction [9,11,14], while others use methods to estimate the energy consumption in virtual machines or servers [2,12,16]. Various input parameters have been used to model the energy consumption: (i) CPU, (ii) cache, (iii) disk, (iv) DRAM, (v) network, and (vi) maximum number of open sockets. In addition, linear regression models, neural networks, and Gaussian mixture models have been utilized to model power consumption.

In this paper, we are interested on estimating the energy consumed by a scientific application running in a data center. Various works [2,3] have shown that the CPU usage is the most important parameter that affecting the power consumption of a computer. In this paper, we aim to improve the prediction accuracy by investigating whether or not other input parameters (e.g., memory usage, memory size, disk size, etc.) can be used to predict the power consumption. Being able to estimate the energy consumption of an application, task scheduling mechanisms can be built to reduce the total energy consumption of a data centers. We are investigating the factors that affect the power consumption of scientific applications. Various machine learning methods are compared in terms of their ability to obtain accurate predictions of the power consumed by an application. Simpler regression models with one and two parameters are proposed.

2   Computational Results

The experiments were performed on a computer with an Intel Xeon CPU E5-2660 v3 (2 CPUs - 10 cores each) and 128 GB of main memory, a clock of 2.6 GHz, an L1 code cache of 32 KB per core, an L1 data cache of 32 KB per core, an L2 cache of 256 KB per core, and an $L3$ cache of 24 MB, running under Centos 7 64-bit.

In this work, we aim to investigate the parameters that affect the power consumed by an application. The parameters that we considered are the following:

– number of cores (nc)
– CPU usage (cu)
– memory size (ms)
– memory usage (mu)
– disk size (ds)
– total number of transfers per second (dt)
– total amount of data written to devices in blocks per second (dw)
– total number of network requests (nn)

- total number of kilobytes received per second (nr)
- total number of kilobytes transmitted per second (nt)

In order to estimate the energy consumption of an application, we used the stress-ng [8] tool to stress CPU, RAM and disk, and the ab [4] tool to stress network. We utilized sar [5] to collect the values of the aforementioned ten input parameters in each second and powerstat [7] to collect the power consumed (Watts). A total of 1,000 runs were performed with different combinations for the number of threads, the memory usage, the disk usage and the network consumed by the application (in this simulation, the application is the stress-ng and the ab tools that stress the CPU, RAM, disk and network). Each experiment was run for 80 s and the instantaneous value of each input and output parameters was stored in each second. Afterwards, we eliminated the first ten and the last ten values and we calculated the average of the remaining 60 values.

We used eight regression methods from scikit-learn to estimate the power consumption:

1. Ordinary least squares linear regression (LinearRegression)
2. Lasso regression (Lasso)
3. Ridge regression (Ridge)
4. Epsilon-support vector regression (SVR)
5. Decision tree regression (DecisionTreeRegressor)
6. Random forest regression (RandomForestRegressor)
7. Regression based on k-nearest neighbors (KNeighborsRegressor)
8. Multi-layer Perceptron regression (MLPRegressor)

We use 70% (700 samples) of data to train each model, and the rest 30% (300 samples) for testing the model. We use 10-fold cross validation to test the accuracy of each model. To evaluate the performance of our model, we use R-squared (R^2, coefficient of determination) that provides an estimate of the strength of the relationship between a regression model and the dependent variable (output). Table 1 presents the results of all regressors using (i) all input parameters (ten parameters), (ii) only the CPU usage (cu) as input (one parameter), and (iii) all combinations of the CPU usage parameter with all other parameters (two parameters). The second column shows the R^2 scores that each regressor achieved with all ten parameters. The third column shows the R^2 scores that each regressor achieves with the cu parameter as single input, while the fourth column shows the best R^2 score with two parameters, one of which is always the CPU usage parameter. All regressors, except from SVR and MLPRegressor, achieve high accuracy when using all parameters as input. The best performing regressor is the RandomForestRegressor with a score of 99.44%. Equation 1 is the best model that was obtained from the three linear regression methods using ten input parameters. As it is obvious, the CPU usage is the most important parameter.

$$Watts = (cu \times 1.14) + (ms \times -1.04e^{-01}) + (ds \times 4.30e^{-01})$$
$$+(dw \times 9.68e^{-06} + (nn \times 2.20e^{-05}) + (nr \times -5.17e^{-05}) + 65.04 \tag{1}$$

Therefore, we investigate the accuracy that can be obtained when using only the CPU usage as input parameter. The accuracy obtained by all regressors, except from SVR and MLPRegressor, is lower than their performance when using all input parameters. However, an accuracy of 97.33% can be obtained. Equation 2 is the best model that was obtained from the three linear regression methods using only the CPU usage as input parameter.

$$Watts = cu \times 1.17 + 64.3 \qquad (2)$$

Finally, we used as input the CPU usage with all other input parameters (nine combinations). The last column in Table 1 shows that very accurate estimations can be made using only two parameters. The disk size (ds) and the total number of transfers per second (dt) are the second most important parameters in predicting the power consumption of an application. Equation 3 is the best model that was obtained from the three linear regression methods using two input parameters.

$$Watts = (cu \times 1.11) + (ds \times 0.49) + 66 \qquad (3)$$

When using only the CPU usage as input parameter, the results for SVR and MLPRegressor are better than when using ten parameters as input. In addition, the rest of the scores is reduced to a very small degree, so CPU usage is the most important input parameter. In most regressors, the error when using one parameter increases by only 2% relative to the error when using ten parameters. When using two parameters, the error increases only by less than 1%. This means that simpler models with one or two parameters can be built to predict the power consumed by an application.

Table 1. R^2 scores for regressors using all (ten), one, and two input parameters.

Regressor	All parameters	One parameter (cu)	Two parameters
LinearRegression	0.9932	0.9725	0.9899
Lasso	0.9939	0.9724	0.9897
Ridge	0.9934	0.9725	0.9899
SVR	−0.1508	0.6479	0.5514
DecisionTreeRegressor	0.9900	0.9689	0.9922
RandomForestRegressor	0.9944	0.9726	0.9938
KNeighborsRegressor	0.9102	0.9733	0.9922
MLPRegressor	−0.4871	0.6281	0.9142

In Fig. 1, we present the scores that each regressor can achieve with different input parameters. The y axis shows the R^2 scores of the regressors and the x axis shows the number of parameters used. Most regressors have similar patterns and their lines overlap because their scores are very close. The KNeighborsRegressor method has a lower accuracy than other regressors when using ten input parameters, but it has a good performance when using one and two input parameters. The SVR and MLPRegressor methods are the worst performers. However, their accuracy scores are significantly improved when using one and two parameters.

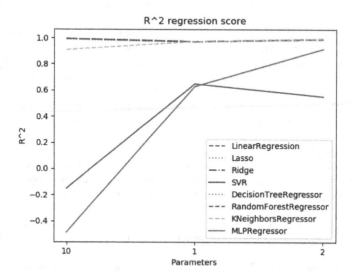

Fig. 1. Accuracy achieved by each regressor using one, two, and ten input parameters.

3   Conclusions

In this paper, we use eight regression methods to predict the power consumption of an application on a computer with 20 cores. Extensive computational results show that the CPU usage is the most important parameter for the energy consumption prediction. We also investigated the accuracy that can be achieved using simpler models. Most regressors are able to achieve a high accuracy score when using one and two parameters. Therefore, simpler models can be utilized to predict the power consumed by an application. In future work, we plan to collect data from different computers and confirm whether or not the models generated on a specific machine can be also used to predict the energy consumption on other machines. We also aim to stress servers with various applications that will be executed concurrently and validate the application of the proposed models. Finally, we will also experiment with tuning the parameters of each regressor in order to further improve their accuracy.

References

1. Avgerinou, M., Bertoldi, P., Castellazzi, L.: Trends in data centre energy consumption under the European code of conduct for data centre energy efficiency. Energies **10**(10), 1470 (2017)
2. Economou, D., Rivoire, S., Kozyrakis, C., Ranganathan, P.: Full-system power analysis and modeling for server environments. In: International Symposium on Computer Architecture. IEEE (2006)
3. Fan, X., Weber, W.D., Barroso, L.A.: Power provisioning for a warehouse-sized computer. ACM SIGARCH Comput. Archit. News **35**, 13–23 (2007)
4. The Apache Software Foundation: ab - apache http server benchmarking tool. https://manpages.ubuntu.com/manpages/disco/en/man1/ab.1.html. Accessed 30 Jan 2020
5. Godard, S.: sar - collect, report, or save system activity information. https://manpages.ubuntu.com/manpages/precise/man1/sar.sysstat.1.html. Accessed 30 Jan 2020
6. Google: Google environmental report 2019. https://services.google.com/fh/files/misc/google_2019-environmental-report.pdf (2019). Accessed 30 Jan 2020
7. King, C.: powerstat - a tool to measure power consumption. http://manpages.ubuntu.com/manpages/xenial/man8/powerstat.8.html (2015). Accessed 30 Jan 2020
8. King, C.: stress-ng - a tool to load and stress a computer system (2016). http://manpages.ubuntu.com/manpages/xenial/en/man1/stress-ng.1.html. Accessed 30 Jan 2020
9. Kumar, J., Singh, A.K.: Workload prediction in cloud using artificial neural network and adaptive differential evolution. Future Gener. Comput. Syst. **81**, 41–52 (2018)
10. Miller, R.: Google's energy story: high efficiency, huge scale. http://www.datacenterknowledge.com/archives/2011/09/08/googles-energy-story-high-efficiency-huge-scale. Accessed 30 Jan 2020
11. Nagothu, K.M., Kelley, B., Prevost, J., Jamshidi, M.: Ultra low energy cloud computing using adaptive load prediction. In: 2010 World Automation Congress, pp. 1–7. IEEE (2010)
12. Öztürk, M.M.: Tuning stacked auto-encoders for energy consumption prediction: a case study. Int. J. Inf. Technol. Comput. Sci. **2**, 1–8 (2019)
13. Rao, G.K.V., Premchand, K.: Scheduling virtual machines across data centres in accordance to availability of renewable sources of energy. Int. J. Eng. Comput. Sci. **5**(10), 18372–18376 (2016)
14. Sinha, A., Chandrakasan, A.P.: Dynamic voltage scheduling using adaptive filtering of workload traces. In: VLSI Design 2001, Fourteenth International Conference on VLSI Design, pp. 221–226. IEEE (2001)
15. Uchechukwu, A., Li, K., Shen, Y.: Energy consumption in cloud computing data centers. Int. J. Cloud Comput. Serv. Sci. **3**(3), 145–162 (2014)
16. Yu, X., Zhang, G., Li, Z., Liangs, W., Xie, G.: Toward generalized neural model for VMS power consumption estimation in data centers. In: ICC 2019–2019 IEEE International Conference on Communications (ICC), pp. 1–7. IEEE (2019)

Automated Tuning of a Column Generation Algorithm

Dario Bezzi[ID], Alberto Ceselli[✉], and Giovanni Righini

Department of Computer Science, University of Milan, Milan, Italy
{dario.bezzi,alberto.ceselli,giovanni.righini}@unimi.it

Abstract. This study concerns the use of automatic classification techniques for the purpose of self-tuning an exact optimization algorithm: in particular, the purpose is to automatically select the critical resource in a dynamic programming pricing algorithm within a branch-and-cut-and-price algorithm for the Electric Vehicle Routing Problem.

Keywords: Column generation · Dynamic programming · Machine learning

1 Introduction

Automated tuning of heuristic algorithms is a well-known research area, whose importance stems from the presence of critical parameters in heuristics and metaheuristics; classical examples are the cooling schedule in simulated annealing, the population size in genetic algorithms and the tabu tenure in tabu search.

However, also exact optimization algorithms, such as branch-and-cut-and-price, are governed by some critical parameters whose value can heavily affect the computational performances, although not the optimality guarantee. In fact, machine learning methods have been successfully applied both in a general context of mathematical programming [20], and specifically in column generation; for instance, in [21] and [22] classification and regression models are used to broaden the application of Dantzig-Wolfe decomposition methods. This study concerns the use of automatic classification techniques for the purpose of self-tuning an exact optimization algorithm: in particular, the purpose is to automatically select the critical resource that must be used in a dynamic programming algorithm to solve the pricing sub-problem within a branch-and-cut-and-price algorithm for an NP-hard routing problem. In fact, preliminary results indicated that performing the right selection makes the difference between converging in seconds, or not reaching optimality in several minutes of computation. We propose two approaches, devising both heuristics, which proceed to partial evaluations of pricing runs, and data driven methods, which instead exploit supervised learning.

Partially funded by Regione Lombardia, grant agreement n. E97F17000000009, Project AD-COM.

ⓒ Springer Nature Switzerland AG 2020
I. S. Kotsireas and P. M. Pardalos (Eds.): LION 14 2020, LNCS 12096, pp. 201–215, 2020.
https://doi.org/10.1007/978-3-030-53552-0_21

The paper is organized as follows. In Sect. 2 we give a formal description of our routing problem; in Sect. 3 we describe the column generation algorithm and the formulation of the pricing sub-problem; in Sect. 4 we give a synthetic description of the bi-directional dynamic programming algorithm which is used to solve it and we discuss the impact of the critical resource selection on the computing time; in Sect. 5 we describe our algorithmic and data driven methods to automatically perform the best critical resource selection; in Sect. 6 we present computational results; conclusions are drawn in Sect. 7.

2   The Problem

The Electric Vehicle Routing Problem (EVRP in the remainder) is a variation of the well-known Vehicle Routing Problem (VRP) in which the fleet is made of electric vehicles (EVs). For a survey on the use of EVs in distribution logistics see Pelletier et al. [14]. The EVRP is by far harder than the classical VRP, both because recharge decisions must be taken in addition to routing decisions and because distance minimization is no longer the only optimization criterion but more complex objective functions must be considered. An up-to-date and comprehensive survey on the EVRP can be found in Keskin, Laporte and Catay [13].

In this paper we consider the EVRP with multiple recharge technologies, first introduced by Felipe et al. [12], where each recharge station may be equipped with one or more recharge technologies and each technology is characterized by a different recharge rate and energy price. Let $\mathcal{G} = (\mathcal{N} \cup \mathcal{R}, \mathcal{E})$ be a given weighted undirected graph whose vertex set is the union of a set $\mathcal{N}$ of N customers and a set $\mathcal{R}$ of R recharge stations. A distinguished station in $\mathcal{R}$ is the depot, where vehicle routes start and terminate. All customer vertices in $\mathcal{N}$ must be visited by a single vehicle; split delivery is not allowed. Each customer $i \in \mathcal{N}$ is characterized by a demand q_i. Multiple visits to stations are allowed as well as partial recharges. A fleet of K identical vehicles with given capacity Q is available; vehicles are equipped with batteries storing up to B energy units. Energy consumption is assumed to be proportional to the distance traveled. The duration of each route is required to be within a given limit T. As opposed to classical vehicle routing problems, where one wants to minimize the overall distance traveled, the objective to be optimized is the overall recharge cost. As the sum of customer demands is a trivial bound to capacity consumption for each route, we set $Q = \sum_{i \in \mathcal{N}} q_i$ if the original value of Q is higher. For more details on the formulation the reader is referred to [7].

3   Branch-and-price

We have developed a branch-and-cut-and-price (BCP) algorithm for the exact optimization of the EVRP with multiple technologies. The BCP algorithm relies upon an extended formulation in which each column in the master problem represents a route that can visit customers as well as recharge stations. The master problem reads as follows.

$$\text{minimize } z = \sum_{r \in \Omega} c_r \theta_r \tag{1}$$

$$\text{s.t.} \sum_{r \in \Omega} x_{ir} \theta_r \geq 1 \qquad \forall i \in \mathcal{N} \tag{2}$$

$$\sum_{r \in \Omega} \theta_r \leq K \tag{3}$$

$$\theta_r \text{ binary} \qquad \forall r \in \Omega \tag{4}$$

A binary variable θ_r corresponds to each feasible route of cost c_r, where Ω indicates the set of feasible routes. The objective function (1) asks for the minimization of the total cost of the selected routes. Covering constraints (2) impose that all customers are visited by at least one route: for this purpose a binary coefficient x_{ir} indicates whether customer i is visited along route r or not. An additional constraint (3) limits the number of routes that can be selected, being K the number of available vehicles.

Such a formulation implies an exponential number of variables θ, one for each feasible route. Therefore its linear relaxation is solved with column generation at every node of a branch-and-bound tree. The pricing sub-problem that must be solved to generate new columns is a Resource Constrained Elementary Shortest Path Problem, where a prize β_i is associated with each customer vertex, being β_i the non-negative dual variable corresponding to constraint $i \in \mathcal{N}$ in the constraint set (2).

The constraints in the pricing sub-problem define feasible routes. In particular, any feasible route respects the following properties: it must correspond to a path in the graph, it must start from the depot and return to it, it must not visit any customer more than once, it must not consume more than Q units of capacity, it must have a duration no longer than T, it must comply with energy constraints and battery capacity limits. The objective of the pricing sub-problem is to compute a route of minimum reduced cost, if any. The reduced cost of a route is equal to its cost decreased by the prizes β_i accumulated when visiting customers.

4 The Dynamic Programming Algorithm

The pricing sub-problem is NP-hard and it is common practice in the VRP literature to address it with dynamic programming. Starting from the depot, nodes are iteratively labelled and suitable dominance rules allow to discard suboptimal labels, thus limiting the combinatorial explosion in the number of labels.

Details of the algorithm are omitted here. The interested reader can refer to the VRP literature (see for instance [6,9–11]) and in particular to [7] and [8] for the EVRP with multiple technologies.

A remarkable speed-up in dynamic programming algorithms to compute constrained shortest paths can be achieved by bi-directional extension of labels [15–17]. This allows to generate shorter paths with respect to mono-directional extension. The reduction in path length usually pays off, because the number of labels grows much faster than linearly with the path length. In bi-directional dynamic programming, path extension proceeds in two opposite directions, forward and backward, and it is fundamental to stop the extension of forward and backward paths as soon as there is the guarantee that the remaining part of any feasible route can be generated in the opposite direction, so that no feasible solution can be lost.

This test is done on the consumption of a *critical resource*: when half of the available amount of the critical resource has been consumed along a path, then its extension is stopped.

In the EVRP, the pricing sub-problem takes into account three main resources: the amount of demand served, because of the constraint on vehicles capacity; the time needed to travel along the path, because of the constraint on the maximum duration of routes; the energy consumption along the path, because of the constraints on battery capacity. The energy cannot be used as a critical resource, because it can be recharged along a route and therefore routes are not guaranteed to be decomposable in two paths along which no more than half of the battery capacity is consumed. This guarantee holds for time and capacity: so, these are the two resources that can be selected as critical.

It may happen that neither time nor capacity are really binding; in these cases bi-directional propagation becomes inefficient. In order to overcome this problem, the number of customers visited in a route is also considered as an additional resource, that becomes critical when the others are not. This prevents the extension of paths when $\lfloor N/2 \rfloor$ customers have already been visited.

The importance of guessing the right critical resource can be seen from Table 1, where we have reported the number of labels examined and the computing time required by the bi-directional dynamic programming algorithm according to the selection of the critical resource (T stands for time, Q stands for capacity, N stands for number of customers). A wrong guess may lead to an increase in computing time of orders of magnitude. It must be remarked that this guess must be repeated as many times as the number of calls to the pricing sub-routine, which typically occurs some dozen times for each node in the branch-and-cut-and-price tree that in turn may contain up to some hundreds thousands nodes.

From these preliminary results one can appreciate the importance of a reliable and fast technique to automatically select the critical resource in order to keep the overall computing time under control.

5   Automatic Selection of the Critical Resource

The task of choosing the right critical resource requires at first a modeling step. In fact, we may define as the *best* critical resource that yielding convergence of the

Table 1. Computing time (sec.) for each resource in single pricing iterations

Dataset	Instance	Iteration	N	Q	T
A	$C101-5$	2	0.0037	0.0021	0.0033
A	$C101-10$	5	4.43	3.76	0.91
A	$C103-15$	1	1.07	1.11	0.61
A	$C103-15$	17	42.8	16.2	1.72
A	$C208-15$	14	5.41	5.98	7.70
B	$10-N10$	1	225	11.5	0.004

pricing algorithm to an optimal solution in the minimum amount of CPU time. Such a definition, however, can hardly be used in the design of a computational method for finding it.

We therefore propose two options. The first one is to consider the number of labels which are produced by extension operations as a proxy for the overall CPU time required to the pricing algorithm to converge. Indeed, the problem of computing the number of labels produced by different choices of the critical resource is as hard as the application of the initial definition; therefore we designed heuristics to provide a reliable estimate of it. Such heuristics still rely on dynamic programming to algorithmically solve a strongly restricted problem at each pricing iteration. We refer to this approach as *partial inspection*, and we detail it in Subsect. 5.1.

The second approach is to assume that features on instance and pricing data exist, which are predictive of the best critical resource. Under this assumption we experimented on data-driven methods, modeling the choice of the best critical resource as a supervised learning one. We considered several pricing problem instances arising during EVRP column generation. For each such an instance we created a data object, measuring and recording several features. We also ran three versions of the pricing algorithm, each using a different critical resource, recording the CPU time they took to reach proven optimality. Finally, we assigned the data object a label corresponding to either T, Q or N, whichever resource was corresponding to the fastest pricing algorithm version. The set of data objects is then used to train classification models, mapping pricing instance features to critical resource labels. The classification model can then be invoked during the optimization of new EVRP instances to predict at each column generation iteration the best critical resource. We detail this approach in Subsect. 5.2.

5.1   Selection by Partial Inspection Algorithms

Intuitively, when the best choice is made for the critical resource, a huge number of label extensions are stopped during each run of the dynamic programming algorithm. We also expect such a phenomenon to become evident already in the early stages of dynamic programming. We experimented with heuristics which

Algorithm 1. Partial Inspection Heuristics.

$\overline{N}, \overline{Q}, \overline{T}, t \leftarrow 0$
Queue $\leftarrow$ {empty label}
while $t <$ L **do**
    $\mathcal{L} \leftarrow$ Queue$[t]$
    if $\mathcal{L}$ is an open label **then**
        for Each node i reachable from $\mathcal{L}$ **do**
            Try to extend $\mathcal{L}$ to i creating $\mathcal{M}$
            if $\mathcal{M}$ is blocked **then**
                if $\mathcal{M}$ blocked by N constraint (more than N/2 customers) **then**
                    $\overline{N} \leftarrow \overline{N} + 1$
                if $\mathcal{M}$ blocked by Q constraint (demand exceeds Q/2) **then**
                    $\overline{Q} \leftarrow \overline{Q} + 1$
                if $\mathcal{M}$ blocked by T constraint (time exceeds T/2) **then**
                    $\overline{T} \leftarrow \overline{T} + 1$
        else
            Add $\mathcal{M}$ to Queue
            Perform dominance
    $t \leftarrow t + 1$
if $\overline{N} = \overline{Q} = \overline{T}$ **then**
    return Q
else
    return $\texttt{argmax}(\overline{N}, \overline{Q}, \overline{T})$

---

perform dynamic programming without necessarily performing a complete extension phase. In fact (a) we forbid the extension of each label consuming more than half of *any* resource (b) we stop once a certain number L of labels have been generated. Once this limited dynamic programming is over, we guess the best critical resource to be that stopping the highest number of extension operations. A complete pseudo-code of our heuristics is given as Algorithm 1.

Note that, given our choice of RCESPP dynamic programming algorithm, a single extension step could fail even if it does not violate any resource constraint; when it happens, no counter is incremented.

The partial inspection algorithm uses Q as a tiebreaker when all of the critical resources block the same number of extensions, which experimentally only happens when $\overline{N} = \overline{Q} = \overline{T} = 0$. This choice is motivated by the fact that the maximum capacity consumption along a route has a trivial bound (we cannot consume more than the sum of customer demands), while the time constraint has not, and therefore we expect in extreme cases $Q/2$ to be more reliable than $T/2$ as a stopping condition.

5.2 Selection by Data Driven Models

The use of heuristics has both appealing and negative features. On one side, they do not require any training, and are ready to be applied to datasets of unknown structure. On the other side, running them consumes time and might therefore increase the overall computing effort; furthermore, they assume the number of labels which are generated in a limited setting to be a reliable proxy for the computing time of a whole pricing iteration, which may or may not be true. In fact, in preliminary experiments we observed that a few instances exist in which not even the overall number of labels is directly correlated with pricing time.

Therefore, we designed the following data driven approach. We assume a dataset of *base instances* to be given. We also assume to perform *simulated runs* on them, performing the full column generation process with *three versions* of the pricing algorithms, one for each possible choice of the critical resource.

For each base instance, and each pricing iteration in its column generation process, we create a data object, which is composed by the following features.

Base Instance Features. We include the number of customers (N), vehicles (K) and recharge stations (R).

Mix of Base Instance Features. We include also the following:

N-normalized Number of customers divided by number of vehicles: $\tilde{N} = N/K$
Q-normalized $\tilde{Q} = \sum_{i \in \mathcal{N}} q_i / Q$
T-normalized $\tilde{T} = \sum_{i \in \mathcal{N}} (s_i + d_{i0}) / T$ (using depot tech for estimating time)

Iteration Features. We include the **pricing iteration counter** as a feature as well, which trivially changes over different pricing iterations of the same base instance.

Iteration and Dual Features. The following values may change over different pricing iterations in non-trivial ways. They are related to the dual solution values. In fact, we argue that the structure of a dual solution might provide insights into the computational behaviour of a pricing algorithm. To give an example, each dual variable having very low value strongly suggests that visiting the associated customer would not pay off; therefore, a high percentage of dual variables having low values suggests optimal routes to be short (thereby making N a bad choice for the critical resource). Let β_i be the dual variables corresponding to constraints (2). In order to obtain promising features, we measure the following:

Beta-mean mean value of β_i
Beta-variance variance of β_i
Low-N Percentage of customers having $\beta_i > 0.1 \sum \beta_i$
Mid-N Percentage of customers having $\beta_i > 0.2 \sum \beta_i$
High-N Percentage of customers having $\beta_i > 0.3 \sum \beta_i$
Low-Q Percentage of customers having $q_i \beta_i > 0.1 \sum (q_i \beta_i)$
Mid-Q Percentage of customers having $q_i \beta_i > 0.2 \sum (q_i \beta_i)$
High-Q Percentage of customers having $q_i \beta_i > 0.3 \sum (q_i \beta_i)$
Low-T Percentage of customers having $(s_i + d_{i0})\beta_i > 0.1 \sum (s_i + d_{i0})\beta_i$
Mid-T Percentage of customers having $(s_i + d_{i0})\beta_i > 0.2 \sum (s_i + d_{i0})\beta_i$
High-T Percentage of customers having $(s_i + d_{i0})\beta_i > 0.3 \sum (s_i + d_{i0})\beta_i$

Class Label. Additionally, after all the three pricers are over, we set the following:

Class either N, Q or T, depending on which pricer is fastest •

5.3 Machine Learning Methods

Then, we exploit the dataset obtained in this way to build classification models with a supervised learning mechanism. That is, we aim to predict class labels (and therefore the best critical resource) by measuring the set of base instance, mix, iteration and iteration dual features. We experimented with many classification models. Indeed, the accuracy is not the only important factor in the choice of the best model, as we may be interested in trading accuracy for ease of embedding in a final branch-and-cut-and-price solver and query efficiency. The following classes were part of our experimental campaign: Bayesian classifier, simple decision tree built by information gain induction heuristics and support vector machines with RBF kernel [2]. These models are indeed all easy to directly implement in custom code. As a comparison term, we experimented with a state-of-the-art random forest model trained by gradient boost, although the embedding of the resulting model in custom code (without calling external libraries) is more involved.

6   Computational Results

Base Instance Datasets. We did our experiments on two datasets.

Dataset A was derived in [18] from the Solomon dataset, by relaxing the time windows constraints: instances have up to 15 customers (the last part of the name indicates the size of each instance) and 5 stations with a single technology. For some instances in this dataset we also modified the number of vehicles with respect to the original value used in [18]. In one case this was done to make the instance feasible, because the original one was not [19]. In some other cases we decreased the number of vehicles to the minimum value for which the instance was known to be feasible [12].

In dataset B, instances have 10 customers, up to 5 vehicles, up to 9 stations and 3 technologies. As a preliminary result of our investigation, we found all instances in B to have T (time) as critical resource, for each pricing iteration, so we report detailed results for this dataset only in Sect. 6.1.

In our experiments we employ a state-of-the-art column generation implementation: the pricing algorithms, as well as the partial inspection heuristics have been implemented in C++; the full column generation algorithm uses SCIP 6.0.2 [1] as a general framework. We set the following time limits: 7200 s for the full column generation process of each base instance, 600 s for each single run of a pricing algorithm 300 s for the label extension phase of every pricing run (that is, we always reserve at least 300s for joining). At each column generation iteration, each pricer might return a set of columns of negative reduced cost (those not hitting the timeout contain an optimal one). In order to keep a reasonable master effort, we keep only at most 100 columns for each pricer (those having minimum reduced cost). That is, at most 300 columns are added to the master problem at each column generation iteration, at least one of which being optimal. We remark that in our experimental setting, each simulation runs three pricing algorithms at each column generation iteration, two of which using a wrong choice of the critical resource, so the overall column generation time in these experiments is not representative of the time required by our procedure in a final implementation (while the time for each pricer does).

The supervised learning models were implemented in R. In particular:

- R Package *e1071* [4] was used to build the Bayesian classifier. Function naive-Bayes() was called with default parameters;
- R Package *rpart* [3] is used to build the decision tree. Function rpart() is called to construct the tree, and the tree with the minimum prediction error is selected;
- R Package *1071* [4] is used to build the support vector machines classifier, calling function svm() with default parameters;
- R Package *xgboost* [5] is used to build the random forest classifier, calling function xgboost() with parameters max.depth = 3, nrounds = 50 (leaving others as default).

The results were obtained using a PC equipped with an i7-6700K 4.0GHz processor and 32 GB of RAM, running Linux Ubuntu 18.

6.1 Profiling Partial Inspection Heuristics

Setup. The heuristic method described in 5 is called twice for every pricing iteration, one with a small limit for the number of labels ("fast" heuristic) and one with a significantly larger limit ("slow" heuristic). "Fast" heuristic sets $L = 500(N + R)$. "Slow" heuristic sets $L = 10000(N + R)$, 20 times more than the "fast" method. The algorithm remains unchanged: labels are tentatively extended until half of one of the resources is consumed, and stopped labels are counted.

In Table 2 we report our results. The table is composed by two blocks, for the slow and fast heuristic respectively. In each block we report the mean ($\pm$ variance) computing time, in seconds (variance values smaller to 10^{-4} are rounded to 10^{-4}), and the accuracy of the heuristic.

One row is included for each of the following sets of base instances:

A-5 Set of instances having 5 customers belonging to dataset A.
A-10 Set of instances having 10 customers belonging to dataset A.
A-15 Set of instances having 15 customers belonging to dataset A.
B Set of all instances belonging to dataset B (having 10 customers).

Each entry represents average values of the instances in the corresponding set.

Table 2. Comparing fast heuristic to slow one

Set	Slow		Fast	
	Time (s)	Accuracy	Time (s)	Accuracy
A-5	0.0008 ± 0.0001	0.96	0.0008 ± 0.0001	0.96
A-10	0.45 ± 3.50	0.66	0.080 ± 0.0037	0.80
A-15	24.11 ± 5339	0.74	0.165 ± 0.010	0.89
B	0.0009 ± 0.0001	0.80	0.0009 ± 0.0001	0.80

Results. For very small instances (like those in set A-5), there are no differences between the two heuristics. When the instance is so small, the label limit L is almost never reached even by fast heuristic, so both of them can extend the whole set of labels (and predict the critical resource with high accuracy). A similar behaviour can be observed for dataset B, where time constraint are so strict that halving them leads to drastically reduce feasible extensions, and both heuristics can proceed until no open label is available. By converse, the accuracy is not as high as A-5 rows: it's possible to predict the wrong label even with a fully extended queue.

For A-10 and A-15 rows, the most surprising result is that the fast heuristic seems more accurate than the slow one. This particular behaviour suggests again that simply counting the stopped labels is not always an accurate proxy. In details, we found that when T is the critical resource, its effect appears clearly already in early extension operations. At the same time, when this is not the case, the partial inspection heuristics often reach termination before hitting any critical consumption of either Q or T. Subsequently, the slow heuristic resorts in predicting N, often being wrong. The fast heuristic, instead, according to the tiebreaking condition, correctly predicts Q.

6.2 Classification in Synthetic Settings

The first experiment is synthetic. It investigates the results obtained by the four classifiers described in Sect. 6 in a basic training-and-testing configuration, and compares them to the partial inspection algorithm described in Sect. 5.

We randomly split the dataset (325 rows), using 2/3 data objects for training and 1/3 for testing. That is, representative pricing iterations of each base EVRP instance are expected to be in both training and testing datasets. Each of the four classifiers described in Sect. 6 split the dataset the same way. Each classifier is permitted to classify according to the features described in Subsect. 5.2. In the original dataset, 133 rows are labeled T, 188 are labeled Q and only 4 are labeled N. The results on the test set are shown in Table 4. Results on the training set are given as a reference in Table 3. Both tables have 6 rows, one for each classification method. Last two rows are included for comparison with partial inspection heuristics. Accuracy (Acc.), Precision (Pr.) and Recall (Rec.)

metrics are reported in each table for each classifier, lying in range $[0-1]$. A "-" symbol means that the classifier never predicted class N.

Table 3. Classification in synthetic settings - training set

Method	Accuracy	Precision			Recall		
		N	Q	T	N	Q	T
Bayesian	0.78	0.10	0.87	0.92	1.00	0.75	0.82
DTree	0.93	0.00	0.95	0.93	–	0.93	0.93
SVM	0.94	0.00	0.95	0.95	–	0.95	0.92
Xgboost	1.00	1.00	1.00	1.00	1.00	1.00	1.00
Slow H.	0.74	1.00	0.57	0.99	0.06	0.99	0.92
Fast H.	0.88	0.67	0.84	0.94	0.13	0.95	0.92

Table 4. Classification in synthetic settings - test set

Method	Accuracy	Precision			Recall		
		N	Q	T	N	Q	T
Bayesian	0.81	0.06	0.90	1.00	1.00	0.76	0.88
DTree	0.97	0.00	1.00	0.96	–	0.95	1.00
SVM	0.95	0.00	0.98	0.94	–	0.94	0.98
Xgboost	0.95	0.00	0.97	0.96	–	0.95	0.98
Slow H.	0.77	0.00	0.61	0.98	–	0.97	0.92
Fast H.	0.85	0.00	0.86	0.85	–	0.86	0.98

Results. A first unexpected result is that classifiers perform in general better than partial inspection heuristics. The Bayesian classifier, despite having the worst accuracy, is the only one who correctly predicts N (in testing). We remark that data objects actually classified N are quite rare (in this random split, only 3 for training and 1 for testing). The other three classifiers give similar performances: none of them predicts N, but other data objects are classified with really high accuracy. It is worth noting that Decision Tree and SVM fully discard N: they have Precision 0 for N even during training.

Features Importance. The Decision Tree structure built by R during this experiment is indeed a simple *if* instruction, which classifies according to the **T-normalized** feature only. Under a certain threshold it predicts Q, otherwise it predicts T (data objects with N critical do exist, but they are discarded as outliers). Still, finding a suitable threshold value requires the statistical evaluation of the training set. From the point of view of branch-and-cut-and-price design, this is an appealing phenomenon: it means that base instances could be classified in preprocessing, without any overhead at runtime.

In order to have a more clear picture, in Table 5 we report the relative importance of each feature in the XGBoost model.

Table 5. Features importance in XGBoost.

Feature	Gain	Cover	Frequence
T-normalized	0.8066	0.163	0.132
Beta-variance	0.0518	0.282	0.236
Beta-mean	0.0396	0.145	0.162
K	0.0267	0.014	0.014
N-normalized	0.0200	0.076	0.078
Mid-N	0.0111	0.015	0.017
Iteration	0.0107	0.067	0.064
N	0.0096	0.022	0.030
R	0.0084	0.020	0.067
Mid-Q	0.0045	0.029	0.040
Low-N	0.0028	0.015	0.030
Low-Q	0.0025	0.023	0.027
High-T	0.0021	0.028	0.023
High-N	0.0017	0.052	0.030
Mid-T	0.0009	0.024	0.017
High-Q	0.0004	0.022	0.017
Low-T	0.0002	0.001	0.003
Q-normalized	0.0002	0.002	0.010

The **T-normalized** feature remains the most important one by far, especially in terms of relative gain it produces when used in a node of a tree. At the same time, the mean and variance of dual variables contribute to the efficacy of the method: such a contribution reflects in the values of cover and frequency. No other feature seems relevant.

6.3   Classification in Realistic Settings

The second experiment compares classifier performances in a realistic operation mode. That is, we split the dataset in blocks, each containing all and only the rows corresponding to a single base EVRP instance. Experiments are always performed without mixing different blocks during training and testing, thereby simulating a setting in which training is performed on a certain dataset, and then the models are used to perform predictions on new, previously unknown base EVRP instances.

Experimental Setup. We employ a leave-one-out approach, iteratively removing one of the blocks described above from the dataset, performing training on the remaining blocks, and testing on the removed one. Since dataset A contains rows corresponding to 36 base EVRP instances, 36 iterations are needed for each classifier, and average results are measured. As before, each classifier is permitted to classify according to the features described in Subsect. 5.2.

Our average results on the testing blocks are reported in Table 7. For reference, in Table 6 we report also the average results during training. Both tables have the same structures of those in the previous experiment.

Table 6. Classification in realistic settings - training

Method	Accuracy	Precision			Recall		
		N	Q	T	N	Q	T
Bayesian	0.86	0.67	0.66	0.66	0.89	0.90	0.90
DTree	0.94	0.64	0.64	0.64	0.95	0.95	0.95
SVM	0.95	0.64	0.64	0.64	0.95	0.95	0.95
Xgboost	1.00	1.00	1.00	1.00	1.00	1.00	1.00
Slow H.	0.75	0.77	0.77	0.78	0.65	0.65	0.65
Fast H.	0.87	0.75	0.75	0.77	0.65	0.65	0.65

Table 7. Classification in realistic settings - testing

Method	Accuracy	Precision			Recall		
		N	Q	T	N	Q	T
Bayesian	0.79	0.70	0.68	0.65	0.70	0.66	0.72
DTree	0.88	0.71	0.75	0.77	0.95	0.86	0.83
SVM	0.85	0.64	0.70	0.78	0.81	0.78	0.74
Xgboost	0.92	0.67	0.73	0.86	0.81	0.87	0.92
Slow H.	0.81	0.64	0.69	0.84	0.74	0.72	0.76
Fast H.	0.89	0.68	0.78	0.85	0.77	0.83	0.84

Results. This experiment further clarifies the behaviour of the methods. Here, all classifiers are able to predict N (at least for some of the 36 runs). As expected, their mean precision over Q and T decreases. However, it does not drop significantly, especially for Bayesian and XGBoost classifiers. Partial inspection heuristics are competitive in this setting. Indeed, this confirms the intuition that such dedicated heuristics could perform better than general purpose classifiers on a real setting.

As an overall final computational insight, we envisage a cascading approach: at first approximation, a choice of critical resource performed during preprocessing according to base instance features only can be enough. If timeouts are observed during the pricing iterations, very fast queries to an external classifier provide accurate predictions in a large share of cases. Finally, if timeouts are still observed, running partial inspection heuristics can be of further help.

7   Conclusions

Our investigation produced interesting insights. First, partial inspection heuristics provide good accuracy in detecting the best critical resource; in particular, their results are accurate even if very tight bounds are imposed to their computing time and number of labels which are allowed to be processed. In fact, we argue that their biasing towards the best critical resource is more crisp in their earlier phases, tending to a more balanced effect as their computation proceeds.

For what concerns the data driven methods, we have designed a few representative features, and analyzed several classification models from the literature, exploiting their values to predict the best critical resource. Some of these features are related only to the base EVRP instance data; others measure properties of the dual values during a single pricing iteration.

In general, data driven models appear superior to partial inspection heuristics in our experiments.

One feature in particular turns out to be always more significant than the others: it is a (rough) estimate of expected time spent in visiting each customer, expressed in relative terms w.r.t. the time limit T. In fact, an unexpected result is the following: very good accuracies can even be reached by correctly estimating a split point over that single feature, and performing classifications using such a single check. Indeed that feature is the only one in which we encoded graph structure, and therefore we expect even better results by a careful design of further similar graph-dependent features.

Including features which depend on dual values, and therefore change at each pricing iteration, proves beneficial when training and testing is performed on different base EVRP instances, thereby indicating that better generalization can be achieved by including them.

Overall, our experiments indicate that integrating simple preprocessing, data driven models and partial inspection heuristics is possible and fruitful.

References

1. Gleixner, A., et al.: The SCIP optimization suite 6.0, July 2018. Available at Optimization Online and as ZIB-Report 18–26. http://nbn-resolving.de/urn:nbn:de:0297-zib-69361. Accessed 28 Jan 2019
2. Han, J., Kamber, M., Pei, J.: Data Mining: Concepts and Techniques, 3rd edn. Morgan Kaufmann Publishers, Waltam (2012)

3. Therneau, T., Atkinson, B.: rpart: recursive partitioning and regression trees. R package version 4.1-15 (2019). https://CRAN.R-project.org/package=rpart. Accessed 22 Jan 2019
4. Meyer, D., Dimitriadou, E., Hornik, K., Weingessel, A., Leisch, F.: e1071: Misc Functions of the Department of Statistics, Probability Theory Group (Formerly: E1071), TU Wien. R package version 1.7-2. https://CRAN.R-project.org/package=e1071. Accessed 22 Jan 2019
5. Chen, T., He, T., Benesty, M.: xgboost: extreme gradient boosting. R package version 0.4-2. https://CRAN.R-project.org/package=xgboost. Accessed 22 Jan 2019
6. Baldacci, R., Mingozzi, A., Roberti, R.: New route relaxation and pricing strategies for the vehicle routing problem. Oper. Res. **59**, 1269–1283 (2011)
7. Bezzi, D.: Algoritmo di ottimizzazione per l'Electric Vehicle Orienteering Problem. Master degree thesis, University of Milan (2017)
8. Bezzi, D., Ceselli, A., Righini, G.: Dynamic programming for the electric vehicle orienteering problem with multiple technologies. In: Odysseus 2018, Cagliari, Italy (2018)
9. Christofides, N., Mingozzi, A., Toth, P.: Exact algorithms for the vehicle routing problem, based on spanning tree and shortest path relaxations. Math. Program. **20**, 255–282 (1981). https://doi.org/10.1007/BF01589353
10. Desaulniers, G., Errico, F., Irnich, S., Schneider, M.: Exact algorithms for electric vehicle-routing problems with time windows. Oper. Res. **64**(6), 1388–1405 (2016)
11. Feillet, D., Dejax, P., Gendreau, M., Gueguen, C.: An exact algorithm for the elementary shortest path problem with resource constraints: application to some vehicle routing problems. Networks **44**, 216–229 (2004)
12. Felipe Ortega, A., Ortuño Sánchez, M.T., Righini, G., Tirado Domínguez, G.: A heuristic approach for the green vehicle routing problem with multiple technologies and partial recharges. Transp. Res. Part E **71**, 111–128 (2014)
13. Keskin, M., Laporte, G., Çatay, B.: Electric vehicle routing problem with time-dependent waiting times at recharging stations. Comput. Oper. Res. **107**, 77–94 (2019)
14. Pelletier, S., Jabali, O., Laporte, G.: Goods distribution with electric vehicles: review and research perspectives. Transp. Sci. **50**(1), 3–22 (2016)
15. Righini, G., Salani, M.: Symmetry helps: bounded bi-directional dynamic programming for the elementary shortest path problem with resource constraints. Discret. Optim. **3**, 255–273 (2006)
16. Righini, G., Salani, M.: New dynamic programming algorithms for the resource-constrained elementary shortest path problem. Networks **51**, 155–170 (2008)
17. Righini, G., Salani, M.: Decremental state space relaxation strategies and initialization heuristics for solving the orienteering problem with time windows with dynamic programming. Comput. Oper. Res. **36**, 1191–1203 (2009)
18. Schneider, M., Stenger, A., Goeke, D.: The electric vehicle-routing problem with time windows and recharging stations. Transp. Sci. **48**, 500–520 (2014)
19. Schneider, M.: Personal Communication (2014)
20. Khalil, E.B.: Machine learning for integer programming. In: Proceedings of the Twenty-Fifth International Joint Conference on Artificial Intelligence (2016)
21. Kruber, M., Lübbecke, M.E., Parmentier, A.: Learning when to use a decomposition. In: Salvagnin, D., Lombardi, M. (eds.) CPAIOR 2017. LNCS, vol. 10335, pp. 202–210. Springer, Cham (2017). https://doi.org/10.1007/978-3-319-59776-8_16
22. Basso, S., Ceselli, A., Tettamanzi, A.: Random sampling and machine learning to understand good decompositions. Ann. Oper. Res. **284**, 501–526 (2018). https://doi.org/10.1007/s10479-018-3067-9

Pool-Based Realtime Algorithm Configuration: A Preselection Bandit Approach

Adil El Mesaoudi-Paul[1], Dimitri Weiß[2], Viktor Bengs[1(✉)], Eyke Hüllermeier[1], and Kevin Tierney[2]

[1] Heinz Nixdorf Institute and Department of Computer Science, Paderborn University, Paderborn, Germany
`viktor.bengs@upb.de`
[2] Decision and Operation Technologies Group, Bielefeld University, Bielefeld, Germany
{`dimitri.weiss,kevin.tierney`}`@uni-bielefeld.de`

Abstract. The goal of automatic algorithm configuration is to recommend good parameter settings for an algorithm or solver on a per-instance basis, i.e., for the specific problem instance being solved. Realtime algorithm configuration is a practically motivated variant of algorithm configuration, in which the problem instances arrive in a sequential manner and high-quality configurations must be chosen during runtime. We model the realtime algorithm configuration problem as an extended version of the recently introduced contextual preselection bandit problem. Our approach combines a method for selecting configurations from a pool of candidates with a surrogate configuration generation procedure based on a genetic crossover procedure. In contrast to existing methods for realtime algorithm configuration, the approach based on contextual preselection bandits allows for the incorporation of problem instance features as well as parameterizations of algorithms. We test our algorithm on different realtime algorithm configuration scenarios and find that it outperforms the state of the art.

Keywords: Realtime algorithm configuration · Contextual preselection bandit

1 Introduction

It is widely known that no single solver produces optimal performance for all types of problem instances [11]. Due to the large space of parameterizations available for most solvers, algorithm designers are forced to tune the parameters of their approaches to provide reasonable performance, a long and arduous process. Recently, automatic algorithm configuration has simplified the search for good parameters by *automatically* identifying and recommending high-quality parameters to algorithm designers and users. Furthermore, these approaches can work in an *instance specific* fashion, providing high-quality parameters specific to the

© Springer Nature Switzerland AG 2020
I. S. Kotsireas and P. M. Pardalos (Eds.): LION 14 2020, LNCS 12096, pp. 216–232, 2020.
https://doi.org/10.1007/978-3-030-53552-0_22

instances being solved [31] including for instances never before seen or envisioned by the algorithm designer.

Traditionally, algorithm configuration research has focused on offline, train-once approaches [2,23,24]. These are based on a static set of instances deemed to be representative of a specific task for which good parameter settings are found and after which the parameters are put into practice. Ignoring the fact that a representative set of instances may not be available for a given problem at the time of algorithm design, train-once methods suffer from changes or *drift* of problem instances. Furthermore, there may be a lack of time for repeated offline training. Online or *realtime* approaches, such as the ReACT and ReACTR systems of [17] and [16], respectively, overcome these issues by choosing a high-quality configuration of the solver during runtime as new instances arrive.

In this paper, we propose an algorithm called Contextual Preselection with Plackett-Luce (CPPL) for realtime algorithm configuration. The latter is based on the contextual preselection bandit problem, a variant of the preference-based multi-armed bandit (MAB) problem [12] recently introduced in [6]. MAB methods have proven successful for the closely related problem of algorithm selection, see e.g. [14,19,22,33,38], but have not yet been applied to algorithm configuration.

The state-of-the-art realtime algorithm configurator ReACTR [16] utilizes the racing principle in which, for a given instance, multiple parameter configurations are run in parallel on multiple CPU cores to see which is best for a given instance. ReACTR gathers information about the performance of parameterizations and applies a ranking mechanism called TrueSkill [21] to decide which choice of parameterizations to select from a pool of options for the next instance. Furthermore, it uses TrueSkill to decide which parameterizations to replace with new ones and how to generate new parameterizations, meaning which individuals to choose as parents for the crossover mechanism included in the generation of new parameterizations. Our goal in this paper is to replace the ranking, choice and generation mechanisms of ReACTR by functions of the CPPL approach while maintaining the framework of the racing principle and parallel execution of ReACTR. Our contributions are as follows:

1. We connect the online contextual preselection bandit setting to the realtime algorithm configuration problem.
2. We introduce the CPPL algorithm for solving realtime algorithm configuration tasks.
3. We provide a novel technique for generating new parameterizations using the surrogate of the CPPL algorithm inspired by the idea of genetic engineering in [1].
4. We show experimentally that CPPL is competitive with the state-of-the-art ReACTR algorithm on different algorithm configuration scenarios.

The paper is organized as follows. In the next section, we give an overview of related work. The online contextual preselection setting and its application to realtime algorithm configuration are presented in Sect. 3. An experimental study is then presented in Sect. 4, prior to concluding the paper in Sect. 5.

2   Related Work

We now provide an overview of the related work in several fields: algorithm configuration, algorithm selection using MABs, and hyperparameter optimization for machine learning algorithms.

2.1   Algorithm Configuration

Approaches for offline algorithm configuration can be broadly categorized into non-model-based, model-based and hybrid approaches. Non-model-based techniques including racing algorithms such as F-Race [9] and its successor Iterated F-Race [10]. In these approaches, parameterizations are run against each other on subsets of instances and those that are performing poorly are eliminated through statistical testing. In the gender-based genetic algorithm (GGA) of [2] and its extension [1] a racing mechanism is also used to run instances in parallel and cut off poor performing parameterizations before they use too much CPU time (in a runtime minimization setting). The ParamILS method [24] employs a focused iterated local search. Sequential Model-based Algorithm Configuration (SMAC) approach by [23] is a model-based method that generates a response surface over parameters of the solver to predict the performance if those parameters were to be used. Finally, [1] can be considered a hybrid approach for configuration, utilizing both a genetic algorithm and a learned forest to select new parameterizations.

Realtime methods include two approaches. The previously mentioned ReACT algorithm [17] stores a set of parameterizations and runs them in parallel on the instances to be solved. Parameterizations are removed if they do not "win" enough of the races with the other parameterizations and are replaced with random parameterizations. ReACT's extension ReACTR [16] enhances ReACT by incorporating the ranking technique of the ranking system TrueSkill and uses a crossover mechanism (as in genetic algorithms) to generate new parameterizations. Both [17] and [16] show that tackling the realtime configuration problem head-on instead of periodically performing a new offline configuration results in better performance.

A recent contribution from [7] introduces the concept of dynamic algorithm configuration, in which the goal is to dynamically configure the parameters of an algorithm while it is running. The problem is modeled as a contextual Markov decision process and reinforcement learning is applied. Algorithm configuration has been examined from a theoretical perspective in several recent works, e.g. [5, 28] resulting in bounds on the quality of configuration procedures.

2.2   Algorithm Selection and Hyperparameter Optimization with Bandits

The K-armed bandit problem is a sequential decision problem where in each trial a learner needs to choose/pull one of the arms of a K-armed slot machine (*bandit*), with each arm having its own reward distribution. After pulling an arm the learner observes its associated reward. The goal of the learner is to maximize the sum of rewards or, equivalently, minimize the total loss defined as the expected difference between the sum of rewards associated with the optimal selection strategy with the best possible reward and the sum of the collected rewards. The main issue affecting the long-term success of a learner in such problems is to find a good trade-off between choosing arms that have turned out to be appealing in the past (exploitation) and trying other, possibly even better arms, but whose appeal is not precisely known so far (exploration).

The problem of algorithm selection can be represented as a bandit problem in a straightforward fashion. The arms are the algorithms, and pulling an arm corresponds to running the associated algorithm on the current problem instance and observing its runtime. The objective is to minimize the total solution time over all problem instances.

Bandit approaches have a long history of success for algorithm selection. In the following, we list some of the works following this approach for different applications. In [40] the authors model the adaptive time warp control problem in the single-agent setting as a MAB problem. A MAB framework for managing and adapting a portfolio of acquisition functions in the context of Bayesian optimization with Gaussian processes is introduced in [22]. In [33] the authors model the problem of scheduling of searches with different heuristics as a MAB problem and use dynamic Thompson sampling (DTS) to solve it. Finally, in [4] a MAB-based learning technique is used to automatically select among several levels of propagation during search for adaptive constraint propagation.

Algorithm configuration is also related to hyperparameter optimization, in which the goal is to find a configuration of hyperparameters that optimize some evaluation criterion based on a given set of hyperparameters associated with some supervised learning task and a search space over them [27]. In hyperparameter optimization, the algorithm is generally a machine learning method rather than an optimization technique. The key difference to algorithm configuration is that hyperparameter optimization generally ignores algorithm runtimes and focuses on improving the quality (i.e., the output) of an approach. Nevertheless, a problem of hyperparameter optimization can be modeled as a bandit problem in the same way.

A number of MAB approaches have also been applied for hyperparameter optimization by casting it into a pure exploration nonstochastic infinite-armed bandit problem. In this way, several well-known bandit algorithms can be leveraged to the hyperparameter optimization problem such as successive halving [27,29], upper confidence bound [39] or Thompson sampling [37].

All the MAB approaches mentioned above use quantitative feedback in the form of absolute numerical rewards (e.g., runtime of a solver or accuracy of a learning method). In contrast, we use a preference-based variant of the MAB problem, in which the feedback is purely qualitative. We merely observe which algorithm performed best among a subset of algorithms on a given problem instance. This setting is a generalization of the dueling bandits problem introduced in [41].

3   Contextual Preselection for Realtime Algorithm Configuration

In this section, we present details of our approach to realtime algorithm configuration. After a brief introduction to this problem in Sect. 3.1, we recall the (contextual) preselection problem with winner feedback in Sect. 3.2. In Sect. 3.3, we then explain how realtime algorithm configuration can be cast as a problem of that kind.

3.1   Realtime Algorithm Configuration

Formally, offline algorithm configuration can be defined as follows [24]. We are given a set of problem instances $\Pi \subseteq \hat{\Pi}$ for which we want to find a good parameter setting λ from a parameter space Λ for an algorithm $\mathcal{A}$. We can evaluate the runtime or quality of some set of parameters using a performance metric $m : \Pi \times \Lambda \to \mathbb{R}$. Without loss of generality, the goal is to find some λ using the instances Π such that $\sum_{\pi \in \hat{\Pi}} m(\pi, \lambda)$ is minimal, i.e., a parameterization λ that performs well on average (across a set of problem instances).

Notice that in offline algorithm configuration we are provided a large set of instances up front and only test the quality of the parameter configuration we find in an offline setting. In contrast, *realtime* algorithm configuration considers Π to be a *sequence* of instances that are solved one after the other. Furthermore, we assume that the underlying distribution of instances $\hat{\Pi}$ is not fixed and may change over time, meaning that the parameters tuned offline may no longer be effective. This setting is a realistic one, for example for logistics companies that must solve routing problems on a daily basis for changing customers or manufacturing companies that have new sets of orders each day. Furthermore, as companies grow, shrink or adjust their business model, $\hat{\Pi}$ could drastically change and require updating.

Figure 1 shows the realtime algorithm configuration process for our approach, and those in the literature. Given a pool of parameterizations λ_1 through λ_n, we must select a limited number according to our parallel computing resources. We run the first instance on the parameterizations and, in the runtime setting, see which one finishes first. The other parameterizations are immediately stopped. We note that in ReACT and ReACTR no information is gained from these runs, i.e., the data we receive is censored. The parameterization pool and a model of how to select the parameterizations is updated and the process is repeated for the next instances.

In this paper, we make use of a recently introduced variant of the multi-armed bandit problem, called the preselection bandit problem [6], which is able to exploit (censored) "winner feedback" as described above, i.e., information about which parameterization among a finite set of parameterizations solved a problem first. Moreover, making use of a *contextualized* extension of preselection bandits [15], we are able to recommend parameterizations of a solver on a *per-instance* basis, i.e., parameterizations that do not only perform well on average but are specifically tailored to the concrete problem instance at hand.

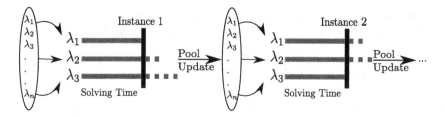

Fig. 1. Realtime algorithm configuration using the racing principle.

3.2 The Contextual Preselection Problem with Winner Feedback

The online contextual preselection problem with winner feedback, as introduced in [15], is a sequential online decision problem, in which, in each time step $t \in \{1, 2, \ldots\}$, a learner is presented a context $\mathbf{X}_t = (\boldsymbol{x}_{t,1} \ldots \boldsymbol{x}_{t,n})$. Each $\boldsymbol{x}_{t,i} \in \mathbb{R}^d$ is associated with one of n different arms, which we shall identify by $\{1, \ldots, n\}$, and contains properties of the arm itself as well as the context in which the arm needs to be chosen. After the context is revealed, the learner selects a subset $S_t \subseteq [n] = \{1, \ldots, n\}$ of $k < n$ distinct arms. Then, it obtains as feedback the top-ranked arm/winner among these arms. In each time step t, the goal of the learner is to select a subset S_t of arms that contains the best possible arm for the current context $\mathbf{X}_t$.

We consider the contextual preselection problem under the (contextualized) Plackett-Luce (PL) model [30,34], a parametrized probability distribution on the set of all rankings over a finite set of n choice alternatives. The PL model is defined by a parameter $\boldsymbol{v} = (v_1, \ldots, v_n)^\top \in \mathbb{R}_+^n$, where each v_i represents the weight or "strength" of the i-th alternative. The probability of a ranking $\boldsymbol{r}$ of the choice alternatives under the PL model is given by

$$\mathbb{P}(\boldsymbol{r} \mid \boldsymbol{v}) = \prod_{i=1}^n \frac{v_{\boldsymbol{r}^{-1}(i)}}{\sum_{j=i}^n v_{\boldsymbol{r}^{-1}(j)}}.$$

Here, a ranking $\boldsymbol{r}$ is a bijective mapping $\boldsymbol{r} : [n] \to [n]$, with $\boldsymbol{r}(k)$ the rank of the kth alternative and $\boldsymbol{r}^{-1}(i)$ the index of the alternative on position i.

The probability that alternative k gets top-ranked among the alternatives in a subset $S \subset [n]$ under the PL model is

$$\mathbb{P}(\boldsymbol{r}_S(k) = 1 \mid \boldsymbol{v}) = \frac{v_k}{\sum_{i \in S} v_i}. \tag{1}$$

In order to integrate context information $\boldsymbol{x}_i \in \mathbb{R}^d$ about the ith choice alternative, the constant latent utility v_i can be replaced by a log-linear function of the arm's context, in a similar way as has been done in [13,36]:

$$v_i = v_i(\mathbf{X}) = \exp\left(\theta^\top \boldsymbol{x}_i\right), \quad i \in \{1, \ldots, n\}, \tag{2}$$

where we summarize the corresponding feature vectors $x_1, \ldots, x_n$ in a matrix $\mathbf{X} \in \mathbb{R}^{d \times n}$. The PL model with context information is then defined as

$$\mathbb{P}(r \mid \theta, \mathbf{X}) = \prod_{i=1}^{n} \frac{v_{r^{-1}(i)}(\mathbf{X})}{v_{r^{-1}(i)}(\mathbf{X}) + \cdots + v_{r^{-1}(n)}(\mathbf{X})} = \prod_{i=1}^{n} \frac{\exp\left(\theta^{\top} x_{r^{-1}(j)}\right)}{\sum_{j=i}^{n} \exp\left(\theta^{\top} x_{r^{-1}(j)}\right)}.$$

The probability for an alternative $k \in S$ to get the top rank among the alternatives in S is analogous to (1) and in particular, the corresponding log-likelihood function for an observation $(k, S, \mathbf{X})$ is

$$\mathcal{L}(\theta \mid k, S, \mathbf{X}) = \theta^{\top} x_k - \log\left(\sum_{j \in S} \exp\left(\theta^{\top} x_j\right)\right). \tag{3}$$

The gradient and Hessian matrix of the log-likelihood function can be computed easily and are used for deriving confidence bounds $c_{t,i}$ on the contextualized utility parameters v_i as specified in (2). The confidence bounds can be written as $c_{t,i} = \omega \cdot I(t, d, x_{t,i})$, where $\omega > 0$ is some suitable constant and I is a function depending on the gradient as well as the Hessian of the log-likelihood function with respect to the observation at time step t. For technical details, we refer to [15].

Tackling the online contextual preselection problem requires estimating the unknown parameter vector θ and solving the well-known exploration-exploitation trade-off. In [15], the former issue is dealt with using the averaged stochastic gradient descent (SGD) method, and the latter is handled by a variant of the upper confidence bounds (UCB) algorithm. Again, we refer to [15] for a detailed description of the CPPL algorithm for the online contextual preselection problem with winner feedback and explain its adaptation to the realtime algorithm configuration problem in the next section.

3.3 Realtime Algorithm Configuration as Contextual Preselection

We model the realtime algorithm configuration problem as an online contextual preselection problem, where each algorithm parameterization is viewed as an arm, and adapt the CPPL algorithm to solve it. We note here that, because of the parametric form of the model in (2), the number of arms (parameterizations) does not need to be finite, as is normally the case in realtime algorithm configuration.

Our algorithm, called CPPL, is described in Algorithm 1. In the following, we denote the features of a parameterization $\lambda \in \Lambda$ resp. a problem instance $\pi \in \Pi$ by $f(\lambda)$ resp. $f(\pi)$. Note that for the considered problem scenario both the features of the solvers as well as the problem instances are of high dimension and reveal high correlations, so that a principal component analysis (PCA) procedure is conducted to reduce the dimensionality of the features as well as their correlation. Also note that we perform the PCA only on a small portion of the instances (20%) and the initial pool of parameterizations, under the assumption that in a real-world setting, at least some data will be available in advance of starting the system. The obtained PCA transformations are then used to transform every new instance and parameterization features during the run of the algorithm. In the experimental study we analyze different parameter settings for the PCA to determine whether this reduces the performance of our approach or not.

The algorithm is initialized with a random parameter vector $\hat{\theta}_0$ and a random set P of n different parameterizations (line 1), which corresponds to the pool of candidates. The algorithm proceeds in discrete time steps $t = 1, 2, \ldots$. In each step, first a new problem instance $\pi \in \Pi$ is observed (line 5) and then a joint feature map of the instance features $f(\pi)$ and each of the parameterizations $f(\lambda_i)$, $i \in \{1, \ldots, n\}$ is built.

The joint feature map we use is a polynomial of the second degree, which consists of all polynomial combinations of the features with degree less than or equal to 2, i.e., it is defined for two vectors $x \in \mathbb{R}^r$ and $y \in \mathbb{R}^c$ as:

$$\Phi(x, y) = \left(1, x_1, \cdots, x_r, y_1, \cdots, y_c, x_1^2, \cdots, x_r^2, y_1^2, \cdots, y_c^2, x_1 y_1, x_1 y_2, \cdots, x_r y_c\right).$$

With this, the estimated skill parameters (see (2)) are computed by using the current estimate of θ and the joint feature map of the parameterization and problem features (line 7). Following the optimism in the face of uncertainty principle, the k most optimistic parameterizations within the pool of candidates are determined by computing the upper confidence bounds on their estimated skill parameter for the given problem instance (line 9). Generally, k corresponds to the number of the available CPU cores in the machine. These k parameterizations are then used (in parallel) to solve the given problem instance resulting in a winner feedback, i.e., the parameterization among the k used which solved the problem instance first (line 10). This winner feedback information is then used in the subsequent step to update the estimate of θ, following a stochastic gradient descent scheme (line 11). After that, poorly performing parameterizations are discarded from the pool of candidates (lines 12–13). To this end, a racing strategy is adopted [32], in which a parameterization is pruned as soon its upper bound on the estimated skill parameter is below the respective lower bound of another parameterization. Note that this differs from the preference model approach based on TrueSkill used in [16]. Finally, the parameterizations discarded in the last step are replaced by generating new ones according to a genetic approach as described in the following (line 14).

Due to the nature of the CPPL model, it is not possible to directly generate parameterizations from the learned model. Nonetheless, we wish to use the learned model to augment our candidate pool with parameterizations that will be effective on future instances. Thus, we implement a crossover-based approach based on the idea of genetic engineering [1]. We generate individuals/parameterizations using a uniform crossover operator on two individuals ranked as the best by the model. To ensure enough diversification of the search for good parameterizations, we introduce mutation of single genes as well as random generation of individuals with a lower probability. All the newly generated parameterizations are then ranked by the learned model and the best

Table 1. Comparison of CPPL and ReACTR regarding the three most important components of the realtime algorithm configuration.

	Discarding parameterizations	Generating parameterizations	Selecting for Runs
CPPL	Racing strategy	Crossover based on model and randomness, evaluation by model	Evaluation by the model
ReACTR	Based on TrueSkill	Crossover based on TrueSkill and randomness	Based on TrueSkill and random choices

individuals are selected. The parameterizations chosen to be terminated by the model are then replaced by the best parameterizations generated. Differences in approaches of solving the three main components of realtime algorithm configuration between CPPL and ReACTR are summarized in Table 1.

Algorithm 1. CPPL$(n, k, \Pi, \alpha, \gamma, \omega, f)$

1: Initialize n random parameterizations $P = \{\lambda_1, \ldots, \lambda_n\} \subset \Lambda$
2: Initialize $\hat{\theta}_0$ randomly
3: $\bar{\theta}_0 = \hat{\theta}_0$
4: **for** $t = 1, 2, \ldots$ **do**
5:     Observe problem instance $\pi \in \Pi$
6:     **for** $j = 1, \ldots, n$ **do**
7:         $\hat{v}_{t,j} = \exp\left(\boldsymbol{x}_{t,j}^\top \bar{\theta}_t\right)$, where $\boldsymbol{x}_{t,j} = \Phi(f(\lambda_j), f(\pi))$
8:     **end for**
9:     Choose S_t as:
$$\underset{S_t \subseteq [n], \; |S_t| = k}{\operatorname{argmax}} \sum_{i \in S_t} \left(\hat{v}_{t,i} + c_{t,i}\right)$$
10:     Run the parameterizations of S_t to solve π and observe the parameterization $w_t \in S_t$ terminating first
11:     Update $\bar{\theta}_t$ by $\bar{\theta}_t = (t-1)\frac{\bar{\theta}_{t-1}}{t} + \frac{\hat{\theta}_t}{t}$ with $\hat{\theta}_t = \hat{\theta}_{t-1} + \gamma t^{-\alpha} \nabla \mathcal{L}\left(\hat{\theta}_{t-1} | w_t, S_t, \mathbf{X}_t\right)$
12:     Let: $K \leftarrow \{\lambda_i \in P \mid \exists \lambda_j \neq \lambda_i \text{ s.t. } \hat{v}_{t,j} - c_{t,j} \geq \hat{v}_{t,i} + c_{t,i}\}$
13:     $\Lambda \leftarrow \Lambda \setminus K$
14:     Generate $|\Lambda| - |K|$ new parameterizations using the genetic approach as described.
15: **end for**

---

4   Computational Results

In this section, we study the following research questions:

- **RQ1:** What effect does the PCA dimension have on the performance of CPPL?
- **RQ2:** What effect does the parameter ω have on the performance of CPPL?
- **RQ3:** How good is the choice provided by CPPL?
- **RQ4:** How does CPPL compare to ReACTR?

4.1 Datasets and Solvers

We first define the datasets and solvers used in our experiments. We consider three solvers: CPLEX [26], CaDiCaL [8] and Glucose [3]. CPLEX is a mixed-integer programming solver, and CaDiCaL and Glucose are satisfiability (SAT) solvers. The type of parameters of each solver are given in Table 2. We excluded categorical parameters from the configuration by CPPL, since a PCA is employed as part of this method, which is not designed for handling non-numerical variables. For ReACTR, which is not restricted in this regard, we do include all parameters in the configuration.

We choose problem instances to emulate industrial problems with drift. For CPLEX, we use the frequency assignment problem generated by a slightly altered approach from [35]. This dataset contains 1,000 problem instances which are generated by setting the number of cells to 5 and the variance of channel requirements per cell to 1.5. The necessary distance between channels is drawn from a normal distribution, and the mean requirement of channels per cell goes from 8 to 18 in 10 stages. To introduce drift into the data, we change the generation parameters after every 10 instances.

For the SAT solvers, we use two datasets. The first dataset contains 1,000 instances generated with the modularity-based random SAT instance generator [20] by setting it to make instances with 10,000 variables, 60,000 clauses, 4 literals per clause, 600 communities and we vary the modularity factor from 0.4 to 0.35 in 10 stages. The second set of 1,000 instances is generated with the power-law random SAT generator [18]. We make instances with 10,000 variables, 93,000 clauses, 4 literals per clause, 18 as the power-law exponent of variables and the power-law exponent of clauses changing as described before from 12.5 to 2.5. The instance features used are based on [25] for MIP and SAT instances. We choose 32 features for MIP and 54 features for SAT instances.

Table 2. Types of parameters being configured in each solver.

Solver	Real	Categorical	Binary
CPLEX	35	54	6
CaDiCaL	64	29	63
Glucose	15	10	92

4.2   RQ1: Choosing the PCA Dimensions

The goal of this experiment is to determine whether PCA is suitable for reducing the dimensionality of the instance and parameter feature input and determine which dimension to use for our problem settings. To this end, we run different experiments with different values for both parameters on 200 modularity-based SAT instances using the glucose solver. The results are given in Table 3, where it can be seen that a value of 3 for the PCA dimension of instance features and 5 for the PCA dimension of algorithm features lead to the best result. The key takeaway from this experiment is that changing the dimensionality of the PCA does not significantly harm CPPL's performance. We thus use these values in further experiments.

Table 3. Overall runtime of glucose in seconds for solving 200 modularity-based SAT instances using different values for the number of PCA dimension of instance features and algorithm features. The results are averaged over 3 repetitions.

# dim. of parameters	# dim. of features				
	3	5	8	10	12
3	1863.17	1861.63	1860.85	1862.08	1865.77
5	1852.27	1864.28	1866.78	1866.87	1865.32
8	1869.94	1870.34	1866.96	1865.35	1865.19
10	1869.14	1865.32	1862.67	1867.56	1866.22
12	1858.52	1863.92	1861.75	1866.75	1866.13

4.3   RQ2: Choosing ω

We again run similar experiments for glucose to determine good values of the parameter ω, which helps determine the confidence intervals $c_{t,i}$. We note here that a smaller value of ω tightens the confidence bounds of contextualized skill parameters, which in turn leads to more parameterizations being discarded, and vice versa for larger values of ω (see line 12 in Algorithm 1). Figure 2 shows the results of several values of ω. Notice that low and high values of ω only have a very slight effect on the performance. However, with $\omega = 0.001$, we get the best cumulative runtime.

Note, that we conducted similar experiments as in Sect. 4.2 for the parameters α and γ where we found the best performance for the values of 0.2 for α and 1 for γ.

4.4   RQ3: Evaluation of the CPPL Choice

We now compare the performance of the CPPL choice with the choice of ReACTR. For this we use the glucose solver on 1000 modular-based SAT

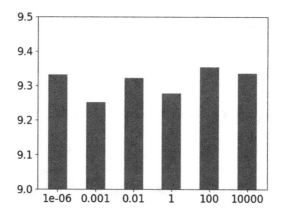

Fig. 2. Average runtime on an instance with CPPL using glucose on 200 modularity-based SAT instances for different values of ω with time in seconds on the y- and ω value on the x-axis.

instances. To have a meaningful comparison, the initial pool of parameterizations and the choice of parameterizations to run on the first instance is fixed to be the same for both approaches. The results given in Fig. 3 show that the CPPL ranking mechanism outperforms the ranking mechanism of ReACTR. We note that even though this result looks somewhat small, without generating new parameters we are dealing with only a randomly generated parameterization pool over the entire run of the algorithm.

4.5 RQ4: Direct Comparison of Performance of CPPL and ReACTR

We now compare the implementations of ReACTR with Glucose, CaDiCaL and CPLEX[1] For each experiment we use 16 Intel Xeon E5-2670 cores running at 2.6 GHz. In all results figures, we display a moving average of the runtimes of the parameterized solvers on the instances. Our results on three different benchmarks are shown in Fig. 4. The experiments with CaDiCaL and Glucose were executed three times and the runtime on each instance was averaged before plotting the comparison. The experiment with CPLEX could only be conducted once due to constrained time.

The computational time needed for ranking the parameterizations, choosing parameterizations that are to be run on the next instance and replacing parameterizations deemed to have bad performance is not included. Although we assume a realtime scenario, the practical applications we have in mind concede enough time between the arrival of instances for adjusting the pool and performing bookkeeping. Although CPPL is significantly more computationally

---

[1] A direct comparison of CPPL and ReACTR is not provided on the Glucose solver with the power-law SAT instance set. Even the first problem instance of this set could not be solved by Glucose within 24 h.

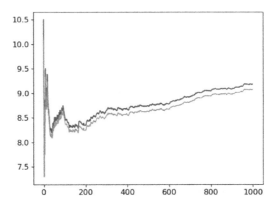

Fig. 3. Rolling average time of CPPL (orange) and ReACTR (blue) both without generating new parameterizations with time in seconds on the y- and number of instances on the x-axis. (Color figure online)

intensive than ReACTR, the time consumed by CPPL for the mentioned operations between instances, for example with CaDiCaL on the power-law SAT instance set, is on average only 0.4 s.

With CaDiCaL (Fig. 4a), CPPL shows considerably better performance than ReACTR on the modular-based SAT instance set. CaDiCaL shows advantageous properties for configuration purposes. The increase in complexity of the problem instances can be overcome by both approaches and the average solution time decreases with both. However, CPPL exploits CaDiCaL's properties further than ReACTR, which amounts in an advantage of approximately 10% of the total solution time compared with ReACTR. Regarding the comparison of the approaches in Fig. 4b, the results show that on the more difficult problem instance set, CPPL achieves better configuration of the solver. The solution time per instance of ReACTR varies in terms of the periodical changes in complexity of the instances. CPPL on the other hand seems to reliably find appropriate parameter configurations and the solution stays on a relatively stable level. The advantage regarding the total solution time of the instance set again amounts to approximately 10%. Figure 4c shows that the increase in difficulty of the instances influences the solution time for both approaches. However, CPPL solves the instances of the dataset about 160 s faster than ReACTR. Considering the average solution time of approximately 10 s per instance for ReACTR, this is an advantage of nearly two percent of the total solution time of the instance set. In Fig. 4d the CPPL approach with CPLEX does not outperform ReACTR. Despite an initial advantage on the instances the performance of the CPPL approach deteriorates. ReACTR outperforms CPPL by around one percent of the total runtime. We note that even if the experiments searching for good parameter values for CPPL showed only very small difference in performance, still choosing all values with best cumulative runtime resulted in a mostly good performance of CPPL.

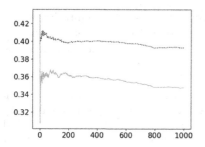

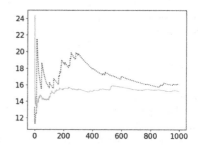

(a) Solver: CaDiCaL; Instances: SAT modular dataset.

(b) Solver: CaDiCaL; Instances: SAT power-law instance SAT set.

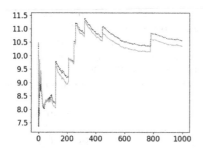

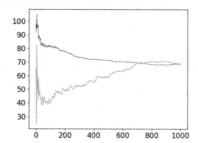

(c) Solver: Glucose; Instances: SAT modular dataset.

(d) Solver: CPLEX; Instances: MIP Frequency Assignment Problem.

Fig. 4. Comparison of CPPL (orange) and ReACTR (dashed blue) using several different solvers and instance sets with time in seconds on the y- and number of instances on the x-axis. (Color figure online)

5   Conclusion and Future Work

In this work, we considered the problem of realtime algorithm configuration and adapted the contextual preselection bandit method with the assumption of a Plackett-Luce ranking probability distribution to solve it, resulting in the CPPL realtime algorithm configurator. The approach allows for incorporating features of problem instances and parameterizations of algorithms and is competitive with the state of the art. Our first experimental results are promising.

In future work, we plan to further elaborate on different components of CPPL in order to fully exploit its potential. The feature engineering part, that is, the joint feature map $\Phi(x, y)$ together with an embedding in a lower-dimensional space, appears to be especially critical in this regard. So far, we used a standard quadratic kernel for Φ and a simple PCA for dimensionality reduction, but there is certainly scope for improvement. Another direction concerns the idea of detecting and reacting to drift in a more active way. So far, an adaptation to changes in the data distribution is only accomplished implicitly through learning in an incremental manner. In the case of abrupt changes, this adaptation is certainly not fast enough. Last but not least, we intend to have a deeper look at the different computational steps of the CPPL algorithm in order to reduce the algorithm's runtime.

Acknowledgements. This work was partly supported by the German Research Foundation (DFG) under grant HU 1284/13-1. This work was also partly supported by the German Federal Ministry of Economics and Technology (BMWi) under grant ZIM ZF4622601LF8. Moreover, the authors would like to thank the Paderborn Center for Parallel Computation (PC^2) for the use of the OCuLUS cluster.

References

1. Ansótegui, C., Malitsky, Y., Samulowitz, H., Sellmann, M., Tierney, K.: Model-based genetic algorithms for algorithm configuration. In: IJCAI (2015)
2. Ansótegui, C., Sellmann, M., Tierney, K.: A gender-based genetic algorithm for the automatic configuration of algorithms. In: International Conference on Principles and Practice of Constraint Programming (CP), pp. 142–157 (2009)
3. Audemard, G.: Glucose and syrup in the SAT race 2015. In: SAT Competition 2015 (2015)
4. Balafrej, A., Bessiere, C., Paparrizou, A.: Multi-armed bandits for adaptive constraint propagation. In: IJCAI (2015)
5. Balcan, M.F., Sandholm, T., Vitercik, E.: Learning to Optimize Computational Resources: Frugal Training with Generalization Guarantees. arXiv preprint arXiv:1905.10819 (2019)
6. Bengs, V., Hüllermeier, E.: Preselection Bandits under the Plackett-Luce model. arXiv preprint arXiv:1907.06123 (2019)
7. Biedenkapp, A., Bozkurt, H.F., Eimer, T., Hutter, F., Lindauer, M.: Dynamic algorithm configuration: foundation of a new meta-algorithmic framework. In: ECAI (2020)
8. Biere, A.: CaDiCaL at the SAT race 2019. In: SAT Race 2019 - Solver and Benchmark Descriptions, p. 2 (2018)
9. Birattari, M., Stützle, T., Paquete, L., Varrentrapp, K.: A racing algorithm for configuring metaheuristics. In: Conference on Genetic and Evolutionary Computation (GECCO), pp. 11–18 (2002)
10. Birattari, M., Yuan, Z., Balaprakash, P., Stützle, T.: F-Race and iterated F-Race: an overview. In: Bartz-Beielstein, T., Chiarandini, M., Paquete, L., Preuss, M. (eds.) Experimental Methods for the Analysis of Optimization Algorithms, pp. 311–336. Springer, Heidelberg (2010). https://doi.org/10.1007/978-3-642-02538-9_13
11. Bischl, B., et al.: ASlib: a benchmark library for algorithm selection. Artif. Intell. **237**, 41–58 (2016)
12. Busa-Fekete, R., Hüllermeier, E., El Mesaoudi-Paul, A.: Preference-Based Online Learning with Dueling Bandits: A Survey. arXiv preprint arXiv:1807.11398 (2018)
13. Cheng, W., Hüllermeier, E., Dembczynski, K.: Label ranking methods based on the Plackett-Luce model. In: ICML, pp. 215–222 (2010)
14. Cicirello, V.A., Smith, S.F.: The max k-armed bandit: a new model of exploration applied to search heuristic selection. In: AAAI (2005)
15. El Mesaoudi-Paul, A., Bengs, V., Hüllermeier, E.: Online preselection with context information under the Plackett-Luce model. arXiv preprint arXiv:2002.04275 (2020)

16. Fitzgerald, T., Malitsky, Y., O'Sullivan, B.J.: ReACTR: realtime algorithm configuration through tournament rankings. In: IJCAI (2015)
17. Fitzgerald, T., Malitsky, Y., O'Sullivan, B.J., Tierney, K.: ReACT: real-time algorithm configuration through tournaments. In: Annual Symposium on Combinatorial Search (SoCS) (2014)
18. Friedrich, T., Krohmer, A., Rothenberger, R., Sutton, A.: Phase transitions for scale-free SAT formulas. In: AAAI, pp. 3893–3899 (2017)
19. Gagliolo, M., Schmidhuber, J.: Algorithm portfolio selection as a bandit problem with unbounded losses. Ann. Math. Artif. Intell. **61**, 49–86 (2011). https://doi.org/10.1007/s10472-011-9228-z
20. Giráldez-Cru, J., Levy, J.: A modularity-based random SAT instances generator. In: IJCAI, pp. 1952–1958 (2015)
21. Guo, S., Sanner, S., Graepel, T., Buntine, W.: Score-based Bayesian skill learning. In: Flach, P.A., De Bie, T., Cristianini, N. (eds.) ECML PKDD 2012. LNCS (LNAI), vol. 7523, pp. 106–121. Springer, Heidelberg (2012). https://doi.org/10.1007/978-3-642-33460-3_12
22. Hoffman, M.D., Brochu, E., de Freitas, N.: Portfolio allocation for Bayesian optimization. In: UAI (2010)
23. Hutter, F., Hoos, H.H., Leyton-Brown, K.: Sequential model-based optimization for general algorithm configuration. In: LION, pp. 507–523 (2011)
24. Hutter, F., Hoos, H.H., Leyton-Brown, K., Stützle, T.: ParamILS: an automatic algorithm configuration framework. J. Artif. Intell. Res. **36**, 267–306 (2009)
25. Hutter, F., Xu, L., Hoos, H.H., Leyton-Brown, K.: Algorithm runtime prediction: methods & evaluation. Artif. Intell. **206**, 79–111 (2014)
26. IBM: CIBM ILOG CPLEX Optimization Studio: CPLEX User's Manual (2016). https://www.ibm.com/support/knowledgecenter/SSSA5P_12.7.0/ilog.odms.studio.help/pdf/usrcplex.pdf
27. Jamieson, K.G., Talwalkar, A.: Non-stochastic best arm identification and hyperparameter optimization. In: AISTATS, pp. 240–248 (2016)
28. Kleinberg, R., Leyton-Brown, K., Lucier, B.: Efficiency through procrastination: approximately optimal algorithm configuration with runtime guarantees. In: IJCAI, vol. 3, p. 1 (2017)
29. Li, L., Jamieson, K., DeSalvo, G., Rostamizadeh, A., Talwalkar, A.: Hyperband: a novel bandit-based approach to hyperparameter optimization. J. Mach. Learn. Res. **18**(1), 6765–6816 (2017)
30. Luce, R.D.: Individual Choice Behavior: A Theoretical Analysis. Wiley, Hoboken (1959)
31. Malitsky, Y., Sellmann, M.: Instance-specific algorithm configuration as a method for non-model-based portfolio generation. In: International Conference on the Integration of Constraint Programming, Artificial Intelligence, and Operations Research (CPAIOR) (2012)
32. Maron, O., Moore, A.W.: Hoeffding races: accelerating model selection search for classification and function approximation. In: NIPS, pp. 59–66 (1993)
33. Phillips, M., Narayanan, V., Aine, S., Likhachev, M.: Efficient search with an ensemble of heuristics. In: IJCAI (2015)
34. Plackett, R.: The analysis of Permutations. J. Roy. Stat. Soc. Ser. C (Appl. Stat.) **24**(1), 193–202 (1975)
35. Santos, H., Toffolo, T.: Python MIP: Modeling Examples (2018–2019). https://engineering.purdue.edu/~mark/puthesis/faq/cite-url/. Accessed 23 Jan 2020

36. Schäfer, D., Hüllermeier, E.: Dyad ranking using Plackett-Luce models based on joint feature representations. Mach. Learn. **107**(5), 903–941 (2018). https://doi.org/10.1007/s10994-017-5694-9

37. Shang, X., Kaufmann, E., Valko, M.: A simple dynamic bandit algorithm for hyperparameter tuning. In: Workshop on Automated Machine Learning at ICML, June 2019

38. St-Pierre, D.L., Teytaud, O.: The Nash and the bandit approaches for adversarial portfolios. In: IEEE Conference on Computational Intelligence and Games, pp. 1–7 (2014)

39. Tavakol, M., Mair, S., Morik, K.: HyperUCB: hyperparameter optimization using contextual bandits (2019)

40. Wang, J., Tropper, C.: Optimizing time warp simulation with reinforcement learning techniques. In: Winter Simulation Conference, pp. 577–584 (2007)

41. Yue, Y., Joachims, T.: Interactively optimizing information retrieval systems as a dueling bandits problem. In: Proceedings of International Conference on Machine Learning (ICML), pp. 1201–1208 (2009)

A Memetic Approach for the Unicost Set Covering Problem

Maxime Pinard[1,2], Laurent Moalic[1,2($\boxtimes$)], Mathieu Brévilliers[1,2],
Julien Lepagnot[1,2], and Lhassane Idoumghar[1,2]

[1] Université de Haute-Alsace, IRIMAS UR 7499, 68100 Mulhouse, France
maxime.pin@live.fr,
{laurent.moalic,mathieu.brevilliers,julien.lepagnot,
lhassane.idoumghar}@uha.fr
[2] Université de Strasbourg, Strasbourg, France

Abstract. The Unicost Set Covering Problem (USCP) is a well-known
$\mathcal{NP}$-hard combinatorial optimization problem. This paper presents a
memetic algorithm that combines and adapts the Hybrid Evolutionary Algorithm in Duet (HEAD) and the Row Weighting Local Search
(RWLS) to solve the USCP. The former is a memetic approach with
a population of only two individuals which was originally developed to
tackle the graph coloring problem. The latter is a heuristic algorithm
designed to solve the USCP by using a smart weighting scheme that prevents early convergence and guides the algorithm toward interesting sets.
RWLS has been shown to be one of the most effective algorithm for the
USCP. In the proposed approach, RWLS is modified to be efficiently used
as the local search of HEAD (for exploitation purpose) on the one hand,
and also to be used as the crossover (for exploration purpose) on the other
hand. The HEAD framework is also adapted to take advantage of the
information provided by the weighting scheme of RWLS. The proposed
memetic algorithm is compared to RWLS on 98 widely-used benchmark
instances (87 from the OR-Library and 11 derived from Steiner triple
systems). The experimental study reports competitive results and the
proposed algorithm improves the best known solutions for 8 instances.

Keywords: Metaheuristics · Memetic algorithms · Unicost set
covering problem

1 Introduction

The Set Covering Problem (SCP) is a fundamental and well-known combinatorial
optimization problem related to a wide range of real-world applications such as
crew scheduling, facility location, city logistic problems, and optimal camera
placement [1,6,9,11,19]. It is one of Karp's well-known $\mathcal{NP}$-complete problems
[18] and have been proved $\mathcal{NP}$-hard in the strong sense, as well as the unicost
version of the problem [15]. The SCP can be defined as follows: given a universe
set $U = \{u_1, \ldots, u_m\}$ and a collection $S = \{s_1, \ldots, s_n\}$ of sets whose union

© Springer Nature Switzerland AG 2020
I. S. Kotsireas and P. M. Pardalos (Eds.): LION 14 2020, LNCS 12096, pp. 233–248, 2020.
https://doi.org/10.1007/978-3-030-53552-0_23

equals the universe U, and where each set in S is associated with a cost, the goal is to find a subset of S that covers all elements in U with minimal total cost.

This paper focuses on the Unicost Set Covering Problem (USCP), which is a special case of the SCP where all sets in S have identical cost, and which is generally considered to be harder to solve [22]. A USCP instance is usually defined as a matrix $A = (a_{i,j})$ of size $m \times n$ such that:

$$\forall i \in \{1, \ldots, m\},\ \forall j \in \{1, \ldots, n\},\ a_{i,j} = \begin{cases} 1 \text{ if } u_i \in s_j \\ 0 \text{ otherwise} \end{cases} \tag{1}$$

A solution is defined by a vector $x = (x_j)$ of size n such that:

$$\forall j \in \{1, \ldots, n\},\ x_j = \begin{cases} 1 \text{ if } s_j \text{ is part of the solution} \\ 0 \text{ otherwise} \end{cases} \tag{2}$$

A solution is valid if and only if:

$$\forall i \in \{1, \ldots, m\},\ \sum_{j \in \{1, \ldots, n\}} a_{ij} x_j \geq 1 \tag{3}$$

And the objective function to minimize is:

$$f(x) = \sum_{j \in \{1, \ldots, n\}} x_j \tag{4}$$

The memetic approach proposed in this article is inspired, on the one hand, by the Hybrid Evolutionary Algorithm in Duet (HEAD) [21] and, on the other hand, by the Row Weighting Local Search (RWLS) algorithm [14].

HEAD is a memetic algorithm that combines the TabuCol algorithm [17] with the Greedy Partition Crossover (GPX, see [13]) to solve the graph coloring problem. As a main feature, HEAD has a population of only two individuals, and experimental results have shown that HEAD is competitive, robust, and fast in comparison with up-to-date state-of-the-art algorithms.

HEAD has also successfully inspired other two-individual evolutionary algorithms, such as in [20] for the sum coloring problem, or to solve the flexible job shop scheduling problem [10], resulting in the improvement of the best known solution for 47 out of 313 benchmark instances.

RWLS is a local search designed to solve the USCP. It explores the search space by using a smart weighting scheme that allows to determine and update the usefulness of the sets of S, i.e. which sets of the solution are the most useless and which available sets would be the most useful. This approach has been assessed on 87 instances from the OR-Library [4] and 4 instances derived from Steiner triple systems, and it has successfully improved 14 best known solutions. To the best of our knowledge, and according to Kritter et al. [19], RWLS is one of the most effective algorithm for solving the USCP.

The main contribution of this paper is the combination and the clever adaptation of HEAD and RWLS in order to solve the USCP. For being used as HEAD local search (for exploitation), RWLS has been modified so that its input can be any solution and any set of weights coming from the last generation in the evolutionary loop (instead of a greedy solution with weights initialized to 1). An adaptation strategy has also been introduced to dynamically adjust the number of search iterations of RWLS for each generation of the memetic approach. Regarding the design of HEAD, the concept of population has been adapted: it still handles two individuals, but it also handles the two corresponding sets of weights resulting from RWLS local search. Two specific crossover operators have thus been designed to generate new starting individuals for the next generation, and to take advantage of the additional information provided by the weighting scheme of RWLS. It is worth noting that RWLS is also used in the crossover (for exploration) to quickly provide new poor solutions by solving a subproblem generated from both current solutions of the population. All these components contribute to the effectiveness of the proposed memetic algorithm, which is assessed on 98 USCP benchmark instances widely-used in the literature.

Sections 2 and 3 briefly explain the HEAD and RWLS algorithms respectively, highlighting the key points needed for a proper understanding of the memetic approach proposed in Sect. 4. Section 5 presents the benchmark testbed and the experimental setting, and it reports the results of the proposed algorithm in comparison with RWLS. Section 6 concludes the paper and gives some perspectives for future work.

2   Hybrid Evolutionary Algorithm in Duet (HEAD)

HEAD is a memetic approach originally designed to address the graph vertex coloring problem (GCP). The main idea of HEAD is that simplicity is probably the key to efficiency. To achieve such a goal of simplicity, the population is reduced to only two individuals. With such a simple population, there is no more selection operator: at each generation, both parents are the two individuals in the population. Nor is there any replacement strategy: at each generation, the two individuals are replaced by the two new children. It can be noticed that with such a small population, there is a risk of premature convergence. Actually, with only two individuals, crossover brings less and less diversity when the two individuals become similar. When the individuals become identical, the algorithm is trapped in a local optimum.

In order to limit the risk of such premature convergence, HEAD introduces the concept of elitist individual. It is the best solution found within a cycle of n generations. This solution is reintroduced in place of one of the 2 individuals at the end of each cycle. This mechanism makes it possible to obtain a very favorable behavior. It is of course interesting in terms of intensity, because it allows to keep a solution with a good fitness value. But it is also interesting in terms of diversity, because the population evolves without the elite solution which is then different from the population.

Algorithm 1 describes the main steps of HEAD, which was originally developed for the GCP. Since then, it has been adapted to work on the scheduling problem [10] with very interesting results: it can thus be expected that HEAD is also well suited to tackle other problems.

Algorithm 1: Pseudo-code of HEAD for the GCP

Input: k, the number of colors; $Iter_{LS}$, the number of *Local Search* iterations; $Iter_{cycle} = 10$, the number of generations into one cycle.

Output: the best k-coloring found: *best*

1 $p_1, p_2, elite_1, elite_2, best \leftarrow$ init() /* **initialize with random** k**-colorings** */
2 *generation, cycle* $\leftarrow 0$
3 **while** $nbConflicts(best) > 0$ *and* $p_1 \neq p_2$ **do**
4      $c_1 \leftarrow Crossover(p_1, p_2)$
5      $c_2 \leftarrow Crossover(p_2, p_1)$
6      $p_1 \leftarrow LocalSearch(c_1, Iter_{LS})$
7      $p_2 \leftarrow LocalSearch(c_2, Iter_{LS})$
8      $elite_1 \leftarrow$ saveBest$(p_1, p_2, elite_1)$ /* **best** k**-coloring of the current cycle** */
9      $best \leftarrow$ saveBest$(elite_1, best)$
10   **if** $generation \% Iter_{cycle} = 0$ **then**
11        $p_1 \leftarrow elite_2$ /* **best** k**-coloring of the previous cycle** */
12        $elite_2 \leftarrow elite_1$
13        $elite_1 \leftarrow$ init()   /* **reset** $elite_1$ **with a random solution** */
14        $cycle + +$
15     $generation + +$

One can notice that the crossover is applied twice on the same parents at the lines 4 and 5. This implies that the crossover must be stochastic.

3 Row Weighting Local Search (RWLS)

RWLS is a heuristic algorithm designed by Gao et al. [14] to solve the USCP. It gathers the three following components together in a local search framework: a weighting scheme which updates the weights of uncovered elements to prevent convergence to local optima, a tabu strategy to avoid possible cycles during the search, and a timestamp method to break ties when prioritizing sets is needed. RWLS main feature is the weighting scheme, which allows to identify hard-to-cover elements of U and helps to rank the sets of S according to their usefulness for covering U. It is implemented by using 2 concepts: the weight for elements of U and the score for sets of S. The weight of an element represents the difficulty to cover this element. It is initialized to 1 and increased by 1 for each iteration of the algorithm with the element left uncovered. The score of a set represents its utility for covering hard-to-cover points in the solution. If the set is part of the solution, its score is the opposite of the sum of the weights of the elements it is the only one to cover in the solution. If it is not part of the solution, its score is the sum of the weights of the elements it would be the only one to cover if added to the solution. This way, at any time of the algorithm, the most useful sets among those available have the highest scores, and the most useful sets among those already selected have the lowest scores. As shown in Algorithm 2, RWLS starts from a greedy solution and then uses these information to improve the current solution by alternately removing and adding sets with the highest scores (i.e. the most useless ones when removing sets of the solution, and the most useful ones when adding new sets in the solution).

Algorithm 2: Pseudo-code of RWLS

---

    input : step_limit
    output: *The best solution encountered*

1 solution ← Greedy()
2 weights ← $[1, ..., 1]$
3 current_step ← 0

4 **while** current_step $\leq$ step_limit **do**
5   **while** IsValid(solution) **do**
6   Update best solution
7   s ← set with the greatest score
8   Remove(solution, s)
9   s ← set in the solution, with the greatest score, not tabu and the oldest if there is a tie
10   Remove(solution, s)
11   e ← random uncovered element
12   s ← set covering e, with the greatest score, that can be added to the solution and the oldest if there is a tie
13   Add(solution, s)
14   Make s tabu
15   Increase weights of uncovered elements
16   current_step ← current_step + 1

---

4 Memetic Algorithm for Set Covering (MASC)

4.1 Algorithm Overview

As in HEAD, the population of the memetic algorithm is composed of two individuals: during a generation, both individuals are processed by parallel RWLS local searches, and a crossover is then applied to get two new individuals (see Fig. 1). However, the proposed algorithm takes into account the resulting RWLS weights to prepare the next generation. It means that the two weight vectors can be considered as a second population handled by HEAD: they are updated during RWLS local searches, and a specific crossover is then applied to get new weight vectors as input for the next generation (see Fig. 1). Obviously, before starting the generation loop, the solution population is greedily initialized and all weights are set to 1 (see Algorithm 3).

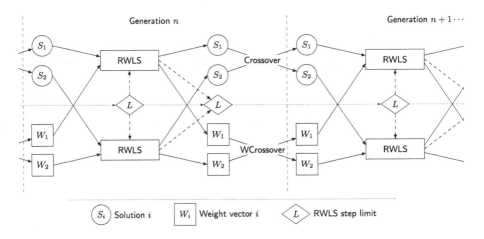

Fig. 1. MASC algorithm main scheme

4.2 Local Search

MASC uses a slightly modified version of RWLS. In the original RWLS algorithm, the input data are the problem instance and the stopping criterion (which is either a fixed maximum number of steps, or a fixed run time). Then an initial solution is greedily constructed and the element weights are initialized to 1. At the end of the optimization loop, the output is the best found solution. In the proposed modified version, the initial solution and the element weights are now also input parameters. Moreover, output data consist here of the best found solution, the element weights when this solution was encountered, and the number of steps needed to reach this solution. These slight modifications allow RWLS to be used as the local search of HEAD: this way, it can start the next generation with information coming from the previous one (i.e. solution and weights).

4.3 Parameter Self-adaptation Strategy

As mentioned above, the number of steps needed to get the best found solution is now part of the output information of RWLS. This allows to define a self-adaptation strategy for RWLS stopping criterion (i.e. the step limit). As observed in Fig. 1, this limit becomes dynamic and it evolves to better fit RWLS need for the current instance, trying to be enough to improve the solution but without loosing time stagnating. The initial limit is the sum of the number of sets and the number of elements of the instance. Then this limit is updated depending on the number of steps at which RWLS found the best solution in the last 10 generations. The exact way the maximum number of steps changes is detailed in Algorithm 3.

Algorithm 3: Pseudo-code of the proposed MASC

---

input : cumulative_step_limit
output: *The best solution encountered*

1 base_steps ← *number of sets + number of rows of the instance*
2 step_limit ← base_steps
3 dynamic_steps ← [base_steps, ..., base_steps]
4 (solution$_1$, solution$_2$) ← GreedySolve()
5 (weights$_1$, weights$_2$) ← [1, ..., 1]
6 cumulative_steps ← 0

7 **while** cumulative_steps $\leq$ cumulative_step_limit **do**
8   (solution$_1$, weights$_1$, steps$_1$) ← RWLS(solution$_1$, weights$_1$, step_limit)
9   (solution$_2$, weights$_2$, steps$_2$) ← RWLS(solution$_2$, weights$_2$, step_limit)
10   cumulative_steps ← cumulative_steps + 2 × step_limit

11   **if** steps$_1$ = 0 or steps$_2$ = 0 **then**
12   (solution$_1$, solution$_2$) ← RandomSolve()
13   step_limit ← 2 × step_limit
14   **else**
15   **if** solution$_1$ = solution$_2$ **then**
16   (solution$_1$, solution$_2$) ← RandomSolve()
17   PushBack(dynamic_steps, steps$_1$ + steps$_2$)
18   PopFront(dynamic_steps)
19   step_limit ← base_steps + Average(dynamic_steps)

20   solution$_{tmp}$ ← solution$_1$
21   solution$_1$ ← SolutionCrossover(solution$_1$, solution$_2$)
22   solution$_2$ ← SolutionCrossover(solution$_2$, solution$_{tmp}$)

23   weights$_{tmp}$ ← weights$_1$
24   weights$_1$ ← WeightCrossover(weights$_1$, weights$_2$)
25   weights$_2$ ← WeightCrossover(weights$_2$, weights$_{tmp}$)

4.4 Crossover Operators

Solution Crossover. The crossover applied to the solutions consists of solving a subproblem of the original instance: this approach is motivated by its successful application to solve the optimal camera placement problem when stated as a USCP [7]. More precisely, not all the sets of S are considered, but only the sets selected by the local search in the two current solutions. This subproblem is then solved by RWLS, which starts with a greedily generated solution and with weights set to 1, and whose step limit is set to the sum of the number of sets and the number of elements in the original instance. The greedy algorithm used here to generate the initial solution iteratively takes the most interesting set and adds it to the solution until it is valid.

Since the crossover is applied twice, exploration would be favoured if RWLS could start with two different initial solutions. For this reason, when ties occurs for selecting the next set during the greedy algorithm, they are solved differently during the first and the second crossover. For the first one, any tie is solved by selecting the first set encountered in the data structure, while in the second one, any tie is solved by selecting the last set encountered.

Weight Crossover. The crossover applied to the weight vectors is based on the following principle: each resulting weight is the maximum weight among the two corresponding weights in the two parent vectors. Given that the weight of an element represents the difficulty to cover this element, this operator gathers together the elements identified as difficult to cover by the two previous RWLS runs. It is also worth noting that this crossover results in the same input weight vector for both RWLS runs of the next generation.

4.5   Premature Convergence Detection

The exploitation part of the algorithm is done by the two parallel local searches while the exploration part is done by the two crossover operators that worsen the solutions and merge the weight information. However, if the solution crossover does not provides a new child solution, then RWLS could fail to improve the solution during the next generation. If it happens, both the best found solution and the weight vector will be left unchanged, inducing a stagnation. A similar problem also arises if the crossover operators do not succeed in escaping an area of the search space that always leads RWLS to the same solution. In addition to that, RWLS can also fail to improve a solution because it would need more iterations (in spite of the self-adaptation strategy described above).

To prevent these problems, as detailed in Algorithm 3, MASC contains stagnation avoiding conditions that replace the solutions with new random ones if RWLS failed to improve a solution or if both solutions are identical. Moreover, RWLS step limit is doubled as soon as a solution is not improved by RWLS.

This strategy replaces elitism of the original HEAD algorithm to avoid premature convergence.

5   Experimental Study

To show the effectiveness of MASC, it has been evaluated on widely-used benchmark testbeds from the literature, and then compared to RWLS.

5.1   Problem Instances

A total of 98 problem instances were used (87 from the OR-Library [4] and 11 derived from Steiner Triple Systems [12]), including the 91 instances used in [14] to assess RWLS. The main characteristics of these problems can be found in Tables 2, 3 and 4: instance name, number of rows (i.e. elements), number of columns (i.e. sets) and the best known solutions (BKS).

OR-Library Instances. Firstly, 70 of them are randomly generated problems: instance sets 4 to 6 are from [2], A to E from [5] and NRE to NRH from [3]. For some of these instances, the optima are known in a non-unicost context but not when converted to USCP instances (by simply ignoring the cost information). Instances from set E are unicost but quite trivial to solve.

Secondly, 10 of them are combinatorial problem instances from [16]: 6 CYC instances (where the number of edges required to hit every 4-cycle in a hypercube has to be minimized) and 4 CLR instances (which require to minimize the number of 4-tuples such that at least one of them is monochromatic whatever the 2-coloring of a given set).

Thirdly, 7 of them are very large-scale instances arising from crew-scheduling for the Italian railway company [8]. These instances are almost-unicost, each column covers at most 12 rows and has a cost of either 1 or 2. These are the only reducible instances of the benchmark testbed: Table 1 details the number of rows and columns removed by the reduction procedures from Sect. 5.2.

Steiner Triple Systems (STS) Instances. The 11 STS problem instances from [12] have a regular structure with each row being covered by exactly 3 columns. Only 4 of them (STS243, STS405, STS729 and STS1215) were used in [14] to demonstrate the effectiveness of RWLS.

5.2 Preprocessing

The two following instance reduction procedures (identified by Beasley [5]) have been used to reduce the 7 large-scale RAIL instances. In the following, *rows* and *columns* denotes the rows and columns of the zero-one matrix A defined in Eq. 1, where the rows are the elements of U, and the columns are the sets of S.

Column Domination. If a column is a subset of another column, this column is said to be dominated by the other: since the problem is unicost, dominated columns can be removed from the instance.

Column Inclusion. If a column is the only one covering a row, it is mandatory to include this column in the solution for it to be valid. Such a column can thus be included by default, which makes the row covered, and then allows to remove the column and the row from the instance.

RWLS requires that there is no row covered by only one column, and column domination can slow down the optimization process. However, it has been observed that removing dominated columns can lead to rows covered by only one column, and that removing rows covered by included columns can lead to dominated columns. Therefore, both reduction procedures have been alternatively applied, starting with the column domination until it is no longer possible to reduce the instance (results are reported in Table 1).

Table 1. Results of the reduction procedures presented in Sect. 5.2. Dominated columns are removed since other columns cover the same rows. Included columns and their corresponding covered rows are also removed, but these columns are still taken into account in the fitness as they are necessarily part of the solution.

Instance	Rows	Columns	Included columns	Covered rows	Dominated columns
RAIL507	507	63 009	8	20	37 842
RAIL516	516	47 311	27	62	8 990
RAIL582	582	55 515	7	17	28 311
RAIL2536	2 536	1 081 841	5	32	290 967
RAIL2586	2 586	920 683	50	155	496 329
RAIL4284	4 284	1 092 610	17	76	436 693
RAIL4872	4 872	968 672	86	321	499 406

5.3   Experimental Setting and Results

MASC has been implemented in C++17 and uses OpenMP to parallelize the 2 concurrent RWLS executions in the algorithm. All experiments were run on the high-performance computing cluster of the University of Strasbourg[1] (which is a heterogeneous computing system).

Tables 2 and 3 present the results of RWLS and MASC on the random USCP instances, and the combinatorial and STS instances, respectively. These tables report the best found solution value, the number of runs finding this best, and the average time over these "successful" runs. Results in boldface indicate that the best known solution (BKS) was reached, and starred results indicate an improvement of the BKS.

RWLS results are taken from [14]. The same experimental setting has thus been used to ensure a fair comparison between RWLS and MASC: 10 runs with different random seeds, a stopping criterion of 3×10^7 steps for the random USCP instances, and 1×10^8 steps allowed for the combinatorial and STS instances.

Regarding these 91 problem instances, MASC gets the same best value as RWLS on 76 instances (including 72 BKS) with similar frequency. The proposed algorithm also improves the best know solution for 8 instances (the best solution value of CYC10 in particular, which is improved from 1798 to 1792 and reached with a good frequency). It is worth noting that RWLS has been run only on 4 of the STS instances [14], but the whole set of 11 STS instances has been considered here: it allows to assess the effectiveness of MASC on the one hand (it has reached all BKS), and to provide the first BKS on instance STS2187 on the other hand.

In [14], the 7 RAIL instances were solved with RWLS-R, a modified version of RWLS that uses Lagrangian relaxation and its dual information to reduce the number of columns. The authors benchmarked RWLS-R with a time limit depending on the instances: 100 s for instances RAIL507, RAIL516 and RAIL586,

---

[1] https://services-numeriques.unistra.fr/les-services-aux-usagers/hpc.html (french).

Table 2. Best solutions found by RWLS and MASC on the random USCP instances

Inst.	Rows	Columns	BKS	RWLS			MASC		
				Best	#Best	Time(s)	Best	#Best	Time(s)
4.1	200	1 000	38	38	10	0.02	38	10	0.26
4.2	200	1 000	37	37	10	0.01	37	10	0.14
4.3	200	1 000	38	38	10	0.01	38	10	0.16
4.4	200	1 000	38	38	10	0.17	38	10	1.43
4.5	200	1 000	38	38	10	0.02	38	10	0.37
4.6	200	1 000	37	37	10	0.13	38	10	0.85
4.7	200	1 000	38	38	10	0.07	38	10	0.34
4.8	200	1 000	37	37	10	0.08	37	10	0.72
4.9	200	1 000	38	38	10	0.02	38	10	0.28
4.10	200	1 000	38	38	10	0.15	38	10	0.85
5.1	200	2 000	34	34	10	0.40	34	10	1.22
5.2	200	2 000	34	34	10	0.10	34	10	0.50
5.3	200	2 000	34	34	10	0.04	34	10	0.28
5.4	200	2 000	34	34	10	0.07	34	10	0.30
5.5	200	2 000	34	34	10	0.06	34	10	0.26
5.6	200	2 000	34	34	10	0.09	34	10	0.39
5.7	200	2 000	34	34	10	0.04	34	10	0.17
5.8	200	2 000	34	34	10	0.17	34	10	0.37
5.9	200	2 000	35	35	10	0.03	35	10	0.28
5.10	200	2 000	34	34	10	0.16	34	10	0.43
6.1	200	1 000	21	21	10	0.02	21	10	0.14
6.2	200	1 000	20	20	10	0.17	20	10	0.48
6.3	200	1 000	21	21	10	0.02	21	10	0.16
6.4	200	1 000	20	20	10	0.48	20	10	1.16
6.5	200	1 000	21	21	10	0.03	21	10	0.15
A.1	300	3 000	38	38	10	320.27	38	10	17.53
A.2	300	3 000	38	38	10	3.46	38	10	1.81
A.3	300	3 000	38	38	10	181.40	38	10	21.71
A.4	300	3 000	37	37	10	6.04	37	10	4.89
A.5	300	3 000	38	38	10	0.42	38	10	0.69
B.1	300	3 000	22	22	10	0.35	22	10	0.43
B.2	300	3 000	22	22	10	0.31	22	10	0.48
B.3	300	3 000	22	22	10	0.68	22	10	0.79
B.4	300	3 000	22	22	10	1.07	22	10	0.92
B.5	300	3 000	22	22	10	0.68	22	10	0.68

(*continued*)

Table 2. (*continued*)

Inst.	Rows	Columns	BKS	RWLS			MASC		
				Best	#Best	Time(s)	Best	#Best	Time(s)
C.1	400	4 000	40	43	10	0.81	43	10	1.00
C.2	400	4 000	40	43	10	1.14	43	10	1.46
C.3	400	4 000	40	43	10	0.73	43	10	2.19
C.4	400	4 000	40	43	10	0.82	43	10	1.47
C.5	400	4 000	43	**43**	10	4.25	**43**	10	3.62
D.1	400	4 000	24	**24**	10	9.73	**24**	10	5.22
D.2	400	4 000	24	**24**	10	285.28	**24**	10	42.44
D.3	400	4 000	24	**24**	10	364.87	**24**	10	56.97
D.4	400	4 000	24	**24**	10	270.87	**24**	10	33.40
D.5	400	4 000	24	**24**	10	346.74	**24**	10	67.62
E.1	50	500	5	**5**	10	0.00	**5**	10	0.05
E.2	50	500	5	**5**	10	0.00	**5**	10	0.07
E.3	50	500	5	**5**	10	0.00	**5**	10	0.05
E.4	50	500	5	**5**	10	0.00	**5**	10	0.04
E.5	50	500	5	**5**	10	0.00	**5**	10	0.04
NRE.1	500	5 000	16	**16**	2	3 441.29	**16**	3	424.55
NRE.2	500	5 000	16	**16**	2	6 437.21	**16**	2	256.20
NRE.3	500	5 000	16	**16**	1	2 935.55	**16**	3	351.32
NRE.4	500	5 000	16	**16**	5	2 925.08	**16**	3	906.27
NRE.5	500	5 000	16	**16**	2	3 576.99	**16**	2	693.12
NRF.1	500	5 000	10	**10**	10	35.09	**10**	10	3.69
NRF.2	500	5 000	10	**10**	10	84.45	**10**	10	4.50
NRF.3	500	5 000	10	**10**	10	71.61	**10**	10	3.64
NRF.4	500	5 000	10	**10**	10	36.48	**10**	10	5.87
NRF.5	500	5 000	10	**10**	10	47.25	**10**	10	6.42
NRG.1	1 000	10 000	61	**61**	10	74.27	**60***	5	191.34
NRG.2	1 000	10 000	61	**61**	10	124.62	**60***	4	166.16
NRG.3	1 000	10 000	61	**61**	10	104.36	**60***	1	143.49
NRG.4	1 000	10 000	61	**61**	10	139.69	**60***	1	231.59
NRG.5	1 000	10 000	61	**61**	10	176.31	**60***	1	109.90
NRH.1	1 000	10 000	34	**34**	10	280.76	**34**	10	268.20
NRH.2	1 000	10 000	34	**34**	10	407.19	**33***	1	682.88
NRH.3	1 000	10 000	34	**34**	10	409.27	**34**	10	499.08
NRH.4	1 000	10 000	34	**34**	10	533.78	**34**	10	197.11
NRH.5	1 000	10 000	34	**34**	10	495.10	**33***	1	190.31

Table 3. Best solutions found by RWLS and MASC on the combinatorial and STS instances

Inst.	Rows	Columns	BKS	RWLS			MASC		
				Best	#Best	Time(s)	Best	#Best	Time(s)
CLR10	511	210	25	**25**	10	0.01	**25**	10	0.07
CLR11	1 023	330	23	**23**	10	0.08	**23**	10	0.07
CLR12	2 047	495	23	**23**	10	0.38	**23**	10	0.12
CLR13	4 095	715	23	**23**	10	3.89	**23**	10	1.07
CYC6	240	192	60	**60**	10	0.00	**60**	10	0.19
CYC7	672	448	144	**144**	10	0.02	**144**	10	0.33
CYC8	1 792	1 024	342	**342**	10	0.30	**342**	10	1.44
CYC9	4 608	2 304	772	**772**	2	266.70	**772**	6	103.95
CYC10	11 520	5 120	1 798	**1 798**	7	663.73	**1 792***	8	267.99
CYC11	28 160	11 264	3 968	**3 968**	1	520.69	**3 968**	1	275.82
STS9	12	9	5	—	—	—	**5**	10	0.05
STS15	35	15	9	—	—	—	**9**	10	0.04
STS27	117	27	18	—	—	—	**18**	10	0.06
STS45	330	45	30	—	—	—	**30**	10	0.17
STS81	1 080	81	61	—	—	—	**61**	10	0.04
STS135	3 015	135	103	—	—	—	**103**	10	2.81
STS243	9 801	243	198	**198**	10	0.09	**198**	10	0.25
STS405	27 270	405	335	**335**	5	321.26	**335**	10	41.40
STS729	88 452	729	617	**617**	10	23.26	**617**	10	10.59
STS1215	245 835	1 215	1 063	**1 063**	1	886.25	**1 063**	1	1971.41
STS2187	796 797	2 187	—	—	—	—	**1 963***	10	1883.39

and 1 000 s for RAIL2536, RAIL2586, RAIL4284 and RAIL4872. For MASC, a larger limit has been used (i.e. 10 000 s for all instances) to try to benefit from the evolutionary loop and the self-adaptation strategy regarding the maximum number of steps allowed in the local search. Table 4 reports the corresponding results, and shows that the proposed algorithm manages to reach only 2 BKS. Actually, even with this time limit, only 2 or 3 generations were realized for the largest RAIL instances, which is not enough for the dynamic RWLS step system to be efficient and for the crossover operators to explore the search space. This highlights that MASC is not well suited for these very large instances, and should thus be further improved to tackle this kind of problem.

Table 4. Best solutions found by RWLS-R and MASC on the `RAIL` instances

Inst.	Rows	Columns	BKS	RWLS-R			MASC		
				Best	#Best	Time(s)	Best	#Best	Time(s)
RAIL507	507	63 009	96	**96**	1	0.06	97	10	320.30
RAIL516	516	47 311	134	**134**	9	0.09	**134**	10	5.84
RAIL582	582	55 515	126	**126**	7	0.08	**126**	1	602.67
RAIL2536	2 536	1 081 841	378	381	4	37.34	386	2	6 188.21
RAIL2586	2 586	920 683	518	520	2	16.18	530	1	5 499.93
RAIL4284	4 284	1 092 610	594	597	3	165.58	614	2	6 549.11
RAIL4872	4 872	968 672	879	882	2	56.34	899	1	4 301.85

6   Conclusion

This paper proposes a memetic approach to solve the USCP. It combines the two-individual HEAD algorithm [21] with the RWLS local search [14]. The HEAD evolutionary framework has been adapted to handle both solutions and weight vectors used by RWLS, and it has lead to the design of two specific crossover operators. RWLS is integrated in HEAD as local search for exploitation purpose on the one hand, and it is also used in the solution crossover to solve a subproblem for exploration purpose on the other hand. In addition to that, an adaptation strategy has been proposed so that the maximum number of RWLS steps can evolve and better fit the needs for each instance.

An experimental study carried out on a set of 98 classical instances (from OR-Library and `STS`) showed the effectiveness of the so-called MASC algorithm, reaching 80 BKS, improving 8 other BKS, and providing the first BKS on instance `STS2187`.

As a future work, it would be interesting to conduct a comprehensive study on the adaptation strategy that automatically adjusts the step limit of RWLS. Other strategies could be considered in addition to the one proposed here, and could be thoroughly compared to the classical static strategy. Another perspective would be to modify and further improve the algorithm to get better results on instances containing a significantly large number of columns but only a few rows (i.e. `RAIL` instances). It can also be planned to test different crossover operators and to analyze the specificity of the instances in order to reach better solutions, especially regarding `NRH` instances (2 improved BKS out of 5) and the two last `STS` instances (whose run times suggest that the algorithm is struggling).

References

1. Balas, E.: A class of location, distribution and scheduling problems: modeling and solution methods. In: Proceedings of the Chinese-U.S. Symposium on Systems Analysis. Wiley Series on Systems Engineering and Analysis. Wiley (1983). ISBN 978-0-471-87093-7

2. Balas, E., Ho, A.: Set covering algorithms using cutting planes, heuristics, and sub-gradient optimization: a computational study. In: Padberg, M.W. (ed.) Combinatorial Optimization. MATHPROGRAMM, vol. 12, pp. 37–60. Springer, Heidelberg (1980). https://doi.org/10.1007/BFb0120886. ISBN 978-3-642-00802-3

3. Beasley, J.E.: A Lagrangian heuristic for set-covering problems. Naval Res. Logist. **37**(1), 151–164 (1990). https://doi.org/10.1002/1520-6750(199002)37:1⟨151::AID-NAV3220370110⟩3.0.CO;2-2

4. Beasley, J.E.: OR-library: distributing test problems by electronic mail. J. Oper. Res. Soc. **41**(11), 1069–1072 (1990). https://doi.org/10.1057/jors.1990.166

5. Beasley, J.: An algorithm for set covering problem. Eur. J. Oper. Res. **31**(1), 85–93 (1987). https://doi.org/10.1016/0377-2217(87)90141-X

6. Boschetti, M., Maniezzo, V.: A set covering based matheuristic for a real-world city logistics problem. Int. Trans. Oper. Res. **22**(1), 169–195 (2015). https://doi.org/10.1111/itor.12110

7. Brévilliers, M., Lepagnot, J., Idoumghar, L., Rebai, M., Kritter, J.: Hybrid differential evolution algorithms for the optimal camera placement problem. J. Syst. Inf. Technol. (2018). https://doi.org/10.1108/JSIT-09-2017-0081

8. Caprara, A., Fischetti, M., Toth, P.: A heuristic method for the set covering problem. Oper. Res. (1999). https://doi.org/10.1287/opre.47.5.730

9. Christofides, N., Korman, S.: A computational survey of methods for the set covering problem. Manag. Sci. **21**(5), 591–599 (1975). https://doi.org/10.2307/2630042

10. Ding, J., Lü, Z., Li, C.-M., Shen, L., Xu, L., Glover, F.: A two-individual based evolutionary algorithm for the flexible job shop scheduling problem. Proceedings of the AAAI Conference on Artificial Intelligence, vol. 33, no. 01, pp. 2262–2271 (2019). https://doi.org/10.1609/aaai.v33i01.33012262

11. Farahani, R.Z., Asgari, N., Heidari, N., Hosseininia, M., Goh, M.: Covering problems in facility location: a review. Comput. Ind. Eng. **62**(1), 368–407 (2012). https://doi.org/10.1016/j.cie.2011.08.020

12. Fulkerson, D.R., Nemhauser, G.L., Trotter, L.E.: Two computationally difficult set covering problems that arise in computing the 1-width of incidence matrices of Steiner triple systems. In: Balinski, M.L. (ed.) Approaches to Integer Programming. MATHPROGRAMM, vol. 2, pp. 72–81. Springer, Heidelberg (1974). https://doi.org/10.1007/BFb0120689. ISBN 978-3-642-00740-8

13. Galinier, P., Hao, J.-K.: Hybrid evolutionary algorithms for graph coloring. J. Comb. Optim. **3**(4), 379–397 (1999). https://doi.org/10.1023/A:1009823419804

14. Gao, C., Yao, X., Weise, T., Li, J.: An efficient local search heuristic with row weighting for the unicost set covering problem. Eur. J. Oper. Res. **246**(3), 750–761 (2015). https://doi.org/10.1016/j.ejor.2015.05.038

15. Garey, M.R., Johnson, D.S.: Computers and Intractability; A Guide to the Theory of NP-Completeness. W. H. Freeman & Co. (1990). http://dl.acm.org/citation.cfm?id=574848

16. Grossman, T., Wool, A.: Computational experience with approximation algorithms for the set covering problem. Eur. J. Oper. Res. **101**(1), 81–92 (1997). https://doi.org/10.1016/S0377-2217(96)00161-0

17. Hertz, A., de Werra, D.: Using tabu search techniques for graph coloring. Computing **39**(4), 345–351 (1987). https://doi.org/10.1007/BF02239976

18. Karp, R.M.: Reducibility among combinatorial problems. In: Miller, R.E., Thatcher, J.W., Bohlinger, J.D. (eds.) Complexity of Computer Computations. The IBM Research Symposia Series, pp. 85–103. Springer, Boston (1972). https://doi.org/10.1007/978-1-4684-2001-2_9. ISBN 978-1-4684-2001-2

19. Kritter, J., Brévilliers, M., Lepagnot, J., Idoumghar, L.: On the optimal placement of cameras for surveillance and the underlying set cover problem. Appl. Soft Comput. **74**, 133–153 (2019). https://doi.org/10.1016/j.asoc.2018.10.025

20. Moalic, L., Gondran, A.: The sum coloring problem: a memetic algorithm based on two individuals. In: 2019 IEEE Congress on Evolutionary Computation (CEC), pp. 1798–1805 (2019). https://doi.org/10.1109/CEC.2019.8789927

21. Moalic, L., Gondran, A.: Variations on memetic algorithms for graph coloring problems. J. Heuristics **24**(1), 1–24 (2018). https://doi.org/10.1007/s10732-017-9354-9

22. Yelbay, B., Birbil, ŞI., Bülbül, K.: The set covering problem revisited: an empirical study of the value of dual information. J. Ind. Manag. Optim. **11**, 575 (2015). https://doi.org/10.3934/jimo.2015.11.575

Dynamic Visual Few-Shot Learning Through Parameter Prediction Network

Nikhil Sathya Kumar$^{(\boxtimes)}$, Manoj Ravindra Phirke, Anupriya Jayapal, and Vishnu Thangam

Imaging CoE, HCL Technologies Ltd., Bangalore, India
{nikhilsathyak,manoj.p,anupriya.j,vishnut}@hcl.com

Abstract. Though machine learning algorithms have achieved great performance when adequate amounts of labeled data is available, there has been growing interest in reducing the volume of data required. While humans tend to be highly effective in this context, it remains a challenge for machine learning approaches. The goal of our work is to develop a visual learning based few-shot system that achieves good performance on novel few shot classes (with less than 5 samples each for training) and does not degrade the performance on the pre-trained large scale base classes and has a fast inference with little or zero training for adding new classes to the existing model. In this paper, we propose a novel, computationally efficient, yet effective framework called Param-Net, which is a multi-layer transformation function to convert the activations of a particular class to its corresponding parameters. Param-Net is pre-trained on large-scale base classes, and at inference time it adapts to novel few shot classes with just a single forward pass and zero-training, as the network is category-agnostic. Two publicly available datasets: MiniImageNet and Pascal-VOC were used for evaluation and benchmarking. Extensive comparison with related works indicate that, Param-Net outperforms the current state-of-the-art on 1-shot and 5-shot object recognition tasks in terms of accuracy as well as faster convergence (zero training). We also propose to fine-tune Param-Net with base classes as well as few-shot classes to significantly improve the accuracy (by more than 10% over zero-training approach), at the cost of slightly slower convergence (138 s of training on a Tesla K80 GPU for addition of a set of novel classes).

Keywords: Param-Net · MiniImagenet · Pascal-VOC · Activations · Few-shot learning

1 Introduction

Current state of the art on semantic segmentation, object detection, image classification and most other learning based tasks rely on deep neural networks. Deep neural networks are high-capacity powerful models which require large amounts

Supported by HCL Technologies Ltd.

I. S. Kotsireas and P. M. Pardalos (Eds.): LION 14 2020, LNCS 12096, pp. 249–263, 2020.
https://doi.org/10.1007/978-3-030-53552-0_24

of annotated data and millions of parameters. Large amounts of supervised training data per concept is required for deep learning algorithms to achieve great performance, and the learning process could take multiple days to weeks using specialized expensive hardware like GPUs. Adapting a deep learning model to recognize a new class includes 2 major steps: 1) Collection of the large scale dataset, 2) Fine-tune the existing model to recognize the new class. Both of these steps are time, memory and resource intensive. If new classes are to be recognized, then typically thousands of training examples are required for training and fine-tuning the model. Sometimes, unfortunately this fine-tuning might result in the model forgetting the initial classes on which it was trained. One of the most important objectives of few-shot learning based algorithms is to adapt the existing models at real-time, to recognize novel classes which were unseen during the initial training phase. The major challenge is that, these novel classes have less than 5 visual examples each for training the model. The performance of state-of-the-art classification models deteriorates when the number of images per new class reduces to less than 10, whereas humans are capable of learning new visual concepts reliably and effortlessly with very few examples. This has inspired scientists to adapt deep learning algorithms to work on few-shot domain, where the main goal is to learn novel concepts using limited number of examples. The main advantage of solving the few-shot problem is that it relies only on very few examples and eliminates or restricts the need to formulate large amount of labeled training data, which is usually a cumbersome and costly process.

In this paper, we propose a novel, computationally efficient, yet effective framework called Param-Net, which is a fusion of the best practices of parameter generation, gradient descent and data augmentation methods. Two publicly available dataset: MiniImageNet and Pascal-VOC have been used in this paper for evaluation and bench-marking. MiniImageNet dataset is the most popular dataset for benchmarking few-shot learning algorithms. Pascal-VOC is the most popular dataset for object detection tasks. Using Pascal-VOC dataset, an attempt can be made to scale the few-shot classification task to few-shot detection task. The dataset is split into: (1) classes that contain adequate number of samples denoted as C_{Base}, this is considered as large scale dataset and (2) classes that contain 1–5 images each which are denoted as C_{Few}, these are the few shot classes. The goal of our work is to devise a visual learning based few-shot system that achieves good performance on novel few shot classes, C_{Few} (with less than 5 samples each) and does not degrade the performance on the large scale base classes C_{Base} and has a fast inference with little or zero training for adding new classes to the existing model.

In neural networks, parameters of a particular class and its activations share a strong relationship and this property is used by Param-Net to predict weights for novel classes. For fair comparison with state-of-the-art approaches, we use a Res-Net based model for extracting the most relevant features/activations of the input images. The activations which are determined prior to the final fully connected layer in the base model, is used as input to the Param-Net, which is a multi-layer transformation function. Param-Net is used to convert activations of a particular class to its corresponding weights. Res-Net as well as Param-Net model is pre-trained on C_{Base}. Param-Net can adapt to novel few shot classes with just a forward pass and zero training as the network is category agnostic.

In this paper, the models are initially tested on MiniImageNet dataset and compared with state-of-the-art few shot algorithms. The proposed Param-Net model outperforms the state-of-the-art methods on few-shot classes, while also not compromising on the efficiency on base classes (C_{Base}). On MiniImageNet dataset, Param-Net achieves an accuracy of 62.69% for 5-way 1-shot learning and 86.14% for 5 shot learning. The inference time of the model to add novel few shot classes is also very low (because of zero-training time), it only takes around 23 ms for adding a novel class on a Tesla K80 GPU. Inspite of this state-of-the-art performance it has been observed that the accuracy of Param-Net on closely similar classes is slightly compromised. To account for this, we suggest fine-tuning the Param-Net with base classes as well as few-shot classes. This significantly improves the accuracy at the cost of slightly slower convergence. Fine-tuned Param-Net achieves an accuracy of: a) 94.23% for 5-way and 5 shot learning on MiniImageNet dataset and b) 87.26% on Pascal-VOC dataset for 20 way 5-shot settings. The fine-tunable version of Param-Net takes around 138 s of training on Tesla K80 GPU for the addition of a set of novel classes for MiniImageNet dataset. Hence with little fine-tuning, Param-Net can be used to adapt a deep learning model to add novel classes with just a single training image.

In "Sect. 2", we describe the techniques widely used and documented in literature to achieve current state-of-the-art results, most of the techniques described in this section are used to benchmark the Param-Net framework. Then in "Sect. 3" we elaborate the proposed Param-Net approach, in "Sect. 4", we discuss the experimental setup, results and benchmarks and in "Sect. 5", we discuss the conclusion and future scope.

2   Related Work

The ideas behind Param-Net has broad prior support in literature, but mostly appear in disjoint or in incompatible problem setting. Research literature on few-shot learning techniques exhibits great diversity. We adapt these concepts into a unified framework for recognition in real-world scenarios. In this section, we focus on methods using the supervised meta-learning paradigm [12], [51], [9] most relevant to ours and compared to in the experiments. We can divide these methods into 5 categories:

Data Generation and Data Augmentation Methods: In [9], a sampling method was proposed that extracts varying sequences of patches by decorrelating an image based on maximum entropy reinforcement learning. This is a form of "learned" data augmentation. In [19], GAN based approach was proposed to address the few shot learning, where GAN allows the few shot classifiers to learn sharper decision boundary, which could generalize better. In [30], a modified auto-encoder was proposed to synthesize new samples for an unseen category just by seeing few examples from it.

Gradient Descent Based Methods: Meta-LSTM [12], treats the model parameters as its hidden states and uses LSTM as a model updater, that not only learns the initial model, but also the update rule. In contrast to Meta-LSTM, MAML [14] only learns an initial update. In MAML, the updating rule is fixed to a stochastic gradient descent (SGD). In [26], a variant of MAML was proposed where only first order gradients were used. In [21], MetaSGD was proposed as an extension of MAML, which learns weight initialization as well as learner's update step size. In [1,2], a modification to MAML was proposed to prevent overfitting. In [2] entropy based and inequality minimization measures were introduced and in [1], Meta-Transfer Learning approach was introduced where it leverages transfer learning and benefits from referencing neuron knowledge in pre-trained deep nets. A framework was proposed in [18] to unify meta learning and gradient based hyperparameter optimization. In [20], Neuron-level adaptation was suggested, to reduce the complexity of weight modification when the connections are dense.

Metric Learning Methods: Siamese Neural Network which uses a two stream convolutional neural network was originally utilized by Koch et al. [30], to learn powerful discriminative representations and then generalized them to unseen classes. Vinyals et al. [13] proposed Matching-Nets and introduced the episodic training mechanism into few-shot learning. Prototypical Network was proposed in [15], which is built upon the matching network [13], uses cosine similarity and 4-layer network. Here, query image is compared with support images using class centroids to eliminate outliers in support set. In [16], a variant of Matching network [13] was proposed and named the Relation-Net. It uses additional network to learn similarity between image through a deep non-linear metric. Relationship of every query-support pair is evaluated using a neural network. As an extension to prototypical network in [15], three light weight and parameter free improvements were proposed in [5]. In [10,25] modifications to Relation-Net was proposed. In [10], images were encoded into feature vectors by an encoding network. In [25], second order descriptors were proposed instead of first order descriptors. Given a new task with its few-shot support set, Garcia et al. [23] proposed to construct a graph where all examples of the support set and a query set are densely connected. There have been modifications proposed to [23], by [27] and [7]. In [27], transductive propagation network was proposed to propagate labels from known labeled instances to unlabeled test instances. In [7], Edge Labeling Graph Neural Network (EGNN) was proposed to predict edge-labels rather than node-labels, this is ideal for performing classification on various tasks without retraining. In [4], local descriptor based image-to-class measure was proposed which was obtained using deep local descriptors of convolutional feature maps.

Parameter Generation Methods: Using attention based mechanism to predict the classification weights of each novel class as a mixture of base classification weights of each novel class, Gidaris et al. in [22] proposed Dynamic-Net to capture class dependencies in the context of few-shot learning. But in Dynamic-Net [22], dependencies were considered between base classes and novel classes. In contrast, the dependencies were considered to exist between all the classes in [6], and these dependencies were proposed to be captured using GNN architectures. But this is computationally more expensive than the simple attention based mechanism proposed in [22]. The episodic formulation proposed in [13], was used by [6], to apply the Denoising Autoencoder framework in the context of few-shot learning, thereby improving the performance of parameter generation, by forcing it to reconstruct more discriminative classification weights. In GNN, input is the labeled training examples and unlabeled test examples of few shot problem and the model is trained to predict the label of test examples. But here, input to GNN is some initial estimate of classification weights of the classes that needs to be learnt and it is trained to reconstruct more discriminative classification weights.

Model Fine-Tuning Methods: Most of the ADAS based models, usually opt for fine-tuning the pre-trained model to add novel classes, but this method works well, only when there is sufficient number of novel class examples for training. The method that has been proposed in this paper is a fusion of the most effective features of Parameter generation, Model fine-tuning, data generation and gradient based methods. In the following section, we shall discuss the architecture of the model followed by the experimental results.

3 Methodology

The datasets are split into large-scale dataset (D_{Base}) and few-shot dataset (D_{Few}), where D_{Base} contains classes which have sufficient number of images for training, whereas D_{Few} contains classes with less than 5 images. C_{Base} refers to the classes present in D_{Base} and C_{Few} refers to the classes present in D_{Few}. There is no overlap between C_{Base} and C_{Few}. The distribution of the dataset into D_{Base} and D_{Few} is illustrated in Table 1.

Table 1. Random distribution of classes from public datasets into large scale and few-shot classes.

Datasets	Number of classes in D_{Base}	Number of classes in D_{Few}
MiniImageNet	80	20
Pascal-VOC	13	7

The goal of our work is to devise a visual learning based few-shot system that achieves good classification performance on novel few shot classes, C_{Few} (with less than 5 samples each) and does not degrade the performance on the large scale base classes C_{Base} and has a fast inference with little or zero training for adding new classes to the existing model.

In a neural network, for a particular class: weights and activations are closely related. In this paper, we propose a novel, computationally efficient, yet effective framework called Param-Net, which is a multi-layer dense regression transformation function to convert the activations of a particular class to its corresponding weights.

Initially, Resnet-101 deep neural network was considered as the base model and was pre-trained on the large scale dataset (D_{Base}). In the base-model, the entire network prior to the final fully connected layer was considered as feature extractor. The final fully connected layer was considered as the classifier network. For an input image 'X_i', the feature extractor will output a "d"-dimensional feature vector $Z_{X_i} = F(X_i)$. The weights "w" of the classifier network consists of "N" classification weights, where "N" is the number of classes in D_{Base}:

$$w = [w_i]_{i=1}^{N} \tag{1}$$

where w_i is the d-dimensional classification weight vector for the ith class. For input image 'X_i and for 'N' classes, the classifier will compute the classification scores as: For input image 'X_i',

$$[s_{1i}, s_{2i}, ..., s_{Ni}] = [Z_{xi}w_1, Z_{xi}w_2, .., Z_{xi}w_N], \tag{2}$$

For an image "X_e", belonging to class "k", the objective of the feature extractor and classifier is to maximize s_{ke}, where $s_{ke} = Z_{xe}w_k$, and minimize $[s_{1e}, s_{2e}, ..., s_{(k-1)e}]$ and $[s_{(k+1)e}, s_{(k+2)e}, ..., s_{Ne}]$. The weights "w" are learnt through back-propagation using the loss function:

$$\frac{\sum\limits_{i=0}^{M} L_i}{M} \tag{3}$$

where,

$$L_i = -log \frac{e^{Z_{X_i} W_{y_i}}}{\sum\limits_{j=1}^{N} e^{Z_{X_i} W_j}} \tag{4}$$

where "M" is the number of images per epoch, for image "X_i": "L_i" is the loss and "y_i" is the annotation label. To adapt the base model to include novel few-shot classes, a transformation layer named Param-Net has been proposed in this paper. The objective of the Param-Net is to predict parameters of a particular class based on its corresponding activations. The activations which are used as input to the Param-Net are determined using the feature extractor network. Parameters of the original base-model classifier network is replaced by the Parameters estimated from the Param-Net which can be denoted as:

$$T(Z_{X_i}) = we_i \tag{5}$$

where T() is the transformation function or the Param-Net, Z_{X_i} is the activation of the image "X_i" and we_i is the estimated parameters for the image "X_i".

During the training phase of the Param-Net, initially a mini-batch of input images is formed from the C_{Base} classes, containing M' images of each class. Hence the size of the mini-batch was N * M', where N is the number of classes. Using the feature extractor network, d-dimensional activation vectors are extracted for the entire mini-batch, containing N classes and M' images for each class. Mean activations are extracted for each class using:

$$Mz_j = \frac{\sum_{k=0}^{M'} Z_{X_{jk}}}{M'} \tag{6}$$

where Mz_j is the mean activation of the batch of images belonging to j^{th} class, M' is the number of images per class in the batch and $Z_{X_{jk}}$ is the activation of the kth image of the jth class.

The d-dimensional mean activations of each of the class is input to the Param-Net to estimate the parameters of the corresponding classes, as illustrated in Fig. 1(a). The Param-Net is a regression network whose input and output dimensions are of the same size, but to reduce overfit, we have posed it as a classification task. After estimating parameters of all the classes individually, they are concatenated into a "d * N" vector.

A mini-batch of training images $X'_1, X'_2, ..., X'_N$, containing an equal sample of all the classes is considered and features are extracted, where X'_1, contains a subset of images belonging to class 1. These features are then convolved with the estimated parameters from the transformation layer and softmax activations are applied. The output is compared with the input annotated labels and the loss function is computed and the gradients are propagated. The loss function from Eq. 4 is modified to:

$$L_i = -log \frac{e^{Z_{x'_i}[T(Mz)]_{y_i}}}{\sum_{j=1}^{N} e^{Z_{x'_i}[T(Mz)]_j}} \tag{7}$$

where, L_i is the loss for the image 'X_i', 'N' is the number of classes, $Z_{x'_i}$ is the activations estimated or features extracted for the image 'X_i' using the feature extraction network, '$[M_z]_{y_i}$' is the mean activation of the actual class that the image 'X_i' belongs to, '$[Mz]_j$' is the mean activation of all the other classes ranging from 1 to N and T[] is the estimated weights from the Param-Net. The entire flow of the network has been illustrated in Fig. 1 (b).

There are three distinct phases in this approach: Training, Parameter predictor and inference. During the training phase, only images from D_{Base} are used to train the transformation layer/Param-Net. None of the images from the D_{Few} are used. In the parameter predictor phase, the images 1–5 of a given class C_{Few} from D_{Few} are considered and passed through the feature extractor and mean activations of the extracted features are determined.

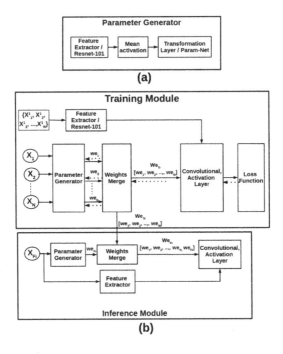

Fig. 1. Illustration of few-shot learning pipeline proposed in this paper using Param-Net based approach with zero-training. (a): Parameter Generator network (b): Training and inference pipeline for the proposed model.

These mean activations are input to the transformation layer to estimate the parameters for the few shot class C_{Few}. These estimated parameters for few-shot classes are concatenated with the estimated parameters of the other base classes. This way, a new class is added to the existing model with zero training. At inference time, for any input image, features are extracted and convolved with the estimated parameters to determine the class of the input image.

We also propose to fine-tune the Param-Net with base classes as well as few-shot classes to significantly improve the accuracy. For addition of a novel class to the existing model, Param-Net is fine-tuned with mean-activations from C_{Base} as well as from C_{Few}.

But the convergence of the fine-tuned model does not require too many epochs nor too many computation cycles because: (1) Param-Net is just a 2-layer dense network, (2) Input to the Param-Net is only mean activations which are of lower dimension, compared to the high-dimension raw input. Hence, with little training, the efficiency of Param-Net can be considerably increased. The modified flow for the proposed Param-Net is illustrated in Fig. 2.

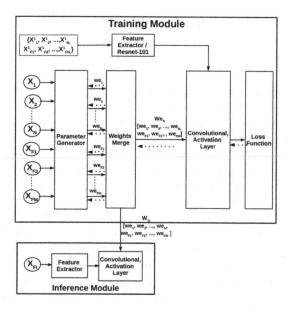

Fig. 2. Illustration of few-shot learning pipeline proposed in this paper using fine-tuning based Param-Net approach with little-training. Training and inference pipeline for the proposed model has been shown with the Param-Net being fine-tuned, every time a novel few-shot class is added to the model.

4   Results

We evaluate the proposed Param-Net extensively in terms of few-shot recognition accuracy and model convergence speed on MiniImageNet and Pascal-VOC datasets. In the following sections, we describe the datasets, detailed experimental setup, comparison to the state-of-the-art methods and an ablation study with respect to different model architectures.

4.1   Dataset

MiniImageNet was proposed by Vinyals et al. in [13], for evaluating few-shot learning based algorithms. Its complexity is high due to the use of images from ImageNet dataset, but requires less resources and cheaper infrastructure than running on the full ImageNet dataset. It is the most popular dataset for benchmarking few-shot learning algorithms. We used the same split proposed by [1] for fair comparison. We consider 5-way classification for both 1-shot and 5-shot.

Pascal-VOC dataset is one of the most popular datasets for object detection task. The logical extension of few-shot learning based algorithms is object detection based few-shot learning. Hence the use of Pascal-VOC helps us to make the model robust to scale to object detection tasks. Param-Net was evaluated on the Pascal VOC for 20-way 1-shot performance.

4.2   Model

For a fair comparison with the state-of-the-art we use pre-trained Resnet-101 as our base model for feature extraction, for MiniImageNet and Pascal-VOC datasets. Resnet-101 was pre-trained on ImageNet and the model was later fine-tuned on D_{Base} dataset. Model fine-tuning was performed by freezing the initial layers and only updating the weights of the final few layers. The layer prior to the final fully connected layer serves as a feature extractor for Param-Net. The dimension of the extracted features is 2048. The learning rate for fine-tuning Resnet-101 was 0.0001 using an Adam optimizer.

For MiniImageNet dataset, the D_{Base} dataset had 80 classes with 600 images each and these images were used for fine-tuning Resnet-101. Of the 600 images, 500 images were used for training and 50 images each for validation and testing. The same distribution of D_{Base} is used for training the Param-Net as well. One of the most important characteristics of Param-Net is that it is class-agnostic, which indicates that the network need not be trained on images from D_{Few}. Adam optimizer was used with a learning rate of 0.001 and it took around 187 s for the Param-Net to be trained on a Tesla K80 GPU. Once the Param-Net is trained, it has the capability to add novel classes with zero-training and the addition of novel classes can be done at real time. It takes only 18 ms to add novel classes to the existing base-classes because Param-Net is just a two layer dense regression network with a low-dimensional input.

In the following sections, we shall discuss the key role that a robust feature extractor plays, in improving the performance of few-shot learning algorithms. We shall also discuss the effect of the number of dense layers in the Param-Net on the quality of the weights generated, in the following sections.

Table 2 shows the comparison between: a) Accuracy of conventional fine-tuning based approach, where Resnet-101 model was fine-tuned and test only using the D_{Base} dataset b) Accuracy of the Param-Net based approach on the D_{Base} dataset, where similarly, the Param-Net was trained and tested only using the D_{Base} dataset and the weights of the classification layer of the Res-Net model was replaced by the weights estimated using the Param-Net. It is evident from the results in Table 2, that the Param-Net is able to achieve comparable performance on the D_{Base} dataset, while achieving state-of-the-art results on D_{Few} dataset as depicted in later sections.

Table 2. Comparative study between: a) ResNet-101 based feature extractor and classifier and b) ResNet-101 based feature extractor with Param-Net based classifier. Both are trained and tested only on D_{Base} dataset.

Dataset	ResNet (Feature extractor + classifier) (%)	Resnet feature extractor + Param-Net classifier (%)
Training Data	94.91	92.36
Validation Data	92.63	91.93
Testing Data	93.14	91.06

4.3   Evaluation Results and Comparisons

In Table 3, we compare Param-Net with state-of-the-art methods of different few-shot techniques. The accuracies of all the techniques have been reported on the test dataset.

Data generation techniques [9,11,19,30]: Different techniques like decorrelation [9], GAN [19], auto-encoder [30] and image deformation network [11] have been used. But these models need to be trained on more than just 5 examples for the generators to generate useful data, otherwise the generators under-perform on few-shot data. Hence the accuracy of the data augmentation based methods is lower than the most of the other approaches.

Metric learning techniques [4,5,7,10,13,15,16,23,25]: In [13], few-shot learning problem was addressed using cosine similarities between the features produced by CNN, which is a very simplistic metric to differentiate between the images. In [15] and [16], instead of cosine similarities, a non-linear similarity score was introduced using neural networks, but the descriptors were of first order. As an extension to [15], different approaches [5,10,25] were proposed like second order descriptors, encoding feature vectors using encoder network, batch folding, few shot localization (fsl) and covariance pooling (cp). But still they under-performed because of the inability of the feature extractors to extract meaningful features. GNN based approach, proposed by [22] under-performed because it uses node-labeling to model intra-cluster similarity and inter-cluster dissimilarity. This was addressed in [7] to achieve better performance. In [4], k-nearest neighbors metric was used, but one drawback of this approach was that it used Conv4 as a feature extractor, which is a weak feature extractor. Most of the mentioned metric learning techniques, learn a feature similarity metric between few-shot training examples and a test (query) example. But, these methods treat each training class independently from one another. Hence, the performance of metric learning frameworks is weakened.

Gradient descent techniques: In [12] and [14], element-wise fine tuning was used, hence inducing overfitting on the designed models, and in [12], LSTM was used to update the initial model as well as the update rule, hence it was time consuming as well, this was addressed in [14], by learning only the initial model. As an update to [14], different solutions were proposed like: using first order gradients [26], joint learning of weight initialization as well as learner's step size [21], entropy based inequality minimization measures [2] and transfer learning and fine-tuning [1]. But the problem with gradient based approaches is that, they require many iterative steps over many examples to perform well.

Parameter generation: In [22], dependencies were considered between novel classes and base classes using simple attention based mechanism. In [6], GNN based techniques were used to differentiate between all the classes, novel and base.

The Param-Net that has been proposed here, is a combination of data augmentation, gradient descent and parameter generation methods. This has resulted in state-of-the-art performance on 1-shot and 5-shot settings for Mini-ImageNet, which is the most popular public dataset for benchmarking few-shot learning algorithms. The existing state-of-the-art methods mainly focus on making relation measure, concept representation and knowledge transfer, but do not pay enough attention to final classification. This issue has been addressed in this paper by posing a regression problem of the Param-Net into a classification task, thereby also reducing the overfit of the model onto the training dataset considerably.

Table 3. 5-way accuracy on MiniImageNet. Blue: Best accuracy.

Method	Algorithm	Models	1-shot (%)	5-shot (%)
Data augmentation	Decorrelation [9]	Conv4	51.03	67.96
	Meta-Gan [19]	Resnet-12	52.71	68.63
	Delta-encoder [30]	VGG-16 (pre)	58.7	73.6
	Image deformation Meta Network [11]	ResNet-18	57.71	74.34
Metric learning	Matching networks [13]	Conv4	43.44	55
	Protonets (PN) [15]	Conv4	49.42	68.2
	RelationNet [16]	Conv4	50.44	65.32
	2nd order similarity network [25]		52.96	68.63
	GNN [23]	Conv-256F	50.33	66.41
	Deep Nearest Neighbor neural network [4]	Conv4	51.24	71.02
	PN+ fsl + CP [5]	Res-Net50		69.45
	Salient-Network [10]	Conv4	57.45	72.01
	Edge-Labeling GNN [7]	Resnet		76.37
Gradient descent	Meta-learning LSTM [12]	Conv-32F	43.56	60
	MAML [14]	Conv-32F	48.7	63
	Reptile [26]	Conv-32F	49.97	65.99
	TAML [2]	Conv-32F	49.4	66.0
	Matasgd [21]	Conv-32F	50.47	64.03
	MTL [1]	Resnet-12	61.2	75.5
Parameter generation	DynamicNet [22]	Conv-4-64	55.45	70.13
	WDAE-ML [6]	WRN-28-10	60.61	76.56
	Our-Param-Net: 2-layer (Resnet)	ResNet-101	**63.31**	**82.29**

Table 4. Results of ablation study

Algorithm	1-shot (%)	5-shot (%)
Resnet-101 + 1 layer Param-Net	61.95	78.95
Resnet-101 + 2 layer Param-Net	63.31	82.29
Nasnet + 2 layer Param-Net	64.69	86.14
Resnet-101 + fine-tune(2 layer Param-Net)	71.18	94.23

Table 3 and Table 4, indicates the importance of a robust feature extractor. In Table 3, the techniques which used Resnet based architecture for feature extraction performed better than the approaches that used Conv-4 or Conv-32 based feature extractors. Similarly, we also experimented using Nas-Net (network based on neural-architecture search) instead of Resnet-101 as feature extractor as shown in Table 4. Nas-Net improved the performance of the algorithm on both 1-shot as well as 5-shot settings. We also conducted experiments to ascertain the contribution of the number of layers in the Param-Net, to the eventual performance of the algorithm on MiniImageNet dataset. It has been observed that a 2-layer dense network performs better than a 1-layer dense network, as has been indicated in Table 4.

In this paper, we also propose an approach where, Param-Net is finetuned with D_{Base} as well as D_{Few}. For every new class that needs to be added, the Param-Net needs to be finetuned. It significantly leads to an increase in the accuracy but with a little extra training time of 138 s for addition of a set of novel classes to the existing model, on a NVIDIA K80 GPU.

5   Conclusion

In this work, we contribute to few-shot learning by developing a novel, computationally efficient framework called Param-Net which achieves top performance in tackling few-shot learning problems.

The main objective of few-shot applications is to add novel classes at real time to the existing model in the presence of less than 5 visual examples. Hence Param-Net has been proposed in this paper. It is a dense transformation layer which converts the activations of a particular class to its corresponding weights. It is pre-trained on large-scale base classes and at inference time it adapts to novel few-shot classes with just a single forward pass and zero or little training as the network is class agnostic.

Extensive comparison with related works indicate that the Param-Net outperforms state-of-the-art algorithms in terms of accuracy (1-shot and 5-shot) and in terms of faster convergence (zero or very-little training). We evaluate the performance of Param-Net on two publicly available datasets: MiniImageNet and Pascal-VOC. MiniImageNet is the most popular dataset for benchmarking few-shot algorithms. Pascal-VOC dataset was used to verify the scalability of Param-Net from few-shot classification task to few-shot detection task.

The future scope of improvement for the proposed Param-Net would be to scale the algorithm to address few-shot detection rather than just few-shot classification problems. The first step to address this challenge has been successfully accomplished by testing the Param-Net on the Pascal-VOC dataset, which is the premier dataset for object detection tasks.

References

1. Sun, Q., Liu, Y., Chua, T.S., Schiele, B.: Meta-transfer learning for few-shot learning. In: Proceedings of the IEEE Conference on Computer Vision and Pattern Recognition, pp. 403–412 (2019)
2. Jamal, M.A., Qi, G.J.: Task agnostic meta-learning for few-shot learning. In: Proceedings of IEEE Conference on Computer Vision and Pattern Recognition (2019)
3. Lifchitz, Y., Avrithis, Y., Picard, S., Bursuc, A.: Dense classification and implanting for few-shot learning. In: Proceedings of the IEEE Conference on Computer Vision and Pattern Recognition, pp. 9258–9267 (2019)
4. Li, W., Wang, L., Xu, J., Huo, J., Gao, Y., Luo, J.: Revisiting local descriptor based image-to-class measure for few-shot learning. In: Proceedings of the IEEE Conference on Computer Vision and Pattern Recognition (2019)
5. Wertheimer, D., Hariharan, B.: Few-shot learning with localization in realistic settings. In: Proceedings of the IEEE Conference on Computer Vision and Pattern Recognition, pp. 6558–6567 (2019)
6. Gidaris, S., Komodakis, N.: Generating Classification Weights with GNN Denoising Autoencoders for Few-Shot Learning. arXiv preprint arXiv:1905.01102 (2019)
7. Kim, J., Kim, T., Kim, S., Yoo, C.D.: Edge-labeling graph neural network for few-shot learning. In: Proceedings of the IEEE Conference on Computer Vision and Pattern Recognition, pp. 11–20 (2019)
8. Li, H., Eigen, D., Dodge, S., Zeiler, M., Wang, X.: Finding task-relevant features for few-shot learning by category traversal. In: Proceedings of the IEEE Conference on Computer Vision and Pattern Recognition (2019)
9. Chu, W.H., Li, Y.J., Chang, J.C., Wang, Y.C.F.: Spot and learn: a maximum-entropy patch sampler for few-shot image classification. In: Proceedings of Conference on Computer Vision and Pattern Recognition (2019)
10. Zhang, H., Zhang, J., Koniusz, P.: Few-shot learning via saliency-guided hallucination of samples. In: Proceedings of the IEEE Conference on Computer Vision and Pattern Recognition, pp. 2770–2779 (2019)
11. Chen, Z., Fu, Y., Wang, Y.X., Ma, L., Liu, W., Hebert, M.: Image deformation meta-networks for one-shot learning. In: Proceedings of the IEEE Conference on Computer Vision and Pattern Recognition (2019)
12. Ravi, S., Larochelle, H.: Optimization as a model for few-shot learning (2016)
13. Vinyals, O., Blundell, C., Lillicrap, T., Wierstra, D.: Matching networks for one shot learning. In: Advances in Neural Information Processing Systems (2016)
14. Finn, C., Abbeel, P., Levine, S.: Model-agnostic meta-learning for fast adaptation of deep networks. In: Proceedings of the 34th International Conference on Machine Learning, vol. 70, pp. 1126–1135, August 2017
15. Snell, J., Swersky, K., Zemel, R.: Prototypical networks for few-shot learning. In: Advances in Neural Information Processing Systems, pp. 4077–4087 (2017)
16. Sung, F., Yang, Y., Zhang, L., Xiang, T., Torr, P.H., Hospedales, T.M.: Learning to compare: relation network for few-shot learning. In: Proceedings of the IEEE Conference on Computer Vision and Pattern Recognition (2018)

17. Mishra, N., Rohaninejad, M., Chen, X., Abbeel, P.: A simple neural attentive meta-learner. arXiv pre-print arXiv:1707.03141 (2017)
18. Franceschi, L., Frasconi, P., Salzo, S., Grazzi, R., Pontil, M.: Bilevel programming for hyperparameter optimization and meta-learning. arXiv preprint arXiv:1806.04910 (2018)
19. Zhang, R., Che, T., Ghahramani, Z., Bengio, Y., Song, Y.: MetaGAN: an adversarial approach to few-shot learning. In: Advances in Neural Information Processing Systems, pp. 2365–2374 (2018)
20. Munkhdalai, T., Yuan, X., Mehri, S., Trischler, A.: Rapid adaptation with conditionally shifted neurons. arXiv preprint arXiv:1712.09926 (2017)
21. Li, Z., Zhou, F., Chen, F., Li, H.: Meta-SGD: learning to learn quickly for few-shot learning. arXiv pre-print arXiv:1707.09835 (2017)
22. Gidaris, S., Komodakis, N.: Dynamic few-shot visual learning without forgetting. In: Proceedings of the IEEE Conference on Computer Vision and Pattern Recognition, pp. 4367–4375 (2018)
23. Garcia, V., Bruna, J.: Few-shot learning with graph neural networks. arXiv preprint arXiv:1711.04043 (2017)
24. Cai, Q., Pan, Y., Yao, T., Yan, C., Mei, T.: Memory matching networks for one-shot image recognition. In: Proceedings of the IEEE Conference on Computer Vision and Pattern Recognition, pp. 4080–4088 (2018)
25. Zhang, H., Koniusz, P.: Power normalizing second-order similarity network for few-shot learning. In: 2019 IEEE Winter Conference on Applications of Computer Vision (WACV), pp. 1185–1193. IEEE, January 2019
26. Nichol, A., Achiam, J., Schulman, J.: On first-order meta-learning algorithms. arXiv preprint arXiv:1803.02999 (2018)
27. Liu, Y., et al.: Learning to propagate labels: transductive propagation network for few-shot learning. arXiv preprint arXiv:1805.10002 (2018)
28. Jiang, X., Havaei, M., Varno, F., Chartrand, G., Chapados, N., Matwin, S.: Learning to learn with conditional class dependencies (2018)
29. Allen, K.R., Shin, H., Shelhamer, E., Tenenbaum, J.B.: Variadic meta-learning by Bayesian nonparametric deep embedding (2018)
30. Koch, G., Zemel, R., Salakhutdinov, R.: Siamese neural networks for one-shot image recognition. In: ICML Deep Learning Workshop, vol. 2, July 2015

An Alternating DCA-Based Approach for Reduced-Rank Multitask Linear Regression with Covariance Estimation

Vinh Thanh Ho$^{(\boxtimes)}$ and Hoai An Le Thi

Computer Science and Applications Department, LGIPM, University of Lorraine, Metz, France
{vinh-thanh.ho,hoai-an.le-thi}@univ-lorraine.fr

Abstract. We investigate a nonconvex, nonsmooth optimization approach based on DC (Difference of Convex functions) programming and DCA (DC Algorithm) for the reduced-rank multitask linear regression problem with covariance estimation. The objective is to model the linear relationship between a multitask response and more explanatory variables by estimating a low-rank coefficient matrix and a covariance matrix. The problem is formulated as minimizing the constrained negative log-likelihood function of these two matrix variables. Then, we consider a reformulation of this problem which takes the form of a *partial* DC program i.e. it is a standard DC program for each variable when fixing the other variable. Next, an alternating version of a standard DCA scheme is developed. Numerical results on many synthetic multitask linear regression datasets and benchmark real datasets show the efficiency of our approach in comparison with the existing alternating/joint methods.

Keywords: DC programming · DCA · Alternating DCA · Reduced-rank multitask linear regression · Covariance estimation

1 Introduction

In this paper, we consider the reduced-rank multitask linear regression problem with covariance estimation (see, e.g., [7]). Given m different tasks with the d-dimensional feature vector denoted $\phi_i \in \mathbb{R}^d$, the corresponding respond denoted $z_i \in \mathbb{R}^m$ is generated using the linear model

$$z_i = X\phi_i + \epsilon_i, \tag{1}$$

where $X \in \mathbb{R}^{m \times d}$ is an unknown matrix whose rows represent the coefficient vector for each task; the error $\epsilon_i \in \mathbb{R}^m$ is assumed from a centered multivariate normal distribution with a covariance matrix $\mathrm{Cov}(\epsilon_i) = (\Theta)^{-1}, \Theta \in \mathbb{R}^{m \times m}$. In most applications of the multivariate regression problem, the errors in the regression model (e.g., ϵ_i in (1)) are distributed with an unknown covariance

© Springer Nature Switzerland AG 2020
I. S. Kotsireas and P. M. Pardalos (Eds.): LION 14 2020, LNCS 12096, pp. 264–277, 2020.
https://doi.org/10.1007/978-3-030-53552-0_25

matrix $Cov(\epsilon_i)$ (see, e.g., [26, 34, 36]). Thus, the estimation of the noise covariance matrix plays an important role. The techniques for estimating the noise covariance matrix for other problems have been developed in many years (see, e.g., [5, 27]).

The objective is to find the matrices X and Θ from n points $\{(z_i, \phi_i)\}_{i=1,\dots,n}$. In the high-dimensional setting, the problem aims to minimize the constrained negative log-likelihood function:

$$\min \left[\frac{1}{n} \sum_{i=1}^{n} (z_i - X\phi_i)^\top \Theta (z_i - X\phi_i) \right] - \log \det(\Theta) \tag{2}$$

$$\text{s.t. } X \in \mathcal{X}, \Theta \in \mathcal{Y},$$

where $\mathcal{X} = \{X \in \mathbb{R}^{m \times d} : \text{rank}(X) \le r\}$ represents the low-rank constraint set and $\mathcal{Y} = \{\Theta \in \mathbb{R}^{m \times m} : \Theta \succeq 0\}$ is the set of positive semi-definite (PSD) matrices.

This problem has many real-world applications ranging from chemometrics (see, e.g., [39]) to imaging neuroscience (see, e.g., [8]), to quantitative finance and risk management [25], to bioinformatics [9], to robotics (see e.g. [4,10]), to cite a few. For instance, in robotics, multivariate regression analysis is applied to evaluate the impact of robotic technique and high surgical volume on the cost of radical prostatectomy [10]. In another robotics application [4], linear regression analysis is performed to quantify the effect of surgeon experience on the operating time for each surgical step in the robotic-assisted laparoscopic prostatectomy procedure. In bioinformatics, the multitask regression algorithms are developed to solve the genomic selection problem in the fields of plant/animal breeding and genetic epidemiology (see [9] for more details).

In general, it is very hard to search globally optimal solutions to the problem (2) due to a double difficulty: first, the objective function of (2) is nonconvex in the variable (X, Θ), and, second, the rank function in the constraint set $\mathcal{X}$ is discontinuous and nonconvex.

For solving the problem (2), some existing approaches used an alternating optimization procedure on the variable (X, Θ). In particular, a classic Alternating Method (AM) will alternate between computing two variables X and Θ at every iteration (see, e.g., [29]). It leads to solving, at each iteration, a reduced-rank regression problem in X (see [1]) and a convex program in Θ. Recently, Ha et al. [7] have proposed an Alternating method using Gradient Descent method (AGD) for solving (2). The AGD method differs from the AM method by the fact that the AGD performs one iteration of the gradient descent method for solving the reduced-rank regression problem. Another approach without computing two variables alternatively is the joint gradient descent (JGD) method [7] which takes one gradient descent step in (X, Θ). All three AM, AGD, and JGD algorithms are described completely in the Appendix.

In this work, we continue using the alternating optimization procedure on the variable (X, Θ). However since the problem (2) is nonconvex in X, we will investigate an alternating approach for solving (2) based on DC (Difference of

Convex functions) programming and DCA (DC Algorithm) (see, e.g., [12,19, 21,30–32] and the list of references in [21]) which are well-known as powerful nonsmooth, nonconvex optimization tools. DCA aims to solve a standard DC program that consists in minimizing a DC function $f = g - h$ (with g, h being convex functions) over a convex set or on the whole space. Here $g - h$ is called a DC decomposition of f, while g and h are DC components of f. The idea of the standard DCA is, at each iteration k, approximating the second DC component h by its affine minorant and then solving the resulting convex subproblem.

Our Contribution. First, we consider a penalized reformulation of the problem (2) which can be expressed as a *partial DC program* i.e. it is a standard DC program in each variable while fixing other variables. Second, we propose an alternating DCA scheme for solving this problem. In particular, at each iteration, we perform one iteration of standard DCA for the corresponding DC program in each variable when fixing the other variable. Finally, we evaluate our alternating approach by comparing with three alternating/joint methods on six synthetic multitask linear regression datasets and eight benchmark real datasets.

The rest of the paper is organized as follows. Section 2 gives a brief introduction to Partial DC programming and Alternating DCA, and then shows how to apply them for solving the penalty problem of (2). Section 3 reports the numerical results on several test problems. Finally, Sect. 4 concludes the paper.

2   Solution Method

DC programming and DCA were introduced by Pham Dinh Tao in a preliminary form in 1985 and have been extensively developed by Le Thi Hoai An and Pham Dinh Tao since 1994. DCA is well-known as an efficient approach in the nonconvex programming framework (see, e.g., [12,19,21,30–32]). In recent years, numerous DCA-based algorithms have been developed for successfully solving large-scale nonsmooth/nonconvex programs in several application areas (see, e.g., [13–16,18,20,22,24,28,33,38] and the list of references in [21]). For a comprehensible survey on thirty years of development of DCA, the reader is referred to the recent work [21].

The standard DCA scheme is described below. Its convergence properties are given completely in, e.g, [30].

Standard DCA scheme

Initialization: Let $x^0 \in \mathbb{R}^p$ be a best guess. Set $k = 0$.

repeat

  1. Calculate $\overline{x}^k \in \partial h(x^k)$.

  2. Calculate $x^{k+1} \in \operatorname{argmin}\{g(x) - \langle x, \overline{x}^k \rangle : x \in \mathbb{R}^p\}$.

  3. $k = k + 1$.

until convergence of $\{x^k\}$.

2.1 A Brief Introduction to Partial DC Programming and Alternating DCA

Now, we briefly introduce partial DC programming and Alternating DCA in [17]. A partial DC (PDC) program takes the form

$$\min F(x, y) := G(x, y) - H(x, y) \text{ s.t. } (x, y) \in \mathbb{R}^p \times \mathbb{R}^q, \tag{3}$$

where G and H are partial convex functions in the sense that they are convex in each variable when fixing all other variables. Such a function F is called a *partial DC function*.

An alternating version of DCA for solving (3) consists in, at the iteration k, *alternatively* computing x^{k+1} and y^{k+1} by performing one iteration of standard DCA for solving the following DC programs in variable x and y, respectively:

$$\min F(x, y^k) := G(x, y^k) - H(x, y^k) \text{ s.t. } x \in \mathbb{R}^p,$$

and

$$\min F(x^{k+1}, y) := G(x^{k+1}, y) - H(x^{k+1}, y) \text{ s.t. } y \in \mathbb{R}^q.$$

This version, named Alternating DCA, is described as follows.

Alternating DCA scheme
Initialization: Let $(x^0, y^0) \in \mathbb{R}^p \times \mathbb{R}^q$ be a best guess. Set $k = 0$.
repeat
   1. Calculate $\overline{x}^k \in \partial_x H(x^k, y^k)$.
   2. Calculate $x^{k+1} \in \operatorname{argmin}\{G(x, y^k) - \langle x, \overline{x}^k \rangle : x \in \mathbb{R}^p\}$.
   3. Calculate $\overline{y}^k \in \partial_y H(x^{k+1}, y^k)$.
   4. Calculate $y^{k+1} \in \operatorname{argmin}\{G(x^{k+1}, y) - \langle y, \overline{y}^k \rangle : y \in \mathbb{R}^q\}$.
   5. $k = k + 1$.
until convergence of $\{(x^k, y^k)\}$.

2.2 A Reformulation of the Problem (2)

Recall that our main goal in this paper is to investigate a nonconvex approach based on DCA for directly solving the nonconvex problem (2). Thus, we need to reformulate the problem (2) as an unconstrained PDC program (3). We penalize the difficult low-rank constraint in $\mathcal{X}$ into the objective function of (2) by using the squared distance function $d_{\mathcal{X}}^2$ as a penalty function (see e.g. [23,35]). Thus, for a given parameter $\alpha > 0$, the problem (2) can be transformed into the following optimization problem

$$\min \frac{1}{n} \sum_{i=1}^{n} (z_i - X\phi_i)^{\top} \Theta(z_i - X\phi_i) - \log \det(\Theta) + \alpha d_{\mathcal{X}}^2(X) \tag{4}$$

$$\text{s.t. } X \in \mathbb{R}^{m \times d}, \Theta \succeq 0.$$

Here the squared distance function $d_{\mathcal{X}}^2$ is defined as $d_{\mathcal{X}}^2(X) := \min_{Y \in \mathcal{X}} \|Y - X\|_F^2$ and $\| \cdot \|_F$ is a Frobenius norm.

Since the PSD constraint $\Theta \in \mathcal{Y}$ in (4) is convex, we use the indicator function $\chi_{\Theta \succeq 0}$, defined as $\chi_{\mathcal{C}}(x) = 0$ if $x \in \mathcal{C}$, $+\infty$ otherwise. Thus, we can derive from (4) the following formulation

$$\min F(X, \Theta) := \frac{1}{n} \sum_{i=1}^{n} (z_i - X\phi_i)^\top \Theta (z_i - X\phi_i)$$
$$- \log \det(\Theta) + \alpha d_{\mathcal{X}}^2(X) + \chi_{\Theta \succeq 0}(\Theta), \tag{5}$$
$$\text{s.t. } X \in \mathbb{R}^{m \times d}, \Theta \in \mathbb{R}^{m \times m}.$$

Note that if (X^*, Θ^*) is a globally optimal solution to the problem (5) and $(X^*, \Theta^*) \in \mathcal{X} \times \mathcal{Y}$, then (X^*, Θ^*) is also a globally optimal solution to the problem (2).

Knowing that the function $d_{\mathcal{X}}^2$ is a DC function with DC decomposition

$$d_{\mathcal{X}}^2(X) = \min_{Y \in \mathcal{X}} \|X - Y\|_F^2 = \|X\|_F^2 - \max_{Y \in \mathcal{X}} (2\langle X, Y \rangle - \|Y\|_F^2),$$

the problem (5) can be rewritten as a partial DC program

$$\min F(X, \Theta) := G(X, \Theta) - H(X, \Theta) \tag{6}$$

where

$$G(X, \Theta) := \frac{1}{n} \sum_{i=1}^{n} (z_i - X\phi_i)^\top \Theta (z_i - X\phi_i) - \log \det(\Theta) + \alpha \|X\|_F^2 + \chi_{\Theta \succeq 0}(\Theta),$$
$$H(X, \Theta) := \alpha \max_{Y \in \mathcal{X}} (2\langle X, Y \rangle - \|Y\|_F^2).$$

Since the function $H(X, \Theta)$ is convex in X and the function $\log \det$ is concave in Θ [3], the functions G and H are partially convex.

2.3   Alternating DCA for Solving the Problem (6)

According to the Alternating DCA scheme in Sect. 2.1, we need to construct two sequences $\{(X^k, \Theta^k)\}$ and $\{(U^k, V^k)\}$ such that

$$U^k \in \partial_X H(X^k, \Theta^k),$$
$$X^{k+1} \in \text{argmin}\{G(X, \Theta^k) - \langle X, U^k \rangle : X \in \mathbb{R}^{m \times d}\}, \tag{7}$$

and

$$V^k \in \partial_\Theta H(X^{k+1}, \Theta^k),$$
$$\Theta^{k+1} \in \text{argmin}\{G(X^{k+1}, \Theta) - \langle \Theta, V^k \rangle : \Theta \in \mathbb{R}^{m \times m}\}. \tag{8}$$

From the subdifferential formula for a maximum function of an infinite number of affine functions [11, 41] and the definition of the function H, we have

$$\partial_X H(X, \Theta) \supset 2\alpha\text{co}\{\text{Proj}_{\mathcal{X}}(X)\}, \quad \partial_\Theta H(X, \Theta) = \{0\}.$$

Here $\text{Proj}_{\mathcal{C}}$ and $\text{co}(\mathcal{C})$ denote, respectively, the projection operator on the set $\mathcal{C}$ and the convex hull of $\mathcal{C}$. Thus, we can choose the subgradients $U^k \in \partial_X H(X^k, \Theta^k)$ and $V^k \in \partial_\Theta H(X^{k+1}, \Theta^k)$ as follows:

$$U^k = 2\alpha W^k, \quad W^k \in \text{Proj}_{\mathcal{X}}(X^k), \quad \text{and} \quad V^k = 0. \tag{9}$$

Solving the convex subproblem (7) amounts to solving the problem

$$\min_{X \in \mathbb{R}^{m \times d}} \left[\frac{1}{n} \sum_{i=1}^{n} (z_i - X\phi_i)^\top \Theta^k (z_i - X\phi_i) \right] + \alpha \|X\|_F^2 - \langle U^k, X \rangle. \tag{10}$$

By setting the derivative of the objective function of the last problem (10) to zero, we can see that its optimal solution X^{k+1} satisfies the Sylvester equation

$$A^k X + X B^k = C^k, \tag{11}$$

where the matrices $A^k \in \mathbb{R}^{m \times m}$, $B^k \in \mathbb{R}^{d \times d}$, and $C^k \in \mathbb{R}^{m \times d}$ are defined as

$$A^k = \alpha(\Theta^k)^{-1}, \quad B^k = \frac{1}{n} \sum_{i=1}^{n} \left(\phi_i \phi_i^\top \right), \quad C^k = \alpha(\Theta^k)^{-1} W^k + \frac{1}{n} \sum_{i=1}^{n} \left(z_i \phi_i^\top \right).$$

Here Z^{-1} denotes an inverse of a matrix Z.

From (8) and the definition of G, Θ^{k+1} is an optimal solution to the convex program

$$\min_{\Theta \succeq 0} \left[\frac{1}{n} \sum_{i=1}^{n} (z_i - X^{k+1}\phi_i)^\top \Theta (z_i - X^{k+1}\phi_i) \right] - \log \det(\Theta). \tag{12}$$

It is easy to check that the problem (12) has a closed-form optimal solution [7] as follows:

$$\Theta^{k+1} = \left(\frac{1}{n} \sum_{i=1}^{n} (z_i - X^{k+1}\phi_i)(z_i - X^{k+1}\phi_i)^\top \right)^{-1}. \tag{13}$$

Finally, the Alternating DCA scheme applied to (6) can be summarized in Algorithm 1 (Al-DCA).

Algorithm 1. Al-DCA: Alternating DCA for solving (6)

Initialization: Let ε be a sufficiently small positive number. Let $X^0 \in \mathbb{R}^{m \times d}$, $\Theta^0 \in \mathbb{R}^{m \times m}$, $\Theta^0 \succeq 0$, $\alpha > 0$. Set $k = 0$.
repeat
   1. Compute $W^k \in \text{Proj}_{\mathcal{X}}(X^k)$.
   2. Compute X^{k+1} by solving the Sylvester equation (11).
   3. Compute Θ^{k+1} using (13).
   4. $k = k + 1$.
until Stopping criteria are satisfied.

The convergence properties of our algorithm are derived from [17] and given in Theorem 1 below.

Theorem 1. *Convergence properties of* Al-DCA
i) Al-DCA *generates the sequence* $\{(X^k, \Theta^k)\}$ *such that the sequence of the objective function values* $\{F(X^k, \Theta^k)\}$ *is decreasing.*
ii) Assume that the sequence $\{(X^k, \Theta^k)\}$ *is bounded and* (X^*, Θ^*) *is its limit point. Then,* (X^*, Θ^*) *is a weak critical point of* $G - H$ *i.e.* $\partial_X G(X^*, \Theta^*) \cap \partial_X H(X^*, \Theta^*) \neq \emptyset$, $\partial_\Theta G(X^*, \Theta^*) \cap \partial_\Theta H(X^*, \Theta^*) \neq \emptyset$.

Remark 1. In numerical experiments, X^* obtained by Al-DCA does often not belong to $\mathcal{X}$. Thus, after running Al-DCA, we propose performing one projection step: projecting X^* into the set $\mathcal{X}$ and then updating Θ^* by (13).

3   Numerical Experiments

Our experiments aim to compare the proposed alternating algorithm Al-DCA with other alternating/joint algorithms for the multitask linear regression problem (2).

Comparative Algorithms. As listed in Sect. 1, we consider three alternating/joint algorithms for solving the problem (2): classic alternating method (AM), alternating method using gradient descent method (AGD) [7], and joint gradient method (JGD) [7] (see the Appendix for more details).

Datasets. We test the four algorithms Al-DCA, AGD, JGD, and AM on six synthetic datasets and eight real datasets.

We generate synthetic datasets using the linear model (1) as described in the works, e.g., [2,7,34,40]. Specifically, the feature vector ϕ_i is drawn independently from a multivariate normal distribution $\mathcal{N}(0, \Sigma_\phi)$ where each element $\Sigma_\phi(i, j) = 0.5^{|i-j|}$. Similarly, the error ϵ_i is also generated from $\mathcal{N}(0, \sigma^2 \Sigma_\epsilon)$ where σ^2 is chosen such that the corresponding signal-to-noise is equal to 1 (see e.g. [2,7]) and Σ_ϵ is defined by the following type: AR(1), denoted ar(ρ_ϵ), with $\Sigma_\phi(i, j) = (\rho_\epsilon)^{|i-j|}$. Here, ρ_ϵ represents a correlation parameter; the larger its value is, the more the degree of dependence of errors would be. The coefficient matrix X is computed as $X = AB$ where the orthonormal matrices $A \in \mathbb{R}^{m \times r}$ and $B \in \mathbb{R}^{r \times d}$ are generated form $\mathcal{N}(0, 1)$. Finally, the respond vector $z_i \in \mathbb{R}^m$ is computed using (1). By setting $r = 3$, $m \in \{10, 20, 60\}$, $d \in \{10, 20, 40\}$, $\rho_\epsilon \in \{0, 0.5\}$, we have six synthetic datasets which are summarized in Table 1. For each synthetic dataset, we generate 50 training samples and 1000 test samples in each run time, and we repeat the whole process 30 times.

As for real datasets, we test on eight benchmark multitask regression datasets[1]. These datasets are collected from various interesting applications and can be found in the recent work of [37] (see the references therein). The parameters of these datasets and the given values of r are provided in Table 3. We split

---

[1] For the detailed descriptions of all datasets, the reader is referred to [37] and the website http://mulan.sourceforge.net/datasets-mtr.html.

each real dataset into a training set containing the first 75% of dataset and a test set containing the rest of dataset.

Comparison Criteria and Stopping Criteria. We are interested in the following aspects: prediction error and CPU time (in seconds) for training the solution (X^*, Θ^*). As for synthetic datasets, the prediction error is defined by the mean squared error (MSE) [2]

$$\text{MSE} = \frac{\sum_{i=1}^{n} \|X\phi_i - AB\phi_i\|_2^2}{nm}, \tag{14}$$

while the relative root mean squared error (RRMSE) on real datasets is used to measure the prediction error of the algorithm on each task and defined as [37]

$$\text{RRMSE} = \sqrt{\frac{\sum_{i=1}^{n} \|\hat{z}_i - z_i\|_2^2}{\sum_{i=1}^{n} \|\bar{z}_i - z_i\|_2^2}}, \tag{15}$$

where $\hat{z}_i$ is a respond vector estimated by the algorithm and $\bar{z}_i$ is the mean value of the respond vectors on the training set. We stop the algorithms if the relative difference between two consecutive points (X^{k-1}, Θ^{k-1}) and (X^k, Θ^k) or between two corresponding objective function values is less than or equal to ε.

Set Up Parameters. Our experiment is performed in MATLAB R2016b on a PC Intel(R) Core(TM) i5-3470 CPU @ 3.20GHz of 8 GB RAM. The MATLAB's sylvester function is used for solving Sylvester equation (11). The projection algorithm Proj on the set $\mathcal{X}$ is given in [6]. The MATLAB's svd function is used for computing the singular value decomposition (SVD) in the Proj algorithm on the set $\mathcal{X}$. All algorithms start with the same point (X^0, Θ^0). The starting point X^0 is set to a zero matrix in $\mathbb{R}^{m \times d}$, and the matrix Θ^0 is computed using (13). To validate the performance of the algorithms on all synthetic/real datasets, we consider the following validation procedure: first we run the algorithm with the different parameters on the training set, then choose the solution (X^*, Θ^*) that provides the best objective function value $F(X^*, \Theta^*)$, and finally evaluate the obtained model using MSE (14) or RRMSE (15) on the test set. The ranges of parameters η_X, η_Θ, and α are defined as: $\alpha \in \{5, 10, 100\}$, $\eta_X \in \{10^{-5}, 10^{-4}, \ldots, 10^2\}$, η_Θ belongs in a geometric sequence from 5 to 400 [7]. The default tolerance is $\varepsilon = 10^{-3}$.

Descriptions of Result Tables. The average MSE and its standard deviation obtained by all comparative algorithms on six synthetic datasets over 30 run times are reported in Table 1. The average results of training time of the algorithms on synthetic datasets are given in Table 2. Table 3 shows the experimental results on real datasets in terms of RRMSE and training time.

Comments on Numerical Results
Synthetic Datasets. We observe from Table 1 that, in terms of MSE, Al-DCA is more efficient than AGD, JGD, and AM. To be specific, Al-DCA is the best

Table 1. Comparative results of Al-DCA, AGD, JGD, and AM in terms of the average of mean-squared-error MSE defined by (14) (upper row) and its standard deviation (lower row) on six synthetic datasets over 30 run times. Bold values indicate the best result.

d	m	Σ_ϵ	Al-DCA	AGD	JGD	AM
10	60	ar(0.0)	**2.44e−02**	2.65e−02	2.64e−02	5.73e−01
			6.47e−03	5.78e−03	5.71e−03	8.94e−01
		ar(0.5)	**2.00e−02**	2.42e−02	2.40e−02	4.34e−01
			6.15e−03	4.64e−03	4.81e−03	6.04e−01
20	10	ar(0.0)	**4.79e−02**	5.83e−02	1.37e−01	5.04e−02
			1.67e−02	2.39e−02	2.24e−02	1.65e−02
		ar(0.5)	**3.30e−02**	5.47e−02	1.33e−01	3.56e−02
			1.24e−02	2.87e−02	2.01e−02	1.10e−02
40	20	ar(0.0)	**6.13e−02**	6.15e−02	6.23e−02	2.34e+00
			8.01e−03	8.07e−03	8.28e−03	9.66e+00
		ar(0.5)	6.35e−02	**6.27e−02**	6.31e−02	5.72e−01
			8.76e−03	8.54e−03	8.42e−03	5.96e−01

Table 2. Comparative results of Al-DCA, AGD, JGD, and AM in terms of the average of training time in seconds (upper row) and its standard deviation (lower row) on six synthetic datasets over 30 run times. Bold values indicate the best result.

d	m	Σ_ϵ	Al-DCA	AGD	JGD	AM
10	60	ar(0.0)	1.52e−02	**4.08e−03**	4.79e−03	5.81e−03
			9.29e−03	1.60e−03	1.62e−03	2.88e−03
		ar(0.5)	1.44e−02	**3.92e−03**	9.79e−03	4.81e−03
			5.35e−03	2.00e−03	3.78e−03	1.34e−03
20	10	ar(0.0)	3.14e−02	4.25e−02	**8.82e−04**	2.32e−03
			1.98e−02	1.65e−02	2.30e−04	1.04e−03
		ar(0.5)	2.06e−02	3.50e−02	**9.33e−04**	1.97e−03
			5.48e−03	1.75e−02	8.11e−05	3.48e−04
40	20	ar(0.0)	1.20e−01	7.35e−02	**1.40e−03**	4.87e−03
			2.60e−02	7.10e−02	6.87e−04	7.55e−04
		ar(0.5)	9.77e−02	6.10e−02	**8.92e−04**	4.64e−03
			1.96e−02	4.54e−02	1.63e−04	2.04e−04

on 5/6 datasets – the ratio of gain of Al-DCA versus AGD, JGD, and AM varies from 0.32% to 39.6%, from 1.60% to 75.1% and from 4.96% to 97.3%, respectively. Moreover, Al-DCA well performs for two model errors (independent, moderately correlated). In terms of training time, all four algorithms run very fast (less than 0.1 s).

Table 3. Comparative results of Al-DCA, AGD, JGD, and AM in terms of the relative root-mean-squared error RRMSE defined by (15) (upper row) and training time in seconds (lower row) on eight real datasets. Bold values indicate the best result.

Name	n	d	m	r	Al-DCA	AGD	JGD	AM
andro	49	30	6	4	**8.42e−01**	3.75e+00	3.75e+00	1.22e+00
					5.17e−01	1.21e−03	**6.32e−04**	4.06e−03
atp7d	296	411	6	4	**3.63e+00**	4.60e+00	4.83e+00	2.74e+04
					2.45e+01	7.93e−03	**3.05e−03**	1.12e−01
oes10	403	298	16	4	**5.18e−01**	1.18e+00	1.26e+00	1.95e+01
					5.41e−01	2.82e−02	**3.56e−03**	7.53e−02
osales	639	376	12	4	**9.89e−01**	1.05e+00	1.05e+00	1.09e+03
					4.71e−02	4.61e−03	**4.50e−03**	1.34e−01
rf2	7679	576	8	6	**1.09e−01**	1.48e+00	1.48e+00	7.70e−01
					2.67e+00	2.54e−02	**2.52e−02**	3.48e−01
scpf	1137	8	3	3	9.88e−01	1.03e+00	1.03e+00	**9.35e−01**
					6.48e−03	**6.22e−04**	6.66e−04	1.65e−03
sf1	323	7	3	3	**9.61e−01**	1.02e+00	1.07e+00	1.24e+00
					1.49e−03	1.40e−03	**4.12e−04**	1.51e−03
wq	1060	16	14	4	9.98e−01	1.01e+00	1.62e+00	**9.14e−01**
					1.48e−02	1.45e−02	**9.50e−04**	2.85e−03

Real Datasets. The error RRMSE obtained by Al-DCA is the best on 6/8 datasets, especially the rf2 dataset with more than 7000 samples. In particular, as for the rf2 dataset, Al-DCA significantly outperforms AGD, JGD and AM with the ratio of gain of 92.6%, 92.6% and 85.8%, respectively. On other datasets, the ratio of gain varies from 1.18% to 77.5%, from 4.07% to 77.5% and from 22.5% to 99.9%. Comparing with AM, Al-DCA is worse on 2/8 datasets with the ratio from 5.36% to 9.19%. In Table 3, training times of Al-DCA are reasonable (less than 1 s on 6/8 datasets and 25 s on the atp7d and rf2 datasets).

4   Conclusions

We have investigated a new approach based on DC programming and DCA for solving the reduced-rank multitask linear regression problem with covariance estimation. An Alternating version of DCA (Al-DCA) has been developed. Numerical results on synthetic/real datasets have turned out that the Al-DCA is more efficient than exiting alternating/joint methods in terms of the prediction error, and runs within a reasonable consuming time. In the future, we plan to show the efficiency of Al-DCA on many other synthetic/real datasets with different model errors as well as various applications.

Appendix: Comparative Algorithms for Solving the Problem (2)

The AM method alternates between computing the variable X and Θ at every iteration. In particular, at iteration k, for fixed Θ, we need to compute X^{k+1}, an optimal solution to the following problem (see, e.g., [1])

$$\min \frac{1}{n} \sum_{i=1}^{n} (z_i - X\phi_i)^\top \Theta^k (z_i - X\phi_i) \text{ s.t. } \text{rank}(X) = r. \qquad (16)$$

Let us denote by Z (resp. Φ) a matrix in $\mathbb{R}^{m \times n}$ (resp. $\mathbb{R}^{d \times n}$) whose each column is a vector z_i (resp. ϕ_i); and define $D^k := (\Phi\Phi^\top)^{(-1/2)} (\Phi Z^\top)(\Theta^k)^{(1/2)}$. A reduced-rank regression estimate X^{k+1} of (16) is given by

$$X^{k+1} = \sum_{t=1}^{r} \lambda_t \left[(1/n)\Phi\Phi^\top \right]^{(-1/2)} u_t v_t^\top (\Theta^k)^{(-1/2)}, \qquad (17)$$

where the sequence $\{\lambda_t\}$ is the singular values of matrix D^k; $\{u_t\}$ and $\{v_t\}$ are the left-hand and right-hand singular vectors of D^k. For fixed X, the AM computes the point Θ^{k+1} using (13) at X^{k+1}. Note that the AM method does not have any parameters.

AM: classic Alternating Method for solving (2)

Initialization: Let ε be a sufficiently small positive number. Let $X^0 \in \mathbb{R}^{m \times d}$, $\Theta^0 \in \mathbb{R}^{m \times m}$, $\Theta^0 \succeq 0$. Set $k = 0$.

repeat
    1. Compute X^{k+1} using (17).
    2. Compute Θ^{k+1} using (13).
    3. $k = k + 1$.
until Stopping criteria are satisfied.

The AGD method differs from the AM method by the fact that the AGD performs one iteration of gradient descent method for solving the convex problem (16). In particular, X^{k+1} is computed as follows [7]:

$$X^{k+1} = \text{Proj}_\mathcal{X} \left(X^k + \frac{2\eta_X}{n} \Theta^k \sum_{i=1}^{n} (z_i - X^k \phi_i)\phi_i^\top \right), \qquad (18)$$

where the step size η_X is a tuning parameter. It is similar for the AM to update the point Θ^{k+1} using (13) at X^{k+1}.

AGD: Alternating method using Gradient Descent method for solving (2)

Initialization: Let ε be a sufficiently small positive number. Let $X^0 \in \mathbb{R}^{m \times d}$, $\Theta^0 \in \mathbb{R}^{m \times m}$, $\Theta^0 \succeq 0$. Set $k = 0$.

repeat

    1. Compute X^{k+1} using (18).

    2. Compute Θ^{k+1} using (13).

    3. $k = k + 1$.

until Stopping criteria are satisfied.

The JGD method does not compute two variables alternatively, but takes one gradient descent step in the joint variable (X, Θ). For estimating X^{k+1}, it is the same as (18), while the point Θ^{k+1} is computed by using gradient descent method for (12) at the point (X^k, Θ^k) as follows [7]:

$$\Theta^{k+1} = \mathrm{Proj}_{\mathcal{Y}} \left(\Theta^k + \eta_\Theta \Delta^k \right), \tag{19}$$

where the step size η_Θ is a tuning parameter and

$$\Delta^k = (\Theta^k)^{(-1)} - \left[\frac{1}{n} \Theta^k \sum_{i=1}^n (z_i - X^k \phi_i)(z_i - X^k \phi_i)^\top \right].$$

JGD: Joint Gradient Descent method for solving (2)

Initialization: Let ε be a sufficiently small positive number. Let $X^0 \in \mathbb{R}^{m \times d}$, $\Theta^0 \in \mathbb{R}^{m \times m}$, $\Theta^0 \succeq 0$. Set $k = 0$.

repeat

    1. Compute X^{k+1} using (18).

    2. Compute Θ^{k+1} using (19).

    3. $k = k + 1$.

until Stopping criteria are satisfied.

References

1. Aldrin, M.: Reduced-Rank Regression, vol. 3, pp. 1724–1728. Wiley, Hoboken (2002)
2. Chen, L., Huang, J.Z.: Sparse reduced-rank regression with covariance estimation. Stat. Comput. 461–470 (2014). https://doi.org/10.1007/s11222-014-9517-6
3. Cover, T.M., Thomas, A.: Determinant inequalities via information theory. SIAM J. Matrix Anal. Appl. **9**(3), 384–392 (1988)
4. Dev, H., Sharma, N.L., Dawson, S.N., Neal, D.E., Shah, N.: Detailed analysis of operating time learning curves in robotic prostatectomy by a novice surgeon. BJU Int. **109**(7), 1074–1080 (2012)
5. Duník, J., Straka, O., Kost, O., Havlík, J.: Noise covariance matrices in state-space models: a survey and comparison of estimation methods - part i. Int. J. Adapt. Control Sig. Process. **31**(11), 1505–1543 (2017)

6. Eckart, C., Young, G.: The approximation of one matrix by another of lower rank. Psychometrika **1**, 211–218 (1936)
7. Ha, W., Foygel Barber, R.: Alternating minimization and alternating descent over nonconvex sets. ArXiv e-prints, September 2017
8. Harrison, L., Penny, W., Friston, K.: Multivariate autoregressive modeling of fMRI time series. NeuroImage **19**, 1477–1491 (2003)
9. He, D., Parida, L., Kuhn, D.: Novel applications of multitask learning and multiple output regression to multiple genetic trait prediction. Bioinformatics **32**(12), i37–i43 (2016)
10. Hyams, E., Mullins, J., Pierorazio, P., Partin, A., Allaf, M., Matlaga, B.: Impact of robotic technique and surgical volume on the cost of radical prostatectomy. J. Endourol. **27**(3), 298–303 (2013)
11. Ioffe, A., Tihomirov, V.: Theory of Extremal Problems. North-Holland (1979)
12. Le Thi, H.A.: Analyse numérique des algorithmes de l'optimisation d. C. Approches locale et globale. Codes et simulations numériques en grande dimension. Applications. Ph.D. thesis, University of Rouen (1994)
13. Le Thi, H.A.: Collaborative DCA: an intelligent collective optimization scheme, and its application for clustering. J. Intell. Fuzzy Syst. **37**(6), 7511–7518 (2019)
14. Le Thi, H.A.: DC programming and DCA for supply chain and production management: state-of-the-art models and methods. Int. J. Prod. Res. 1–37 (2019). https://doi.org/10.1080/00207543.2019.1657245
15. Le Thi, H.A., Ho, V.T.: Online learning based on online DCA and application to online classification. Neural Comput. **32**(4), 759–793 (2020)
16. Le Thi, H.A., Ho, V.T., Pham Dinh, T.: A unified DC programming framework and efficient DCA based approaches for large scale batch reinforcement learning. J. Glob. Optim. **73**(2), 279–310 (2018). https://doi.org/10.1007/s10898-018-0698-y
17. Le Thi, H.A., Huynh, V.N., Pham Dinh, T.: Alternating DC algorithm for partial DC programming (2016). Technical report, University of Lorraine
18. Le Thi, H.A., Nguyen, M.C.: Self-organizing maps by difference of convex functions optimization. Data Mining Knowl. Discov. (2), 1336–1365 (2014). https://doi.org/10.1007/s10618-014-0369-7
19. Le Thi, H.A., Pham Dinh, T.: The DC (difference of convex functions) programming and DCA revisited with DC models of real world nonconvex optimization problems. Ann. Oper. Res. **133**(1–4), 23–46 (2005)
20. Le Thi, H.A., Pham Dinh, T.: Difference of convex functions algorithms (DCA) for image restoration via a Markov random field model. Optim. Eng. **18**(4), 873–906 (2017). https://doi.org/10.1007/s11081-017-9359-0
21. Le Thi, H.A., Pham Dinh, T.: DC programming and DCA: thirty years of developments. Math. Program. **169**(1), 5–68 (2018). https://doi.org/10.1007/s10107-018-1235-y
22. Le Thi, H.A., Pham Dinh, T., Le, H.M., Vo, X.T.: DC approximation approaches for sparse optimization. Eur. J. Oper. Res. **244**(1), 26–46 (2015)
23. Le Thi, H.A., Pham Dinh, T., Ngai, H.V.: Exact penalty and error bounds in DC programming. J. Glob. Optim. **52**(3), 509–535 (2012)
24. Le Thi, H.A., Ta, A.S., Pham Dinh, T.: An efficient DCA based algorithm for power control in large scale wireless networks. Appl. Math. Comput. **318**, 215–226 (2018)
25. Lee, C.L., Lee, J.: Handbook of Quantitative Finance and Risk Management. Springer, Heidelberg (2010). https://doi.org/10.1007/978-0-387-77117-5

26. Lee, W., Liu, Y.: Simultaneous multiple response regression and inverse covariance matrix estimation via penalized Gaussian maximum likelihood. J. Multivariate Anal. **111**, 241–255 (2012)
27. Nez, A., Fradet, L., Marin, F., Monnet, T., Lacouture, P.: Identification of noise covariance matrices to improve orientation estimation by Kalman filter. Sensors **18**, 3490 (2018)
28. Ong, C.S., Le Thi, H.A.: Learning sparse classifiers with difference of convex functions algorithms. Optim. Methods Softw. **28**(4), 830–854 (2013)
29. Ortega, J., Rheinboldt, W.: Iterative Solution of Nonlinear Equations in Several Variables. Academic Press (1970)
30. Pham Dinh, T., Le Thi, H.A.: Convex analysis approach to DC programming: theory, algorithms and applications. Acta Mathematica Vietnamica **22**(1), 289–355 (1997)
31. Pham Dinh, T., Le Thi, H.A.: DC optimization algorithms for solving the trust region subproblem. SIAM J. Optim. **8**(2), 476–505 (1998)
32. Pham Dinh, T., Le Thi, H.A.: Recent advances in DC programming and DCA. In: Nguyen, N.-T., Le-Thi, H.A. (eds.) Transactions on Computational Intelligence XIII. LNCS, vol. 8342, pp. 1–37. Springer, Heidelberg (2014). https://doi.org/10.1007/978-3-642-54455-2_1
33. Phan, D.N., Le Thi, H.A., Pham Dinh, T.: Sparse covariance matrix estimation by DCA-based algorithms. Neural Comput. **29**(11), 3040–3077 (2017)
34. Rothman, A.J., Levina, E., Zhu, J.: Sparse multivariate regression with covariance estimation. J. Comput. Graph. Stat. **19**(4), 947–962 (2010)
35. Smith, A.E., Coit, D.W.: Constraint-handling techniques - penalty functions. In: Handbook of Evolutionary Computation, pp. C5.2:1–C5.2.6. Oxford University Press (1997)
36. Sohn, K.A., Kim, S.: Joint estimation of structured sparsity and output structure in multiple-output regression via inverse-covariance regularization. In: Lawrence, N.D., Girolami, M. (eds.) Proceedings of the Fifteenth International Conference on Artificial Intelligence and Statistics. Proceedings of Machine Learning Research, vol. 22, pp. 1081–1089. PMLR, La Palma (2012)
37. Spyromitros-Xioufis, E., Tsoumakas, G., Groves, W., Vlahavas, I.: Multi-target regression via input space expansion: treating targets as inputs. Mach. Learn. **104**(1), 55–98 (2016). https://doi.org/10.1007/s10994-016-5546-z
38. Tran, T.T., Le Thi, H.A., Pham Dinh, T.: DC programming and DCA for enhancing physical layer security via cooperative jamming. Comput. Oper. Res. **87**, 235–244 (2017)
39. Wold, S., Sjöström, M., Eriksson, L.: PLS-regression: a basic tool of chemometrics. Chem. Intell. Lab. Syst. **58**(2), 109–130 (2001)
40. Yuan, M., Lin, Y.: Model selection and estimation in the Gaussian graphical model. Biometrika **94**(1), 19–35 (2007)
41. Zălinescu, C.: Convex Analysis in General Vector Spaces. World Scientific (2002)

PSO-Based Cooperative Learning Using Chunking

Malek Sarhani[✉] and Stefan Voß

Institute of Information Systems, University of Hamburg, Hamburg, Germany
{malek.sarhani,stefan.voss}@uni-hamburg.de

Abstract. Bio-inspired optimization consists of drawing inspiration from the behavior and internal functioning of physical, biological and social systems to design enhanced optimization algorithms. Our aim in this paper is to enhance the performance of particle swarm optimization (PSO), when optimizing the feature selection (FS) problem, by combining two bio-inspired approaches which are chunking and cooperative learning. The experimental results show that our hybrid wrapper-filter approach gives competitive results in terms of the predictive accuracy on the training set.

Keywords: Particle swarm optimization · Cooperative learning · Chunking · Feature selection

1 Introduction

Nowadays, with the appeal of artificial intelligence, a challenging problem is to stimulate principles of real life and turn them into algorithmic solutions that can be applied to address real-life issues. The ultimate goal is to come up with simple, easy-to-implement ideas that are adequate for complex real-world problems in which traditional approaches have difficulties to be applied successfully.

In particular, over the last few decades, many researchers have tried to advance optimization in this way. The idea of these approaches, which fall under the umbrella of bio-inspired (or nature-inspired) optimization, is to draw inspiration from the behavior and internal functioning of physical, biological and social systems to design enhanced optimization algorithms with potential value in real-world applications [6]. In many cases, these metaheuristics [5] rely either on evolutionary computation or swarm intelligence.

By analyzing the state of the literature on this subject, we can notice that there is a large variety of similar methods which have been defined in different forms or applied in related fields without cross-referencing them. Readers are referred to [19] for more details and clarifications on these issues that challenge research in the field of metaheuristics. Therefore, instead of re-iterating the existing knowledge in a different form, a line of research that could make a

Supported by Alexander von Humboldt Foundation.

I. S. Kotsireas and P. M. Pardalos (Eds.): LION 14 2020, LNCS 12096, pp. 278–288, 2020.
https://doi.org/10.1007/978-3-030-53552-0_26

real contribution is to properly and newly combine ideas already approved and to take advantage of different concepts.

In this paper, we are interested in two wily nature-inspired learning approaches which could be utilized, while solving complex optimization problems: which are chunking and cooperation. The former, which is inspired by human behavior, consists in grouping basic information units allowing to build a higher level solution. Regarding the latter, we can note here that swarm intelligence approaches are by principle based on cooperation, and that most bio-inspired algorithms belong to the category of swarm intelligence [6] (even if some of them were criticized for instance in [19]).

These two concepts have been defined in different forms, and were proposed as an alternative to the classical approach which consists of decomposing a large problem into sub-components, solve each one using a basic approach, and then concatenate them to find the final solution. Such approaches were integrated in some optimization algorithms. In this paper, we focus on PSO which is nowadays among the most adopted ones. In fact, owing to its simple concept and high efficiency, PSO has become a widely adopted optimization technique and has been used successfully to solve many optimization problems, including machine learning (ML) optimization and function optimization. But, unlike continuous optimization, less attention has been paid to assessing the relevance of PSO for discrete optimization problems. In addition, many of the proposed approaches have shown certain weaknesses and have been applied only to discrete problems of small dimensionality.

In this work, we are interested in whether our extension of PSO will work well in combinatorial optimization. In particular, we are interested in the problem of FS which is among the most important optimization problems in ML. This problem could be defined as the process of selecting a subset of the most relevant features (also named attributes or predictors) for the use in the model construction. FS is a combinatorial optimization problem with 2^n possible combinations, where n is the number of original features. That is, the search space grows exponentially with the number of features, and therefore FS is still a problematic issue especially on high-dimensional data due to its huge search space.

Researchers have proposed a large number of FS methods for prediction problems, which can be classified into wrapper and filter approaches [16]. While a wrapper method evaluates the goodness of a feature subset using a prediction algorithm, a filter method is based solely on the intrinsic characteristic of the training data. Therefore, wrappers can usually obtain better prediction performance than filters, but with higher computation time. Therefore, our contribution in this paper is to propose a combination of these two approaches to combine the advantages of both approaches.

The rest of the paper is organized as follows: in the next section, the related works are outlined. In Sect. 3, we expose our proposed approach. Section 4 presents the experiments. Finally, the conclusion and the perspectives are depicted.

2   Literature Review

In this section, we begin by presenting the notions of chunking and cooperative learning, and then we focus on the use of PSO in FS.

2.1   Chunking and Cooperative Learning

On the one hand, the purpose of chunking is to limit the cost of computation. It was first introduced in psychology [13] and then extended into ML [9]; its first introduction in optimization was in [23]. There were few attempts to apply this work (e.g. [22] and [24]). In this paper, we aim to use this idea of chunking as the basis to include cooperative learning in PSO.

On the other hand, regarding cooperative learning, it should first be noted that the fundamental principle of PSO is based on learning through cooperation and sharing of knowledge: each particle participates in the evolution of the population and in the improvement of the solution. However, in the canonical version, all the particles follow the best at each iteration, then the algorithm could be easily trapped in a local optimum. Therefore, several different learning mechanisms have been introduced to tackle this issue. The most commonly used learning concept is the comprehensive learning which has been introduced in [11] in which each particle learns from another particle chosen according to a learning probability. We refer to [1] for other examples of learning approaches that were adopted in PSO. However, as further illustrated in [2], in these learning approaches, as for the canonical PSO, it is possible that some vector components have moved closer to the optimal solution, while others have moved away from it. As long as the effect of the enhancement outweighs the impact of the weakening components, these PSO would consider the new vector as an overall enhancement, although some vector components may have shifted further from the solution. Therefore, another form of learning was introduced in [2] under the name of cooperative PSO.

We can notice in this part that most of the adopted approaches for cooperative PSO are tailored to continuous optimization. In contrast, less attention has been paid to understanding how learning and cooperative approaches can be adapted to combinatorial optimization. An example of such approaches is given in [7]. In this paper, a discrete adaptation of the cooperative PSO was used to solve the problem of field programmable gate array placement. Their idea consists of the placement of the I/O and logic block being optimized by different swarms. In [12], a cooperative multi-swarm PSO algorithm was adopted to solve the problem of minimizing the electricity payments, the consumer's dissatisfaction, and the carbon dioxide emissions.

2.2   PSO for FS

Nowadays, FS has become an essential technique in data pre-processing especially on high-dimensional data in different ML applications [10]. Typical FS methods are greedy sequential forward selection (SFS) and backward FS methods (SBS). However, such greedy approaches are prone to be stuck in local optima, especially now in the big data era. A global search technique is then needed to explore this huge search space better, and therefore metaheuristics have gained much attention in solving this problem. Concerning the use of PSO for FS, it has been applied for FS in different forms; the most used form is the native one proposed by [8] as presented in the following section. This approach has been applied, for instance, in [20] and [25].

With regard to the concepts outlined above, to the best of our knowledge, the unique paper which used such kind of cooperation in swarm intelligence for FS is [21]. However, their cooperative design adopted the first ant colony system to detect optimal cardinality features and the second to select the most relevant based on this information. When dealing with the problem of FS, we can notice that most of the papers that were interested in the concept of chunking defined it under the term of clustering. For instance, in [18], the features are divided into clusters by using a graph-theoretic clustering method. Their idea is to form a set of clusters of independent features. Another graph-based approach was proposed in [15]. In this paper, the authors adopted the community detection algorithm defined in [3] to divide the features into clusters.

3 The Proposed Approach

In this section, we start by showing how PSO could be applied to the problem of FS, and then exposing how we have incorporated both the concepts of chunking and cooperation to improve it.

3.1 PSO for FS

PSO was initially designed for continuous optimization problems, but was extended in several binary and discrete forms to deal with combinatorial optimization problems such as FS. In this paper, we adopt the traditional and most known discrete and binary version which was proposed in [8]. In this binary PSO form, the update of the population of particles $t+1$ is done according to Eqs. 1, 2 and 3, concerning the inertia weight, and it is updated as in Eq. 4 (more details can be found in [8]).

$$v_i^j(t) = wv_i^j(t-1) + c_1 r_1(p_i^j(t-1) - x_i^j(t-1)) + c_2.r_2.(p_g^j(t-1) - x_i^j(t-1)) \quad (1)$$

where $v_i^j(t)$ and $x_i^j(t)$ correspond to the jth dimension of velocity and position vectors of the particle i. $p_i^j(t-1)$ represents the j dimension of the best previous position of particle i. $p_g^j(t-1)$ represents the j dimension of the best previous position among all particles in the population. r_1 and r_2 are two independently uniformly distributed random variables. c_1 and c_2 are the acceleration factors and w is the inertia weight.

$$v_i^j(t) = sig(v_i^j(t)) = \frac{1}{1 + e^{-v_i^j(t)}} \quad (2)$$

$$x_i^j(t) = \begin{cases} 1 & \text{if } rand_i \leq sig(v_i^j(t)) \\ 0 & else \end{cases} \tag{3}$$

where sig is the sigmoid function used to transform continuous values into binary ones, and $rand_i$ is a uniform random variable in the interval [0,1].

$$w = w_{max} - \frac{t}{t_{max}}(w_{max} - w_{min}) \tag{4}$$

t and t_{max} are, respectively, the current iteration and the maximum number of iterations.

PSO could be integrated with a wrapper approach as in Algorithm 1 ([17]). In the following section, we show our contribution to this algorithm.

3.2 FS by Combining Chunking and Cooperative Learning

In this part, our objective is to bridge the gap between the different concepts highlighted in the previous section. That is, we propose an insight into how to adopt cooperative learning in PSO by using chunking/clustering in FS.

In the conventional PSO presented in the previous section, each individual in the swarm represents a complete solution vector. But, in our proposed PSO-based cooperative learning, instead of having one swarm (of n particles) trying to find the optimal d-dimensional vector, the vector is split into its clusters of features that we can consider independent of the others. In other words, a solution vector of the selected features is a combination of the different solutions provided by each swarm according to the same principle of the cooperative PSO [2]. However, a challenging issue is to assign each particle to its appropriate swarm in order to obtain swarms that optimize an independent sub-problem. In this paper, we associate with each swarm a cluster of features for which their F-Correlation with the remaining features could be neglected, and could then be considered independent of the others. For this, we adopt the idea defined in [18] which we summarize in the next paragraph.

In [18], after computing the F-Correlation for each pair of features, the feature set is considered as a vertex set (graph), where each feature is considered as a vertex and the weight of the edge between vertices f_i' and f_j' is their F-Correlation, the concept of a minimum spanning tree (MST) was adopted for grouping features, as it does not suppose that the points are grouped around the centers. More concretely, after having connected all the vertices so that the sum of the weights of the edges is minimal, the clusters are formed by removing the edges whose weights are smaller than the T-Relevance of both features with the class. We must note that we are based here on the consideration of [18] that the highly correlated features are assembled in a cluster; that is why we defined our assumption of independence of different clusters.

In Algorithm 2, we define the different steps of the proposed approach.

As we can see above, the approach is simple and consists mainly of combining the approaches explored before. In fact, unless it is necessary, it is generally recommended to define the approaches in the simplest form. Our proposed approach is based on two concepts which are chunking/grouping and cooperative

---

Algorithm 1: PSO for FS [17]

Data: Training set $(X, Y) = (x_1, y_1), ..., (x_m, y_m)$, c_1, c_2 (PSO parameters), l population size

Initialization: Initialize the vector positions $(p_1, ..., p_n$ where $p_i = p_i^1, ..., p_i^d \ \forall i = 1, ..., n)$ and velocities $(v_1, ..., v_n$ where $v_i = v_i^1, ..., v_i^d \ \forall i = 1, ..., n)$ of particles, the fitness fit , $pbest$ and $gbest$;

Initialize the scalars inertia weight (w) and $f(gbest)$) ;

Set t value to 1 ;

while *number of iterations* $t \leq t_{max}$ **do**

    for $i = 1$ *to* n **do**

        for $j = 1$ *to* d **do**

            Compute the velocity $v_i^j(t)$ according to the Eqs. (1) and (2);

            Compute the position $x_i^j(t)$ according to the Eq. (3) ;

        end

        Evaluate the fitness $fit_i(t)$ of the particle i using ML prediction ;

        if $(fit_i(t) \leq f(p_i(t)))$ **then**

            $(p_i(t) \leftarrow x_i(t))\&(f(p_i(t)) \leftarrow fit_i(x_i(t)))$

        end

        if $(fit_i(t) \leq f(p_g(t)))$ **then**

            $(p_g(t) \leftarrow x_i(t))\&(f(gbest) \leftarrow fit_i(x_i(t)))$

        end

    end

    Update the inertia weight w according to the Eq. (4) ;

    $t \longrightarrow t + 1$

end

Return p_g ;

for $k = 1$ *to* d **do**

    if $(p_g(k) = 1)$ **then**

        $(f'(l) \leftarrow f(k))$;

        $l \leftarrow l + 1$

    end

end

for $k = i$ *to* n **do**

    $x_i' = (x_{i(f_1')}, ..., x_{i(f_d')})$;

end

Result: A training subset of $(x_1', y_1), ..., (x_m', y_m)$ of d* features $(f_1', ..., f_d')$ (where $x_i' = (x_{i(f_1')}, ..., x_{i(f_d')})$

---

learning, and our contribution is to newly combine them to better address the problem of FS.

Algorithm 2: Cooperative PSO with chunking for FS

Data: Same of Algorithm 1
Initialization: positions and velocities of particles ;
Divide dataset into a training set and a test set ;
Set k value to 1 ;
calculate the F-Correlation for each pair of features ;
Determine a minimum spanning tree ;
Generate d clusters by removing edges using T-Relevance [18] ;
for $k = 1$ to d **do**
  | Run Algorithm 1 ;
end
Concatenate the selected features to obtain the final solution ;
Evaluate the solution on both training and test set ;
Result: The chosen subset and parameter values based on the best solution in
       the training set and the corresponding fitness values on the training
       and test sets

---

4   Experiment

4.1   Parameter Setting

In this experiment, on the one hand, 80% of the sample are used in training while the remaining 20% of the sample are used in testing. Concerning the ML approach adopted in the wrapper part of our approach, a support vector machine (SVM) approach [4] is chosen as it has shown good performance in both ML classification and regression tasks. That is, SVM is adopted in this paper in order to measure the predictive power of the subset. To implement PSO for FS, we have adopted the "PySwarms" research toolkit [14]. Concerning PSO parameters, we consider $c_1 = 0.5$ and $c_2 = 0.5$; the number of particles is set to 30. We have chosen these values, which are the default parameters provided in "PySwarms", as they enable a balance between exploration and exploitation. More details on the impact of these parameters can be found in [1]. The maximum number of iterations is set to 10. We must note that in this paper, as a first experiment, we have considered the numbers of chunks/clusters in Algorithm 2 as 2. In other words, if the T-Relevance-based condition is detected, the feature set is clustered for only once.

To evaluate our approach, we used publicly available datasets from the UCI ML repository. More information on these datasets can be found in Table 1.

To measure the prediction accuracy of the algorithm, we have adopted for classification problems the classification accuracy, while for the regression problem, we have used the mean percentage error. We compare our approach with the PSO-based wrapper approach (with the same design adopted in [17]). The algorithms were executed on a computer equipped with an Intel i7-9750H and 16 GB of RAM.

Table 1. Benchmark dataset

	Data name	Features	Samples	Target	Type
1	Boston	13	506	Real 5. - 50	Regression
2	Breast cancer	30	569	2 classes	Classification
3	Ionosphere	34	351	2 classes	Classification
4	Lung cancer	56	32	2 classes	Classification
5	Sonar	60	208	2 classes	Classification
6	Digits	60	1797	10 classes	Classification
7	Madelon	100	4400	2 classes	Classification
8	Semeion	256	1593	2 classes	Classification
9	Voice rehabilitation	309	126	2 classes	Classification
10	Speech	756	754	2 classes	Classification

4.2 Results

For each one of the two optimizations algorithms: (our approach and the PSO-based wrapper approach), we display in Table 2 the results obtained for the different datasets, which consists of the prediction accuracy on training and test sets (based on the measurements mentioned above).

Table 2. Comparison of the prediction on training and test sets

	Our approach		PSO-based approach	
	Training acc.	Test acc.	Training acc.	Test acc
Boston	0.1291	1.0207	0.1291	1.0207
Breast-cancer	**0.980**	0.763	0.943	**0.886**
Ionosphere	**0.907**	0.901	0.896	0.901
Lung cancer	**0.840**	0.429	0.480	0.429
Sonar	0.512	0.619	**0.530**	**0.714**
Digits	**1.000**	0.508	0.999	**0.714**
Madelon	0.500	0.492	0.500	0.492
Semeion	**0.651**	0.352	0.590	**0.426**
Voice rehabilitation	1.000	0.720	1.000	0.720
Speech	1.000	0.429	1.000	0.429

We can see that the proposed approach outperforms the PSO-based approach in terms of the accuracy on the training set. Indeed, with the exception of the "Sonar" dataset, and other datasets such as "Voice Rehabilitation" and "Speech" in which both approaches give the same results (as the best result is trivial), our approach gives better results on the training set in the remaining datasets, which is more representative of the efficiency of the algorithms. That is, both optimization algorithms try to optimize the training accuracy, by selecting the most relevant features, so as to have the best ML prediction possible. We can conclude then that if the features are clustered in an appropriate way, PSO may be able to optimize the SVM performance in a smaller number of iterations (in our case 10). But, this performance also depends on other factors (e.g. SVM parameters) and this is the reason for the difference in performance for the different data.

On the other hand, we can see that the PSO-based approach gives better results in many cases on the test set, based on the selected features.

Moreover, to show a more in-depth comparison of the two approaches, we depict in Table 3 a comparison of the CPU time (in seconds) and the number of selected features.

Table 3. Comparison of the CPU time and the number of features

	Our approach		PSO-based approach	
	CPU time	Nb. of Features	CPU time	Nb. of Features
Boston	2.2207	5	**1.466**	5
Breast-cancer	**1.7131**	11	1.7272	12
Ionosphere	0.7679	14	**0.5984**	13
Lung cancer	1.2077	26	**1.0588**	29
Sonar	1.9025	20	**1.8045**	29
Digits	**22.1199**	37	22.6183	29
Madelon	25.2395	46	**15.1205**	47
Semeion	2.3154	212	**1.9778**	212
Voice rehabilitation	1.4800	142	**1.3682**	153
Speech	30.0225	383	**25.0361**	379

We can see that both approaches have given a close CPU time depending on each case study. The same remark applies to the number of features. We can note in particular that for the "Semeion" dataset, even if the number of selected features is equal, they have selected different features, and hence different results are obtained in Table 2. On the other hand, for "Voice Rehabilitation" and "Speech" datasets, even if the number of selected features is different, the two approaches have given similar results.

5   Conclusion

In this paper, we have introduced a novel extension of PSO based on an hybridization of two bio-inspired concepts, which are chunking and cooperative learning, adapted to the problem of FS. The experiment consisted of a first investigation of the efficiency of the approach and has shown that it could be more suitable than a PSO-based wrapper approach in terms of the predictive accuracy on the training set.

Further research should attempt to examine the proposed approach in the largest datasets available in the literature, and to compare it with the state of the art of FS approaches. Moreover, statistical tests will be useful to provide a more detailed comparison of the different approaches.

References

1. Aoun, O., Sarhani, M., El Afia, A.: Particle swarm optimisation with population size and acceleration coefficients adaptation using hidden Markov model state classification. Int. J. Metaheuristics **7**(1), 1–29 (2018). https://doi.org/10.1504/ijmheur.2018.091867
2. van den Bergh, F., Engelbrecht, A.: A cooperative approach to particle swarm optimization. IEEE Trans. Evol. Comput. **8**(3), 225–239 (2004). https://doi.org/10.1109/TEVC.2004.826069
3. Blondel, V.D., Guillaume, J.L., Lambiotte, R., Lefebvre, E.: Fast unfolding of communities in large networks. J. Stat. Mech.: Theory Exp. **2008**(10), P10008 (2008). https://doi.org/10.1088/1742-5468/2008/10/p10008
4. Boser, B.E., Guyon, I.M., Vapnik, V.N.: A training algorithm for optimal margin classifiers. In: Proceedings of the Fifth Annual Workshop on Computational Learning Theory - COLT 1992. ACM Press (1992). https://doi.org/10.1145%2F130385.130401
5. Caserta, M., Voß, S.: Metaheuristics: intelligent problem solving. In: Maniezzo, V., Stützle, T., Voß, S. (eds.) Matheuristics, vol. 10, pp. 1–38. Springer, Heidelberg (2009). https://doi.org/10.1007/978-1-4419-1306-7_1
6. Dokeroglu, T., Sevinc, E., Kucukyilmaz, T., Cosar, A.: A survey on new generation metaheuristic algorithms. Comput. Ind. Eng. **137**, 106040 (2019). https://doi.org/10.1016/j.cie.2019.106040
7. El-Abd, M., Hassan, H., Anis, M., Kamel, M.S., Elmasry, M.: Discrete cooperative particle swarm optimization for FPGA placement. Appl. Soft Comput. **10**(1), 284–295 (2010). https://doi.org/10.1016/j.asoc.2009.07.011
8. Kennedy, J., Eberhart, R.: A discrete binary version of the particle swarm algorithm. In: IEEE International Conference on Systems, Man, and Cybernetics. Computational Cybernetics and Simulation. IEEE (1997). https://doi.org/10.1109/icsmc.1997.637339
9. Laird, J.E., Rosenbloom, P.S., Newell, A.: Chunking in soar: the anatomy of a general learning mechanism. Mach. Learn. **1**(1), 11–46 (1986). https://doi.org/10.1007/bf00116249
10. Lessmann, S., Voß, S.: Feature selection in marketing applications. In: Huang, R., Yang, Q., Pei, J., Gama, J., Meng, X., Li, X. (eds.) ADMA 2009. LNCS (LNAI), vol. 5678, pp. 200–208. Springer, Heidelberg (2009). https://doi.org/10.1007/978-3-642-03348-3_21
11. Liang, J., Qin, A., Suganthan, P., Baskar, S.: Comprehensive learning particle swarm optimizer for global optimization of multimodal functions. IEEE Trans. Evol. Comput. **10**(3), 281–295 (2006). https://doi.org/10.1109/tevc.2005.857610

12. Ma, K., Hu, S., Yang, J., Xu, X., Guan, X.: Appliances scheduling via cooperative multi-swarm PSO under day-ahead prices and photovoltaic generation. Appl. Soft Comput. **62**, 504–513 (2018). https://doi.org/10.1016/j.asoc.2017.09.021

13. Miller, G.A.: The magical number seven, plus or minus two: some limits on our capacity for processing information. Psychol. Rev. **63**(2), 81–97 (1956). https://doi.org/10.1037/h0043158

14. Miranda, L.J.V.: PySwarms: a research toolkit for particle swarm optimization in Python. J. Open Source Softw. **3**(21), 433 (2018). https://doi.org/10.21105/joss.00433

15. Moradi, P., Rostami, M.: Integration of graph clustering with ant colony optimization for feature selection. Knowl.-Based Syst. **84**, 144–161 (2015). https://doi.org/10.1016/j.knosys.2015.04.007

16. Raza, M.S., Qamar, U.: Introduction to feature selection. In: Qamar, U., Raza, M.S., et al. (eds.) Understanding and Using Rough Set Based Feature Selection: Concepts, Techniques and Applications, pp. 1–25. Springer, Singapore (2019). https://doi.org/10.1007/978-981-32-9166-9_1

17. Sarhani, M., Afia, A.E., Faizi, R.: Facing the feature selection problem with a binary PSO-GSA approach. In: Amodeo, L., Talbi, E.-G., Yalaoui, F. (eds.) Recent Developments in Metaheuristics. ORSIS, vol. 62, pp. 447–462. Springer, Cham (2018). https://doi.org/10.1007/978-3-319-58253-5_26

18. Song, Q., Ni, J., Wang, G.: A fast clustering-based feature subset selection algorithm for high-dimensional data. IEEE Trans. Knowl. Data Eng. **25**(1), 1–14 (2013). https://doi.org/10.1109/tkde.2011.181

19. Sörensen, K.: Metaheuristics-the metaphor exposed. Int. Trans. Oper. Res. **22**(1), 3–18 (2013). https://doi.org/10.1111/itor.12001

20. Unler, A., Murat, A.: A discrete particle swarm optimization method for feature selection in binary classification problems. Eur. J. Oper. Res. **206**(3), 528–539 (2010). https://doi.org/10.1016/j.ejor.2010.02.032

21. Vieira, S.M., Sousa, J.M., Runkler, T.A.: Two cooperative ant colonies for feature selection using fuzzy models. Expert Syst. Appl. **37**(4), 2714–2723 (2010). https://doi.org/10.1016/j.eswa.2009.08.026

22. Voß, S., Gutenschwager, K.: A chunking based genetic algorithm for the Steiner tree problem in graphs. In: Pardalos, P., Du, D.Z. (eds.) Network Design: Connectivity and Facilities Location, DIMACS Series in Discrete Mathematics and Theoretical Computer Science, vol. 40, pp. 335–355. Princeton, AMS (1998)

23. Woodruff, D.L.: Proposals for chunking and tabu search. Eur. J. Oper. Res. **106**(2–3), 585–598 (1998). https://doi.org/10.1016/s0377-2217(97)00293-2

24. Woodruff, D.L.: A chunking based selection strategy for integrating meta-heuristics with branch and bound. In: Voß, S., Martello, S., Osman, I.H., Roucairol, C. (eds.) Meta-Heuristics: Advances and Trends in Local Search Paradigms for Optimization, pp. 499–511. Springer, Heidelberg (1999). https://doi.org/10.1007/978-1-4615-5775-3_34

25. Xue, B., Zhang, M., Browne, W.N.: Particle swarm optimisation for feature selection in classification: novel initialisation and updating mechanisms. Appl. Soft Comput. **18**, 261–276 (2014). https://doi.org/10.1016/j.asoc.2013.09.018

The Problem of the Hospital Surgery Department Debottlenecking

Alexander A. Lazarev[1], Darya V. Lemtyuzhnikova[1,3],
Alexander S. Mandel[1,2], and Nikolay A. Pravdivets[1]

[1] Institute of Control Sciences, 65 Profsoyuznaya street, 117997 Moscow, Russia
pravdivets@ipu.ru
[2] Lomonosov Moscow State University, 1 Leninskie Gory, Moscow 119991, Russia
[3] Moscow Aviation Institute, 4 Volokolamskoe highway, Moscow 125993, Russia

Abstract. We consider the dynamic patient scheduling for the hospital surgery department with electronic health records. Models for increasing the throughput of the surgery are proposed. It is based on classical intellectual optimization problems, such as the assignment problem, the scheduling problem, and the forecasting problem. Various approaches to solving the proposed problem are investigated. The formalization of the surgery planning problem of the large medical hospital surgery department is considered.

Keywords: Dynamic patient scheduling · Health care programs · Electronic information card · Scheduling problem

1 Introduction

High-tech medical care (HTMC) is medical care with the use of high technologies for the treatment of complex diseases. It includes both treatment and diagnostic services that are performed in a specialized hospital. There is a list of HTMC operations that can be done for Russian citizens for free. If the required operation is in the list, the order of further actions is as follows: visit the General Practitioner; consult and get a referral for analysis and diagnostic tests; pass the necessary tests; go to a doctor again with the results of the tests; get a referral from the doctor for a commission; pass the commission; visit a doctor to get a referral for hospitalization; undergo second testing for hospitalization; admission to the hospital registration in the hospital, setting the date of the operation. In total, the entire process of preparing for a free operation can take up to six months. If the operation is needed urgently, patient can agree to a paid operation, and then apply for a compensation. Each organization has its own special aspects of the surgery unit. In the daily activities of a multi-specialty surgical hospital with a large bed capacity, the capacity of the surgery unit can

Partially supported by RFBR (project 20-58-S52006) and the Basic Research Program of the National Research University Higher School of Economics.

I. S. Kotsireas and P. M. Pardalos (Eds.): LION 14 2020, LNCS 12096, pp. 289–302, 2020.
https://doi.org/10.1007/978-3-030-53552-0_27

often become a limiting factor for the intensification of the whole hospital's functioning. This problem can be particularly acute if individual surgery rooms are specialized and equipped to perform surgical interventions of a dedicated profile (for example, neurosurgery). The Burdenko Neurosurgery Center is a neurosurgical clinic that provides care to patients with diseases of the central and peripheral nervous system. Optimizing the operation of the surgery department of the Burdenko Neurosurgery Center will increase its throughput and improve conditions for patients.

The scheduling practices are being discussed in health care programs of different countries. Patients undergo a series of medical analysis before being eligible for elective surgery. At the same time various aspects of implementation and improvement of medical Electronic Information Card (EIC) are studied and analyzed in many countries, such as technological [1–3], legal [4], managerial [5,6]. In Russia the situation is different [7–9] because of the process of organizing medical care is quite complicated. There are three main areas of research in medical processes automation: electronic medical records in various versions; decision support based on clinical guidelines; clinical process management .

An interesting research is being conducted in the Kaliningrad [10,11] on the use of EIC in solving diagnostic problems and organizing consultations. The most significant analytical information systems (AIS) are the following.

- AIS in the president's polyclinic. The 1st version of it was developed in the 70s and 80s in Institute of Control Sciences of Russian Academy of Sciences.
- The system developed at the Burdenko Neurosurgical Center [12–15].
- Regional information and analytical medical system (an example can be found in [16]).

However, the majority of potential users have insufficient skills to work with modern information technologies [17]. This significantly complicates the implementation and wide distribution of the technologies. Among the restraints to the widespread introduction of EIC, researchers include: unwillingness of many ordinary doctors to spend additional time working on a computer; immetodical approach to automation; bad automation experience obtained earlier; inflated (or low) expectations of the opportunities provided by AIS and EIC; psychological problems; fear of appearing incompetent; unwillingness to provide primary information openly; laziness and illiteracy. All of these interferences inhibit the widespread use of medical AIS and EIC. But this is surmountable. In the last few years, a team of scientists from the Institute of Control Sciences of RAS and the Burdenko Neurosurgical Center creates a more modern EIC focused on solving management problems [18–20].

Features of the treatment process organization in the Burdenko Neurosurgical Center are dependent of the following two main circumstances that creates additional problems during the organization of the treatment process.

1. The system of financing of the treatment is complex and quite confusing.
2. Patients come to Burdenko Center from all over Russia. Forming the input flow of patients is a hard and time-consuming work.

The main source of information and expert opinions for building the concept was expert information received from the specialists of the Burdenko Neurosurgical Center: the chief physician, heads of surgical departments, doctors of the emergency department, etc. Working with experts and analyzing the information received from them requires a special technique. The issue is that different experts have different (sometimes diametrically opposite) opinions on many key aspects of the treatment process organization. It happens because their opinions depend on their position in the organization and the functions they perform. Institute of Control Sciences of RAS has developed a method of collective multivariate expertise for working with experts in a specific language they understand. The method has many modifications for different types of problems, but the principle of cross-discussion is common to all modifications. But the Burdenko Neurosurgical Center has another source of information: a carefully designed and efficient information system that provides electronic storage of medical records and centralized access to the medical information. For each patient, time moments of main treatment events are recorded. This information was not used before this project. In the process of developing the concept, this information have been used to build a clearly arranged picture of the treatment process, to identify the real workload of surgical departments and the surgery units, the available reserves, which allow to check and clarify the opinions of the specialists about the bottlenecks of the system and possible ways to improve it.

The paper is organized as follows. In Sect. 2, we reviewed existing solution methods of health scheduling problems. In Sect. 3 scheme of the hospitalization process in the Burdenko Neurosurgical Center. The initial data of the mathematical model are discussed in Sect. 4. Mathematical model we consider in Sect. 5. Section 6 reviews the experimental results. And some remarks in Conclusion.

2   Health Scheduling

Due to modern technology, medical care is becoming automated. Therefore, the study of optimizing service processes is relevant. There are many publications on health scheduling. Let's consider some articles for the investigated problem of optimizing surgery rooms.

The article [21] is devoted to the study of the schedule of surgical operations and the appointment of surgeons in the operating room focusing on elective patients with different urgency. Long waiting times can increase the urgency of the patient and lead to complications. The goal is to maximize the sum of the urgency values assigned to each operation. The average duration of surgery was obtained from hospital data. Since it takes a long computational time to solve a large-scale model problem, the authors developed a local search algorithm based on a simple heuristic to solve the problem. A simple heuristic (SH) is developed to schedule the surgeries based on their urgency value. An urgency value will determine the assignment of the surgeries to the operating rooms on each days. The SH start by sequencing the waiting list. Authors sequence the patient with high urgency value over patient with lower urgency value. Finally, they will scheduled the patients based on the list. After executing the SH, an initial schedule is obtained. Next you need to improve the schedule in terms of efficiency through LS. This heuristic is developed to assist the hospital by giving priority to the urgent patient based on their condition. LS consider the searching of solutions in the neighborhood and it will move from one neighborhood to another to find a better solution. The iteration will continue until no better solution can be found. If the best solution is found, it will replaced the current solution with the best solution. Based on the results, both heuristics are very good in reducing the large running time of the ILP model. The Local Search is better since it significantly reduce the computational time while giving a good solution which is as close as the optimal solution from the model.

The article [22] presents an integer programming model for scheduling operating rooms. This model tries to minimize the total weighted time to start operations. To calculate the weight, three age groups of patients are taken into account. The setup time for this group of patients depends on the sequence and the duration of the operation is analyzed using fuzzy logic. The proposed model is solved using a hybrid algorithm. When the operation is completed, it takes time to prepare the operating room for the next operation (cleaning the room, changing doctors, replacing equipment). Installation time plays a very important role in choosing the operating room schedule. The main contribution of this study is that the installation time between operations depends on the sequence, and the duration of the surgical operation depends on fuzzy numbers. The scheduling decision for operating room (OR) includes the assignment of surgical operations to ORs, and the operation sequence in each OR. Different types of each operations may need various resources and equipment. Some resources and equipment for each surgical operation can be used among several ORs, some of them are dedicated to particular ORs. Each surgical operation is constrained by the resources and equipment associated with the OR. Solution procedure has two steps: first, need to choose initial scheduled based on Weighted Shortest Processing Time because the objective of proposed model relates to total weighted so this procedure for initial solution is decent option. Second, we tested some local search heuristics, which focus on solution improvement by swapping two different surgeries between OR-days (a two-exchange), or by moving one surgery to another OR (a one exchange). The algorithm is implemented using the software package MATLAB version 8.4. After the launch of the model, acceptable results were achieved. The paper [23] proposes a multi-step approach and a typing priority rule to generate the initial sequence for "bin-packing". Case studies show that the PTD (Priority-Type-Duration) rule is superior to the LPT (Longest Processing Time) rule based on the cost of operating planning. In the proposed model, patients are divided into 5 groups according to priority. At the planning stage, the interest is in defining a set of planned activities for resource allocation. In this study, N elective cases with different priorities and different types of operations are selected from the waiting list. A set of costs is defined as a measure for evaluating operational cost planning, for example, fixed costs, overtime costs, downtime costs and installation costs. A bin-packing model maximizes utilization and minimizes the idle time, which consequently affects the cost at the planning phase. Since the priority is the most important factor for performing a surgery, authors first sequence surgeries according to their relative priorities. There are five groups. The second step is to group surgeries according to surgery types within each priority group. The third step is to sequence surgeries in each subgroup by the LPT rule based on their durations. After obtaining the initial sequence, authors assign surgeries to surgery room from the head of sequence (the highest priority) to the tail of sequence (the lowest priority), while they avoid combining different surgery types into one surgery room. If there is still some remaining time in the surgery room after assigning all patients, need to search for a compatible patient from lower priority groups with the same surgery type. If there is no compatible cases, they leave the remaining time idle.

Paper [24] focuses on the block operating room planning strategy. In such a strategy, each specialty receives several operating blocks during a certain planning period, in which it can distribute its surgical cases. The planning problem is further complicated by the change in the duration of surgical cases, which reduces the use of operating rooms. The proposed model includes trying to optimize the use of operating rooms without canceling and increasing overtime. The objective function includes patient waiting time, operating downtime and overtime. Patients belong to a waiting list, where they are registered at the moment they are referred. A subset of patients is selected who will be operated on in the planning horizon under consideration and assigned to weeks and operating blocks, while ensuring that the capacity of each block is not exceeded. The objective function is aimed at minimizing the overall penalty due to delays in service of patients. Authors propose an approach combining offline and online decisions. The offline solutions are applied and modified online so as to manage patients who have been cancelled and must be rescheduled and newly patient arrivals. Uncertainty in surgery duration must be considered in the offline step, so as to reduce the number of cancelled patients: Authors apply a cardinality-constrained robust optimization approach to model the off-line scheduling problem. Tests on a set of real-based instances are carried on. They apply the proposed two-step approach on a set of randomly generated scenarios in order to assess its behavior in managing patients to be rescheduled and new arrivals. Beside, we evaluate the benefit of applying a robust solution rather than a non-robust one in the off-line step.

This model can be solved using stochastic programming and various heuristic methods. In [25], a problem of scheduling of the surgical department of a Chinese hospital is considered. Emergency and planned patients can be served in the same operating rooms. There are fixed time slots for planned operations as well as flexible slots for unscheduled operations. Authors propose a simulation-optimization approach consisting of two models. For a two-stage stochastic optimization model, uncertain arrival times of emergency patients are represented by a set of scenarios. The discrete event simulation model is designed to eliminate the uncertainties associated with the duration of the operation and the length of stay in the hospital, as well as to verify the basic schedule of the operation developed using the stochastic model. A simulation model is also used to generate scenarios. The resulting system shows good results both on emergency waiting time and on the stability of the planned operations schedule.

3   Scheme of the Hospitalization Process

The goal is to increase throughput due to two factors: reducing gaps in schedule and increasing the number of resources (bed stock, IDs of surgery rooms). We can use information system of the Burdenko Neurosurgical Center and doctors' expert evaluation to solve the problem. Experts identify subproblems such as hospitalization, surgical department manipulations, and monitoring of surgery rooms.

The problem is divided into three subproblems. The first is the problem of allocating specialists to the appropriate rooms at a certain time. The second problem is to create a schedule for receiving patients for surgery. The third is the problem of predicting downtimes of surgery rooms. Let's consider the facts that determine the features of the problem:

1. Hospitalization. The main focus is on patients who are admitted to the hospital for their planned hospitalization and are registered in the queue. In addition, there are patients who are served on a commercial basis. As soon as a patient leaves the hospital, employees look through the queue list and select the next patient. The choice depends on the department, where a vacant bed is appeared. The person from the queue whose diagnosis corresponds to the service in the department is being selected.

   The constraints of the problem are formed out of the following facts.
   - Patients come from different regions of the Russian Federation as soon as beds are released in 10 specialized surgical departments.
   - All the patient documents are checked.
   - Comorbidities are checked.
   - There is a mandatory list of medical analysis and diagnosing tests.
   - Appearing of unscheduled patients, as well as those who arrived without a call or their representatives.
   - The anesthesiologist consults patients only from 14:00 to 15:00.
   - Necessary time for monitoring the course of the disease.

   The following factors affect the delay in hospitalization.
   A. The difference between the call for hospitalization and the actual arrival of the patient. It is assumed that calling a patient means reserving a bed. We assume that the reserve has a time limit.
   B. Difference between patient arrival and hospitalization. It consists of the time spent on checking documents, the comorbidities, conducting a mandatory list of medical analysis and diagnosing tests, as well as the need to consult an anesthesiologist and other specialists.

2. The surgical department.
   - The patient's stay includes a pre-operation procedures, one or more operations, and restorative treatment.

- Each surgical department has 30 beds.
- When a patient is undergoing surgery or intensive care, a new patient cannot be placed in their bed.
- Necessary time for monitoring the course of the disease.
3. Monitoring of surgeries.
    - The medical center has 14 main surgery rooms, including rooms numbered R1 – R4 (usually less busy) for operations that require x-ray inspection.
    - Each surgery department has priority for using its own surgery room, but if there is free time, it provides the surgery room to other surgery departments.
    - Longer operations are to be started at the beginning of the day.

The rooms. Each department contains a set of rooms. The room can be: registry, consulting, patients, surgery and emergency. Capacity of registry room is determined by the number of available receptionists. In consulting rooms the doctor can consult one patient at a time and doctors can accept patients from different departments. Patients rooms have 30 beds for each department, wards are assigned to certain departments. Only one patient can be in a surgery room at a time. There are 14 surgery rooms, including 4 with x-ray equipment. Each surgery room is assigned to a specific department, but can be provided for the work of another department. In emergency room also can be only one patient at a time. Many rooms of different departments may overlap in certain cases.

The Specialists. The specialist can be: a consulting doctor, a doctor performing a surgery, a GP who is supervising the patient, a receptionist. In cases where several specialists are working with the patient at the same time, the responsible doctor will be designated as a specialist. For example, in the case of an operation, this is the doctor who performs the operation directly. Many specialists from different departments may overlap in certain cases.

The Patients. Each patient is assigned to a specific department upon arrival, from which they can be transferred to other departments during treatment. For each patient, there is a disease (group of diseases) that corresponds to a set of procedures: observation, consultation, operation, resuscitation, delivery of the analysis. We will also include here the procedures necessary for receiving each patient: registration (this also includes checking documents), arriving (time from receiving the call to arriving at the hospital).

Time for each procedure has a fixed upper bound. The set of procedures may change depending on the preceding procedures. For example, an operation may be canceled due to a specialist's consultation, or an unplanned emergency procedures may occur after the operation. After certain procedures are completed, the service may be terminated prematurely if:

- the patient did not collect all the documents (this is checked at the registry just after arrival),
- the patient's analysis and diagnosing tests do not allow the operation to be performed in the near future,
- the consultation states that the operation is contraindicated,

– the patient is being transferred to another hospital,
– the patient died.

The precedence relations between the procedures are set in the form of an acyclic oriented graph. Based on the problem condition, a specific set and sequence of procedures are typical for each department.

4 Data

Burdenko Neurosurgical Center information system includes information about a patient's hospitalization, the principles of his treatment, the work of the department, including occupied beds, the work of the surgical department, etc.

For each patient, information about their operations is generated in the system. This creates a table that consists of the following columns: number of the patient; date; the host department; surgery room ID; complexity category of operation; start of operation; end of operation.

Based on this table, we can conclude that usage periods of surgery rooms usually have gaps, which greatly reduces the efficiency of surgery rooms.

According to examples for each operation we know its duration, complexity category and the department in which it should be performed. The relationship of all entered parameters of the operation is not defined. Thus, the connection between the complexity of the operation and the time of its execution is not known. By using these example we created the frequency dictionary of various parameters and used it in the generation.

In the generated data the Center received a random number of patients n with a 15 min frequency. The value n is distributed normally with parameters μ and σ^2: $n \sim N(\mu, \sigma^2)$. By varying the values of μ and σ^2, it is possible to set the admission density of patients. The department and complexity category for each patient was set randomly based on a weighted selection procedure. That is, the more often the complexity category was encountered in the initial data, the greater will be the probability of its generation. The time of operation was also set on the basis of a weighted selection, but from those values that were usual of the given complexity and for the given department.

A pattern could not be identified for preliminary stay. Thus, the time of preliminary stay of patients awaiting for hospitalization is an random value, evenly distributed at $[T_0, T_1]$ interval.

5 Monitoring of Surgeries

Consider the problem of planning patients $J = \{1, \ldots, n\}$ in the operating room $I = \{1, \ldots, 10\}$. Each patient $j \in J$ has time r_j, after which the patient can start the operation, and processing time of operation p_j. Each patient is tied to a specific operating room $w_j \in I$. We introduce the decision variable X_{ij}^t, which is equal 1 if the patient j starts operation in operating room $i \in I$ in time $t \in l$, and equal 0 otherwise. l - is the upper time limit. A set of times S represents a off-hours. In daytime mode, operations are carried out only from 9:00 to 18:00. As the objective function, we take the total completion time of all operations.

We consider two models: a model of a twenty-for-hour clinic (*Model A*) and a model of a clinic with 'standard' working day from 9:00 to 18:00 (*Model B*).

Model A

Objective function:

$$\sum_{j \in J} C_j \rightarrow \min_{X_{ij}^t} \tag{1}$$

Subject to:

$$\sum_{t=0}^{l} X_{w_j j}^t = 1, \qquad \forall j \in J; \tag{2}$$

$$\sum_{j \in J} \sum_{h=max(0,t-p_j)}^{t-1} X_{ij}^h \le 1, \qquad \forall i \in I, \quad \forall t \in l; \tag{3}$$

$$C_j = \sum_{i \in I} \sum_{t=0}^{l-1} (t + p_j) X_{ij}^t, \qquad \forall j \in J; \tag{4}$$

$$C_j \ge r_j + p_j, \qquad \forall j \in J. \tag{5}$$

Model B is model A with the following constraints added:

$$C_j - p_j \ge 9 : 00, \qquad \forall j \in J; \tag{6}$$

$$C_j \le 18 : 00, \qquad \forall j \in J. \tag{7}$$

An objective function is to minimize a total completion time of all operations. A set of constraints (2) ensures that each patient will be operated in a pre-assigned operating room. Constraint set (3) allows at most one job to be processed at any time on any machine. Constraint sets (4) and (5) determines the completion time of operations. Constraints (6,7) define the working hours of the surgery rooms.

6   Computational Experiments

We conducted experiments on the real-like data for models A and B.

Experiment 1. The objective and constraints of model A are represented by formulas (1–5). Operating rooms are supposed to work around the clock. The results of the experiments are presented in the Figs. 1 and 2. In the Fig. 1 the department number is plotted on the x-axis, the objective function value is plotted on the y-axis at different densities of patient penetration. The figure shows that with a patient density of 25 patients per hour, a significant increase in the value of the objective function for the 7th department occurs. For the density of patient admission of 5 people per hour, an increase in the objective

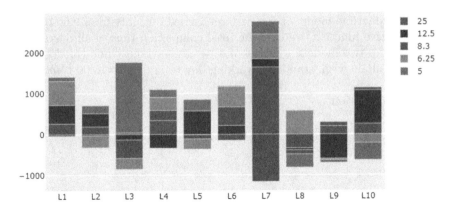

Fig. 1. Histogram of objective function values for different densities, around the clock.

function in the 3rd department is visible. The solution seems to be to compensate for the load on the busiest branch of the department by transferring the load to a less busy department.

The Fig. 2 shows a Gantt chart for a patient density of 25 patients per hour. Departments are on the y-axis. The picture shows that operating rooms loaded with different intensities.

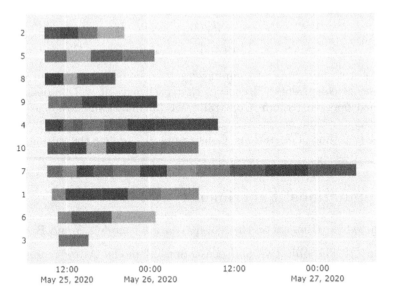

Fig. 2. Gantt chart for round-the-clock operation

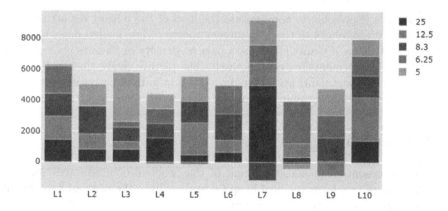

Fig. 3. Histogram of objective function values for different densities, 9:00 to 18:00.

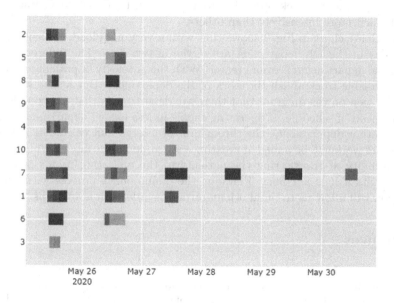

Fig. 4. Gantt chart for working 9:00 to 18:00

Experiment 2. The objective function and constraints of model B are represented by formulas (1–7). Operating rooms are supposed to work from 9:00 to 18:00 and to be closed at night. The results are presented in the Figs. 3 and 4. Figure 3 on the x-axis contains number of the department, on the y-axis contains the value of the objective function for the department, the color of the column indicates the density of patients per hour. We can see a significant increase in the objective function value for 7th department at the patient arrival density of 25 patients per hour, which indicates an overload of the department. For an arrival density of 5 patients per hour, we can see the increase of the objective function in the 3rd department. It is possible to compensate the load of the busiest department by transporting it's patients to the free surgery rooms of less loaded departments.

The Fig. 4 shows a Gantt chart for a patient density of 25 patients per hour for model B. Departments are on the y-axis. The picture shows that with a high load in the 7th department, up to five days delay in patient care can occur. The 7th department should be unloaded at the expense of other departments.

7   Conclusion

This paper presents a formal statement of the problem of predictive planning of surgery units in a large medical hospital and outlines the ways of its optimal solution. Experiments were carried out on the real-like data, which were generated on the basis of data provided by Burdenko Institute. An experiment for operating round-the-clock operating rooms showed that some departments are loaded much more intensively than others.

The Gantt chart in Fig. 2 shows that with a very high load, patients arrived in one day in the 7th department are served in two days. The difference in the load of the departments is even greater. With this intensity of patient admission, it is impossible to establish the work of the department. But at the same time it can be seen on the diagram that there are departments that are complete the service ahead of schedule. If operating rooms in less loaded departments will be used for other departments, the throughput of a large unit can be increased.

Further work will be aimed at creating a schedule for the model, which implies the possibility of transferring the patients to the operating rooms of another departments if it is not busy. Planning the maintenance process with the new condition will allow us to build a long-term schedule with fewer gaps.

References

1. Adam, D., Milner, J., Quereshy, F.: Dynamic patient scheduling for multi-appointment health care programs. Prod. Oper. Manag. **27**(1), 58–79 (2018)
2. Gunter, T.D., Terry, N.P.: The emergence of national electronic health record architectures in the United States and Australia: models, costs, and questions. J. Med. Internet Res. **7**(1), 1–13 (2005)
3. Blumenthal, D.: Launching HITECH. New England J. Med. **362**(5), 382–385 (2010)
4. Evans, D.C., Nichol, W.P., Perlin, J.B.: Effect of the implementation of an enterprise-wide Electronic Health Record on productivity in the Veterans Health Administration. Health Econ. Policy Law **1**(2), 163–169 (2006)
5. Hansen, T., Onders, R.: Rural Practice Redesigns Care Processes to Allow Multidisciplinary Teams to Leverage Electronic Health Record, Leading to Better Screening of Medically Underserved, pp. 35–41
6. Kierkegaard, P.: Electronic health record: wiring Europe's healthcare. Comput. Law Secur. Rev. **27**(5), 503–515 (2011)
7. Nazarenko, G.I., Osipov, G.S.: Fundamentals of the Theory of Medical Technological Processes. Part 1, 144 p. FIZMATLIT, Moscow (2005). (in Russian)

8. Nazarenko, G.I., et al.: Development of ontology of technological maps for managing patients in a multi-profile hospital when modeling medical technological processes. Artif. Intell. Decision-Making **2**, 68–77 (2014). (in Russian)

9. Molodchenkov, A.I.: Application of AQ-algorithm for personification of medical and diagnostic processes. Theory and Practice of System Analysis: Proceedings of the First all-Russian Scientific Conference of Young Scientists. Rybinsk: RGATA im. P. A. Solovyova, vol. 1, pp. 79–84 (2010). (in Russian)

10. Kolesnikov, A.V., Rumovskaya, S.B., Listopad, S.V., Kirikov, I.A.: System analysis in solving complex diagnostic problems. In: Kanta, I. (ed.) System Analysis and Information Technologies: Proceedings of the VI International Conference SAIT-2015 in 2 volumes, vol. 1, pp. 157–167. Publishing House of BFU, Kaliningrad (2015). (in Russian)

11. Kirikov, I.A., Kolesnikov, A.V., Listopad, S.V., Rumovskaya, S.B.: "Virtual consultation" - instrumental environment for supporting complex diagnostic decision-making. Inform. Appl. **10**(3), 81–90 (2016). (in Russian)

12. Shifrin, M.A., Kalinina, E.E., Kalinin, E.D.: Electronic medical history of the research Institute of neurosurgery. N. N. Burdenko: concept, development, implementation. In: Proceedings of the Interregional Conference "Problems of Development and Implementation of Information Systems in Health Care and Health Insurance, Krasnoyarsk, pp. 269–275 (2000). (in Russian)

13. Shifrin, M., Kalinina, E., Kalinin, E.: EPR project for N.N. Burdenko neurosurgical institute: goals, technology, results, future trends. Br. J. Healthcare Comput. Inf. Manag. **20**(7), 18–21 (2003)

14. Shifrin, M., Kalinina, E., Kalinin, E.: Sustainability view on EPR system of N.N. Burdenko neurosurgical institute. In: Proceedings of the 12th World Congress on Health (Medical) Informatics, Part 2, pp. 1214–1216. IOS Press (2007)

15. Shifrin, M.: On a seamless transition from a running EPR system to a new one. In: Seamless Care - Safe Care Proceedings of the EFMI Special Topic Conference, pp. 219–222. IOS Press (2010)

16. Regional information and analytical medical system (RIAMS) "Promed". http://swan.perm.ru/elektronnoe_zdravoohranenie/riams_promed. (in Russian)

17. What should be an electronic medical card and why doctors do not want it. http://www.ristar.ru/docnoemk. (in Russian)

18. Chernyavsky, A.L., Dorofeyuk, A.A., Dorofeyuk, Y.A., Mandel, A.S., Pokrovskaya, I.V.: Analysis of the process of hospitalization of patients in a large clinic by methods of collective multivariate examination. Manag. Large Syst. **64**, 151–186 (2016)

19. Chernyavsky, A.L., Dorofeyuk, A.A., Mandel, A.S., Pokrovskaya, I.V., Spiro, A.G.: The use of multi-agent systems in the management of medical institutions. Inf. Technol. Comput. Syst. **4**, 92–100 (2016). (in Russian)

20. Dorofeyuk, A.A., Dorofeyuk, Y.A., Mandel, A.S., Spiro, A.G., Chernyavsky, A.L., Shifrin, M.A.: The concept of an intelligent patient routing subsystem in a patient-oriented management system of a large clinic. Sensors Syst. **10**, 3–8 (2016). (in Russian)

21. Rashid, N.S.A., et al.: Local search heuristic for elective surgery scheduling considering patient urgency. Discov. Math. (Menemui Matematik) **41**(2), 46–56 (2019)

22. Lahijanian, B., Zarandi, M.H.F., Farahani, F.V.: Proposing a model for operating room scheduling based on fuzzy surgical duration. In: 2016 Annual Conference of the North American Fuzzy Information Processing Society (NAFIPS), pp. 1–5. IEEE (2016)

23. Abedini, A., Ye, H., Li, W.: Operating room planning under surgery type and priority constraints. Procedia Manuf. **5**, 15–25 (2016)
24. Addis, B., et al.: Operating room scheduling and rescheduling: a rolling horizon approach. Flexible Serv. Manuf. J. - **28**(1–2), 206–232 (2016)
25. Bai, X., et al.: Day surgery scheduling and optimization in large public hospitals in China: a three-station job shop scheduling problem (2020)

Learning Optimal Control of Water Distribution Networks Through Sequential Model-Based Optimization

Antonio Candelieri[1]([⊠]) [iD], Bruno Galuzzi[2] [iD], Ilaria Giordani[2] [iD],
and Francesco Archetti[2] [iD]

[1] Department of Economics, Management and Statistics,
University of Milano-Bicocca, 20126 Milan, Italy
`antonio.candelieri@unimib.it`
[2] Department of Computer Science Systems and Communication,
University of Milano-Bicocca, 20126 Milan, Italy

Abstract. Sequential Model-based Bayesian Optimization has been successfully applied to several application domains, characterized by complex search spaces, such as Automated Machine Learning and Neural Architecture Search. This paper focuses on optimal control problems, proposing a Sequential Model-based Bayesian Optimization framework to learn optimal control strategies. The strategies are synthetized by pressure-based rules, whose parameters are the design variables of the optimization problem whose black-box objective is the energy cost. A Bayesian optimization framework is presented which handles a quite general formalization of the control problem including multiple constraints, also black box. Relevant results on a real-life Water Distribution Network are reported, comparing different possible choices for the proposed framework.

Keywords: Sequential Model-based Bayesian optimization · Optimal control · Water distribution networks

1 Introduction

Sequential Model-based Bayesian Optimization (SMBO) is a sample-efficient strategy for global optimization (GO) of black-box, expensive and multi-extremal functions [1–4], where the solution of the problem is traditionally constrained to over a box-bounded search space X:

$$\min_{x \in X \subset \mathbb{R}^d} f(x) \tag{1}$$

SMBO has been successfully applied in several domains, ranging from design problems (new materials, drugs, software, structural design) to robotics, control and finance (a brief overview about application domains is provided in Chap. 7 of [5]).

In the Machine Learning (ML) community, it recently became the standard strategy for Automated Machine Learning (AutoML) [6] and Neural Architecture Search

© Springer Nature Switzerland AG 2020
I. S. Kotsireas and P. M. Pardalos (Eds.): LION 14 2020, LNCS 12096, pp. 303–315, 2020.
https://doi.org/10.1007/978-3-030-53552-0_28

(NAS) [7], which are usually characterized by a search space being more complex than a box-bounded domain. More precisely, x can consists of mixed (continuous, integer, categorical) components as well as "conditional", where conditional means that the value of a component x_i depends on the value of another component x_j, with $i \neq j$. An example of complex search space in AutoML is related to the optimization of ML pipelines, such as the one presented in [8].

Starting from the SMBO advances in the ML domain, we consider in this paper an optimal control problem sharing many characteristics with AutoML. More precisely, we addressed the optimal definition of control rules regulating the ON/OFF switching of pumps in a Water Distribution Network (WDN). The objective is the minimization of the energy-related costs while guaranteeing the supply of the water demand.

We remark that the problem is black-box because the evaluation of both the objective function and the constraints is based on a hydraulic simulator: moreover the constraints on decision variables make the search-space complex (i.e., analogously to AutoML, decision variables are both mixed – numeric and discrete – and possibly conditional).

The rest of the paper is organized as follows: in Sect. 2, the methodological background about SMBO and optimization of operations in WDNs is presented. Section 3 provides the mathematical formulation of the optimal control problems, along with the proposed solution. Section 4 defines the experimental setting and Sect. 5 summarizes the results obtained. Finally, conclusions and discussion on advantages and limitations of the proposed approach are provided.

2 Background

2.1 Sequential Model-Based Bayesian Optimization

To solve problem (1), SMBO uses two key components: a *probabilistic surrogate model* of $f(x)$, sequentially updated with respect to new function evaluations, and an *acquisition function* (aka *infill criterion* or *utility function*), driving the choice of the next promising point x where to evaluate $f(x)$ while dealing with the exploitation-exploration dilemma. A typical choice for the probabilistic surrogate model is a Gaussian Process (GP) [9] (in this case, SMBO is also known as GP-based optimization or Bayesian Optimization [5, 10]). An alternative probabilistic surrogate model is a Random Forest (RF) [11], an ensemble learning approach which, by construction, can deal with mixed and conditional components of x, making RFs more well-suited than GPs to solve problems with these characteristics.

The probabilistic surrogate model – whichever it is – should provide an estimate of $f(x)$ along with a measure of uncertainty about such an estimate, with $x \in X$. These two elements are usually the mean and standard deviation of the prediction provided by the probabilistic surrogate model, denoted by $\mu(x)$ and $\sigma(x)$, respectively.

With respect to the acquisition function, several alternatives have been proposed, implementing different mechanisms to balance exploitation and exploration (i.e., $\mu(x)$ and $\sigma(x)$, respectively) [5, 10]. In this paper we focused on a subset of acquisition functions, reported in the experimental setting section.

Due to the sequential nature of SMBO, at a generic iteration n we can denote the set of function evaluations performed so far by $D_{1:n} = \{(x^{(i)}, y^{(i)})\}_{i=1,\ldots,n}$, where $y^{(i)} = f(x^{(i)}) + \varepsilon$, and $\varepsilon \sim \mathcal{N}(\mu_\varepsilon, \sigma_\varepsilon)$ in the case of a noisy objective function.

The probabilistic surrogate model is learned at every iteration, providing the updated $\mu^{(n)}(x)$ and $\sigma^{(n)}(x)$. Then, $x^{(n+1)}$, is chosen by solving the auxiliary problem:

$$x^{(n+1)} = \underset{x \in X \subset \mathbb{R}^d}{\mathrm{argmax}}\, \alpha^{(n)}(x) \tag{2}$$

where $\alpha^{(n)}(x)$ is the acquisition function, typically $\alpha^{(n)}(x, \mu^{(n)}(x), \sigma^{(n)}(x))$. This auxiliary problem is usually less expensive than the original one, and can be solved by gradient-based methods (e.g., L-BFGS) – in the case that the analytical form of $\mu^{(n)}(x)$ and $\sigma^{(n)}(x)$ is given (i.e., when a GP is used as probabilistic surrogate model) – or GO approaches (e.g., DIRECT, Random Search and its recent variants, evolutionary meta-heuristics, etc.) – in the case that $\mu^{(n)}(x)$ and $\sigma^{(n)}(x)$ are also black-box (i.e., when a RF is used as a probabilistic surrogate model).

Then, the objective function is evaluated at $x^{(n+1)}$, leading to the observation of $y^{(n+1)}$ and the update $D_{1:n+1} = D_{1:n+1} \cup \{(x^{(n+1)}, y^{(n+1)})\}$. The process is iterated until some termination criterion is achieved, such as a maximum number of function evaluations has been performed.

2.2   Constrained SMBO

Real life optimization problems have most often constraints making the search space more complex than simply box-bounded [10] and the "vanilla" SMBO not well suited for solving them. In constrained SMBO, the problem (1) can be rewritten as:

$$\min_{x \in X \subset \mathbb{R}^d} f(x)$$
$$g_i(x) \leq 0\, i = 1, \ldots, n_g \tag{3}$$

Solving approaches can be categorized depending on the nature of the constraints: they can be known a-priori and given in analytical form or, on the contrary, they are unknown (aka hidden, black-box). With respect to the first case, several approaches have been proposed in the GO community [12–14], while the second case is more related to simulation-optimization and AutoML [15–19].

A further consideration, with respect to unknown constraints, is that the objective function could be not computable in association with the violation of one or more constraint, leading to the global optimization of partially defined functions [20–22]. Recently, a two-stage approach has been proposed in [23], using Support Vector Machine (SVM) to estimate the portion of the box-bounded search space where the objective function is defined (aka computable), depending on a set of unknown constraints. In the second stage a constrained Bayesian Optimization task is performed on the estimated feasible region. This paper makes use of this approach.

3 Problem Definition and Solution Approach

3.1 Optimization of Operations in Water Distribution Networks

Optimization of WDNs' operations has been an active research field in the last decades. Optimal pump operation, aimed to minimize energy related costs due to pumping water, has been one of the most relevant topics. A systematic review on WDNs' operations optimization has been recently provided in [24], where approaches for optimal pumps management are categorized into: *(i) explicit control* of pumps by times to operate and *(ii) implicit control* by pumps' pressures, flows or speeds, or tanks levels. Although *explicit control* solutions were the most frequently adopted, the optimization problem (also known as Pump Scheduling Optimization, PSO) could be characterized by a huge number of decision variables in the case that the WDN has many pumps and/or times to operate (e.g., decisions about pump activation every hour on a daily horizon).

Most of the *explicit control* solutions proposed use meta-heuristics, mainly evolutionary strategies, such as in [25–27]. However, contrary to SMBO, these strategies are not sample efficient, requiring a huge number of hydraulic simulation runs to identify an optimal pump schedule. More recently, an SMBO approach to PSO has been initially proposed in [28] and then extended in [18] to include unknown constraints on the hydraulic feasibility of the pump schedules proposed by SMBO.

On the other hand, *implicit control* strategies allow reduction of the number of decision variables, but make more complex the search space, due to the introduction of further constraints on and conditions among decision variables. Another important advantage offered by implicit control solutions is that they do not require specification of times to operate; they usually work by applying simple (control) rules depending on the values of collected measurements. Thus, time to operate is given by the data acquisition rate instead of prefixed timestamps as in *explicit control* solutions.

3.2 Learning Optimal Control Rules as a Black-Box Optimization Problem

We consider the case of an *implicit control* solution, where pumps are controlled depending on the associated pressure values. In the simplest case, control for a given pump is defined by two different thresholds, x_1 and x_2, and the following rule:

> IF (pump's pressure $<x_1$ AND pump is OFF) THEN pump is switched ON
> ELSE
> IF (pump's pressure $> x_2$ AND pump is ON) THEN pump is switched OFF

This means that the pump is activated if its pressure is lower than a minimum threshold, x_1, it is deactivated if its pressure exceeds a maximum threshold, x_2, and remains in the current status (ON/OFF) otherwise. Clearly, x_1 and x_2 are the decision variables to optimize with respect to the minimization of energy cost, constrained to water demand satisfaction. A graphical representation of this kind of simple control for a single pump is reported in Fig. 1.

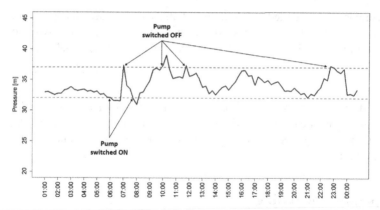

Fig. 1. A schematic representation of implicit pump control based on thresholds (red dotted lines) on pressure (in blue). If pressure goes below/over the lower/upper threshold the pump is switched ON/OFF, respectively. Pressure value could not change immediately after the pump switch because it also depends on the status of the other pumps in the WDN (Color figure online)

It is important to highlight that both energy costs and water demand satisfaction (as well as any other relevant constraints related to the hydraulic behavior of the WDN, such as min/max tanks levels) are black-box, because they can be only evaluated after having fixed the values of the decision variables. Moreover, an analytical constraint must be added, modelling that the minimum threshold x_1 cannot be greater or equal than the maximum one x_2. As follows, we define the optimization problem in the more general case consisting of more than a pair of thresholds. This situation is quite common in real-life WDNs, having more than one pump and/or requiring different control thresholds over the day (e.g., during morning and evening) for a given pump:

$$\min_{x \in X \subset \mathbb{R}^d} f(x)$$

$$
\begin{array}{lll}
x_i \in S_i & i = 1, \ldots, 2\tau & (c_1) \\
x_j - x_{j+\tau} \leq 0 & j = 1, \ldots, \tau & (c_2) \\
g(x) = 0 & & (c_3)
\end{array}
\tag{4}
$$

where $f(x)$ is the energy cost associated to the control rule defined by the x_i value, S_i is the set of possible values for the thresholds, $S_i = \{s_1, \ldots, s_{N_j}\}$, τ is the number of thresholds pairs to be set up (leading to $d = 2\tau$) and $g(x)$ is related to the hydraulic feasibility: it is unknown/black-box and makes $f(x)$ partially defined. Thus, both $f(x)$ and $g(x)$ are black-box and are computed via hydraulic software simulation, typically over a simulation horizon of a day. The open-source EPANET 2.0 is the most widely adopted tool for simulating the hydraulic behavior of a pressurized WDN, so that the search for the optimal values of the control thresholds is sequentially performed on the software model of the WDN. A single simulation run, referred to a specific set up of the thresholds, involve computational costs; SMBO is a sample-efficient strategy to identify an optimal set up within few simulation runs (i.e., function evaluations).

Thresholds are modelled as discrete variables to consider the resolution of the monitoring sensors (i.e., in the case study analyzed in this paper, measurements are acquired with a resolution of 0.5[m]). In the case of continuous variables, constraint c_1 turns into $x_i \in \left[\min S_i, \max S_i \right]$. According to (4), the optimal definition of an implicit control strategy, based on pressure values, shares common characteristics with AutoML: decision variables are discrete (c_1) and conditional (c_2) – such as many Machine and Deep Learning algorithms' hyperparameters – and $g(x)$ is black-box – such as a constraint on resources (i.e., memory usage) for a trained Machine/Deep Learning algorithm.

4   Experimental Setting

4.1   Case Study Description

The case study considered in this paper refers to a WDN in Milan, Italy, supplying water to three different municipalities: Bresso (around 20'000 inhabitants), Cormano (around 26'000 inhabitants) and Cusano-Milanino (around 19'000 inhabitants). The overall WDN consists of 7418 pipes, 8493 junctions, 14 reservoirs, 1381 valves, 9 pumping stations with 14 pumps overall. Piezometric level of the WDN ranges in 136 to 174 m (average: 148 m). Moreover, this WDN is also interconnected with the WDNs of other three municipalities (namely, Paterno Dugnano, Sesto San Giovanni and Cinisello Balsamo). The hydraulic software models of these further municipalities were not available, the hydraulic behavior at the interconnections was modelled through three reservoirs with levels varying over time according to historical data about the flow from the WDN to the other three municipalities and vice versa (Fig. 2).

Fig. 2. The three municipalities considered in the study (on the left) and the hydraulic software model, developed in EPANET, of the associated WDN (on the right)

4.2   SMBO Setting

In this section we provide all the details about the setting of our experiments, organized in two different sub-sections. The first one provides all the details about the SMBO process applied to problem (4). In the second, we decided to relax the constraint related to the discreteness of the decision variables. The aim is to evaluate which could be the difference between the optimal solution identified for the problem (4) and a more optimistic one, in the hypothetical case that the numerical precision of the actual control could be finer. In both cases, RF have been used as probabilistic surrogate model due to the presence of conditional decision variables.

Since the actual global optimizer x^* is unknown, we cannot use performance measures such as regret [29] or Gap metrics [30], but just looking at the best value observed over SMBO iterations, the so called "best seen":

$$y^{+(n)} = \min_{i=1,\ldots,n} \left\{ f\left(x^{(1)}\right), \ldots, f\left(x^{(n)}\right) \right\}$$

Finally, it is important to highlight that, in this study, we have evaluated all the control strategies identified through SMBO by simulating them over the same "test day". This means that we have considered an unnoisy setting, so $y^n = f\left(x^{(n)}\right)$, for every $n = 1, \ldots, N$ and with N the maximum number of function evaluations.

4.2.1 RF-Based SMBO

As mentioned in Sect. 4.1, the WDN has 14 pumps, overall. However, 4 are only used to support supply during peak-hours. They are controlled by time and will not be part of the optimization. With respect to the other 10 pumps, 8 of them requires the identification of optimal control thresholds which can be different during the day (i.e., 06:00–23:00) and the night (i.e., 23:00–06:00). This means that we have optimized 2 thresholds for 2 pumps and 4 thresholds for 8 pumps, leading to 36 decision variables overall (i.e., thresholds x_i) for the problem (4) – that is $\tau = 18$. It is important to highlight that the number of decision variables is significantly higher in the case of *explicit control*: the optimization of hourly-based schedules on the same case study would require 240 decision variables (that is 10 pumps time 24 h).

The possible discrete values for all the *lower* thresholds, that are the sets $S_{i=1,\ldots,\tau}$, range from 21[m] to 32[m], with a step of 0.5[m] (i.e., 23 possible values). The possible discrete values for all the *upper* thresholds, that are the sets $S_{i=\tau+1,\ldots,2\tau}$, range from 26[m] to 44[m], with a step of 0.5[m] (i.e., 23 possible values). These two sets instantiate the constraint (c_1) of the problem (4). Due to the nature of the problem, involving discrete and conditional decision variables, the most suitable probabilistic surrogate model is a RF. Initialization of the probabilistic surrogate model was performed by randomly sampling 10 initial vectors of control thresholds ("initial design"). More precisely, a Latin Hypercube Sampling (LHS) procedure has been applied. Remaining budget (i.e., function evaluations) has been set to 200.

We decided to compare three different acquisition functions, namely Lower Confidence Bound (LCB), Expected Improvement (EI) [5, 10] and Augmented Expected Improvement (AEI) [31] – the last usually replaces EI in the noisy setting. Although in this paper we solve the case study deterministically (i.e., in the noise-free setting), we have decided to include AEI just to evaluate how much the assumption of working in a noisy setting – while the problem is noise-free – could affect the final solution. We plan to use AEI to extend our approach to the noisy setting by considering the water demand as a random variable, instead of known (or predicted) a priori.

$$LCB(x) = \mu(x)^{(n)} - \beta^{(n)}\sigma(x)^{(n)}$$

$$EI(x) = \begin{cases} \left(y^+ - \mu(x)^{(n)}\right)\Phi(Z) + \sigma(x)^{(n)}\phi(Z) & \text{if } \sigma(x)^{(n)} > 0 \\ 0 & \text{otherwise} \end{cases}$$

$$AEI(x) = \begin{cases} \left(y^+ - \mu(x)^{(n)}\right)\Phi(Z) + \sigma(x)^{(n)}\phi(Z)\left(1 - \dfrac{\sigma_\varepsilon}{\sqrt{\sigma_\varepsilon^2 + \left(\sigma(x)^{(n)}\right)^2}}\right) & \text{if } \sigma(x)^{(n)} > 0 \\ 0 & \text{otherwise} \end{cases}$$

where $\beta^{(n)}$ manages the exploitation-exploration trade-off, $Z = \frac{y^+ - \mu(x)^{(n)}}{\sigma(x)^{(n)}}$ and y^+ is the best seen up to n. Then, $x^{(n+1)}$ is selected by minimizing LCB and maximizing EI and AEI.

Since we are using a RF as probabilistic surrogate model, the acquisition functions are also black-box. A global-local method has been used to solve the auxiliary problem (2) and identify the next promising $x^{(n+1)}$. More precisely, the global-local method used is known as "focus-search" [32]: it can handle with numeric, discrete and mixed search spaces, also involving conditional variables. Other approaches, also recent evolutionary methods such as Covariance Matrix Adaptation Evolution Strategy (CMA-ES) [33] are suitable for dense spaces, but the conditionality of the search spaces makes their usage still problematic. Focus-search is an adaptive Random Search strategy: it starts from a large set of random points where the acquisition function is evaluated. Then, it shrinks the search space around the current best point and perform a new random sampling of points within the "focused space". The shrinkage operation is iteratively performed until a maximum number of iterations and the entire procedure can be restarted multiple times to mitigate the risk to converge to a local optimum. Finally, the best point over all restarts and iterations is returned as the solution of the auxiliary problem (2). The R package mlrMBO [32] provides an implementation of focus-search.

Although different acquisition functions have been used, all the associated SMBO processes started from the same initial design. Furthermore, to mitigate the effect of randomness, we have performed 20 experiments with 20 different initial designs.

4.2.2   RF Based SMBO with Relaxation of the Discreteness Constraint

In this experiment we have relaxed the problem (4) by removing the constraint about the discreteness of the decision variable (c_1). This makes the initial box-bounded search space continuous, even if it remains complex due both to the presence of conditional decision variables (c_2) and the black-box hydraulic feasibility constraint (c_3). The rest of the experimental setup is identical to what reported in the previous sub-section.

5 Results

This section summarizes the most relevant results. Figure 3 shows how the "best seen" changes over function evaluations: solid lines are the averages over 20 different runs, while the shaded areas represent the standard deviations (almost 0). The first value, at iteration 0, is the best seen observed within the initial design.

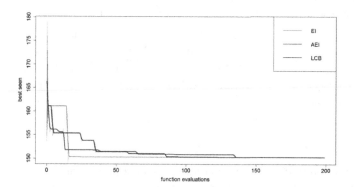

Fig. 3. Best seen over function evaluations of RF-based SMBO using three acquisition functions: EI (red), AEI (green) and LCB (blue). Solid lines and shaded areas represent, respectively, mean and standard deviation (that is almost 0) of the best seen over 20 different runs (Color figure online)

With respect to the second experiment – related to the relaxation of the discreteness constraint (c_1) – Fig. 4 shows how the "best seen" changes over function evaluations. Visualization is limited to the first 20 evaluations – out of the overall 200 – because, already after two function evaluations, no further improvements have been obtained.

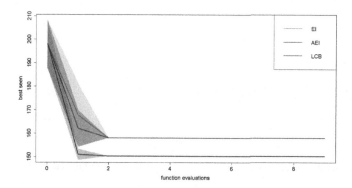

Fig. 4. Best seen over function evaluations of RF-based SMBO with relaxation of the discreteness constraint. Comparison between three acquisition functions: EI (red), AEI (green) and LCB (blue). Solid lines and shaded areas represent, respectively, mean and standard deviation (that is almost 0) of the best seen over 20 different runs (Color figure online)

Finally, we have evaluated the improvement, in terms of energy costs reduction, provided by SMBO with respect to the energy cost implied by the current pressure-based control operated by the water utility, that is 332,30€/day. In Table 1, the best cost over 20 runs has been selected for every acquisition function and separately for the two types of experiments described in Sect. 4.2.1 and 4.2.2.

Table 1. Optimal energy costs obtained via SMBO and associated costs reduction with respect to the current cost implied by the current pressure-based control operated by the water utility

	Original Problem (4)			Relaxation of (c_1)		
	EI	AEI	LCB	EI	AEI	LCB
Energy cost [€]	*150.29*	150.32	150.31	158.12	154.91	150.42
Cost reduction w.r.t. the control strategy currently operated [€]	*182.01*	181.98	181.99	174.18	177.39	181.88

The relaxation of discreteness constraint does not provide any improvement. This could be due to the use of RF, which can result less effective on continuous variables than discrete ones, also depending on the smoothness of the objective function. Probably, in the second experiment, the approach was not able to escape from some plateau, within the maximum number of function evaluations allowed.

As an example, we report in Fig. 5 the activation pattern of two pumps, A and B, according to the control currently operated by the WDN ("curr" suffix) versus the new activation implied by the new control optimized through SMBO ("opt" suffix).

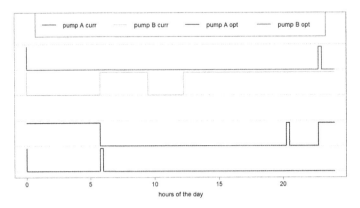

Fig. 5. Activation of two pumps according to current and optimized implicit control

6   Conclusions and Discussion

We have presented a SMBO approach for solving optimal control problems characterized by black-box objective functions and complex, partially unknown, search spaces. A general formalization of the problem was provided along with an instantiation on a specific real-life application, that is the optimal control of pumps in water distribution networks. The use of a hydraulic simulation software, EPANET, makes both objective function and constraints – related to hydraulic feasibility of the identified control rules – black-box. Using SMBO to search for an optimal *implicit* control allowed us to work with a dimensionality which is significantly lower than the one required by the (more widely adopted) *explicit* controls. A more realistic experimentation should consider different "simulation days", characterized by random water demands whose empirical distribution is generated from historical data. This requires evaluating the robustness of the implicit control rules proposed by SMBO and to move towards a "distributionally robust" SMBO.

Acknowledgements. This study has been partially supported by the Italian project "PerFORM WATER 2030" – programme POR (Programma Operativo Regionale) FESR (Fondo Europeo di Sviluppo Regionale) 2014–2020, innovation call "Accordi per la Ricerca e l'Innovazione" ("Agreements for Research and Innovation") of Regione Lombardia, (DGR N. 5245/2016 - AZIONE I.1.B.1.3 – ASSE I POR FESR 2014–2020) – CUP E46D17000120009.

We greatly acknowledge the DEMS (Department of Economics, Management and Statistics) Data Science Lab for supporting this work by providing computational resources.

References

1. Kushner, H.J.: A new method of locating the maximum point of an arbitrary multipeak curve in the presence of noise. J. Basic Eng. **86**(1), 97–106 (1964)
2. Močkus, J.: On Bayesian methods for seeking the extremum. In: Marchuk, G.I. (ed.) Optimization Techniques 1974. LNCS, vol. 27, pp. 400–404. Springer, Heidelberg (1975). https://doi.org/10.1007/3-540-07165-2_55
3. Zhigljavsky, A., Zilinskas, A.: Stochastic Global Optimization, vol. 9. Springer, Heidelberg (2007). https://doi.org/10.1007/978-0-387-74740-8
4. Hutter, F., Hoos, H.H., Leyton-Brown, K.: Sequential model-based optimization for general algorithm configuration. In: Coello, C.A.C. (ed.) LION 2011. LNCS, vol. 6683, pp. 507–523. Springer, Heidelberg (2011). https://doi.org/10.1007/978-3-642-25566-3_40
5. Archetti, F., Candelieri, A. (eds.): Bayesian Optimization and Data Science. SO. Springer, Cham (2019). https://doi.org/10.1007/978-3-030-24494-1
6. Hutter, F., Kotthoff, L., Vanschoren, J.: Automated Machine Learning-Methods, Systems, Challenges. Springer, Heidelberg (2019). https://doi.org/10.1007/978-3-030-05318-5
7. Elsken, T., Metzen, J.H., Hutter, F.: Neural architecture search: a survey. J. Mach. Learn. Res. **20**(55), 1–21 (2019)
8. Candelieri, A., Archetti, F.: Global optimization in machine learning: the design of a predictive analytics application. Soft. Comput. **23**(9), 2969–2977 (2018). https://doi.org/10.1007/s00500-018-3597-8
9. Williams, C.K., Rasmussen, C.E.: Gaussian Processes for Machine Learning. MIT Press, Cambridge (2006). 2(3)

10. Frazier, P.I.: Bayesian optimization. In: INFORMS TutORials in Operations Research, pp. 255–278 (2018)
11. Ho, T.K.: Random decision forests. In: Proceedings of the 3rd International Conference on Document Analysis and Recognition, pp. 278–282 (1995)
12. Sergeyev, Y.D., Pugliese, P., Famularo, D.: Index information algorithm with local tuning for solving multidimensional global optimization problems with multiextremal constraints. Math. Program. **96**(3), 489–512 (2003)
13. Paulavičius, R., Žilinskas, J.: Advantages of simplicial partitioning for Lipschitz optimization problems with linear constraints. Optim. Lett. **10**(2), 237–246 (2014). https://doi.org/10.1007/s11590-014-0772-4
14. Strongin, R.G., Sergeyev, Y.D.: Global Optimization with Non-convex Constraints: Sequential and Parallel Algorithms, vol. 45. Springer, Berlin (2013)
15. Digabel, S.L., Wild, S.M.: A taxonomy of constraints in simulation-based optimization. arXiv preprint arXiv:1505.07881 (2015)
16. Gramacy, R.B., Lee, H.K.M., Holmes, C., Osborne, M.: Optimization under unknown constraints. Bayesian Stat. **9**, 229 (2012)
17. Bernardo, J., et al.: Optimization under unknown constraints. Bayesian Stat. **9**(9), 229 (2011)
18. Candelieri, A., Archetti, F.: Sequential model based optimization with black-box constraints: feasibility determination via machine learning. In: AIP Conference Proceedings, vol. 2070, no. 1, p. 020010. AIP Publishing LLC (2019)
19. Hernández-Lobato, J.M., Gelbart, M.A., Hoffman, M.W., Adams, R.P., Ghahramani, Z.: Predictive entropy search for Bayesian Optimization with unknown constraints. In: Proceedings of the 32nd International Conference on Machine Learning, p. 37 (2015)
20. Strongin, R.G., Sergeyev, Y.D.: Global Optimization with Non-convex Constraints: Sequential and Parallel Algorithms. Kluwer Academic Publishers, Dordrecht (2000)
21. Bachoc, F., Helbert, C., Picheny, V.: Gaussian process optimization with failures: classification and convergence proof. HAL id: hal-02100819, version 1 (2019)
22. Sacher, M., et al.: A classification approach to efficient global optimization in presence of non-computable domains. Struct. Multidisc. Optim. **58**(4), 1537–1557 (2018). https://doi.org/10.1007/s00158-018-1981-8
23. Antonio, C.: Sequential model based optimization of partially defined functions under unknown constraints. J. Glob. Optim. **573**, 1–23 (2019). https://doi.org/10.1007/s10898-019-00860-4
24. Mala-Jetmarova, H., Sultanova, N., Savic, D.: Lost in optimisation of water distribution systems? A literature review of system operation. Environ. Model Softw. **93**, 209–254 (2017)
25. Bene, J.G., Selek, I., Hos, C.: Comparison of deterministic and heuristic optimization solvers for water network scheduling problems. Water Sci. Technol. Water Supply (2013)
26. Castro Gama, M.E., Pan, Q., Salman, S., Jonoski, A.: Multivariate optimization to decrease total energy consumption in the water supply system of Abbiategrasso (Milan, Italy). Environ. Eng. Manag. J. (EEMJ) **14**(9) (2015)
27. Castro Gama, M., Pan, Q., Lanfranchi, E.A., Jonoski, A., Solomatine, D.P.: Pump scheduling for a large water distribution network. Milan, Italy. Procedia Eng. **186**, 436–443 (2017)
28. Candelieri, A., Perego, R., Archetti, F.: Bayesian optimization of pump operations in water distribution systems. J. Glob. Optim. **71**(1), 213–235 (2018). https://doi.org/10.1007/s10898-018-0641-2
29. Srinivas, N., Krause, A., Kakade, S.M., Seeger, M.W.: Information-theoretic regret bounds for Gaussian process optimization in the bandit setting. IEEE Trans. Inf. Theory **58**(5), 3250–3265 (2012)

30. Huang, D., Allen, T.T., Notz, W.I., Zheng, N.: Global optimization of stochastic black-box systems via sequential Kriging meta-models. J. Glob. Optim. 3(34), 441–466 (2006)
31. Picheny, V., Wagner, T., Ginsbourger, D.: A benchmark of kriging-based infill criteria for noisy optimization. Struct. Multidiscip. Optim. 48(3), 607–626 (2013)
32. Bischl, B., Richter, J., Bossek, J., Horn, D., Thomas, J., Lang, M.: mlrMBO: A modular framework for model-based optimization of expensive black-box functions. arXiv preprint arXiv:1703.03373 (2017)
33. Hansen, N., Müller, S.D., Koumoutsakos, P.: Reducing the time complexity of the derandomized evolution strategy with covariance matrix adaptation (CMA-ES). Evol. Comput. 11(1), 1–18 (2003)

Composition of Kernel and Acquisition Functions for High Dimensional Bayesian Optimization

Antonio Candelieri[1]($\boxtimes$) (iD), Ilaria Giordani[2] (iD), Riccardo Perego[2] (iD), and Francesco Archetti[2] (iD)

[1] Department of Economics, Management and Statistics, University of Milano-Bicocca, 20126 Milan, Italy
antonio.candelieri@unimib.it
[2] Department of Computer Science, Systems and Communication, University of Milano-Bicocca, 20126 Milan, Italy

Abstract. Bayesian Optimization has become the reference method for the global optimization of black box, expensive and possibly noisy functions. Bayesian Optimization learns a probabilistic model about the objective function, usually a Gaussian Process, and builds, depending on its mean and variance, an acquisition function whose optimizer yields the new evaluation point, leading to update the probabilistic surrogate model. Despite its sample efficiency, Bayesian Optimization does not scale well with the dimensions of the problem. Moreover, the optimization of the acquisition function has received less attention because its computational cost is usually considered negligible compared to that of the evaluation of the objective function: its efficient optimization is also inhibited, particularly in high dimensional problems, by multiple extrema and "flat" regions. In this paper we leverage the additivity – aka separability – of the objective function into mapping both the kernel and the acquisition function of the Bayesian Optimization in lower dimensional subspaces. This approach makes more efficient both the learning/updating of the probabilistic surrogate model and the optimization of the acquisition function. Experimental results are presented for a standard test function and a real-life application.

Keywords: Bayesian optimization · Gaussian processes · Additive functions

1 Introduction

Bayesian Optimization (BO) [1, 2] has become the reference method for the global optimization of a black box function, whose evaluations are expensive and possibly noisy. BO maintains a probabilistic surrogate model – usually a Gaussian Process (GP) – of the objective function, updated depending on data as we observe them, namely function evaluations [3–6]. GP-based BO has two key design factors: the choice of the GP kernel – which sets an assumption about the hypothetical smoothness of the black-box function to optimize – and the acquisition function whose optimization provides the next point for the evaluation of the objective function, then exploited to update the GP model. BO has been generalized to the constrained case [7–9],

© Springer Nature Switzerland AG 2020
I. S. Kotsireas and P. M. Pardalos (Eds.): LION 14 2020, LNCS 12096, pp. 316–323, 2020.
https://doi.org/10.1007/978-3-030-53552-0_29

including black-box constraints and more recently partially defined objective functions or non-computable domains [10]. BO is sample efficient when applied to low dimensional problem, up to 10 or 20, but already midsize problems are challenging. This difficulty is of course exacerbated when the evaluations of the objective function are very expensive, like in most simulation-optimization problems, hyperparameter optimization of large-scale Machine Learning applications (e.g., Automated Machine Learning [11], Neural Architecture Search [12] and complex Machine Learning pipelines [13]). Many approaches have been suggested to mitigate the computational load of high dimensional BO, such as random embeddings [14] and the exploitation of the *additivity* of the objective function. Additive structure is often assumed in high dimensional GP-based regression, even if $f(x)$ is not additive: in particular, in small sample situations, fitting a simpler model to the data gives better results. This is exactly the case in which BO is the method of choice since querying the black-box objective function – and in case constraints – 1 is expensive. The global structure of the problem is managed through the kernel: [15, 16] describe methods to search over possible kernel compositions starting with basic kernels. A major step was taken in [17] and successively developed in [18] by introducing, based on the Bochner's theorem, a Fourier Features approximation which is shown to yield an error decreasing exponentially with the number of features, while allowing a reduction in the complexity of the kernel matrix inversion.

In this paper we focus on a strategy which leverages additivity – aka separability – of the objective function into the decomposition of the kernel and the acquisition function. The main contributions of this paper are:

- To show that the function additivity can be leveraged to *(i)* reduce the cost of updating the GP and *(ii)* improve the performance of different acquisition functions.
- To show, in a realistic problem, that also the case of black-box constraints fits naturally into the proposed framework.
- To compare the proposed additive BO with the standard one providing results on a benchmark test function and a real-life case study.

2   Background

The global optimization problem is usually defined as:

$$\min_{x \in \mathcal{X} \subset \mathbb{R}^d} f(x)$$

where the search space $\mathcal{X}$ is generally box-bounded. A GP is a stochastic process given by a collection of random variables, any finite number of which have a joint Gaussian distribution. The random variables consist of function values $f(x)$ at different locations x within the domain $\mathcal{X} \subseteq \mathbb{R}^d$. A GP is fully specified by its mean and covariance functions, respectively denoted by $\mu(x)$ and the *kernel* $k : \mathcal{X} \times \mathcal{X} \rightarrow \mathbb{R}$. Many kernels are available, such as Squared Exponential (SE) and Matérn among others [1, 2]. The SE kernel has been used in this paper:

$$k(x, x') = e^{-\frac{\|x-x'\|^2}{2\ell^2}}$$

Let $D_{1:t} = \{(x_i, y_i)\}_{i=1,\dots,t}$, where $y_t = f(x_t) + \varepsilon$, and $\varepsilon \sim \mathcal{N}(\mu_\varepsilon, \sigma_\varepsilon^2)$ in the case of a noisy objective function. Given the set of observations $D_{1:t}$, the posterior distribution of the GP is given by:

$$P(y_{t+1}|D_{1:t}, x_{t+1}) = \mathcal{N}\left(y_{t+1}|\mu_t(x_{t+1}), \sigma_t^2(x_{t+1})\right)$$

where

$$\mu_t(x_{t+1}) = \mathbf{k}_t^T \left[\mathbf{K}_t + \sigma_\varepsilon^2 I\right]^{-1} \{y_1, \dots, y_t\}$$

$$\sigma_t^2 = k(x_{t+1}, x_{t+1}) - \mathbf{k}_t^T \left[\mathbf{K}_t + \sigma_\varepsilon^2 I\right]^{-1} \mathbf{k}_t$$

with $\mathbf{K}_{t,[i,j]} = k(x_i, x_j)$ and $\mathbf{k}_t = [k(x, x_1), \dots, k(x, x_t)]$.

The kernel's parameter, that is the length-scale ℓ in the case of SE kernel, is usually fit on data via Maximum Likelihood Estimation (MLE). It has been recently demonstrated [19] that SE kernel could be inadequate for many functions, specifically for any dense sequence of points $x_1, x_2, \dots, x_N$ with $N \to \infty$. However, according to the goal of this study, the length of the sequences generated through BO are small, making less relevant this issue.

The acquisition function is the mechanism to implement the trade-off between exploration and exploitation in BO. This trade-off is also addressed in deterministic global optimization methods, such as via local tuning in Lipschitz global optimization [20]. Any acquisition function aims to guide the search of the optimum towards points with potentially low values of objective function either because the prediction of $f(x)$ is low or the uncertainty is high (or both). While *exploiting* means to consider the region of the search space providing more chance to improve the current best solution (with respect to the current surrogate model), *exploring* means to move towards less explored regions. Many acquisition functions have been proposed, such as Probability of Improvement, Expected Improvement, Confidence Bound (Upper/Lower Confidence Bound for maximization/minimization problems, respectively), Entropy Search, Predictive Entropy Search, Knowledge Gradient and Thompson Sampling (TS) – a brief review is provided in [2]. Each acquisition offers its own blend of the GP's mean and variance. One of the most widely adopted is the Upper/Lower Confidence Bound (UCB/LCB), respectively for solving maximization/minimization problems. The next promising point, where to evaluate the objective function, is obtained by solving the following auxiliary problem ($\pm$ refer to UCB and LCB, respectively):

$$x_{t+1} = \underset{x \in D}{\operatorname{argmax}} \, \mu_t(x) \pm \beta_t^{1/2} \sigma_t(x)$$

where β_t satisfies the converge criteria analysed in [21]. Although UCB/LCB shows some theoretical deficiency from the point of view of rational decision theory [22], we did not experienced this deficiency in our study.

Thompson Sampling (TS) is based on a different approach: when used in BO, TS searches for the minimizer of a function drawn from the GP posterior and that minimizer will be chosen as the location where the next function evaluation takes place. In [23], an analysis of TS for GP is presented along with the proposal to use an ε-greedy strategy to mitigate the intrinsic exploitation bias of TS.

3 Additive Functions and Basic Decomposition in BO

Additive GP based models assume that the objective function $f(x)$ is a sum of functions $f^{(j)}$ defined over low-dimensional components:

$$f(x) = \sum_{j=1}^{G} f^{(j)}\left(x^{(j)}\right)$$

where each $x^{(j)}$ belongs to a low dimensional subspace $\mathcal{X}^{(j)} \subseteq D$ and G denote the number of these components. We assume in this paper that $\mathcal{X}^{(j)} \cap \mathcal{X}^{(l)} = \emptyset$ if $l \neq j$.

The concept of additive functions translates to GPs, where the stochastic process is a sum of stochastic processes, each one having low dimensional indexing. The effective dimensionality of the model is defined as the largest dimension among all additive groups, $\bar{d} = max_{j \in [G]} \dim\left(\mathcal{X}^{(j)}\right)$.

Under the additive assumption, the kernel and the mean function of a GP decompose similarly to the GP's stochastic process. Specifically, $k(x, x') = \sum_{j=1}^{G} k^{(j)}\left(x^{(j)}, x'^{(j)}\right)$ and $\mu(x) = \sum_{j=1}^{G} \mu^{(j)}\left(x^{(j)}\right)$. Consequently, BO can be applied, independently, on each subspace, while at the same time including the cross correlation of additive groups through the observations $D_{1:t}$ restores the global structure of the objective function.

Preliminary we provide some relevant notations. A kernel can be viewed as the inner product in a Reproducing Kernel Hilbert Space (RKHS) $\mathcal{H}$ equipped with a feature map $\varphi : \mathcal{X} \to \mathcal{H}$, such that $k(x, x') = \langle \varphi(x), \varphi(x') \rangle_{\mathcal{H}}$. In the case that $\mathcal{H}$ is separable, the inner product can be approximated leading to:

$$k(x, x') = \langle \varphi(x), \varphi(x') \rangle_{\mathcal{H}} \approx \phi(x)^{\mathsf{T}} \phi(x')$$

with $\phi(x) : \mathcal{X} \to \mathbb{R}^m$ a finite-dimensional feature map, with m defining the level of approximations: the greater m the better the approximation. For stationary kernels, Bochner's theorem implies that a suitable m-dimensional feature map can be constructed using a set of random Fourier features.

Thus, the covariance matrix can be approximated as $K_t \approx \mathbf{\Phi}(X_t)^{\mathsf{T}} \mathbf{\Phi}(X_t)$, with $X_t = \{x_1, \ldots, x_t\}$ and $\mathbf{\Phi}(X_t) = [\phi(x_1), \ldots, \phi(x_t)]^{\mathsf{T}}$.

Let define $v_t = \left[\mathbf{\Phi}(X_t)^{\mathsf{T}} \mathbf{\Phi}(X_t), + \sigma_\varepsilon^2 I\right]^{-1} \mathbf{\Phi}(X_t)\{y_1, \ldots, y_t\}$, then the approximated GP's mean and variance can be computed as: $\tilde{\mu}_t(x) = \mathbf{\Phi}(x)^{\mathsf{T}} v_t$ and $\tilde{\sigma}_t(x)^2 = \sigma_\varepsilon^2 \mathbf{\Phi}(x)^{\mathsf{T}}$

$[\boldsymbol{\Phi}(X_t)^\mathsf{T}\boldsymbol{\Phi}(X_t), +\sigma_\varepsilon^2 I]^{-1}\boldsymbol{\Phi}(x)$, when $\|\boldsymbol{\Phi}(x)\|_2 = 1$ (which is true for Random as well as Quadrature Fourier Features approximations [18]).

Following, we report the algorithm for additive BO with TS as acquisition function

---

Algorithm 1

1: **for** $t = 1, \ldots, T$ **do**
2:     Update ν_t and $\boldsymbol{\Phi}(X_t)^\mathsf{T}\boldsymbol{\Phi}(X_t) + \sigma_\varepsilon^2 I$
3:     Sample $\theta_t \sim \mathcal{N}(\nu_t, [\boldsymbol{\Phi}(X_t)^\mathsf{T}\boldsymbol{\Phi}(X_t) + \sigma_\varepsilon^2 I]^{-1})$
4:     **for** $j = 1, \ldots, G$ **do**
5:         Find $x_{t+1} = \text{argmax}_{x \in \mathcal{X}} \theta_t^{(j)\mathsf{T}} \boldsymbol{\Phi}^{(j)}(x^{(j)})$
6:     **endfor**
7:     Evaluate $f(x_{t+1}) = \sum_{j=1}^{G} f^{(j)}(x_{t+1}^{(j)})$
8: **endfor**

In the case that a different acquisition function is used, such as EI or UCB/LCB, step 3 is not needed and step 5 is modified with the associated equation.

4 Experiments

4.1 Benchmark Function

The first experiment is related to a well-known test function, namely the Styblinski-tang [24]:

$$f(x) = \frac{1}{2}\sum_{i=1}^{d}\left(x_i^4 - 16x_i^2 + 5x_i\right)$$

where d is the dimensions of the search space, which is the hypercube $x_i \in [-5;5]$ for all $i = 1, \ldots, d$. In our experiment we considered $d = 10$, where the maximum number of dimensions for each group of decision variables is $\bar{d} = 1$.

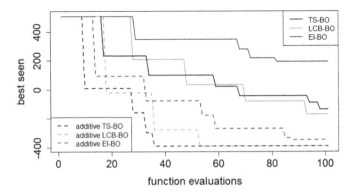

Fig. 1. Best-seen over function evaluations; comparison between additive BO and "standard" BO with three acquisition functions (i.e., TS, LCB and EI)

Figure 1 shows the best value of the objective function observed along the BO function evaluations (aka "best seen"). All the additive BO implementations resulted both more effective (i.e., they provided a better optimal solution) and more efficient (i.e., they achieve the optimal solution in a lower number of function evaluations) than "standard" BO, where "standard" refers to traditional BO algorithm working on the whole search space. Moreover, TS and LCB resulted better than EI, irrespectively to standard and additive BO.

4.2 Control Problems

Here we consider the Pump Scheduling Optimization (PSO) of an urban Water Distribution Network (WDN), using a benchmark WDN, namely AnyTown [25], powered by 4 variable speed pumps. The goal of PSO is to minimize the energy cost associated to the pump schedule, that is defined as the status (or speed) of the pumps at given times (e.g., hours of the day). Status refers to ON/OFF pumps, while speed refers to variable speed pumps.

$$minC = \Delta t \sum_{t=1}^{T} c_E^t \sum_{k_v=1}^{N_v} \gamma x_k^t Q_k^t \frac{H_k^t}{\eta_k}$$

where C is the total energy cost, Δt is the time step, C_E^t is the energy price at time t, γ is the specific weight for water, $Q_{k_j}^t$ is the water flow provided by the pump k_j at time t, $H_{k_j}^t$ is the head loss on pump k_j at time t, η_{k_j} is the efficiency of pump k_j, and P_i^t is the pressure at the control node i at time t, N_v is the number of variable speed pumps.

The decision variables are $x_{k_v}^t \in (0, 1)$, where each one represents the speed of pump k_v at time t. Both energy cost and hydraulic feasibility of the a given pump schedule are computed via software simulation, more specifically through the tool EPANET. Indeed, even if the objective function is clearly additive, inter-temporal relations exist from a time step to the next: all the underlying equations, related to water mass balance, pressure variation, etc., are solved by EPANET.

In the case of non-additive BO, the problem has been solved in [25], where the search space of pump schedules had dimensionality $d = 96$ (i.e., 4 pump's speeds for each hour of the day). By using additivity, we solved 24 problems with dimensionality 4 each (i.e., $\bar{d} = 4$). Table 1 summarizes the preliminary results obtained on this benchmark study, compared to those obtained via "standard" BO (i.e., non-additive) and previously reported in [25]. Results confirm what observed on the test function experiment: additive BO provides better optimal solutions with less computational time.

Table 1. PSO of a WDN: results on the AnyTown benchmark, standard vs additive BO

Strategy	Energy cost ($)	Iteration number	Overall clock time
BO-LCB [25]	653.55	356	21256.48 [s]
BO-EI [25]	609.17	290	23594.36 [s]
Additive BO-LCB	605.23	344	12331.65 [s]
Additive BO-EI	653.55	276	10775.44 [s]
Additive BO-TS	**604.51**	281	11331.31 [s]

5   Conclusions

Although the results are preliminary, exploiting the additivity property inherent in the objective function allowed to boost BO performance in our experiments.

This improvement is observed for several acquisition functions but is more significant for Thompson Sampling than LCB and EI. We observe experimentally that the additive algorithm convergence, proved for the cumulative regret, is no-regret also for the performance metric based on best seen.

Significantly this result translates from the test function to the real-life case of the control of a water distribution network. The resulting optimization problem has close to 96 variables: the additive algorithm brings it to a dimension (i.e., $\bar{d} = 4$) where BO is very sample efficient.

Acknowledgments. This study has been partially supported by the Italian project "PerFORM WATER 2030" – programme POR (Programma Operativo Regionale) FESR (Fondo Europeo di Sviluppo Regionale) 2014–2020, innovation call "Accordi per la Ricerca e l'Innovazione" ("Agreements for Research and Innovation") of Regione Lombardia, (DGR N. 5245/2016 - AZIONE I.1.B.1.3 – ASSE I POR FESR 2014–2020) – CUP E46D17000120009.

We greatly acknowledge the DEMS Data Science Lab for supporting this work by providing computational resources.

References

1. Frazier., P.I.: Bayesian optimization. In: INFORMS TutORials in Operations Research, pp. 255–278 (2018)
2. Archetti, F., Candelieri, A.: Bayesian Optimization and Data Science. Springer, Cham (2019). https://doi.org/10.1007/978-3-030-24494-1
3. Krige, D.G.: A statistical approach to some basic mine valuation problems on the Witwatersrand. J. Chem. Metal. Min. Soc. South Africa **52**, 119–139 (1951)
4. Kushner, H.J.: A new method of locating the maximum point of an arbitrary multipeak curve in the presence of noise. J. Basic Eng. **86**(1), 97–106 (1964)
5. Močkus, J.: On Bayesian methods for seeking the extremum. In: Marchuk, G.I. (ed.) Optimization Techniques IFIP Technical Conference. LNCS, pp. 400–404. Springer, Heidelberg (1975). https://doi.org/10.1007/978-3-662-38527-2_55
6. Zhigljavsky, A., Zilinskas, A.: Stochastic Global Optimization, vol. 9. Springer, Boston (2007). https://doi.org/10.1007/978-0-387-74740-8
7. Gramacy, R.B., Lee, H.K.M., Holmes, C., Osborne, M.: Optimization under unknown constraints. Bayesian Stat. **9**, 229 (2012)
8. Candelieri, A., Archetti, F.: Sequential model based optimization with black-box constraints: feasibility determination via machine learning. In: AIP Conference Proceedings, vol. 2070, no. 1, p. 020010. AIP Publishing LLC, February 2019
9. Hernández-Lobato, J.M., Gelbart, M.A., Hoffman, M.W., Adams, R.P., Ghahramani, Z.: Predictive entropy search for Bayesian Optimization with unknown constraints. In: Proceedings of the 32nd International Conference on Machine Learning, vol. 37 (2015)
10. Candelieri, A: Sequential model based optimization of partially defined functions under unknown constraints. J. Global Optim. 1–23 (2019)

11. Hutter, F., Kotthoff, L., Vanschoren, J.: Automated Machine Learning-Methods, Systems, Challenges. Automated Machine Learning. Springer, Cham (2019). https://doi.org/10.1007/978-3-030-05318-5
12. Elsken, T., Metzen, J.H., Hutter, F.: Neural architecture search: a survey. J. Mach. Learn. Res. **20**(55), 1–21 (2019)
13. Candelieri, A., Archetti, F.: Global optimization in machine learning: the design of a predictive analytics application. Soft. Comput. **23**(9), 2969–2977 (2018). https://doi.org/10.1007/s00500-018-3597-8
14. Wang, Z., Hutter, F., Zoghi, M., Matheson, D., de Feitas, N.: Bayesian optimization in a billion dimensions via random embeddings. J. Artif. Intell. Res. **55**, 361–387 (2016)
15. Duvenaud, D.K., Nickisch, H., Rasmussen, C.E.: Additive Gaussian processes. In: Advances in Neural Information Processing Systems, pp. 226–234 (2011)
16. Duvenaud, D., Lloyd, J.R., Grosse, R., Tenenbaum, J.B., Ghahramani, Z.: Structure discovery in nonparametric regression through compositional kernel search. arXiv preprint arXiv:1302.4922 (2013)
17. Wilson, A., Adams, R.: Gaussian process kernels for pattern discovery and extrapolation. In: International Conference on Machine Learning, pp. 1067–1075, February 2013
18. Mutny, M., Krause, A.: Efficient high dimensional Bayesian optimization with additivity and quadrature Fourier features. In: Advances in Neural Information Processing Systems, pp. 9005–9016 (2018)
19. Zhigljavsky, A., Žilinskas, A.: Selection of a covariance function for a Gaussian random field aimed for modeling global optimization problems. Optim. Lett. **13**(2), 249–259 (2019). https://doi.org/10.1007/s11590-018-1372-5
20. Sergeyev, Y.D.: An information global optimization algorithm with local tuning. SIAM J. Optim. **5**(4), 858–870 (1995)
21. Srinivas, N., Krause, A., Kakade, S.M., Seeger, M.W.: Information-theoretic regret bounds for gaussian process optimization in the bandit setting. IEEE Trans. Inf. Theory **58**(5), 3250–3265 (2012)
22. Žilinskas, A., Calvin, J.: Bi-objective decision making in global optimization based on statistical models. J. Global Optim. **74**(4), 599–609 (2018). https://doi.org/10.1007/s10898-018-0622-5
23. Basu, K., Ghosh, S.: Analysis of Thompson sampling for Gaussian process optimization in the bandit setting. arXiv preprint arXiv:1705.06808 (2017)
24. Jamil, M., Yang, X.S.: A literature survey of benchmark functions for global optimization problems. arXiv preprint arXiv:1308.4008 (2013)
25. Candelieri, A., Perego, R., Archetti, F.: Bayesian optimization of pump operations in water distribution systems. J. Global Optim. **71**(1), 213–235 (2018). https://doi.org/10.1007/s10898-018-0641-2

A Pareto Simulated Annealing for the Integrated Problem of Berth and Quay Crane Scheduling at Maritime Container Terminals with Multiple Objectives and Stochastic Arrival Times of Vessels

Abtin Nourmohammadzadeh$^{(\boxtimes)}$ (iD) and Stefan Voß (iD)

Institute of Information Systems (IWI), University of Hamburg, Hamburg, Germany
{abtin.nourmohammadzadeh,stefan.voss}@uni-hamburg.de,
https://www.bwl.uni-hamburg.de/en/iwi

Abstract. Efficient planning and scheduling of operations at congested seaside container terminals are issues of extreme importance because of the ever growing worldwide demand for container shipments. In this paper, the two main problems of berth and quay crane scheduling are integrated in a novel mathematical model. It is assumed that the arrival times of vessels are stochastic and can take any value that exists within a specific interval. The presented model includes three objectives. They are the minimisation of weighted deviations from the target berthing locations and times as well as departure delays. In the first solution attempt, an ϵ-constraint method is used which employs an exact solver. Since the problem has high complexity and cannot be solved in large scales with an exact solver, a Pareto Simulated Annealing (PSA) algorithm is designed for it. It is proved that this metaheuristic can provide better non-dominated solutions in much shorter times compared to the ϵ-constraint approach. Furthermore, the advantage of integrating the berth and quay crane scheduling is examined by comparing the results with the case that these two problems are processed separately.

Keywords: Multiobjective optimisation · Maritime container terminal · Berth and quay crane scheduling · Mathematical modelling · Pareto simulated annealing

1 Introduction

Containerisation provides reliable and standardised means of transportation which leads to shorter transit times and the possibility of using multiple modalities. Moreover, it reduces the shipping as well as handling costs [24]. According to [2], the global container port throughput in the world has raised from 560 million 20-foot equivalent units (TEUs) in 2010 to 753 million in 2017. In addition, the increase in the worldwide containerised trade volume was more than

© Springer Nature Switzerland AG 2020
I. S. Kotsireas and P. M. Pardalos (Eds.): LION 14 2020, LNCS 12096, pp. 324–340, 2020.
https://doi.org/10.1007/978-3-030-53552-0_30

40% in the decade 2007–2016. Respecting this considerable growth, container terminals are faced with larger quantities of throughput and becoming busy and congested more and more. This clearly shows the significance of efficient planning and scheduling of operations at a seaside container terminal. Operations in a container terminal can be grouped based on the location into: seaside operations, yard operations, and land-side operations [21,23] and [18].

The berth scheduling problem (BSP), also in less detail called berth allocation problem (BAP), is recognised as the most important operational problem at the seaside area [6]. Usually it is sought to reduce the process time of vessels at the berth. This strongly depends on the assignment and scheduling of the available quay cranes, which work on the vessels. This problem is called quay crane assignment problem (QCAP) and its more complete version which determines the start and finish time of each crane on each vessel is known as quay crane scheduling problem (QCSP). One other goal can be berthing of the vessels with minimum deviation from their target berthing times. Furthermore, it is important to assign the vessels to the berthing locations which are possibly less distant from their desired berthing locations.

Regarding the connections between BSP and QCSP, they can be integrated as one problem called berth and quay crane scheduling problem (BQCSP). The advantage of this integration is verified in our work by means of some numerical comparisons. A comprehensive mathematical model with the components of both problems is developed. This novel model includes three objectives of minimising the weighted sum of distances from the desired (target) berthing locations, the weighted sum of temporal deviations from the target berthing times and the weighted sum of the departure (completion time) delays of all vessels. A real aspect in such problems at container terminals is the variability of inputs, which has not been taken into account enough in the previous works. Nonetheless, our work considers this aspect by solving a version of the BQCSP which includes stochastic arrival times for vessels.

In the first step, a classical solution method is applied with the aid of an exact solver. This method is called ϵ-constraint. Since the problem has a high computational complexity, in the next step, a metaheuristic called Pareto simulated annealing (PSA) is adapted and employed.

This paper is organised as follows: Sect. 2 gives a brief categorisation of some related works. The model is introduced in Sect. 3. Section 4 describes the solution methodologies. The computational efforts and experiments as well as the comparisons of results are covered in Sect. 5. Finally, Sect. 6 draws the conclusions of our research and gives some directions for its extension.

2   Related Work

A variety of researches have been conducted on the BAP and the QCAP so far. They considered different versions of the problems and variable assumptions. Some works considered an integrated model (BAQCAP) to solve these two inter-related problems simultaneously. [11] considers vessels' fuel consumption and emissions in the BAQCAP and applies a novel non-linear multi-objective mixed-integer programming model. [9] proposes an integrated heuristics-based solution methodology that tackles a continuous BAQCAP. A GRASP-based metaheuristic is applied to the problem in [20] aiming at minimisation of the total waiting time elapsed to serve all vessels. [25] formulates the problem as a binary integer linear model that is later extended by incorporating the quay crane scheduling problem. In [29], an online model is given considering a hybrid berth which consists of three adjacent small berths together with five quay cranes. [12] uses a mixed integer linear programming model including multiple objectives. [16] considers the problem with a single objective of minimising a weighted sum of the waiting times, the deviations from desired locations and the departing delays. [17] presents a mathematical model for the BQCSP which encompasses all associated operations and constraints. A significant assumption in their problem is that preemption is allowed in the quay crane tasks.

Some others works considered the berth and quay crane assignment or both separately (are, e.g., [4,7,10,13,15,23,26–28] and [22]). The uncertainty or dynamics in the inputs are respected in a small number of researches such as [26–28] and [13]. There are a couple of multiobjective models (e.g. in [11,26,28] and [13]), and the application of heuristic and metaheuristic solution methodologies is observable (for example in [13,16,20,22,27] and [7]) in the previous literature.

3 Mathematical Model

In this section, our multiobjective mathematical model is explained in three subsections of assumptions, notations and formulations.

3.1 Assumptions

Our model is built based on some assumptions, which can also be called the characteristics of the BQCSP focused in this work. We try to make these characteristics conform with the reality as much as possible while also following the goal of simplification. The berth, where vessels dock at, is assumed to be continuous and has a given length. The vessels are assumed to be of three sizes (small, medium and large). The arrival times of vessels are stochastic in the sense that an interval is considered for this parameter. A linear and symmetric probability density function (PDF) is defined as shown in Fig. 1. An average value is determined for the stochastic parameter a shown as $E(a)$ and the parameter values are between $min(a) = 0.75E(a)$ and $max(a) = 1.25E(a)$. The maximum probability density $max[f(a)]$ is at the average value and its minimum is $min[f(a)] = \frac{max[f(a)]}{6}$ happening at $min(a)$ and $max(a)$. Regarding the probability rules we have:

$$\int_{min(a)}^{max(a)} f(a)da = min[f(a)].(max(a) - min(a)) + (max[f(a)] - min[f(a)]).$$

$$(\frac{max(a) - min(a)}{2}) = \frac{1}{6}max[f(a)](0.5E(a)) + \frac{5}{6}max[f(a)](0.25E(a)) =$$

$$E(a)(\frac{1}{12}max[f(a)] + \frac{5}{24}max[f(a)]) = E(a)(\frac{7}{24}max[f(a)]) = 1 \quad (1)$$

Therefore, by having $E(a)$, we can calculate $min(a)$, $max(a)$, $min[f(a)]$ and $max[f(a)]$.

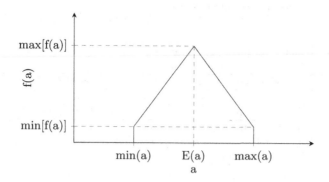

Fig. 1. Probability density function for arrival times of vessels $f(a)$

3.2   Notations

Our mathematical model contains the following notations:

Sets

V	The set of arriving vessels $i, j \in V$
QC	The set of quay cranes $k, l, m \in QC$

Parameters

H	The planning horizon
BL	The length of the berth
TT_i	Target berthing time (arrival time) of vessel i. This is stochastic and its average is called $E(TT_i)$
TB_i	Target berthing location of vessel i
RT_i	The required quay crane hours to serve vessel i
L_i	The length of vessel i
C_i	The importance coefficient of vessel i
$Qmax_i$	The maximum number of cranes which can work on vessel i at the same time

Decision Variables

b_i	The berthing location of vessel i
sv_i	The berthing time of vessel i
fv_i	The time that vessel i leaves the berth
sc_{ki}	The time that quay crane k begins its work on vessel i
fc_{ki}	The time that quay crane k finishes its work on vessel i
q_{ki}	Binary variable $= 1$, if crane k works on vessel i
y_{ij}	Binary variable $= 1$, if vessel j docks physically after vessel i along the berth
z_{ij}	Binary variable $= 1$, if vessel j docks chronologically (in terms of time) after vessel i
u_{kli}	Binary variable $= 1$, if cranes k and l work on vessel i at the same time
w_{kli}	Binary varaible $= 1$, if crane l begins its work on vessel i after the start time of crane k
m_{kij}	Binary variable $= 1$, if crane k works on vessel i before j
ΔB_i	The deviation from the target berthing location of vessel i
ΔTS_i	The deviation from the trarget berthing time of vessel i
ΔTF_i	The departure delay from an early departure time calculated according to the case that vessel i is served with the maximum number of cranes

3.3   Formulations

Objectives

$$Min\ Z_1 = \sum_{i \in V} C_i \Delta B_i \tag{2}$$

$$Min\ Z_2 = \sum_{i \in V} C_i \Delta TS_i \tag{3}$$

$$Min\ Z_3 = \sum_{i \in V} C_i \Delta TF_i \tag{4}$$

Constraint

$$\sum_{k \in QC} q_{ki}(fc_{ki} - sc_{ki}) \geq RT_i \qquad i \in V \tag{5}$$

$$w_{kli} + w_{lki} = 1 \qquad i \in V; k, l \in QC \tag{6}$$

$$q_{ki}q_{li}w_{kli}(sc_{li} - fc_{ki})(sc_{li} - sc_{ki}) \leq u_{kli}H \qquad i \in V; k, l \in QC \tag{7}$$

$$-q_{ki}q_{li}w_{kli}(sc_{li} - fc_{ki})(sc_{li} - sc_{ki}) \leq (1 - u_{kli})H \qquad i \in V; k, l \in QC \tag{8}$$

$$q_{ki}q_{li}w_{lki}(sc_{ki} - fc_{li})(sc_{ki} - sc_{li}) \leq u_{lki}H \qquad i \in V; k,l \in QC \tag{9}$$

$$-q_{ki}q_{li}w_{lki}(sc_{ki} - fc_{li})(sc_{ki} - sc_{li}) \leq (1 - u_{lki})H \qquad i \in V; k,l \in QC \tag{10}$$

$$\frac{1}{6}\sum_{k \in QC}\sum_{l \in QC/k}\sum_{m \in QC/k,l} u_{kli}u_{mli}(2 + u_{kmi}) \leq Qmax_i \qquad i \in V \tag{11}$$

$$\Delta B_i \geq TB_i - b_i \qquad i \in V \tag{12}$$

$$\Delta B_i \geq b_i - TB_i \qquad i \in V \tag{13}$$

$$\Delta TS_i \geq TT_i - sv_i \qquad i \in V \tag{14}$$

$$\Delta TS_i \geq sv_i - TT_i \qquad i \in V \tag{15}$$

$$fv_i \geq q_{ki}fc_{ki} \qquad i \in V, k \in QC \tag{16}$$

$$\Delta TF_i \geq fv_i - (sv_i + \frac{RT_i}{Qmax_i}) \qquad i \in V \tag{17}$$

$$b_j + BL(1 - y_{ij}) \geq b_i + l_i \qquad i,j \in V \tag{18}$$

$$sv_j + (1 - z_{ij})H \geq fv_i \qquad i,j \in V \tag{19}$$

$$y_{ij} + y_{ji} + z_{ij} + z_{ji} \geq 1 \qquad i,j \in V \tag{20}$$

$$(kq_{ik} - lq_{jl})(b_i - b_j) \geq 0 \qquad i,j \in V; k,l \in QC \tag{21}$$

$$sc_{kj} + (1 - m_{kij})H \geq fc_{ki} \qquad i,j \in V; k \in QC \tag{22}$$

$$m_{kij} + m_{kji} \leq 1 \qquad i,j \in V; k \in QC \tag{23}$$

$$m_{kij} + m_{kji} \leq q_{ki} \qquad i,j \in V; k \in QC \tag{24}$$

$$m_{kij} + m_{kji} \leq q_{kj} \qquad i,j \in V; k \in QC \tag{25}$$

Three objectives are considered for our BQCSP. Equation (2) and (3) formulate the sum of weighted deviations of all vessels from the target berthing locations and times, respectively. The third objective, (4), is the sum of weighted delays of all vessels from their ideal departure times. Constraint (5) implies satisfying the required crane-hours of each vessel. Constraints (6)–(10) together have the function of setting $u_{kli} = 1$ if and only if the working times of cranes k and l overlap and otherwise $u_{kli} = 0$. Respecting the maximum available number of cranes which can work on each vessel is guaranteed by constraint (11). The deviation from the target berthing location is calculated by relations (12) and (13). Likewise, (14) and (15) calculate the deviation from the target berthing times. (16) states that the departure time of each vessel is equal to the latest finish time of all quay cranes which work on it. Constraint (17) calculates the departure delay according to a desirable process time which is based on the assignment of maximum possible number of cranes to the vessel. Constraint (18) is for respecting the vessel's length at the berth and assures that no more than one vessel can dock in each point at the same time. Constraint (19) ensures that if vessel j chronologically berths after i ($z_{ij}=1$), its berthing time is after the departure time of i. Regarding any pair of vessels i and j, they must not collide each other. It means that they berth at different locations by respecting their lengths or they berth at different times considering the process time of the former. This is guaranteed by (20). Constraint (21) prevents cranes from crossing each other. Each crane cannot serve more than one vessel at a time which is enforced by (22)–(25).

4   Solution Methodologies

In multiobjective optimisation each solution corresponds to more than one objective value. Therefore, solutions can not be easily sorted based on the objective function value. Instead, a trade-off between the objectives should be considered. Here we are involved with the concept of dominance. It is said that in the presence of n objectives, a solution x dominates another solution y, $x \prec_d y$, if the following conditions hold:

1) $z_i(x) \leq z_i(y), \forall i \in 1, 2, ..., n$
2) $\exists j \in 1, ..., n : z_j(x) < z_j(y)$

The goal in multiobjective optimisation is to find a set of globally non-dominated solutions as the best ones instead of only one optimal solution. These solutions are known as Pareto optimal and the set is called Pareto front.

There are various solution methodologies for multiobjective problems. Some of them are classical and some in the category of evolutionary algorithms. Due to some drawbacks of classical methods such as long computational times or the necessity of numerous runs to obtain a set of non-dominated solutions, Multi-Objective Evolutionary Optimisation Algorithms (MOEAs) are used as alternative approaches for large-sized instances.

In this research, a practical MOEA called Pareto Simulated Annealing (PSA) is adapted to be applicable to our BQCSP with regard to the high computational complexity of this problem. The performance of this evolutionary approach is tested by comparing its results with those of an ϵ-Constraint method which uses an exact solver.

4.1 ϵ-Constraint

The ϵ-constraint is one promising classical multi-objective optimisation method, which is proposed by [5]. Unlike some other classical methods, it does not have any problem in finding solutions on non-convex parts of the Pareto curve. In this method, one objective is chosen out of all (n) objectives to be minimised. The other objectives are constrained to be less than or equal to some given values. Mathematically expressed, if $f_i(x)$ is chosen for minimisation, we have the following problem $P(\epsilon_i)$:

$$min\ f_i(x)$$
$$f_j(x) \leq \epsilon_j,\ \forall j \in \{1, ..., n\}\backslash i.$$

Each time that the problem is solved by different values of ϵ_i, one Pareto optimal solution may be found. Therefore, we need multiple solving attempts to search for a set of non-dominated Pareto solutions. In the case of our BQCSP, each of the three objectives can be selected to be minimised and the other two are restricted by two values.

4.2   Pareto Simulated Annealing

Pareto simulated annealing (PSA), proposed in [8], is a multiple criteria metaheuristic that uses the general concept of the classical single objective Simulated Annealing (SA) as well as several specific notions related to multicriteria.

The basic SA proposed by [14] begins with an initial solution (s_0) and temperature (T_0). Then in each iteration a specific number of neighbourhood searches are done. Each time the fitness of a neighbouring solution is compared with that of the current solution. If the neighbouring solution has a better fitness, the current solution is replaced with it, otherwise this happens by a probability according to the fitness difference and the current temperature. The probability is calculated as $e^{\frac{-(Z_{new}-Z)}{T}}$, where Z_{new} and Z are the objective function value of the new and the current solution; T is the current temperature. The smaller the difference and the higher the temperature is, the bigger is the replacement probability. If a specific number of neighbours (NS) have been investigated, the SA starts a new iteration by reducing the temperature based on a plan, for example, dividing the current temperature by a constant value CT. By going forward iteration by iteration, the chance of moving to a new solution becomes lower. Finally, after meeting a termination criterion, the algorithm stops. For example, if the algorithm has exceeded the maximum allowable number of its consecutive unsuccessful iterations (MQ) or its execution time limit (MET) is over.

Now, in the PSA algorithm some new ideas which take the multiobjective aspect are added or replace some concepts of the single objective SA. The specific concepts that PSA uses are as follows:

- *Generating solutions or agents:*
  Initially, a set G of generating solutions with a fixed cardinality $|G| = \phi$ is randomly produced to provide a sufficiently large search space for PSA. With each of the generating solutions, a separate weight vector $\lambda^s = (\lambda_1^s, ..., \lambda_n^s)$ is associated such that $\lambda_i^s \in [0,1]$ and $\sum_{i=1}^n \lambda_i^s = 1$. Manipulating these weights provides a good dispersion of the outcome Pareto front and makes the solutions of PSA more representative. The generating solutions are treated as "spy agents" which can work almost independently while exchanging information about their position.
- *Aggregation function-based acceptance probabilities:*
  In PSA, when we move from solution s to its neighbour s' due to having multiple objectives to be taken into account and considering the concept of domination, one of these mutually exclusive cases can happen: 1) s' dominates or is equivalent to s 2) s' is dominated by s 3) s' is non-dominated with respect to s. In the first case, the new solution is not worse than the current one. Thus it should be accepted with probability one. In the second case, the new solution is worse than the current one (s is considered as potentially Pareto-optimal) and should be accepted with a probability less than one in order to avoid getting trapped in local optima. In the third case, s and s' are incomparable (and initially non-dominated). Hence, the probability of accepting the new solution s' for a problem with n objectives is defined as:

$$P(s, s', \lambda, T) = min(1, exp(- \max_{i=1,...,n} \frac{\lambda_i[(z_i(s') - z_i(s)]}{T})) \qquad (26)$$

  where T is the current temperature.
- *Management of generating solutions or repulsion:*
  A degree of repulsion α is determined as a very small positive value close to zero. The weight vector λ^s associated with a given agent s is modified in order to increase the probability of moving it away from its closest neighbour agent $\tilde{s}$ which is non-dominated with respect to s. This is achieved by increasing the weights of those objectives for which s is better than $\tilde{s}$ and decreasing the weights of the objectives for which s is worse than $\tilde{s}$. This can be stated as below:

$$\lambda_i^s = \begin{cases} \lambda_i^s + \alpha, & z_i(s) \leq z_i(\tilde{s}) \\ \lambda_i^s - \alpha, & otherwise. \end{cases} \qquad (27)$$

  In the space of normalised objectives, the metric distance between solutions which is used to determine the closest neighbour $\tilde{s}$ from s is:

$$\sum_{i=1}^n (z_i(s) - z_i(\tilde{s}))^2 \qquad (28)$$

  Moreover, we need to normalise λ_i^s in each iteration to satisfy: $\sum_{i=1}^n \lambda_i^s = 1$
- *Updating the set of potentially Pareto optimal solutions:*
  The set of potentially Pareto optimal solutions ρ is empty in the beginning of the algorithm. It is updated each time after a new non-dominated solution

is generated. Updating the set of potentially Pareto optimal solutions with solution s requires: (1) adding s to ρ if no solution in ρ dominates s and (2) removing all solutions dominated by s from ρ.

The pseudocode of the PSA is given as Algorithm 1.

Algorithm 1: The applied Pareto Simulated Annealing (PSA)

Data: Problems' inputs and PSA parameters (ϕ, T_0, CT, NS, α, MQ, and MET)

Result: A set of high quality non-dominated solutions

1 - Generate a set G of agents at random, $|G| = \phi$
2 - Generate λ^s for $s \in G$
3 - $T = T_0$
4 - $q = 0$. For the number of consecutive unsuccessful iterations.
5 **while** $q < MQ$ *and time* $\leq MET$ **do**
6   - For each agent find its closest non-dominated agent according to the distance calculated by (28).
7   - Update λ^s based on (27).
8   **for** $s \in S$ **do**
9   -$n = 1$
10   **while** $n \leq NS$ **do**
11   - Construct s' as a neighbouring solution of s.
12   - If s does not dominate s', update the set of non-dominated solutions ρ with s'.
13   - Replace s with s' according to (26).
14   - $n = n + 1$
15   **end**
16   **end**
17   - If no new non-dominated solution has been found during the whole iteration, set $q = q + 1$; otherwise set $q = 0$
18   - Decrease the temperature and calculate $T = \frac{T}{CT}$.
19 **end**
20 - Output the set ρ

4.3   Application to the BQCSP

Solution Encoding. Solutions are encoded as matrices with $|V|$ (number of vessels) columns each containing the data of one vessel. The number of rows is $2 + 2|QC|$. The first two rows are for berthing locations and berthing times, where the values are encoded in the interval [0,1]. The next $|QC|$ rows are for the starting time and the last $|QC|$ rows are for the finish time of QCs on vessels of the corresponding rows. The contents of these two sections are generated in the interval [-2,1]. If a cell has a negative value, it means that the QC is not assigned to the corresponding vessel and values from 0 to 1 are decoded to times in the planning horizon.

Neighbourhood. A neighbour solution (s') of a solution (s) is defined as one, which is in nbn cells different from s. Therefore, we randomly choose nbn cells and change their values to other allowable random values.

Constraints-Handling. The constraints of the problem are respected during the process of solution generation. By this way, firstly, a feasible solution is found among a population of randomly generated solutions. Then the variables are randomly prioritised. We begin from the variable with the highest priority, consider its value to be unknown and find a feasible interval for it according to the given values of the other variables. Then the corresponding value coming from the encoded structure (matrix) explained above is decoded to a real amount within the found interval. This process is repeated for the next variables until all variables are reassigned with new feasible values. The method is applied to a subset of the variables to construct a feasible neighbouring solution in the PSA.

Normalisation of the Objectives. An objective value z_j is normalised to $znorm_j$ as:

$$znorm_j = \frac{z_j - Zmin_j}{Zmax_j - Zmin_j} \qquad Zmin_j = 0; \; j = 1,2,3 \qquad (29)$$

Where $Zmax_j$ and $Zmin_j$ are the maximum and minimum value of objective j. $Zmax_j$ for the three objectives are calculated as below:

$$Zmax_1 = \sum_{i \in V} C_i[\gamma_i(BL - TB_i) + (1 - \gamma_i)TB_i] \qquad \gamma_i = \begin{cases} 1, & TB_i \leq \frac{BL}{2} \\ 0, & TB_i > \frac{BL}{2} \end{cases}. \qquad (30)$$

$$Zmax_2 = \sum_{i \in V} C_i[\beta_i(H - TT_i) + (1 - \beta_i)TT_i] \qquad \beta_i = \begin{cases} 1, & TT_i \leq \frac{H}{2} \\ 0, & TT_i > \frac{H}{2} \end{cases}. \qquad (31)$$

$$Zmax_3 = \sum_{i \in V} C_i[H - (TT_i + \frac{RT_i}{Qmax_i})] \qquad (32)$$

5   Computational Experiments

The experiments are done on a Core(TM) i7 computer with 3.10GHz CPU and 16 GB of RAM. PYTHON is used for programming the models and algorithms. In addition, we apply GUROBI [1] for exact solution required in the ϵ-constraint.

5.1   Generating Test Problems

The required test problems are randomly generated according to the following patterns. $BL = 100$; $|QC| = 10$; $H = 6|V|$; $E(TT_i) = U[0, H - \frac{RT_i}{Qmax_i}]$; $TB_i = U[0, BL - l_i]$.

The data of vessels are based on their size. They are shown in Table 1.

Table 1. Technical specifications and cost rates

Class	Proportion of frequency	l_i	RT_i	$Qmax_i$	C_i
Small	0.5	15	10	2	1
Medium	0.3	25	20	3	2
Large	0.2	35	30	4	3

5.2   Evaluation Metrics

The metrics used to evaluate the performance of our multiobjective methods
are quality metric (QM) and Hypervolume (HV) [30]. To calculate QM, the
final non-dominated sets obtained by all the methods are merged together. Subsequently, we find the solutions which are not dominated by any other solution existing in this pool and put them in a set called TNS. For each method:
$QM = \frac{\text{The number of the method's solutions in } TNS}{\text{The number of solutions in } TNS}$.

HV, which can indicate both the accuracy and diversity, is used the most
in the multiobjective works [19]. It is an unary metric that measures the size of
the objective space covered by the set of non-dominated solutions found by the
method under evaluation based on a reference point. Regarding the normalised
objectives, [1,1,1] is considered as our reference point for the calculation of HV.

5.3   Parameter Tuning

The parameters of our PSA are set by the response surface method (RSM) [3],
which is a design of experiments approach. The tuning is done for each problem
size (number of vessels) separately based on a half design duo to so many factors.
The response factor is HV. An interval is given for each parameter value in the
beginning, which is determined according to the experience. These intervals and
the values set by RSM are shown in Table 2. The parameter as used in the last
column is the number of cells or elements in the solution matrix. The solution
time limit (MET) is set to 30 min or 1800 s.

Table 2. Parameter setting of PSA by RSM

Nr. of vessels	ϕ	NS	α	MQ	T_0	CT	nbn
Interval	$[10^2, 10^3]$ integer	[5,30] integer	[0.25,0.75]	[5,20] integer	$[10^2, 10^5]$ only multiples of 10	[1,10] integer	$[0.1 \times as,$ $0.3 \times as]$ integer
10	202	7	0.43	12	10^3	5	26
20	320	10	0.45	15	10^4	5	61
50	484	15	0.46	15	10^4	4	140
100	618	18	0.51	16	10^4	2	264
200	967	26	0.56	18	10^5	2	538

5.4   Results

Test instances including 10, 20, 50, 100 and 200 vessels are built according to the explained patterns. For the implementation of the ϵ-constraint method, the third objective (z_3) is considered to be minimised, and z_1 and z_2 are constrained by the combination of eleven ϵ_1 and ϵ_2 values which are scattered in equal distances within [0,1], i.e. they are 0, 0.1, 0.2, 0.3,..., 1. This means that we have altogether $11 \times 11 = 121$ runs for the ϵ-constraint to find distinct non-dominated solutions. A time limit of 10 min is considered for each run.

The PSA is applied to each of the instances with 20 replications because solutions can be different each time. As the arrival times of the vessels are stochastic, the values of the objectives are calculated according to stochastic optimisation by the reference to averages regarding the probability distributions. Figure 2 depicts the number of non-dominated solutions obtained by the two methods.

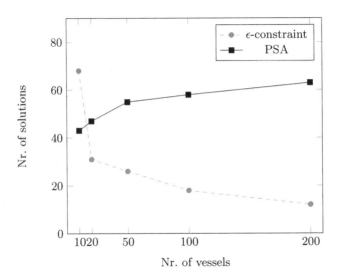

Fig. 2. The number of non-dominated solutions found by the methods

In the following, the outcomes of the approach that the berth and quay crane scheduling are separately solved by the PSA and then the results are added together are compared with the integrated solutions. The QM and HV metrics for the three approaches are depicted in Fig. 3 and 4.

Figure 5 illustrates the average execution time of only the PSA method because except for the smallest case with 10 vessels, the exact solver cannot find any optimal solution in any of the ϵ-constraint runs for larger instances. The total execution time of the ϵ-constraint for the smallest instance is over 21000 s, while the average elapsed time of the PSA is under 270s even for the largest instance including 200 vessels.

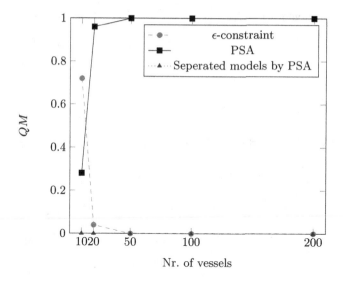

Fig. 3. Quality Metric (QM) of the methods

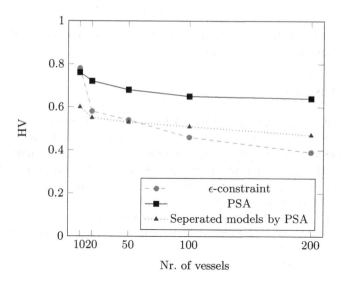

Fig. 4. Hypervolume (HV) metric of the methods

As it is evident from the results, the integration of the two problems leads to better solutions which dominate the solutions obtained by the separated models (regarding QM) and have altogether higher HV. In addition, we observe that our solution methodology can locate more non-dominated solutions which have better evaluation metrics in comparison to the ϵ-constriant. The reason is

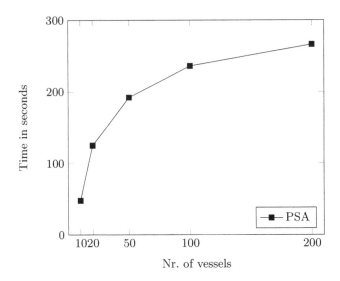

Fig. 5. Execution time of the methods

that the exact solver in the ϵ-constraint cannot find any non-dominated solution within our time limit in many runs.

6   Conclusions

We presented a novel integrated model for the BQCSP with stochastic arrival times for vessels and solved the problem by a modified PSA algorithm. This method provides good non-dominated solutions regarding the evaluation metrics. In addition, the advantage of integration is verified by comparing the results with the outcomes of the separated BSP and QCSP added together. For future research, the model can be extended by adding other related problems and their elements to it. Other non-deterministic parameters and their behaviour can be found and suitable solution methodologies can be devised to tackle the problem.

References

1. Gurobi optimization. https://www.gurobi.com
2. UNCTAD 2018: Review of maritime transport, united nations conference on trade and development (2018)
3. Box, G.E.P., Draper, N.R.: Response Surfaces, Mixtures, and Ridge Analyses. 2nd edn. Wiley-Interscience
4. Buhrkal, K., Zuglian, S., Ropke, S., Larsen, J., Lusby, R.: Models for the discrete berth allocation problem: a computational comparison. Transp. Res. Part E: Log. Transp. Rev. **47**(4), 461–473 (2011)
5. Chankong, V., Haimes, Y.Y.: Multiobjective Decision Making: Theory and Methodology. Elsevier Science, New York (1983)

6. Correcher, J.F., Alvarez-Valdes, R., Tamarit, J.M.: New exact methods for the time-invariant berth allocation and quay crane assignment problem. Eur. J. Oper. Res. **275**(1), 80–92 (2019)
7. Correcher, J.F., Van den Bossche, T., Alvarez-Valdes, R., Vanden Berghe, G.: The berth allocation problem in terminals with irregular layouts. Eur. J. Oper. Res. **272**(3), 1096–1108 (2019)
8. Czyżak, P., Jaszkiewicz, A.: Pareto simulated annealing. In: Fandel, G., Gal, T. (eds.) Multiple Criteria Decision Making, pp. 297–307. Springer, Heidelberg (1997). https://doi.org/10.1007/978-3-642-59132-7_33
9. Elwany, M.H., Ali, I., Abouelseoud, Y.: A heuristics-based solution to the continuous berth allocation and crane assignment problem. Alexandria Eng. J. **52**(4), 671–677 (2013)
10. Gutierrez, F., Lujan, E., Asmat, R., Vergara, E.: A fully fuzzy linear programming model for berth allocation and quay crane assignment. In: Simari, G.R., Fermé, E., Gutiérrez Segura, F., Rodríguez Melquiades, J.A. (eds.) IBERAMIA 2018. LNCS (LNAI), vol. 11238, pp. 302–313. Springer, Cham (2018). https://doi.org/10.1007/978-3-030-03928-8_25
11. Hu, Q.-M., Hu, Z.-H., Du, Y.: Berth and quay-crane allocation problem considering fuel consumption and emissions from vessels. Comput. Ind. Eng. **70**, 1–10 (2014)
12. Idris, N., Zainuddin, Z.M.: A simultaneous integrated model with multiobjective for continuous berth allocation and quay crane scheduling problem. In: 2016 International Conference on Industrial Engineering, Management Science and Application (ICIMSA), pp. 1–5 (2016)
13. Karafa, J., Golias, M.M., Ivey, S., Saharidis, G.K.D., Leonardos, N.: The berth allocation problem with stochastic vessel handling times. Int. J. Adv. Manuf. Technol. **65**(1–4), 473–484 (2012)
14. Kirkpatrick, S.: Optimization by simulated annealing: quantitative studies. J. Stat. Phys. **34**, 975–986 (1984)
15. Legato, P., Mazza, R.M., Gullì, D.: Integrating tactical and operational berth allocation decisions via simulation-optimization. Comput. Ind. Eng. **78**, 84–94 (2014)
16. Nourmohammadzadeh, A., Hartmann, S.: An efficient simulated annealing for the integrated problem of berth allocation and quay crane assignment in seaside container terminals. In: 7th Multidisciplinary International Conference on Scheduling: Theory and Applications (MISTA 2015), pp. 154–164 (2015)
17. Abou Kasm, O., Diabat, A., Cheng, T.C.E.: The integrated berth allocation, quay crane assignment and scheduling problem: mathematical formulations and a case study. Ann. Oper. Res. 1–27 (2019). https://doi.org/10.1007/s10479-018-3125-3
18. Rashidi, H., Tsang, E.P.K.: Novel constraints satisfaction models for optimization problems in container terminals. Appl. Math. Model. **37**(6), 3601–3634 (2013)
19. Riquelme, N., Von Lücken, C., Baran, B.: Performance metrics in multi-objective optimization. In: 2015 Latin American Computing Conference (CLEI), pp. 1–11 (2015)
20. Rodriguez-Molins, M., Salido, M.A., Barber, F.: A GRASP-based metaheuristic for the berth allocation problem and the quay crane assignment problem by managing vessel cargo holds. Appl. Intell. **40**(2), 273–290 (2013)
21. Stahlbock, R., Voß, S.: Operations research at container terminals : a literature update. OR Spectr. **30**, 1–52 (2008)
22. Tavakkoli-Moghaddam, R., Makui, A., Salahi, S., Bazzazi, M., Taheri, F.: An efficient algorithm for solving a new mathematical model for a quay crane scheduling problem in container ports. Comput. Indu. Eng. **56**(1), 241–248 (2009)

23. Ting, C.-J., Wu, K.-C., Chou, H.: Particle swarm optimization algorithm for the berth allocation problem. Expert Syst. Appl. **41**(4), 1543–1550 (2014)

24. Türkoğulları, Y.B., Taşkın, Z.C., Aras, N., Altınel, İ.K.: Optimal berth allocation, time-variant quay crane assignment and scheduling with crane setups in container terminals. Eur. J. Oper. Res. **254**(3), 985–1001 (2016)

25. Türkoğulları, Y.B., Taşkın, Z.C., Aras, N., Altınel, İ.K.: Optimal berth allocation and time-invariant quay crane assignment in container terminals. Eur. J. Oper. Res. **235**(1), 88–101 (2014)

26. Xi, X., Liu, C., Lixin, M.: A bi-objective robust model for berth allocation scheduling under uncertainty. Transp. Res. Part E: Logist. Transp. Rev. **106**, 294–319 (2017)

27. Xu, D., Li, C.-L., Leung, J.Y.-T.: Berth allocation with time-dependent physical limitations on vessels. Eur. J. Oper. Res. **216**(1), 47–56 (2012)

28. Zhen, L., Chang, D.-F.: A bi-objective model for robust berth allocation scheduling. Comput. Indu. Eng. **63**(1), 262–273 (2012)

29. Zheng, F., Qiao, L., Liu, M.: An online model of berth and quay crane integrated allocation in container terminals. In: Lu, Z., Kim, D., Wu, W., Li, W., Du, D.-Z. (eds.) COCOA 2015. LNCS, vol. 9486, pp. 721–730. Springer, Cham (2015). https://doi.org/10.1007/978-3-319-26626-8_53

30. Zitzler, E., Thiele, L.: Multiobjective evolutionary algorithms: a comparative case study and the strength Pareto approach. IEEE Trans. Evol. Comput. **3**(4), 257–271 (1999)

HotelSimu: Simulation-Based Optimization for Hotel Dynamic Pricing

Andrea Mariello, Manuel Dalcastagné$^{(\boxtimes)}$, and Mauro Brunato

Department of Information Engineering and Computer Science, University of Trento,
Via Sommarive 9, 38123 Povo, TN, Italy
andrea.mariello@alumni.unitn.it,
{m.dalcastagne,mauro.brunato}@unitn.it

Abstract. Exact and approximated mathematical optimization methods have already been used to solve hotel revenue management (RM) problems. However, to obtain solutions which can be solved in acceptable CPU times, these methods require simplified models. Approximated solutions can be obtained by using simulation-based optimization, but existing approaches create empirical demand curves which cannot be easily modified if the current market situation deviates from the past one. We introduce HotelSimu, a flexible simulation-based optimization approach for hotel RM, whose parametric demand models can be used to inject new information into the simulator and adapt pricing policies to mutated market conditions. Also, cancellations and reservations are interleaved, and seasonal averages can be set on a daily basis. Monte Carlo simulations are employed with black-box optimization to maximize revenue, and the applicability of our models is evaluated in a case study on a set of hotels in Trento, Italy.

Keywords: Revenue management · Dynamic pricing · Simulation-based optimization

1 Introduction

Information technology drastically changed how people plan travels and accomodations. In fact, tools such as online travel agencies or price comparison websites are now extensively used [31], and hotels are no longer forced to sell their rooms only through traditional intermediaries. Also, many hotels have already adopted RM techniques to manage their availability of rooms, in order to maximize their revenue.

Optimization problems related to hotel RM are usually expressed following two approaches: capacity control [3,5,8,10,11,15,21,22], where the decision variable is the amount of offered supply, and dynamic pricing [4,6,20,34,35], where the price is the decision variable. In both cases, several mathematical optimization methods have already been proposed to maximize revenue [12,19,28]. Many of these formulations assume that demand is independent from the chosen policy. More complex scenarios, where demand is influenced by other factors (e.g.,

© Springer Nature Switzerland AG 2020
I. S. Kotsireas and P. M. Pardalos (Eds.): LION 14 2020, LNCS 12096, pp. 341–355, 2020.
https://doi.org/10.1007/978-3-030-53552-0_31

price), are more difficult to handle and closed-form solutions are rarely available [9]. Demand is usually considered as a known deterministic function or as a stochastic function following a known distribution family with unknown parameters. Also, if stochastic cancellations are considered, the CPU time for solving the problem tends to grow exponentially and approaches like dynamic programming are effective only in specific cases [23]. A possible solution to mitigate the complexity of the model is approximated dynamic programming [3,8,35], where the problem is partitioned into simpler subproblems. Nonetheless, approximated models cannot provide exact solutions for realistic scenarios because of the large number of possible states [27].

The approximate maximization of revenue can be achieved using simulation-based optimization [7,13]. The analytical model is substituted with a simulator of many inter-related processes like reservations, cancellations, no-shows, walk-ins. Then, black-box optimization is used to find the policy which maximizes the revenue. An effective technique to maximize revenue and simulate different stochastic aspects of the hotel booking scenario is Monte Carlo simulation [26]. The generation of reservations and cancellations leads to a distribution of possible revenues, and the expected value of the distribution is considered as the variable to be maximized. For example, in [6,34] a Monte Carlo approach is employed to simulate demand as the result of many stochastic processes, and in [6] the effect that price has on demand is also considered.

In this paper, we present a flexible simulation-based optimization approach for hotel RM based on dynamic pricing. We simulate demand using a novel set of parametric models based on the RIM quantifiers [32], whose parameters are daily statistics which can be estimated from data. Our models allow to change the curves parametrically, redistributing demand along the booking horizon, without requiring any change of advance historical data. In fact, bookings and cancellations associated to each day are distributed along the booking horizon with a non-homogeneous Poisson process, where demand expectations of each day are defined by our parametric models. The hotel manager can inject new information in the system, adapting pricing policies to the mutated conditions of the market. For the optimization, we use an efficient implementation of the Covariance Matrix Adaptation Evolution Strategy (CMA-ES)[16][1]. We position after the work of [6,34], on which we build, to provide a simpler way for the hotel manager to run *what-if* analyses. Furthermore, reservation requests and cancellations are not grouped into disjoint sets of events like in [34], but occur in an interleaved way. The structure of the remainder of this paper is as follows. Section 2 describes HotelSimu, and defines more in detail the parametric models used for the simulations. Section 3 provides some details about the optimization algorithm, and Sect. 4 shows the applicability of our models to a set of hotels in Trento, Italy. Results show that our approach leads to an average revenue increase similar to that of other dynamic pricing strategies, even though only aggregated data has been used. Finally, Sect. 5 provides the main implications of our work for the hotel manager and briefly describes possible extensions.

---

[1] Code available at http://beniz.github.io/libcmaes.

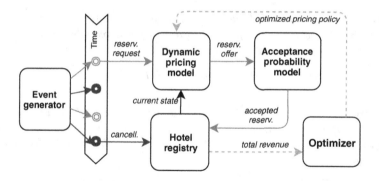

Fig. 1. HotelSimu overview. Reservation requests and cancellations are interspersed. The state of the hotel after one complete simulation is used by the optimizer to compute the total revenue and adjust the pricing policy.

2   Simulation Methods

The main components of HotelSimu are shown in Fig. 1. An event generator simulates the reservation requests and the cancellations. A registry stores the information about the state of the hotel, in particular accepted reservations and room availability. A dynamic pricing model proposes an offer for each reservation request, and an acceptance probability model simulates the stochastic process by which customers accept or discard reservation offers. An optimizer searches for the optimal pricing policy to maximize revenue.

2.1   Definitions

Let us now define the main concepts and the notation used throughout the paper.

Definition 1. *A reservation request (RR) is an event characterized by the following features. The reservation day (RR_{res}), which is the day the request occurs. The arrival day (RR_{arr}), which is the day the customer arrives at the hotel. The length of stay (RR_{los}), which is the number of nights reserved. The size (RR_{size}), which is the number of rooms reserved.*

Definition 2. *A reservation offer (RO) is an admissible reservation request (for which there is room availability) characterized by the price (RO_{price}) proposed by the hotel, which depends on the features of RR.*

Definition 3. *An accepted reservation or simply reservation (R) is a reservation offer accepted by the customer. It is registered on the hotel registry and it effectively changes room availability.*

Definition 4. *The acceptance probability of a reservation offer ($Pr_{accept}(RO)$) is the probability that a customer accepts RO and the proposed price, and therefore is equal to the probability that RO is registered on the book.*

Definition 5. *The state of the hotel $S(t)$ is defined as the state of the booking registry at time t, which corresponds to the historical records up to t as well as the set of reservations for future arrival days that are in the registry at time t.*

Definition 6. *Given two days identified by $i, j \in \{0, 1, 2, \ldots\}$, the number of days between i and j, or their distance, is $d(i, j) = d(j, i) = |i - j| \geq 0$.*

Definition 7. *Given a reservation R, the time-to-arrival of R is $R_{TTA} = d(R_{res}, R_{arr})$. If $R_{TTA} = 0$, a customer makes a reservation on the arrival day or arrives at the hotel with no reservation and we refer to the customer as a walk-in user.*

Definition 8. *The booking time window or booking horizon (BH) is the maximum time-to-arrival allowed by the hotel.*

Definition 9. *A cancellation (C) is characterized by the cancellation day (C_{day}), which is the day the event occurs, and the reservation (C_{res}), which is the reservation on the book that is canceled by the customer. When a reservation is canceled, it is removed from the hotel registry and the associated rooms can be booked by other customers.*

Definition 10. *The cancellation probability, t days before arrival of a reservation R $(Pr_{cancel}(R, t))$, is the probability that the customer associated with R cancels it exactly t days before arrival, with $t \in [0, R_{TTA}]$. According to this definition, the probability that R is canceled within its lifetime is*

$$Pr_{cancel}(R) = \sum_{t \in [0, R_{TTA}]} Pr_{cancel}(R, t). \tag{1}$$

Definition 11. *The reservation requests horizon (RH) is the set of all the reservation days to be simulated. It corresponds to the values that each R_{res} can assume during the simulation.*

Definition 12. *The arrivals horizon (AH) is the set of all possible arrival days. It corresponds to the values that each R_{arr} can assume during the simulation.*

Definition 13. *The optimization horizon (OH) is the set of arrival days for which there is the need of an optimal dynamic pricing policy to maximize revenue.*

For each simulated reservation day $r \in$ RH, a random sequence of C_r cancellations and R_r reservation requests is generated. Each reservation request is associated with an arrival day $a \in$ AH following or coinciding to r ($a \succeq r$), and each cancellation is associated with a registered reservation. The proposal of a price depends on a reservation request and on the state of the hotel at the moment the event occurs. Once a price has been proposed to the customer, a reservation is accepted according to the acceptance probability model. It is then registered into the hotel registry and, if a cancellation does not occur until the end of the simulation, it is considered in the evaluation of the total revenue to be passed to the optimizer. As concerns the optimization, one objective function evaluation corresponds to the average total revenue of several simulation runs, with respect to the reservations recorded in the registry within the OH.

2.2   Simulation of Reservation Requests

Let $\mathcal{R}_r^a$, $r \in \mathrm{RH}$, $a \in \mathrm{AH}$, be the number of reservation requests generated on day r that are associated with arrival day a. The total number of requests generated within RH and associated with one arrival day is therefore given by:

$$\mathcal{R}^a = \sum_{\substack{r \in \mathrm{RH} \\ r \preceq a}} \mathcal{R}_r^a, \tag{2}$$

where $\preceq$ describes the relation *precedes or coincides to*. The expected total number of reservation requests associated with one arrival day can be seen as the result of several independent processes, which occur on each simulated day within the BH of an arrival day:

$$\mathbb{E}[\mathcal{R}^a] = \Lambda(a) = \sum_{i=0}^{\mathrm{BH}} \lambda(i, a), \tag{3}$$

where $\lambda(i, a)$ is the expected number of reservation requests occurring i days before the arrival day a. If historical data are available, one can estimate directly $\lambda(i, a)$ for each i and a. To avoid the computational load of a point-wise estimation, and to facilitate what-if analyses, we define each $\lambda(i, a)$ by the following parametric model:

$$\begin{aligned} \lambda_\alpha(i, a) &= \Lambda(a) \times Q_\alpha(i, \mathrm{BH}) \\ &= \Lambda(a) \times \left(\left(\frac{\mathrm{BH} + 1 - i}{\mathrm{BH} + 1} \right)^\alpha - \left(\frac{\mathrm{BH} - i}{\mathrm{BH} + 1} \right)^\alpha \right), \end{aligned} \tag{4}$$

with $i = 0, 1, \ldots, \mathrm{BH}$, $a \in \mathrm{AH}$, and for any parameter $\alpha > 0$. The expression of $Q_\alpha(i, \mathrm{BH})$ is similar to that of the RIM quantifiers proposed in [32], after reflection and translation. We use $Q_\alpha(i, \mathrm{BH})$ because:

– they define a function with discrete domain and continuous values;
– they sum up to 1:

$$\sum_{i=0}^{BH} Q_\alpha(i, \mathrm{BH}) = 1,$$

for any $\alpha > 0$ and therefore can represent a discrete probability distribution or a normalized curve;
– they can model different reservation scenarios through α, from a constant curve ($\alpha = 1$) to increasing and decreasing curves (see Fig. 2);
– they provide a simple way of finding α from the ratio of walk-in users with respect to the total number of reservations, that is, $Q_\alpha(0, \mathrm{BH})$.

In the current implementation, we assume that the reservation requests follow a non-homogeneous Poisson process with an expected value given by our

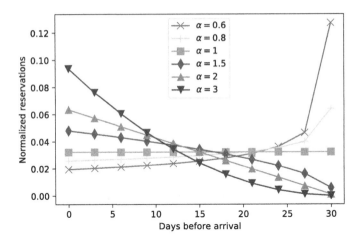

Fig. 2. $Q_\alpha(i, \text{BH})$ for $\text{BH} = 30$ and for different values of α.

parametric model, so $\mathcal{R}^a \sim \text{Poisson}(\Lambda(a))$. Therefore, reservation requests are generated for each simulated day according to the following model:

$$\begin{cases} \mathcal{R}_r^a \sim \text{Poisson}(\lambda_\alpha(i, a)) & \text{if } i \leq \text{BH}, \\ \mathcal{R}_r^a = 0 & \text{otherwise.} \end{cases} \tag{5}$$

Poisson processes are usually chosen to model arrival processes [14] and, in our context, they can represent the arrival of reservation requests with a minimum set of parameters. In [34], a binomial distribution is used, with additional constraints on the variance of samples in order to set the success probability and the number of trials. However, a binomial distribution converges to a Poisson distribution when the number of trials (e.g., customers generating requests) grows. Removing the limit on the pool of customers that can generate new reservations makes the model more realistic, since the number of possible customers is usually unbounded and independent from the capacity of the hotel. For the estimation of $\Lambda(a)$, we assume that it is possible to estimate the expected number of reservation requests for a specific arrival day that are accepted by the customers and not canceled ($\mathcal{R}_{\text{accept}}^a$). Similarly, we assume that one has access to the expected number of reservation requests for a specific arrival day that are accepted by the customers and canceled ($\mathcal{R}_{\text{cancel}}^a$). $\mathcal{R}_{\text{accept}}^a$ can be approximated by the expected number of arrivals, while $\mathcal{R}_{\text{cancel}}^a$ can be seen as the expected number of cancellations.

HotelSimu includes also a model of the acceptance probability $\text{Pr}_{\text{accept}}(RO)$. A model of probabilities (possibly one for each admissible input) can be estimated from data retrieved by an online booking platform, where one can keep track of users that search for a room and decide to finalize the reservation or leave the website. One can also estimate the expected acceptance probability $\mathbb{E}[\text{Pr}_{\text{accept}}(RO)]$ as the expected fraction of reservation requests that are finalized by the users after the search. Therefore, the expected total number of reservation requests (accepted or rejected) associated with one arrival day can be estimated as follows:

$$\mathbb{E}[\mathcal{R}^a] = \Lambda(a) \approx \frac{\mathcal{R}^a_{\text{accept}} + \mathcal{R}^a_{\text{cancel}}}{\mathbb{E}[\text{Pr}_{\text{accept}}(RO)]}. \tag{6}$$

2.3   Simulation of Nights and Rooms

Let $nights^a$ be the expected number of nights for a reservation associated with an arrival day a. Analogously, $rooms^a$ is the expected number of rooms. $max\text{-}nights^a$ and $max\text{-}rooms^a$ represent the limits imposed by the hotel manager. Since each reservation request includes at least one night and one room, we model the discrete probability distribution of the number of additional nights/rooms as follows:

$$\text{Pr}(X - 1 = k) = \int_{\frac{k}{\max(X)}}^{\frac{k+1}{\max(X)}} \frac{(1 - x)^{\frac{\max(X)}{\text{avg}(X)-0.5} - 2}}{\text{B}(1, \frac{\max(X)}{\text{avg}(X)-0.5} - 1)} dx, \tag{7}$$

where X is the number of nights/rooms, $X - 1$ is the number of additional nights/rooms, $\max(X)$ is either $max\text{-}nights^a$ or $max\text{-}rooms^a$, and $\text{avg}(X)$ is either $nights^a$ or $rooms^a$. $k = 0, 1, \ldots, \max(X) - 1$, and $\text{B}(\alpha, \beta)$ is the Beta function with parameters α and β.

The previously defined distribution is a discrete analogue of a (continuous) Beta distribution with $\alpha = 1$ and $\beta = \frac{\max(X)}{\text{avg}(X)-0.5} - 1$. The value of α is chosen so as to have a distribution with an exponential-decay profile, which is similar to the distribution seen in [34]. β is chosen so as to have an expected value approximately equal to $\text{avg}(X) - 1$. This is achieved by imposing the equality of the expected value of the (continuous) Beta distribution, which is $\frac{\alpha}{\alpha+\beta}$, to the expected number of additional nights/rooms rescaled to $[0, 1]$, which is $\frac{\text{avg}(X)-0.5}{\max(X)}$. We consider a correction of 0.5 to account for the discretization error and to position rescaled expected values in the middle of the discretization interval. Experiments show that the maximum error between the expected values and the empirical averages of the discrete analogue with $\max(X) = 5$ is at most 0.33, for expected values equal to $0, 0.1, 0.2, \ldots, \max(X) - 1$.

Even though modeling the length of stay or the number of rooms as Bernoulli or Poisson processes provides a simple and exact way of imposing the expected value, it is not applicable to our context, which cannot be reduced to a coin toss or to an arrival process. In the literature, the Beta distribution is often used to model unknown probability distributions, with shapes that can be controlled by the parameters α and β. By building a discrete analogue of a Beta distribution, we can exploit its macroscopic features and to obtain a realistic model of the variable of interest. A similar model can be defined also for $group$ reservations, which usually follow a different distribution from that of the length of stay of $normal$ reservations. This can be easily achieved by considering a different value for $\text{avg}(X)$. By following (7), an instance of the random variable X, which is either RR_{los} or RR_{size}, is generated as $X = 1 + \lfloor Y \times \max(X) \rfloor$, where $Y \sim \text{Beta}(1, \frac{\max(X)}{\text{avg}(X)-0.5} - 1)$.

2.4   Simulation of Cancellations

Under the same assumptions of Sect. 2.2, and by analogy to (1), the probability that a reservation is canceled during its lifetime can be seen as the summation of the probabilities that a reservation is canceled exactly on a specific day within its lifetime:

$$\mathrm{Pr}_{\mathrm{cancel}}(R) = \Omega(a) = \sum_{i=0}^{R_{\mathrm{TTA}}} w(i,a), \tag{8}$$

where $w(i,a)$ is the probability that R is canceled exactly i days before the arrival day a, with i within its lifetime. We define each $w(i,a)$ by the following parametric model:

$$
\begin{aligned}
w_\alpha(i,a) &= \Omega(a) \times Q_\alpha(i, R_{\mathrm{TTA}}) \\
&= \Omega(a) \times \left(\left(\frac{R_{\mathrm{TTA}} + 1 - i}{R_{\mathrm{TTA}} + 1} \right)^\alpha - \left(\frac{R_{\mathrm{TTA}} - i}{R_{\mathrm{TTA}} + 1} \right)^\alpha \right),
\end{aligned}
\tag{9}
$$

with $i = 0, 1, \ldots, R_{\mathrm{TTA}}$, $a = R_{\mathrm{arr}}$, and for any parameter $\alpha > 0$. In this context one can also find α from the fraction of cancellations that occur on the last day $(Q_\alpha(0, R_{\mathrm{TTA}}))$, which includes the so-called *no-shows*. $\Omega(a)$ can be estimated as follows:

$$\Omega(a) \approx \frac{\mathcal{R}^a_{\mathrm{cancel}}}{\mathcal{R}^a_{\mathrm{cancel}} + \mathcal{R}^a_{\mathrm{accept}}}, \tag{10}$$

with an arrival day $a = R_{\mathrm{arr}}$. In HotelSimu, different stochastic cancellation scenarios can be simulated by changing $w_\alpha(i,a)$ through $\Omega(a)$ and α.

3   Optimizing the Noisy Simulator Function

Since Monte Carlo simulation employs stochastic processes, the performance of each solution corresponds to a distribution of results. The expected value of the distribution is used as an approximation of the objective function to be optimized, so the optimization operates in the presence of noise.

In the literature, multiple works tested diverse heuristic algorithms on noisy functions, and they have shown that population-based approaches like CMA-ES are a good choice to optimize noisy functions [1,2,24,25]. In fact, instead of relying only on a single solution, at each iteration CMA-ES combines a subset of its candidate solutions in order to direct the search in the most promising direction. By combining multiple solutions located in a restricted area of the search space, the impact of noise is decreased due to an *implicit averaging* effect [1,33]. Moreover, to further reduce the effect of noise on the optimization, we compute the performance of each solution as the mean of the outcome of multiple simulations. From probability theory, one knows that the effect of noise can be reduced by evaluating multiple times each solution [33]. More precisely, CMA-ES is an evolutionary optimization algorithm in which a multivariate normal distribution $N(\mu_t, M_t)$ is used to sample solutions, where t defines the iteration of the algorithm. At each iteration, the mean μ_t defines the center of the distribution, while the covariance matrix M_t determines shape and orientation of the ellipsoid corresponding to $N(\mu_t, M_t)$. Also, a step size σ_t controls the spread of the distribution as a percentage of the search space. Iteratively, CMA-ES follows the following steps. First, a population of λ solutions is sampled from $N(\mu_t, M_t)$. Second, candidates are evaluated and ranked according to the respective evaluations. Third, the best $\lfloor \frac{\lambda}{2} \rfloor$ results are used to update μ_t and M_t, in order to move the search towards the most promising search direction. Fourth, σ_t is increased or decreased according to the length of the so-called evolution paths. Evolution paths are weighted vector sums of the last points visited by the algorithm. They provide information about the correlations among points, and they are used to find the direction recently followed by the optimization. If consecutive steps are going in the same direction, the same distance could be covered by longer steps and the current path is too long. If consecutive steps are not going in the same direction, single steps tend to cancel each other out and so the current path is too short.

CMA-ES is executed with $\lambda = 4 + \lfloor 3 \log d \rfloor$ and $\sigma = 0.5$, where d is the dimensionality of the objective function and $\sigma \in (0, 20]$. The size of the population is the one suggested by the authors of [16], who have also tested that CMA-ES with this population size is a robust and fast local search method [18]. We experimented even other parameter settings of the algorithm, but because of space limitations of this publication we present only the preliminary results obtained with the mentioned settings. Also, all the standard stopping criterias of CMA-ES are active [17]. Each time a stopping criteria is triggered, the algorithm is restarted from another randomly generated point in the search space, with a new population of the same size.

4   Results

In the following experiments, we show how our models can be used to search for the optimal pricing policies that maximize the total revenue of a set of hotels of different sizes. We assume there is only one category of rooms, and that at least historical data about final demand is available. However, if advance historical data is also available and empirical demand curves can be estimated, our models can be calibrated using optimization algorithms [29].

4.1   Setup of the Experiments

We consider a monotonically decreasing reservation curve with 40% of the customers treated as walk-in users, calibrating our models according to the reservation models estimated from historical data in [34]. The goodness of this choice is also confirmed by data collected by the Italian Institute of Statistics (Istat) on the features of trips[2], which show that approximately 40% of the interviewed

---

[2] http://dati.istat.it/?lang=en;section:Communication,culture,trips/Trips/Tripsand theircharacteristics.

people travel without booking. As a consequence, it is reasonable to assume that the remaining 60% of the reservations is monotonically distributed in the BH in a decreasing fashion as moving away from the walk-in day. We also assume that the maximum number of cancellations occurs on the last day, and we fix this number to 40% of the total number of cancellations. BH is fixed to 180 days, the maximum number of nights for one reservation to 10, and the maximum number of rooms to 4. As concerns the pricing policy, we use the model proposed in [6], which is based on a set of multipliers that leads to an increase or decrease in the average price according to the features of a reservation request. The multipliers vary around 1, and each multiplier changes the reference price according to the value it assumes: a value lower than 1 corresponds to a discount and a value larger than 1 is a price increase. We assume that RO_{price} corresponds to the unit price for 1 room and 1 night, where the unit price proposed to the customer is computed as follows:

$$RO_{\text{price}} = price^a \cdot \xi(RR_{\text{TTA}}, RR_{\text{los}}, RR_{\text{size}}, S, \Delta, \eta), \tag{11}$$

where $price^a$ is the expected unit price for customers arriving on day a, and $\xi(\cdot)$ is a function of the reservation request features and of the hotel registry, with average value equal to 1. This function smoothly adjusts the price within the interval $[(1 - \Delta)price^a, (1 + \Delta)price^a]$, with a slope proportional to η:

$$\xi(RR_{\text{TTA}}, RR_{\text{los}}, RR_{\text{size}}, S, \Delta, \eta) = \xi(t, l, s, S, \Delta, \eta) = \tag{12}$$
$$= (1 - \Delta) + 2\Delta \cdot \Phi(\eta \cdot (M_T(t)M_L(l)M_S(s)M_C(S) - 1)).$$

$\Phi(\cdot)$ is the cumulative distribution function of the standard normal distribution, and $M_T(\cdot)$, $M_L(\cdot)$, $M_S(\cdot)$ and $M_C(\cdot)$ are functions (or *multipliers*) of the time-to-arrival, the length of stay, the number of rooms and the remaining hotel capacity at the moment the reservation request is generated, respectively. As concerns the parameters of the multipliers, we set $T_0 = 30$ and $C_0 = L_0 = G_0 = 1.6$. Also, $\eta = 3$ and $\Delta = 0.6$ in order to propose prices with a maximum increase/decrease of 60% with respect to $price^a$.

The effect on the room demand of changing the unit price is modeled by the acceptance probability, which we define similarly to [34]. When the proposed price is equal to the average price of reservations with the same arrival day, the acceptance probability is set to 0.5, to model the absence of any preference about accepting or rejecting the reservation. With prices fixed to the average values, the expected number of accepted reservations is equal to half of the total number of reservation requests. The expected percentage of accepted reservations increases when the price decreases and decreases otherwise. This phenomenon, called *price elasticity*, is modeled by the following function:

$$\Pr_{\text{accept}}(RO) = 1 - \Phi(\rho \cdot (RO_{\text{price}} - price^a)), \tag{13}$$

where $\Phi(\cdot)$ is the cumulative distribution function of the standard normal distribution, and ρ is a parameter that controls the slope of the function and allows us to consider different price elasticity scenarios. In the experiments, ρ is chosen so that $\Pr_{\text{accept}}(RO) \approx 1$ when there is a discount of at least 50% and $\Pr_{\text{accept}}(RO) \approx 0$ when the price increases of at least 50%.

Table 1. Characteristics of hotels used for the tests, and results. Arrivals, occupancy (as room-nights) and revenue after optimization are expressed as percentage increase, where maximum and minimum values are in bold. Optimization total CPU time and single-run simulation CPU time are defined in seconds.

Hotel	Rooms	Price (€)	Arrivals	Occupancy	Revenue	Optimization	Simulation
01	52	120.00	48.2 ± 0.5	47.6 ± 0.6	18.4 ± 0.6	11000 ± 109.0	2.0 ± 0.012
02	34	69.50	50.6 ± 0.6	50.6 ± 0.7	20.4 ± 0.7	10800 ± 82.33	1.91 ± 0.006
03	136	290.00	51.6 ± 0.3	52.6 ± 0.3	21.6 ± 0.3	18400 ± 258.5	3.54 ± 0.030
04	46	153.33	44.0 ± 0.5	44.2 ± 0.6	17.8 ± 0.6	9910 ± 97.61	1.78 ± 0.008
05	113	136.675	$\mathbf{55.5 \pm 0.3}$	$\mathbf{55.2 \pm 0.4}$	$\mathbf{23.1 \pm 0.4}$	16700 ± 241.2	3.19 ± 0.031
06	37	74.00	46.9 ± 0.6	45.8 ± 0.6	17.8 ± 0.6	9570 ± 87.87	1.69 ± 0.012
07	9	39.00	$\mathbf{38.2 \pm 1.1}$	$\mathbf{37.7 \pm 1.3}$	$\mathbf{12.8 \pm 1.2}$	7370 ± 17.70	1.25 ± 0.010
08	22	216.50	43.2 ± 0.7	42.0 ± 0.8	18.0 ± 0.8	7870 ± 35.67	1.35 ± 0.004
09	14	66.50	42.0 ± 0.9	41.8 ± 1.0	17.7 ± 1.0	7740 ± 32.26	1.3 ± 0.010
10	19	82.67	41.8 ± 0.8	40.5 ± 0.9	17.3 ± 0.9	8650 ± 46.21	1.49 ± 0.008

We empirically show the applicability of HotelSimu to 10 hotels in Trento, Italy. We selected representative hotels from the official open data of the Province of Trento[3], as reported in Table 1. The information on the average arrivals and the average number of nights per reservation is taken from the Statistics Institute of the Province of Trento (Ispat)[4]. No information is available about the average number of rooms per reservation, so we assumed it to be equal to 1. We disaggregated data on arrivals and mapped them onto each hotel according to their capacity, under the assumption that bigger hotels usually register more arrivals than smaller hotels. We use real aggregated data on tourists and different hotels to simulate time series of reservations and cancellations, and we consider these time series as a baseline to be compared to the outcome of the optimization.

In the experiments RH starts on July 1st, 2017, and ends on December 31st, 2018. AH starts on July 1st, 2017, and ends on January 31st, 2019. OH starts on January 1st, 2018, and ends on December 31st, 2018.

The optimization has a budget of 300 iterations (for a maximum running time of 5/6 h). Each iteration retrieves the total revenue as the average on 20 simulation runs, all with the same parameter configuration, for a total of 6000 simulations within one optimization run. The optimization is repeated 10 times. Each experiment is started from an initial solution which has been generated by a uniform distribution defined over the search space. Tests have been run on

---

[3] http://dati.trentino.it/dataset/esercizi-alberghieri.

[4] http://www.statistica.provincia.tn.it, section "Annuari del Turismo".

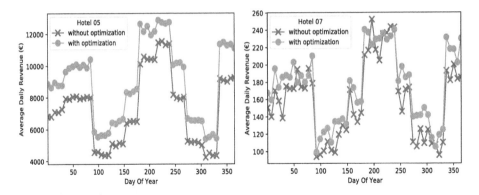

Fig. 3. Average daily revenue for Hotel 05 and Hotel 07 (one value per week).

virtual machines using a KVM hypervisor (1 per hotel), each one with 512 MB of RAM and 1 CPU (1 core) at 2.1 GHz.

4.2 Results on Arrivals, Occupancy and Revenue

In Table 1 we report the results on customer arrivals, occupancy and total revenue as the percentage increase led by the optimized pricing model with respect to the configuration with the multipliers equal to 1. Results are expressed in terms of averages and standard errors, and they are statistically significant according to the two-tailed unequal variances t-test [30], with a significance level $\alpha = 0.01$. A unit of occupancy corresponds to the so-called *room-night*, which is a room occupied for one night.

Results are promising for all the hotels, with a minimum of 12.8% increase in revenue, 37.7% in occupancy and 38.2% in arrivals. The maximum increase in revenue is reached for Hotel 05, with a value of 23.1%. The minimum values are reached for small hotels, where the limited number of rooms leads to fewer arrivals and then relatively low revenues. In this context, there is also more variability, since the hotel can become full with few reservations, thus leading to the rejection of more requests. Experiments suggest that higher revenues can be obtained for medium and big hotels, where the system exploits the capacity of the hotel to increase the number of arrivals. The time series of the average daily revenue during the year of interest for the best and worst scenarios are reported in Fig. 3. For Hotel 05, it is evident that the time series produced by the optimized model is significantly higher than that produced without optimization. In this case, there is less chance of having a loss in revenue because of an optimistic configuration found during the optimization process. For Hotel 07, the two time series are not significantly different because of the higher uncertainty caused by the small dimension of the hotel. This leads to higher risk and to the possibility of having a loss (with probability ≈ 0.03), as it is evident from the distribution of the increase in revenue in Fig. 4. These results are in accordance with the expected behavior of non-homogeneous Poisson distributions, whose coefficient

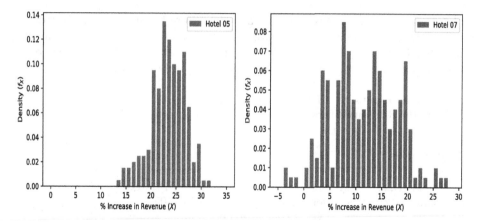

Fig. 4. Estimated distributions of increase in revenue after optimization for Hotel 05 and Hotel 07.

of variation decreases as the expected value increases. In the context of hotel demand, this property implies that for smaller hotels, which can accommodate a limited number of guests and therefore are characterized by less arrivals, the coefficient of variation is higher than that of large hotels. As a consequence, the increased variability for small hotels leads to higher risk of losses, as empirically shown by our results.

5   Conclusions

In this work we proposed HotelSimu, a flexible simulation-based optimization approach which can be used for maximizing the revenue of hotels. Since the output of the simulations is noisy, we optimized the noisy simulator function by using CMA-ES, a population-based algorithm which has already been studied in the literature and proved to be effective in noisy scenarios. Furthermore, we aggregated the outcome of multiple simulations in order to use the expected value to further reduce the effects of the noise on the optimization.

HotelSimu models stochastic arrivals and cancellations in an interleaved fashion, considering several characteristics of reservation requests in order to propose dynamic prices. Furthermore, it models the effect that price variations have on demand (price elasticity). Our models, based on the RIM quantifiers, allow the hotel manager to adapt pricing policies to dynamic market conditions, and to analyze different booking scenarios by changing a compact set of meaningful parameters. Seasonal averages can be set even on a day-by-day basis, thus allowing the hotel manager to adapt the pricing policy to special events and to consider monthly as well as weekly seasonal effects.

The case study shows that our parametric models lead to results similar to other dynamic pricing models in the literature, while relying only on aggregated data. The average revenue increase is $\approx 19\%$ with respect to the original pricing policies, and the risk of losses is absent for medium-big hotels and limited for small hotels, with a maximum loss probability of ≈ 0.03. Moreover, experiments show that HotelSimu can simulate one year and a half in ≈ 2 s on average on a low-end machine. Also, a complete optimization can be run within one night.

Acknowledgements. A. Mariello would like to thank H. T. Nguyen for the useful advice on the RIM quantifiers.

References

1. Arnold, D.V.: Noisy Optimization with Evolution Strategies, vol. 8. Springer, Cham (2012). https://doi.org/10.1007/978-1-4615-1105-2

2. Arnold, D.V., Beyer, H.G.: A comparison of evolution strategies with other direct search methods in the presence of noise. Comput. Optim. Appl. **24**(1), 135–159 (2003)

3. Aydin, N., Birbil, S.: Decomposition methods for dynamic room allocation in hotel revenue management. Eur. J. Oper. Res. **271**(1), 179–192 (2018)

4. Aziz, H.A., Saleh, M., Rasmy, M.H., Elshishiny, H.: Dynamic room pricing model for hotel revenue management systems. Egypt. Inform. J. **12**(3), 177–183 (2011)

5. Baker, T.K., Collier, D.A.: A comparative revenue analysis of hotel yield management heuristics. Decis. Sci. **30**(1), 239–263 (1999)

6. Bayoumi, A.E.M., Saleh, M., Atiya, A.F., Aziz, H.A.: Dynamic pricing for hotel revenue management using price multipliers. J. Revenue Pricing Manag. **12**(3), 271–285 (2013)

7. Bertsimas, D., De Boer, S.: Simulation-based booking limits for airline revenue management. Oper. Res. **53**(1), 90–106 (2005)

8. Bertsimas, D., Popescu, I.: Revenue management in a dynamic network environment. Transp. Sci. **37**(3), 257–277 (2003)

9. Bitran, G., Caldentey, R.: An overview of pricing models for revenue management. Manuf. Serv. Oper. Manag. **5**(3), 203–229 (2003)

10. Bitran, G.R., Mondschein, S.V.: An application of yield management to the hotel industry considering multiple day stays. Oper. Res. **43**(3), 427–443 (1995)

11. Choi, T.Y., Cho, V.: Towards a knowledge discovery framework for yield management in the Hong Kong hotel industry. Int. J. Hosp. Manag. **19**(1), 17–31 (2000)

12. Denizci Guillet, B., Mohammed, I.: Revenue management research in hospitality and tourism: a critical review of current literature and suggestions for future research. Int. J. Contemp. Hosp. Manag. **27**(4), 526–560 (2015)

13. Figueira, G., Almada-Lobo, B.: Hybrid simulation-optimization methods: a taxonomy and discussion. Simul. Model. Pract. Theory **46**, 118–134 (2014)

14. Grube, P., Núñez, F., Cipriano, A.: An event-driven simulator for multi-line metro systems and its application to Santiago de Chile metropolitan rail network. Simul. Model. Pract. Theory **19**(1), 393–405 (2011)

15. Guadix, J., Cortés, P., Onieva, L., Muñuzuri, J.: Technology revenue management system for customer groups in hotels. J. Bus. Res. **63**(5), 519–527 (2010)

16. Hansen, N.: The CMA evolution strategy: a comparing review. In: Lozano, J.A., Larrañaga, P., Inza, I., Bengoetxea, E. (eds.) Towards a New Evolutionary Computation. Studies in Fuzziness and Soft Computing, vol. 192, pp. 75–102. Springer, Heidelberg (2006). https://doi.org/10.1007/3-540-32494-1_4

17. Hansen, N.: Benchmarking a BI-population CMA-ES on the BBOB-2009 function testbed. In: Proceedings of the 11th Annual Conference Companion on Genetic and Evolutionary Computation Conference: Late Breaking Papers, pp. 2389–2396. ACM (2009)

18. Hansen, N., Ostermeier, A.: Completely derandomized self-adaptation in evolution strategies. Evol. Comput. **9**(2), 159–195 (2001)

19. Ivanov, S.: Hotel Revenue Management: From Theory To Practice. Zangador (2014)

20. Kleywegt, A.J.: An optimal control problem of dynamic pricing. School of Industrial and Systems Engineering, Georgia Institute of Technology 2001 (2001)

21. Lai, K.K., Ng, W.L.: A stochastic approach to hotel revenue optimization. Comput. Oper. Res. **32**(5), 1059–1072 (2005)

22. Liu, S., Lai, K.K., Wang, S.: Booking models for hotel revenue management considering multiple-day stays. Int. J. Revenue Manag. **2**, 78–91 (2008)

23. McGill, J.I., Van Ryzin, G.J.: Revenue management: research overview and prospects. Transp. Sci. **33**(2), 233–256 (1999)

24. Nissen, V., Propach, J.: On the robustness of population-based versus point-based optimization in the presence of noise. IEEE Trans. Evol. Comput. **2**(3), 107–119 (1998)

25. Rana, S., Whitley, L.D., Cogswell, R.: Searching in the presence of noise. In: Voigt, H.M., Ebeling, W., Rechenberg, I., Schwefel, H.P. (eds.) Parallel Problem Solving from Nature - PPSN IV, vol. 1141, pp. 198–207. Springer, Heidelberg (1996). https://doi.org/10.1007/3-540-61723-X_984

26. Stefanovic, D., Stefanovic, N., Radenkovic, B.: Supply network modelling and simulation methodology. Simul. Model. Pract. Theory **17**(4), 743–766 (2009)

27. Subulan, K., Baykasoğlu, A., Eren Akyol, D., Yildiz, G.: Metaheuristic-based simulation optimization approach to network revenue management with an improved self-adjusting bid-price function. Eng. Econ. **62**(1), 3–32 (2017)

28. Vives, A., Jacob, M., Payeras, M.: Revenue management and price optimization techniques in the hotel sector: a critical literature review. Tourism Econ. **24**, 720–752 (2018)

29. Vock, S., Enz, S., Cleophas, C.: Genetic algorithms for calibrating airline revenue management simulations. In: 2014 Proceedings of the Winter Simulation Conference, pp. 264–275 (2014)

30. Welch, B.L.: The generalization of student's problem when several different population variances are involved. Biometrika **34**(1/2), 28–35 (1947)

31. Xiang, Z., Magnini, V.P., Fesenmaier, D.R.: Information technology and consumer behavior in travel and tourism: insights from travel planning using the Internet. J. Retail. Consum. Serv. **22**, 244–249 (2015)

32. Yager, R.R.: Quantifier guided aggregation using OWA operators. Int. J. Intell. Syst. **11**(1), 49–73 (1996)

33. Jin, Y., Branke, J.: Evolutionary optimization in uncertain environments-a survey. IEEE Trans. Evol. Comput. **9**(3), 303–317 (2005)

34. Zakhary, A., Atiya, A.F., El-Shishiny, H., Gayar, N.E.: Forecasting hotel arrivals and occupancy using Monte Carlo simulation. J. Revenue Pricing Manag. **10**(4), 344–366 (2011)

35. Zhang, D., Weatherford, L.: Dynamic pricing for network revenue management: a new approach and application in the hotel industry. INFORMS J. Comput. **29**(1), 18–35 (2016)

Heuristic Search Strategies for Noisy Optimization

Manuel Dalcastagné[(✉)]

Department of Information Engineering and Computer Science, University of Trento,
Via Sommarive 9, 38123 Povo, TN, Italy
m.dalcastagne@unitn.it

Abstract. Many real-world optimization problems are subject to noise, and making correct comparisons between candidate solutions is not straightforward. In the literature, various heuristics have been proposed to deal with this problem. Most studies compare evolutionary strategies with algorithms which propose candidate solutions deterministically. This paper compares the efficiency of different randomized heuristic search strategies, and also extends randomized algorithms non based on populations with a statistical analysis technique in order to deal with the presence of noise. Results show that this extension can outperform population-based algorithms, especially with higher levels of noise.

Keywords: Noisy optimization · Heuristics · Statistical analysis

1 Introduction

In many real-world optimization problems, the evaluation of candidate solutions is affected by noise. Possible sources of noise include physical measurement limitations, or the stochastic component employed in simulations. Similarly, in machine learning, the diversity of data used to train and test models adds a layer of uncertainty to the problem. Different models are usually compared using cross-validation approaches, but comparisons are not guaranteed to be correct. In noisy scenarios, since the true value of the objective function is distorted, making correct comparisons between candidate solutions is not straightforward. If the noise is too high with respect to the difference between the true values of two candidates (*signal*), and so the signal-to-noise ratio is too low, comparisons done using a single evaluation per solution might be wrong.

In order to deal with noise, various heuristics have been proposed and studied [2,4,5,17,27]. In particular, many studies employ variants of evolutionary algorithms, which adopt a set of candidate solutions (*population*) subject to local perturbative search and stronger diversification means, often described with terms derived from genetics. Since these algorithms iteratively employ a population to explore the search space and propose new solutions, they are considered to be robust to the presence of noise [2,21]. Multiple works compared various heuristic algorithms with evolutionary strategies, and they have shown

© Springer Nature Switzerland AG 2020
I. S. Kotsireas and P. M. Pardalos (Eds.): LION 14 2020, LNCS 12096, pp. 356–370, 2020.
https://doi.org/10.1007/978-3-030-53552-0_32

that population-based approaches are a good choice to optimize noisy functions [2,4,27]. However, these studies mostly compare evolutionary strategies with algorithms which propose candidate solutions deterministically. But, according to the results of [4], when information about gradients is not available and the objective function is noisy, randomized algorithms might be an effective choice.

Apart from the search policy, which defines how the search space is explored and new solutions are proposed, also other building blocks are necessary to build effective noisy optimization strategies. In fact, to be efficient in this context, algorithms must deal with the presence of noise. In general, this can be achieved in two ways: by increasing the strength of the signal, or by reducing the effect of noise. In randomized algorithms, the signal can be improved by adapting the search region according to the signal-to-noise ratio. Multiple variants of this strategy have been studied in the field of evolutionary algorithms [2,5]. It has been shown that the way in which the search region is adapted during the optimization has a relevant impact. On the other hand, the effect of noise can be reduced by evaluating multiple times each solution [8,9,13,28,29]. Also, if a heuristic is population-based, the impact of noise can be decreased by incrementing the number of candidates (*size*) of the population [3,5,15]. This effect, called *implicit averaging*, has been studied using normally distributed noise with zero mean and different values of constant standard deviations [16,17,22]. Also, [17] shows that increasing indiscriminately the population size can be counterproductive, but without providing an explanation for this behavior. As will be shown empirically, the effect of implicit averaging depends on the type and the amount of noise in the objective function. It is also worth noticing that, in order to deal with noise, randomized algorithms can be extended with statistical analysis techniques. An example is given by simulated annealing [32], which has been extended by adapting the number of samples per solution based on some statistical analysis [1,7]. However, studies which compare diverse randomized algorithms extended with statistical methods are missing in the literature, and this work is a first step in this direction.

This study aims at comparing the efficiency of different heuristic search strategies in the presence of noise, and to investigate the effects that different components of these strategies have on the performance. Differently from previous studies, all the heuristic search strategies employed in this study are randomized, and algorithms not based on populations are extended using a reactive sample size scheme proposed by [14]. The rest of the paper is structured as follows. Section 2 states more formally the noisy optimization problem and gives an overview of the reactive sample size scheme. Section 3 outlines the heuristic search algorithms which have been used in the experiments, and comments the components of the algorithms which are analyzed empirically. Section 4 defines the experiments and analyzes the results.

2   Noisy Optimization

Let F be a stochastic function that models a real world problem. The output of F depends on some decision variables x and on a random vector ξ that represents the stochasticity of the problem. The expectation of F is defined as

$$f(x) = \mathbb{E}[F(x, \xi)] \tag{1}$$

and it can be estimated by using a sample $\xi_1, ..., \xi_n$ of independent identically distributed (i.i.d.) realizations of the random vector ξ, in order to compute the Sample Average Approximation (SAA) of (1) as

$$\hat{f}_n(x) = \frac{1}{n} \sum_{i=1}^{n} F(x, \xi_i). \tag{2}$$

If the sample $\xi_1, ..., \xi_n$ is i.i.d., by the Law of Large Numbers, as n approaches infinity $\hat{f}_n(x)$ converges to $f(x)$ and so $\hat{f}_n(x)$ is an unbiased estimator of $f(x)$. Moreover, if the variance of F is finite, by the Central Limit Theorem $\hat{f}_n(x)$ asymptotically follows a normal distribution with mean $f(x)$ and variance σ^2/n where σ^2 is the variance of F. As a consequence, the accuracy of the estimation increases with sample size n, but this also increments the computational burden (see also [26,31]). The problem might be defined in the constraints-defined region Θ in which x can assume values as

$$\min_{x \in \Theta} f(x). \tag{3}$$

The SAA defined in (2) can be used as objective function by heuristic optimization techniques, in order to optimize $f(x)$. The presence of noise might require large samples in order to obtain sufficiently accurate estimates, so comparing the performance of different configurations is not straightforward. Given any configuration x, $f(x) - \hat{f}_n(x)$ defines an error $\epsilon_n(x)$ that goes to 0 only in the limit of n going to infinity. As a consequence, when comparing two configurations x_1 and x_2, the difference $\hat{f}_n(x_1) - \hat{f}_n(x_2)$ is not sufficient to decide which configuration has a better average. If the signal $|f(x_1) - f(x_2)|$ is lower than the noise $|\epsilon_n(x_1) - \epsilon_n(x_2)|$, the signal-to-noise ratio is too low and the comparison might be not significant.

2.1 A Reactive Sample Size Algorithm

In this work, in order to deal with the presence of noise, heuristic techniques which do not use a population are extended with a reactive sample size algorithm [14] based on paired t-tests and indifference-zone (IZ) selection. IZ selection is a concept commonly used in ranking and selection (R&S) algorithms. These methods aim at selecting, in a statistically significant manner, the best solution x^* which performs better among a finite set of k possibilities. In R&S methods based on IZ selection, the target is to select the best configuration x^* among a finite set of k configurations, where x^* is better than all other configurations in the set by at least δ and the probability of correct selection (PCS) is $1 - \alpha > 0$, where α is the probability of making an error of type I. δ is called the IZ parameter, and it defines the minimum difference in means considered to be worth detecting. More information about R&S can be found in [10,23,25].

The algorithm works as follows. Given a pair of configurations $\{x_1, x_2\}$ to be compared, paired evaluations are obtained by evaluating x_1 and x_2 using the same ξ_i. Then, these evaluations are used to compute the paired t-test statistic. In fact, as observed by [24], using the same realizations helps to reduce the effect of noise. The correlation among pairs of evaluations reduces the variance with respect to an unpaired statistic. Also, the scheme assumes that F is normally distributed and that its variance is finite.

The algorithm reactively decides, in an online manner, the sample size to be used for each comparison done during the optimization. Also, all the evaluations of solutions previously visited during the search are kept in memory, to avoid the waste of computational budget if a configuration has to be compared multiple times. Significant differences are detected by considering the relationship between probabilities α and β of making an error of type I and type II. To remind the reader, given a null hypothesis H_0 and an alternative hypothesis H_1, α is the probability to reject H_0 when H_0 is true and β is the probability to fail to reject H_0 when H_0 is false. To compute β, a significant difference in means for which H_0 is assumed to be false and H_1 to be true has to be defined. The value for which H_1 is assumed to be true is $\delta_{observed}$, which corresponds to $\hat{f}_n(x_1) - \hat{f}_n(x_2)$. So, given a paired sample, the minimum sample size n that should be used to test a one-tailed hypothesis with error probabilities α and β can be computed. See [14] for more details.

Also, in real world problems, one might not be interested to correctly detect very small differences between means. If x_1 and x_2 have a very similar performance and $\delta_{observed}$ is smaller than a certain user-defined δ, the comparison is done heuristically by considering only the values of $\hat{f}_n(x_1)$ and $\hat{f}_n(x_2)$. The value of δ is expressed as a percentage of the current best solution, because in many cases the user does not know *a priori* the best possible result which can be obtained by the optimization.

3   Optimization Algorithms

The optimization algorithms employed in the experiments are Random Search (RS), the Reactive Affine Shaker (RAS) [11] and the Covariance Matrix Adaptation Evolutionary Strategy (CMA-ES) [19,20]. RS is a simple stochastic local search algorithm which is often used as baseline for comparisons, while RAS and CMA-ES are more advanced stochastic schemes which adapt step size and direction of the search during the optimization.

In RS, a new candidate solution x_{new} is sampled from an interval defined in a neighborhood of the current best solution $x_{current}$, according to a uniform distribution. A step size σ is used to define, as a percentage of the intervals which define Θ along each dimension, the boundaries of the local search region located around $x_{current}$. Consequently, diverse step sizes correspond to search policies with different levels of locality. A step size of 1 would make the search global, and the optimization would correspond to pure random search.

In RAS, a local search region is adapted by an affine transformation. The aim is to scout for local minima in the attraction basin where the initial point falls. The step size σ and the direction of the search region are adapted in order to maintain heuristically the largest possible movement per function evaluation. The search occurs by generating points in a stochastic manner, with a uniform probability in the search region, following a *double shot* strategy. A single displacement Δ is generated, and two specular points $x_{current} + \Delta$ and $x_{current} - \Delta$ are considered for evaluation. An evaluation is successful if the objective function value in at least one of the two candidates is better than $\hat{f}(x_{current})$. The search region is modified according to the outcome of the comparisons. It is compressed if both comparisons are unsuccessful, and it is expanded otherwise. In both cases, the search region is modified according to an expansion factor $\rho > 1$.

CMA-ES [19,20] is an evolutionary optimization paradigm in which configurations are sampled from a multivariate normal distribution $N(\mu_t, M_t)$, where t defines the iteration of the algorithm. At each iteration, the mean μ_t defines the center of the distribution, the covariance matrix M_t determines shape and orientation of the ellipsoid corresponding to $N(\mu_t, M_t)$, and a step size σ_t controls the spread of the distribution as a percentage of the intervals which define each dimension of Θ. The ellipsoid is the local search region used by the algorithm to explore the search space and propose candidate solutions. Iteratively, CMA-ES follows four steps. First, it samples a fixed number λ of new configurations from $N(\mu_t, M_t)$, creating a population. Second, candidates are evaluated and ranked according to the quality of the evaluations. Third, the best $\lfloor \frac{\lambda}{2} \rfloor$ results are used to update $N(\mu_t, M_t)$, in order to move the search towards the most promising search direction. Fourth, σ_t is increased or decreased according to the length of the so-called evolution paths, in order to maximize the expected improvement of the optimization. This last step is explained more in detail in the following subsection. Also, CMA-ES has been extended with an uncertainty handling (UH) method, to deal with possible noise in the objective function [18]. In this version of CMA-ES, referred as UH-CMA-ES, the uncertainty is measured by rank changes among members of the population. Once each solution of the population has been evaluated and ranked, a few additional evaluations are taken and the population is ranked again. By doing so the algorithm tries to estimate the amount of noise in the evaluations, in order to increase σ_t and prevent the signal-to-noise ratio from becoming too low.

3.1   A Note on CMA-ES Step-Size Adaptation

The adaptation of σ_t, also called cumulative step-size adaptation, is based on the evolution paths mentioned in Sect. 3. An evolution path is a weighted vector sum of the last points successively visited by the algorithm. It provides information about the correlations between points, and it can be used to detect the direction of consecutive steps taken by the optimization. If consecutive steps are going in the same direction (scalar product greater than zero), the same distance could be covered by longer steps and the current path is too long. If consecutive steps are not going in the same direction (scalar product lower than zero), single steps tend to cancel each other out and so the current path is too short. Therefore, to make successive steps more efficient, σ_t is changed accordingly. The step size determines the signal strength used by CMA-ES to estimate the direction of the gradient. If the steps of the algorithm are very small, the signal is also likely low and therefore the signal-to-noise ratio becomes small as well. Also, it has been shown that in noisy optimization the cumulative step-size adaptation may result in premature convergence [6].

4 Experiments

Before showing the results about the performance of different randomized algorithms in various noisy scenarios and higher dimensions, a preliminary study on CMA-ES is presented. In order to investigate the effects of implicit averaging, the algorithm is tested using populations larger than the standard size proposed by its authors.

4.1 Benchmarking Functions and Noise Models

In order to test diverse heuristic strategies in the presence of noise, Sphere and Rastrigin functions have been extended with multiple types and levels of noise. As in other works in the literature, both functions are optimized in $[-5.12, 5.12]^d$, where d is the number of dimensions. To evaluate the impact of noise on the optimization, a standard practice in the literature is to extend deterministic functions by introducing multiplicative or additive noise. In the case of multiplicative noise, a percentage ϵ of $f(x)$ is added to $f(x)$ according to a displacement generated using a standard normal distribution:

$$f(x, \epsilon) = f(x) + f(x) \cdot \epsilon \cdot N(0, 1). \tag{4}$$

This kind of noise is typical of devices which take physical measurements, like speed cameras, where values are guaranteed to be accurate up to a certain percentage of the measured quantity. However, as the optimization proceeds towards lower values, the noise decreases. This means that, as the optimization approaches the global optimum x^*, it is easier to move into the right direction and, if $f(x^*) = 0$, there is almost no noise in proximity of the global optimum.

Although such a situation is true in many real-world scenarios, there exist other problems where the noise does not always go to zero as the global optimum (if any) is approached. As examples, consider the optimization of simulation models, or the tuning of the hyperparameters of machine learning algorithms. In this case, the noise can be simulated by adopting additive noise. However, determining the amount of noise to add is up to the practitioner. In fact, additive noise is usually normally distributed with zero mean and constant standard deviation σ_ϵ. Since this kind of perturbations does not depend on the signal, the signal-to-noise ratio might cause problems only when approaching the minimum and its effects are going to be very different from function to function.

To avoid these drawbacks, a possibility is to define additive noise as normally distributed with zero mean and dynamic standard deviation. Since the step size used by randomized algorithms impacts the strength of the signal, in order to test harder noise scenarios it makes sense to set the noise level according to the step size. So, given a point x in Θ and a percentage ϵ, the dynamic standard deviation σ_i along each dimension i is computed as follows. Compute lower bound $l_i = x_i - \epsilon \cdot \Theta_i$ and upper bound $u_i = x_i + \epsilon \cdot \Theta_i$, where Θ_i is the interval in which the function is defined along dimension i. Then, find the minimum m and the maximum M among $\{f(l_i), f(x_i), f(u_i)\}$. Finally, $\sigma_i = |m - M|$ and the additive noise model is defined as

$$f(x, \epsilon) = f(x) + \sum_{i=1}^{N} N(0, \frac{\sigma_i}{k}), \tag{5}$$

where N is the dimensionality of the function and k is a constant used to control the amount of noise. In the experiments, $\epsilon = 0.1$ and $k \in \{1, 2, 3, 6\}$. Therefore, while using this model of noise, the distortion of the signal is set according to the maximum signal which can be detected by an algorithm which adopts a fixed step size (like RS). With $k = 6$, 99.7% of the noise is generated within the intervals which define the local search region. With $k = 3, k = 2, k = 1$ the same is true respectively for $81.86\%, 68, 27\%, 34.14\%$ of the noise.

4.2   Setup

Each experiment is based on 100 macroreplications, where each optimization process has a budget (number of function evaluations) of 5000. Initial solutions are generated according to a uniform distribution defined on the interval of each dimension of Θ. In each experiment, algorithms start the optimization from a randomly generated solution x_0 and consider a local search region of the same size defined around x_0. Then, the local search region is iteratively modified according to the algorithm. Apart from CMA-ES and UH-CMA-ES, the algorithms are employed in two versions: with a naive scheme which uses 1 evaluation for each solution, and the reactive scheme proposed by [14]. The acronym of each algorithm is preceded with N in the former case and with R in the latter. As suggested in [14], the values of the parameters are set as $\alpha_{req} = 0.1$, $\beta_{req} = 0.4$ and $\delta = 0.01$.

In the first set of experiments on CMA-ES, restarts are not considered. In fact, the standard implementation of CMA-ES includes various stopping criterias and restart policies [20], but they have been deactivated in order to improve the analysis of the different components of the algorithm. When activated, all algorithms use global restarts based on a single stopping criterion: if the current best solution does not improve by at least 10% in $k = 500$ function evaluations, a restart is done. An exception is given by RAS, which possibly needs to be restarted because of its double-shot strategy. In fact, if x_0 is generated nearby the boundaries of the search space, the double shot strategy might be unable to generate a valid configuration.

RS uses $\sigma \in \{0.1, 0.2\}$, while RAS employs $\sigma \in \{0.1, 0.2\}$ and $\rho = 2$. CMA-ES and UH-CMA-ES adopt $\sigma \in \{1.0, 2.0\}$, because the library used to implement the algorithm defines $\sigma \in (0, 10]$[1]. Also, $\lambda = 4 + \lfloor 3 \log d \rfloor$, where d is the dimensionality of the objective function. The values of ρ and λ are the ones suggested respectively by the authors of [11,20].

4.3   Average Loss Signal per Iteration

In a noisy scenario, in order to understand how the algorithm behaves with populations of different sizes, a possible way to proceed is to measure the magnitude of the error made when estimating the gradient. To do so, at each iteration, the population of candidate solutions $p = \{x_1, ..., x_m\}$ is ranked in two ways. Firstly, according to the noisy ranking $\hat{r}$ based on $\hat{f}_1(x_1), ..., \hat{f}_1(x_m)$. Secondly, following the noiseless ranking r defined by $f(x_1), ..., f(x_m)$. Then, the signal loss L is defined as the difference between the signal of these two rankings, where the signal of a ranking is the sum of the absolute differences among the ordered values used for ranking. More formally, in the case of $\hat{r}$ and r,

$$\hat{s}(\hat{r}, p) = \sum_{i=1}^{m-1} |\hat{f}_1(x_i) - \hat{f}_1(x_{i+1})| \tag{6}$$

and

$$s(r, p) = \sum_{i=1}^{m-1} |f(x_i) - f(x_{i+1})|, \tag{7}$$

where the set $\{1, ..., m-1\}$ is ordered respectively according to $\hat{r}$ and r. Therefore,

$$L(\hat{r}, r, p) = \hat{s}(\hat{r}, p) - s(r, p) \tag{8}$$

and the average signal loss per iteration is defined as

$$\mathbb{E}(L) = \frac{1}{N} \sum_{i=1}^{N} L(\hat{r}_i, r_i, p_i), \tag{9}$$

where N is the number of iterations of the algorithm. By following this procedure, the optimization estimates the gradient according to $\hat{r}$ and it is possible to measure by how much the optimization goes in the wrong direction. Also, by comparing $\hat{r}$ and r, the number of misrankings among each population's candidates can be computed, as well as the average misrankings' percentage $\mathbb{E}(M)$ of the whole optimization.

4.4   Results When Using Larger Populations in CMA-ES

This set of experiments is based on the bidimensional Sphere function. As it is possible to see in Table 1, $\mathbb{E}(L)$ keeps increasing as the population grows. Larger populations can potentially detect more signal, but are unable to do so. In fact,

---

[1] https://github.com/beniz/libcmaes.

Table 1. Performance of CMA-ES with $\sigma = 1.0$ and $\lambda \in \{6, 36, 60\}$ on the Sphere function in 2 dimensions. Macrocolumns show respectively the results obtained when using multiple levels of constant additive noise, multiplicative noise and dynamic additive noise. In each macrocolumn, the average best results $f(x^*)$ are in bold and all values based on $f(x)$ are normalized by the number of dimensions.

			Constant additive			Multiplicative				Dynamic additive			
λ	N	σ_ϵ	$\mathbb{E}(M)$	$\mathbb{E}(L)$	$f(x^*)$	ϵ	$\mathbb{E}(M)$	$\mathbb{E}(L)$	$f(x^*)$	k	$\mathbb{E}(M)$	$\mathbb{E}(L)$	$1f(x^*)$
6	833	0.1	81.92	0.0048	0.02	0.1	13.37	0.0032	**0.00**	6	82.28	0.0392	**0.15**
6	833	1.0	82.55	0.0512	0.22	0.2	24.34	0.0099	**0.00**	3	82.39	0.1219	**1.45**
6	833	2.0	82.74	0.0986	0.42	0.3	33.82	0.0264	**0.28**	2	82.61	0.2362	**3.26**
6	833	3.0	82.85	0.1476	0.75	0.4	43.78	0.0526	**2.29**	1	82.50	1.1282	9.10
36	138	0.1	93.26	0.0138	0.01	0.1	53.48	0.0400	**0.00**	6	95.72	0.1639	0.35
36	138	1.0	95.72	0.1506	0.05	0.2	70.86	0.0662	**0.00**	3	95.85	0.5872	2.26
36	138	2.0	95.98	0.3336	0.11	0.3	79.03	0.1368	0.38	2	95.87	1.3667	4.22
36	138	3.0	96.20	0.4303	0.14	0.4	84.30	0.2789	4.63	1	96.22	4.1927	**9.07**
60	83	0.1	93.75	0.0188	**0.00**	0.1	66.15	0.0508	**0.00**	6	97.10	0.2641	0.40
60	83	1.0	97.05	0.2078	**0.04**	0.2	80.16	0.1322	**0.00**	3	97.25	0.8849	2.49
60	83	2.0	97.50	0.3776	**0.08**	0.3	86.12	0.2235	0.58	2	97.33	2.1270	5.24
60	83	3.0	97.59	0.6161	**0.12**	0.4	90.09	0.4825	6.56	1	97.64	5.7731	10.04

as Figs. 1a, 1b, 1c show, larger populations contribute to mantain a larger step size and so to make wider steps. However, as the amount of noise increases, misrankings are going to happen first among similarly ranked candidates and then also between solutions ranked farther away from each other. Also, the magnitude of the errors increases with the noise. Consequently, larger populations tend to lose more signal and to guide the optimization farther away from the true direction of the gradient.

The effect of the misrankings depends on the amount of noise. Even if $\mathbb{E}(M)$ increases with the population size, it is not implied that the performance of the optimization deteriorates. For example, Table 1 shows that in the case of constant additive noise, larger populations obtain better results. This happens because the signal-to-noise ratio is sufficiently high to avoid misrankings which would guide the optimization towards a significantly wrong direction. The amount of noise starts creating problems only when approaching the minimum, as shown in Fig. 2. With this amount of noise, even N-RS is able to perform comparably well. In contrast, in the case of dynamic additive noise, the signal-to-noise ratio is approximately the same throughout the search space and results are expected to deteriorate much more, as confirmed by the results in Table 1. Even in the case of multiplicative noise, larger populations possibly worsen the performance of the optimization. Therefore, when the signal-to-noise ratio is low and misrankings happen among further positions, increasing the population size or the number of parents is not going to significantly improve the robustness of the optimization.

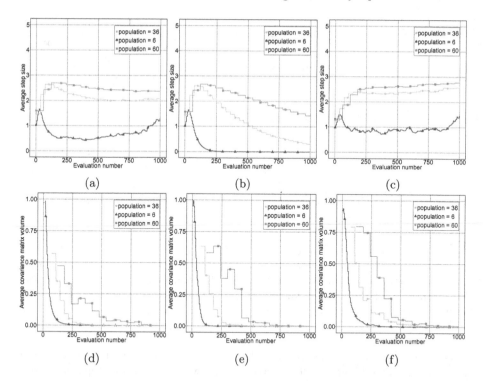

Fig. 1. Each row shows respectively average variation of CMA-ES step size (1a) and CMA-ES covariance matrix volume (1a), with $\sigma = 1.0$ and $\lambda \in \{6, 36, 60\}$. Each column refers respectively to a particular case of the diverse types of noise used in Table 1. More precisely, they show the cases with constant additive noise with $\sigma = 1.0$ (first column), multiplicative noise with $\epsilon = 0.2$ (second column) and dynamic additive noise with $k = 3$ (third column).

In this case, it is preferable to increase the sample size used to estimate each candidate solution.

It is also worth noticing the premature convergence of the covariance matrix. Figures 1d, 1e, 1f show that the volume of the covariance matrix goes to zero in the first part of the optimization. After that, CMA-ES is no longer able to propose significantly different solutions.

4.5   Results When Using Larger Sample Size

These experiments are based on both Sphere and Rastrigin functions, with $d \in \{2, 10\}$. Results in Tables 2 and 3 show that, with high levels of noise, a simple optimization algorithm such as R-RS performs better than more complex algorithms like RAS, CMA-ES or UH-CMA-ES. Without increasing the sample size of estimators, using a population of solutions is not able to compete with single-point algorithms which adapt the sample size of estimators according to empirical evidence. Furthermore, in this context, UH-CMA-ES might even

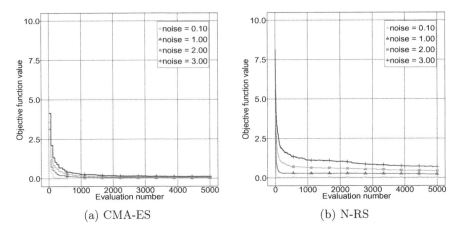

(a) CMA-ES   (b) N-RS

Fig. 2. Average convergence of CMA-ES with $\sigma = 1.0$ and $\lambda = 60$, and N-RS with $\sigma = 0.1$ on the bidimensional Sphere function, with multiple levels of constant additive noise. Lines represent the mean noiseless value of the best configuration found during the optimization (which is not known by the optimizers). Also, all evaluations are normalized by the number of dimensions.

worsen the performance. As shown in Fig. 3d, increasing the step size according to observed misrankings provides better results when the initial step size is very low with respect to the distortion caused by the noise.

Figure 3a shows that a larger step size can improve the efficiency of the optimization. On average, compared solutions correspond to more different estimators and a lower sample size is required to make statistically significant comparisons. However, since a larger step size also implies reducing the sampling granularity, the optimization enhances the global search phase and the convergence speed tends to decrease. For the same reason, as shown in Fig. 3b, the effectiveness of the double-shot strategy in a noisy environment is questionable. Although such an approach can be a good strategy for deterministic optimization, on each iteration very similar configurations are compared, the signal-to-noise ratio tends to be low and the sample size required to make statistically significant comparisons increases. Furthermore, as the search region is compressed, this effect is further enhanced.

In noisy scenarios, step-size adaptation mechanisms and adaptations of the search space are potentially counterproductive. In deterministic functions, compressions of the search region usually lead to a better exploitation of the local structure of the objective function. However, because of the presence of noise, decisions to compress the search region might be wrong and therefore the optimization might prematurely converge to false local optima.

In larger dimensions, the situation changes in the case of Rastrigin function with lower levels of noise. However, it is expected that a population-based algorithm performs better than single-point algorithms in the case of a multimodal function like Rastrigin. Combining a set of candidate solutions at each iteration

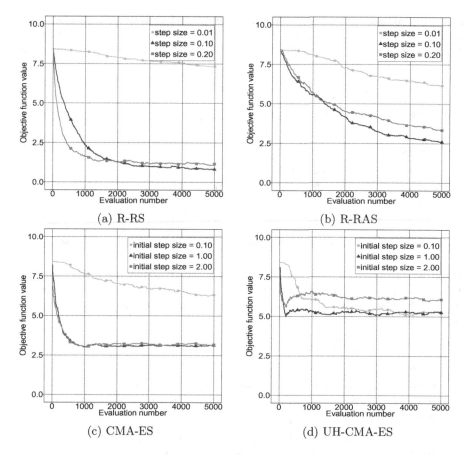

(a) R-RS

(b) R-RAS

(c) CMA-ES

(d) UH-CMA-ES

Fig. 3. Average convergence of the optimizers on the Sphere function, with $d = 10$ $k = 3$. The results obtained by using a very small step size are also shown.

gives the algorithm the ability to adapt to the local topology of the objective function, reducing the risk to get stuck in local optima.

5   Conclusions

This paper investigated different components of diverse heuristic strategies in the context of noisy optimization. A preliminary study on the bidimensional Sphere function showed how the *implicit averaging* effect of population-based algorithms does not always improve the optimization as the size of the population is increased, and analyzed how different amounts of noise change the impact of this effect on the optimization. Randomized algorithms not based on populations have been extended with a statistical analysis technique [14] to deal with the presence of noise, and they have been compared with CMA-ES and UH-CMA-ES.

Table 2. Average best objective function value found by different optimizers, with different levels of dynamic additive noise added to the Sphere function. In each column, the best results are in bold and all values are normalized by the number of dimensions.

Optimizer	σ	$d = 2$				$d = 10$			
		$k = 1$	$k = 2$	$k = 3$	$k = 6$	$k = 1$	$k = 2$	$k = 3$	$k = 6$
R-RS	0.1	**1.14**	**0.29**	**0.15**	**0.05**	4.29	1.75	**0.81**	**0.29**
R-RS	0.2	1.25	0.45	0.26	**0.05**	**3.77**	**1.74**	1.16	0.51
R-RAS	0.1	4.12	1.37	0.78	0.12	6.85	4.27	2.59	0.83
R-RAS	0.2	3.41	1.35	0.55	**0.05**	7.63	4.63	3.34	0.70
CMA-ES	1.0	13.95	5.72	2.45	0.44	11.98	7.22	3.13	0.69
CMA-ES	2.0	14.76	5.00	2.14	0.41	12.17	7.47	3.17	0.71
UH-CMA-ES	1.0	14.71	5.09	2.15	0.45	12.96	9.86	5.26	1.02
UH-CMA-ES	2.0	15.46	5.66	2.03	0.38	12.79	10.33	6.09	0.98

Table 3. Average best objective function value found by different optimizers, with different levels of dynamic additive noise added to the Rastrigin function. In each column, the best results are in bold and all values are normalized by the number of dimensions.

Optimizer	σ	$d = 2$				$d = 10$			
		$k = 1$	$k = 2$	$k = 3$	$k = 6$	$k = 1$	$k = 2$	$k = 3$	$k = 6$
R-RS	0.1	**1.87**	**0.58**	**0.30**	**0.13**	10.94	8.07	7.29	6.60
R-RS	0.2	2.19	0.77	0.48	0.25	**10.30**	**7.24**	6.40	5.93
R-RAS	0.1	7.68	5.65	5.10	3.68	13.20	9.55	8.47	6.97
R-RAS	0.2	6.61	4.77	4.61	2.86	13.98	10.16	8.35	6.88
CMA-ES	1.0	16.85	6.62	3.58	1.03	18.83	12.37	**6.21**	2.24
CMA-ES	2.0	17.48	6.16	2.85	0.98	19.59	13.31	6.46	**2.13**
UH-CMA-ES	1.0	16.79	4.73	2.98	1.11	19.89	15.94	11.02	3.95
UH-CMA-ES	2.0	17.67	5.21	2.11	0.95	19.71	16.57	12.64	3.91

Results in Sect. 4.4 confirm the findings of [17]. The analysis provides an explanation about the reason for which larger populations do not always improve the optimization, and a higher sample size should be preferred when the signal-to-noise ratio is too low. Furthermore, these results also agree with [2]: in the presence of noise, step length control mechanisms are crucial to the performance of the optimization. If optimization methods are extended with statistical analysis techniques such as [14], the resolution at which the search space is explored matters significantly. With lower step sizes, solutions correspond to more similar function evaluations, and the sample size required to statistically determine a difference increases.

Future work aims at extending current results in order to consider other derivative-free optimization strategies and more benchmarks. Also, it would be worth investigating the performance of more global search policies, which iteratively compare configurations located farther away in Θ and search more locally in different parts of Θ only if sufficient empirical evidence to do so is observed. Approaches like CoRSO [12] or Bayesian Optimization [30] could be good choices.

References

1. Ahmed, M.A., Alkhamis, T.M.: Simulation-based optimization using simulated annealing with ranking and selection. Comput. Oper. Res. **29**(4), 387–402 (2002)
2. Arnold, D.V.: Noisy Optimization with Evolution Strategies, vol. 8. Springer, Cham (2012)
3. Arnold, D.V., Beyer, H.G.: Investigation of the (μ, λ)-es in the presence of noise. In: Proceedings of the 2001 Congress on Evolutionary Computation (IEEE Cat. No. 01TH8546), vol. 1, pp. 332–339. IEEE (2001)
4. Arnold, D.V., Beyer, H.G.: A comparison of evolution strategies with other direct search methods in the presence of noise. Comput. Optim. Appl. **24**(1), 135–159 (2003)
5. Beyer, H.G.: Evolutionary algorithms in noisy environments: theoretical issues and guidelines for practice. Comput. Methods Appl. Mech. Eng. **186**(2–4), 239–267 (2000)
6. Beyer, H.G., Arnold, D.V.: Qualms regarding the optimality of cumulative path length control in CSA/CMA-evolution strategies. Evol. Comput. **11**(1), 19–28 (2003)
7. Branke, J., Meisel, S., Schmidt, C.: Simulated annealing in the presence of noise. J. Heuristics **14**(6), 627–654 (2008)
8. Branke, J., Schmidt, C.: selection in the presence of noise. In: Cantú-Paz, E., et al. (eds.) GECCO 2003. LNCS, vol. 2723, pp. 766–777. Springer, Heidelberg (2003). https://doi.org/10.1007/3-540-45105-6_91
9. Branke, J., Schmidt, C., Schmeck, H.: Efficient fitness estimation in noisy environments. In: Proceedings of the 3rd Annual Conference on Genetic and Evolutionary Computation, pp. 243–250. Morgan Kaufmann Publishers Inc. (2001)
10. Branke, J., Chick, S.E., Schmidt, C.: Selecting a selection procedure. Manag. Sci. **53**(12), 1916–1932 (2007)
11. Brunato, M., Battiti, R.: RASH: A Self-adaptive Random Search Method, pp. 95–117. Springer, Heidelberg (2008)
12. Brunato, M., Battiti, R.: CoRSO (collaborative reactive search optimization): blending combinatorial and continuous local search. Informatica **27**(2), 299–322 (2016)
13. Cantú-Paz, E.: Adaptive sampling for noisy problems. In: Deb, K. (ed.) GECCO 2004. LNCS, vol. 3102, pp. 947–958. Springer, Heidelberg (2004). https://doi.org/10.1007/978-3-540-24854-5_95
14. Dalcastagné, M., Mariello, A., Battiti, R.: Reactive Sample Size for Heuristic Search in Simulation-based Optimization. arXiv e-prints arXiv:2005.12141 (2020)
15. Fitzpatrick, J.M., Grefenstette, J.J.: Genetic algorithms in noisy environments. Mach. Learn. **3**(2–3), 101–120 (1988)
16. Goldberg, D.E., Deb, K., Clark, J.H., et al.: Genetic algorithms, noise, and the sizing of populations. Complex Syst. Champaign **6**, 333–333 (1992)

17. Hammel, U., Bäck, T.: Evolution strategies on noisy functions how to improve convergence properties. In: Davidor, Y., Schwefel, H.-P., Männer, R. (eds.) PPSN 1994. LNCS, vol. 866, pp. 159–168. Springer, Heidelberg (1994). https://doi.org/10.1007/3-540-58484-6_260

18. Hansen, N., Niederberger, A.S.P., Guzzella, L., Koumoutsakos, P.: A method for handling uncertainty in evolutionary optimization with an application to feedback control of combustion. IEEE Trans. Evol. Comput. **13**(1), 180–197 (2009)

19. Hansen, N.: The CMA evolution strategy: a comparing review. In: Lozano, J.A., Larrañaga, P., Inza, I., Bengoetxea, E. (eds.) Towards a New Evolutionary Computation. Studies in Fuzziness and Soft Computing, vol. 192, pp. 75–102. Springer, Heidelberg (2006). https://doi.org/10.1007/3-540-32494-1_4

20. Hansen, N.: Benchmarking a bi-population CMA-ES on the BBOB-2009 function testbed. In: Proceedings of the 11th Annual Conference Companion on Genetic and Evolutionary Computation Conference: Late Breaking Papers, pp. 2389–2396. ACM (2009)

21. Hansen, N., Niederberger, A.S., Guzzella, L., Koumoutsakos, P.: A method for handling uncertainty in evolutionary optimization with an application to feedback control of combustion. IEEE Trans. Evol. Comput. **13**(1), 180–197 (2008)

22. Harik, G., Cantú-Paz, E., Goldberg, D.E., Miller, B.L.: The gambler's ruin problem, genetic algorithms, and the sizing of populations. Evol. Comput. **7**(3), 231–253 (1999)

23. Hong, L.J., Nelson, B.L., Xu, J.: Discrete Optimization via Simulation, pp. 9–44. Springer, New York (2015)

24. Jian, N., Henderson, S.: An introduction to simulation optimization. In: Yilmaz, L., Chan, W.K.V., Moon, I., Roeder, T.M.K., Macal, C., Rossetti, M.D. (eds.) Proceedings of the 2015 Winter Simulation Conference, pp. 1780–1794 (2015)

25. Kim, S.H., Nelson, B.L.: Chapter 17 selecting the best system. Simul. Handb. Oper. Res. Manag. Sci. **13**, 501–534 (2006)

26. Kim, S., Pasupathy, R., Henderson, S.G.: A Guide to Sample Average Approximation, pp. 207–243. Springer, New York (2015)

27. Nissen, V., Propach, J.: On the robustness of population-based versus point-based optimization in the presence of noise. IEEE Trans. Evol. Comput. **2**(3), 107–119 (1998)

28. Sano, Y., Kita, H.: Optimization of noisy fitness functions by means of genetic algorithms using history of search with test of estimation. In: Proceedings of the 2002 Congress on Evolutionary Computation. CEC 2002 (Cat. No. 02TH8600), vol. 1, pp. 360–365. IEEE (2002)

29. Schmidt, C., Branke, J., Chick, S.E.: Integrating techniques from statistical ranking into evolutionary algorithms. In: Rothlauf, F., et al. (eds.) EvoWorkshops 2006. LNCS, vol. 3907, pp. 752–763. Springer, Heidelberg (2006). https://doi.org/10.1007/11732242_73

30. Shahriari, B., Swersky, K., Wang, Z., Adams, R.P., De Freitas, N.: Taking the human out of the loop: a review of bayesian optimization. Proc. IEEE **104**(1), 148–175 (2015)

31. Shapiro, A., Dentcheva, D., Ruszczynski, A.: Lectures on Stochastic Programming: Modeling and Theory. 2nd edn. Society for Industrial and Applied Mathematics, Philadelphia (2014)

32. Van Laarhoven, P.J., Aarts, E.H.: Simulated annealing. In: Simulated annealing: Theory and applications, pp. 7–15. Springer, Dordrecht (1987). https://doi.org/10.1007/978-94-015-7744-1_2

Uncertainty of Efficient Frontier in Portfolio Optimization

Valery A. Kalygin[✉] and Sergey V. Slashchinin

Laboratory of Algorithms and Technologies for Network Analysis,
National Research University Higher School of Economics, Nizhny Novgorod, Russia
vkalyagin@hse.ru

Abstract. Portfolio optimization is a large area of investigation both in theoretical and practical setting since the seminal work by Markowitz where a mean-variance model was introduced. From optimization point of view, the problem of optimal portfolio in mean-variance setting can be formulated as convex quadratic optimization under uncertainty. In practice one needs to estimate parameters of the model to find an optimal portfolio. Error in the parameter estimation generates error in the optimal portfolio. It was observed that the out of sample behavior of obtained solution is not in accordance with what is expected. Main reason for this phenomena is related with the estimation of means of stock returns, estimation of covariance matrix being less important. In the present paper we study uncertainty of identification of efficient frontier (Pareto optimal portfolios) in mean-variance model. In order to avoid the estimation of means of returns we use CVaR optimization method by Rockafellar and Uryasev. First we prove, that for a large class of elliptical distributions efficient frontier in mean-variance model is identical to the trajectory of CVaR optimal portfolios with the change of the confidence level. This gives an alternative way to recover efficient frontier in mean-variance model. Next we conduct a series of numerical experiments to test the proposed approach. We show that proposed approach is competitive with existing methods.

Keywords: Optimization under uncertainty · Portfolio optimization · Efficient frontier · Mean-variance optimization · CVaR optimization

1 Introduction

Financial optimization is an attractive fields in optimization which is marked by presence of uncertainty, such as economic factors, returns on financial instruments, future prices of goods, etc. One of the most known problems in this field is portfolio optimization initiated by the seminal work by Markowitz [15]. The objective is to distribute in optimal way the capital between stocks in the financial market. Markowitz approach, also called mean-variance approach, has become very popular and continue to be useful our days [9,14]. However, practical application of this approach hurts a serious obstacle. Since Markowitz's

© Springer Nature Switzerland AG 2020
I. S. Kotsireas and P. M. Pardalos (Eds.): LION 14 2020, LNCS 12096, pp. 371–376, 2020.
https://doi.org/10.1007/978-3-030-53552-0_33

framework assumes that mean vector and covariances of returns are known, in practice it requires estimation of parameters from historical data. This can result in estimation errors or bias and hence non-optimal solutions and poor out-of-sample performance. Impact of estimation error in optimal portfolios calculation can be so important that in some cases trivial equal weights portfolio can be competitive with optimal portfolios [4,16]. Different aspects of the impact of estimation errors on optimal portfolio were investigated in [2,3]. Various techniques to improve estimations of means and covariances were developed in the literature [8,11,12]. However, as it was shown in [10] these improvements don't eliminate the "bias" of optimal portfolio obtained by estimations from the expected optimal portfolio.

In the present paper we study uncertainty of identification of efficient frontier of optimal portfolios in mean-variance setting. For a large class of distributions (elliptical distributions) we suggest a different approach for efficient frontier identification which is based directly on observations and does not involve parameter estimations. More precisely, we suggest to use conditional value at risk (CVaR) optimization for the identification of efficient frontier. In order to do it we prove that for elliptical distributions mean-variance efficient frontier is identical to the trajectory of CVaR optimal portfolios. This allows to apply CVaR portfolio optimization technique developed by Rockafellar and Uryasev [17,18] to efficient frontier identification. We show by experiments that proposed approach is competitive with other approaches.

2   Basic Definitions and Notations. Problem Statement

Suppose we have N stocks on the stock market. Let $R = (R_1, R_2, \ldots, R_N)$ be a random vector of stock returns with some multivariate distribution. Denote by $\mu = (\mu_1, \mu_2, \ldots, \mu_N)$ the vector of means of R, $\mu_i = E(R_i)$, and by $\Sigma = (\sigma_{i,j})$ the covariance matrix of R, $\sigma_{i,j} = Cov(R_i, R_j)$. Portfolio is defined by the vector $w = (w_1, w_2, \ldots, w_N)$ of the proportions of the capital investment, i.e. w_i is the proportion of the capital invested in the stock i. Portfolio return is the random variable $R(w) = \sum_{i=1}^{N} w_i R_i$, with the mean $E(R(w)) = \sum_{i=1}^{N} w_i \mu_i = w\mu'$ and variance $\sigma^2(R(w)) = \sum \sum \sigma_{i,j} w_i w_j = w\Sigma w'$. In what follows we assume that $w_i \geq 0$, $i = 1, 2, \ldots, N$ (short sales are forbidden). We are interested in portfolios related with two-objectives optimization problem

$$
\begin{aligned}
E(R(w)) &= w\mu' \to \max, \\
\sigma(R(w)) &= w\Sigma w' \to \min, \\
s.t. \ \ w_i &\geq 0, \ \ \textstyle\sum_{i=1}^{N} w_i = 1
\end{aligned}
$$

Efficient (optimal) portfolio is a Pareto optimal solution of this problem, i.e. for efficient portfolio it is impossible to simultaneously improve both objectives: efficiency and risk. Efficient frontier is defined in the coordinates (σ, E) as the set of points $(\sigma(R(w)), E(R(w)))$ where w are efficient portfolios. It is known that efficient frontier is a concave curve which is bounded in the case $w_i \geq 0$,

$i = 1, 2, \ldots, N$. To find the efficient frontier one can solve a family of one-objective optimization problems ($\gamma \geq 0$):

$$w\mu' - \gamma w \Sigma w' \to \max,$$
$$s.t. \quad w_i \geq 0, \quad \sum_{i=1}^{N} w_i = 1,$$

where γ is usually called risk aversion coefficient. This is convex quadratic optimization problem and one can use existing effective algorithms to solve it. It can be proved that all portfolios from efficient frontier can be obtained as solutions of this optimization problem for different values of $\gamma \geq 0$.

However, in practice vector of means μ and covariance matrix Σ are not known and to identify the efficient frontier on can use estimates $\hat{\mu}$, $\hat{\Sigma}$ and solve associated one-objective problems. This "obvious" approach has some drawback. It results in estimation errors or bias of the obtained efficient frontier from the true efficient frontier [10]. It is interesting in this case to find a way for efficient frontier identification, which uses observed data directly and avoids estimations of μ, and Σ (data driven approach). In the present paper we propose to use for this purpose CVaR optimization technique initiated by Rockafellar and Uryasev [17,18].

Suppose that loss Y has the density distribution function, which we denote by $p(y)$. Conditional value at risk for the loss Y at the confidence level β is defined by (see more details in [19])

$$\beta - CVaR: \quad \phi_\beta(Y) = E(Y/Y \geq \alpha_\beta) = \frac{1}{(1 - \beta)} \int_{y \geq \alpha_\beta} p(y)dy$$

where α_β is the Value at Risk (VaR) associated with the loss Y:

$$\beta - VaR: \quad \alpha_\beta(Y) = \min\{\alpha : \Psi(\alpha) \geq \beta\}, \text{ where } \Psi(\alpha) = P(Y \leq \alpha)$$

By definition, CVaR is the expected value of loss under condition that loss is larger than α_β. In the case of CVaR portfolio optimization the loss is defined by $Y = -R(w)$ and optimization problem is $\beta - CVaR(-R(w)) \to \min$, $w_i \geq 0$, $\sum w_i = 1$.

3   Efficient Frontiers. Theoretical Results

In this Section we show that for elliptical distributions the mean-variance efficient frontier can be recovered by CVaR optimal portfolios. Elliptically contoured distributions (or simply elliptical distributions) are known to be useful in many applications and especially in finance [7]. Random vector X belong to the class of elliptically contoured distributions if its density function has the form [1]:

$$f(x; \mu, \Lambda) = |\Lambda|^{-\frac{1}{2}} g\{(x - \mu)'\Lambda^{-1}(x - \mu)\} \tag{1}$$

where $\Lambda = (\lambda_{i,j})_{i,j=1,2,\ldots,N}$ is positive definite symmetric matrix, $g(x) \geq 0$, and

$$\int_{-\infty}^{\infty} \cdots \int_{-\infty}^{\infty} g(y'y)dy_1 dy_2 \cdots dy_N = 1$$

This class is a natural generalization of the class of Gaussian distributions. Many properties of Gaussian distributions have analogs for elliptical distributions, but this class is much larger, in particular it includes distributions with heavy tails, such as multivariate Student distributions. For detailed investigation of elliptical distributions see [1,7]. It is known that if $E(X)$ exists then $E(X) = \mu$. One important property of elliptical distribution X is the connection between covariance matrix of the vector X and the matrix Λ. Namely, if covariance matrix exists one has

$$\sigma_{i,j} = Cov(X_i, X_j) = C \cdot \lambda_{i,j} \tag{2}$$

where C is some constant. In particular, for Gaussian distribution one has $Cov(X_i, X_j) = \lambda_{i,j}$. For multivariate Student distribution with ν degree of freedom $(\nu > 2)$ one has $\sigma_{i,j} = \nu/(\nu - 1)\lambda_{i,j}$.

Theorem 1. *Let random vector of returns R has elliptical distribution. Then any mean variance efficient portfolio is exactly CVaR optimal portfolio for an appropriate choice of the parameter β.*

Sketch of the Proof. The proof uses general properties of elliptical distributions [1], description of CVaR optimal portfolios from [17], and specific properties of CVaR risk measure [5].

This theoretical results gives a possibility to use CVaR optimization technique for identification of the efficient frontier from observations. We model observations of stock returns as a sample $R(t)$, $t = 1, 2, \ldots, T$ of the size T from distribution of R. Observed values of the stocks return R_i are denoted by $r_i(t)$. It is shown in [17] that calculation of $\beta - CVaR$ optimal portfolio can be reduced to the solution of the following LP problem, where all observations $r_i(t)$ are involved:

$$\alpha + \frac{1}{T(1 - \beta)} \sum_{t=1}^{T} u_t \to \min$$

subject to $(t = 1, 2, \ldots, T)$

$$x \in X, \; u_t \geq 0, \; x_1 r_1(t) + x_2 r_2(t) + \cdots + x_N r_N(t) + \alpha + u_t \geq 0$$

where u_t are dummy variables. This LP problem can be solved by any LP solver. It is possible to improve the quality (uncertainty) of obtained solution by an appropriate use of machine learning scenario generation technique [6].

4   Efficient Frontiers. Experimental Results

We test our approach by simulations using a real stock market data of daily stock returns from S&P100 index. The collected data contains returns from January 2014 to November 2018 for 96 stocks. From these data we estimated distribution parameters (vector of means and covariance matrix) and then use these parameter in the generator to generate samples of independent identically distributed random vectors from multivariate elliptical distributions. The experiments were conducted as follows.

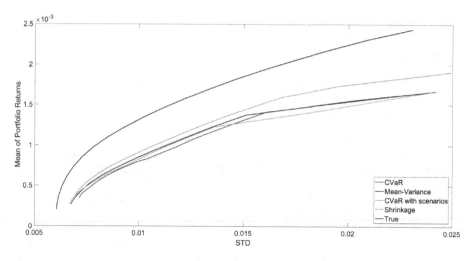

Fig. 1. Efficient frontiers obtained by different approaches, samples generated from Student t distribution

1. Select elliptical distribution for data generator
2. Generate sample of T observations
3. For each model, solve the corresponding optimization problems to construct the efficient frontier of portfolios
4. Measure out-of-sample characteristics of obtained portfolios using true parameters of the underlying distribution
5. Repeat steps 2–4 100 times to compute more stable estimation of out-of-sample characteristics of selected portfolios
6. Construct efficient frontier

We compare 4 different approaches for identification of efficient frontier: mean-variance (standard estimation of means and covariance matrix), shrinkage estimator for covariance matrix from [13], CVaR optimization, CVaR optimization with generated scenarios [6]. The results show that CVaR approach competitive with other approaches especially in the case of heavy tail distributions. This is illustrated by the Fig. 1, where efficient frontiers obtained by different approaches are presented.

5   Concluding Remarks

A new data driven approach for identification of mean-variance efficient frontier of investment portfolios is proposed. This approach is based on CVaR optimization. It is shown that proposed approach is competitive with other approaches. This first observation needs further investigation.

Acknowledgement. The article was prepared within the framework of the Basic Research Program at the National Research University Higher School of Economics (HSE) and is partly supported by RFFI grant 18-07-00524.

References

1. Anderson, T.W.: An Introduction to Multivariate Statistical Analysis, 3rd edn. Wiley-Interscience, New York (2003)
2. Best, M.J., Grauer, R.R.: On the sensitivity of mean-variance-efficient portfolios to changes in asset means: some analytical and computational results. Rev. Financial Stud. **4**(2), 315–342 (1991)
3. Chopra, V.K., Ziemba, W.T.: The effect of errors in means, variances, and covariances on optimal portfolio choice. J. Portfolio Manag. **19**, 6–11 (1993)
4. DeMiguel, V., Garlappi, L., Uppal, R.: Optimal versus naive diversification: how inefficient is the 1/n portfolio strategy? Rev. Financial Stud. **22**(5), 1915–1953 (2009)
5. Embrechts, P., McNeil, A., Straumann, D.: Correlation and dependence in risk management: properties and pitfalls. In: Dempster, M. (ed.) Risk Management Value at Risk and Beyond. Cambridge University Press, Cambridge (2002)
6. Guastaroba, G., Mansini, R., Speranza, M.G.: On the effectiveness of scenario generation techniques in single-period portfolio optimization. Eur. J. Oper. Res. **192**(2), 500–511 (2009)
7. Gupta, F.K., Varga, T., Bodnar, T.: Elliptically Contoured Models in Statistics and Portfolio Theory. Springer, Heidelberg (2013). https://doi.org/10.1007/978-1-4614-8154-6
8. Jorion, P.: Bayes-Stein estimation for portfolio analysis. J. Financ. Quantitative Anal. **21**(3), 279–292 (1986)
9. Kolm, P.N., Tütüncü, R., Fabozzi, F.: 60 Years of portfolio optimization: Practical Challenges Curr. Trends. Eur. J. Oper. Res. **234**(2), 356–371 (2014)
10. Kalyagin, V.A., Slashchinin, S.V.: Impact of error in parameter estimations on large scale portfolio optimization. In: Demetriou, I.C., Pardalos, P.M. (eds.) Approximation and Optimization. SOIA, vol. 145, pp. 151–184. Springer, Cham (2019). https://doi.org/10.1007/978-3-030-12767-1_9
11. Ledoit, O.: Portfolio Selection: Improved Covariance Matrix Estimation (1994)
12. Ledoit, O., Wolf, M.: Improved estimation of the covariance matrix of stock returns with an application to portfolio selection. J. Empirical Finance (2003)
13. Ledoit, O., Wolf, M.: Honey, I Shrunk the Sample Covariance Matrix (2004)
14. Lyuu, Y.-D.: Financial Engineering & Computation: Principles, Mathematics, Algorithms. Cambridge University Press, Cambridge (2002)
15. Markowitz, H.: Portfolio selection. J. Finance **7**, 77–91 (1952)
16. Raffinot, T.: Hierarchical clustering-based asset allocation. J. Portfolio Manag. **44**(2), 89–99 (2017)
17. Rockafellar, R.T., Uryasev, S.: Optimization of conditional value-at-risk. J. Risk **2**(3), 21–40 (2000)
18. Rockafellar, R.T., Uryasev, S.: Conditional value-at-risk for general loss distributions. J. Bank. Finance **26**(7), 1443–1471 (2002)
19. Sarykalin, S., Serraino, G., Uryasev, S.: Value-at-risk vs. conditional value-at-risk in risk management and optimization. In: INFORMS Tutorial (2008). https://doi.org/10.1287/educ.1080.0052. ISBN 978-1-877640-23-0

Learning to Configure Mathematical Programming Solvers by Mathematical Programming

Gabriele Iommazzo[1,2], Claudia D'Ambrosio[1], Antonio Frangioni[2(✉)], and Leo Liberti[1]

[1] CNRS LIX Ecole Polytechnique, 91128 Palaiseau, France
{giommazz,dambrosio,liberti}@lix.polytechnique.fr
[2] Dip. di Informatica, Università di Pisa, 56127 Pisa, Italy
frangio@di.unipi.it

Abstract. We discuss the issue of finding a good mathematical programming solver configuration for a particular instance of a given problem, and we propose a two-phase approach to solve it. In the first phase we learn the relationships between the instance, the configuration and the performance of the configured solver on the given instance. A specific difficulty of learning a good solver configuration is that parameter settings may not all be independent; this requires enforcing (hard) constraints, something that many widely used supervised learning methods cannot natively achieve. We tackle this issue in the second phase of our approach, where we use the learnt information to construct and solve an optimization problem having an explicit representation of the dependency/consistency constraints on the configuration parameter settings. We discuss computational results for two different instantiations of this approach on a unit commitment problem arising in the short-term planning of hydro valleys. We use logistic regression as the supervised learning methodology and consider CPLEX as the solver of interest.

Keywords: Mathematical programming · Optimization solver configuration · Hydro Unit Commitment

1 Introduction

Mathematical Programming (MP) is a formal language for describing optimization problems; once a problem is modelled by a MP formulation, an off-the-shelf solver can be used to solve it. Off-the-shelf solvers must be general enough to encompass a significant family of problems, and yet fast enough that sufficiently large-scale instances will be solved in reasonable time. By the usual trade-off

This paper has received funding from the European Union's Horizon 2020 research and innovation programme under the Marie Sklodowska-Curie grant agreement n. 764759 "MINOA".

I. S. Kotsireas and P. M. Pardalos (Eds.): LION 14 2020, LNCS 12096, pp. 377–389, 2020.
https://doi.org/10.1007/978-3-030-53552-0_34

between generality and efficiency, implementing a good solver is extremely hard. Today's most successful solvers, such as e.g. IBM-ILOG CPLEX [17], meet these specifications by actually embodying a corpus of different solution algorithms, each with their own (often very large) set of algorithmic options [18]. The default values for these options are usually chosen so that the solver will perform reasonably well on a large instance library, but for many problem classes performances can be improved by carefully tuning the solver configuration. Identifying good solver parameters for a given problem instance, however, is a difficult art, which requires a considerable experience in solver usage, and an in-depth hands-on knowledge of the application giving rise to the considered MP formulation.

Automatic configuration of algorithmic parameters is an area of active research, going back to the foundational work in [26]. Many approaches are based on sampling algorithmic parameter values, testing the solver performance, and performing local searches in order to find the parameters that most improve performance [10,14,16,22]. An algorithmic configuration method, derived from [16] and specifically targeted to MP solvers (including CPLEX), is described in [15]. All of these methods learn from a set of instances the best configuration for a similar set of instances; the configuration provided is not "per-instance" but common to all of the instances in the same problem class. This is different to the approach investigated in the present paper, which aims at providing a specific configuration for each given instance. The per-instance approach is necessary whenever the solver performance on instances of a problem class varies much, as in the case of our specific application. A more theoretical approach to choosing provably optimal parameter values based on black-box complexity, and limited to evolutionary algorithms, is given in [9] and references therein. Many artificial intelligence methodologies have been applied to this problem, see e.g. [23].

While the previously cited methodologies—and the one proposed in this paper—try to learn the best parameter values of an algorithm before launching it to solve a new instance, other approches learn on-the-fly, during its execution. For instance, the CPLEX Automatic Tuning Tool [17, Ch. 10], accompanying the corresponding solver, runs it on an instance (or a set thereof) several times, within a user-decided time limit, testing a specific parameter setting at each run, and saves the configuration providing the best algorithmic performance. Another on-the-fly methodology is presented, e.g., in [3], where one or more parameters of a tabu-search heuristic are adjusted, during its execution and in function of its behaviour, until a good configuration is learned.

In this work we present a new two-phase approach to automatic solver configuration. In the first phase, called the *Performance Map Learning Problem* (PMLP), we use a supervised Machine Learning (ML) methodology in order to learn the relationships between the features f of an instance ι, the solver configuration c and the performance quality p of the solver configured by c on the instance ι. Formally, p is defined as a function $p : (f, c) \rightarrow \mathbb{R}$ measuring the integrality gap achieved by the solver within a certain time limit. We propose two different variants of the PMLP. In the Performance-as-Output (PaO) one, we learn an approximation $\bar{p}(f, c)$ of the performance function. In the Performance-as-Input (PaI) one, we instead learn an approximation $\bar{C}$ of the map $C : (f, r) \to c$, where c is any configuration allowing to obtain a required performance level $r \in \mathbb{R}$ for the instance ι. In the second phase we use the learnt information (either $\bar{p}$ or $\bar{C}$) to define the *Configuration Space Search Problem* (CSSP), a constrained optimization problem, which is different for the two variants PaO and PaI. The input of the CSSP is an encoding of the performance map as well as the features of the instance to be solved. Its constraints encode the logical dependency and compatibility conditions over the configuration parameters: this ensures feasibility of the produced configuration. The objective function of the CSSP is a proxy to the performance map: optimizing it over the constraints yields a good configuration for the given instance.

Our approach is therefore capable of handling configuration spaces having arbitrarily complex logical conditions. This overcomes a weakness in previous learning-based approaches to the algorithm configuration problem, as acknowledged e.g. by [15,16]. To see how this weakness might adversely impact a solver configuration methodology, consider the following naive approach: learn the map C using a supervised ML method, then ask the trained method to output $C(f, 1)$ (1 being the best possible performance) for a new, unseen instance encoded by f. Unfortunately, this approach would fail over most off-the-shelf supervised ML methodologies, which are unable to reliably enforce dependency and compatibility constraints on the output configuration. Some attempts have been made to overcome this issue. For instance, the authors of ParamILS [16], that performs local searches in configuration space, declare that their algorithm supports the encoding of dependence/compatibility constraints on feasible parameters configurations. Unfortunately, we were unable to find the precise details of this encoding. Other approaches, used in learning-based optimization try to directly integrate a constrained optimization problem in a neural network, embedding it into the gradient computations of the back-propagation pass [11,29] or into an individual layer [2]. However, they are not generalizable to any ML algorithm and/or MP. We make one last introductory remark about the parameter search space: obviously, our approach can help configure any subset of solver parameters; in order to reduce the time spent in constructing the training set, a judicious choice would consider a reasonably small subset of parameters that are thought to have a definite impact on the problem at hand.

The rest of this paper is organized as follows. In Sect. 2 we formally introduce the notation and the main ingredients of our approach. In Sect. 3 we discuss both variants of the PMLP, and in Sect. 4 we discuss the corresponding CSSP. Finally, in Sect. 5 we report computational experiments and we draw some conclusions.

2   Notation and Preliminaries

2.1   The Training Set

Let C be the set of valid solver configurations. We assume for simplicity that $C \subseteq \{0, 1\}^s$, although extension to integer and continuous numerical parameters is clearly possible. Since every subset of the unitary hypercube can be described by means of a polytope [25, Cor. 1], we assume that its representation as a set of linear inequalities in binary variables, say

$$C = \{ c \in \{0, 1\}^s \mid Ac \le d \} \tag{1}$$

is known. In practice, deriving A and d from the logical conditions on the parameters can be assumed to be easy.

Let $\mathscr{I}$ be an optimization problem consisting of an infinite number of instances. In order to be able to use a ML approach, we have to encode each instance $\iota \in \mathscr{I}$ by a feature vector $f_\iota \in \mathbb{R}^t$ for some fixed $t \in \mathbb{N}$. This is surely possible at least by restricting ourselves to some subset of $\mathscr{I}' \subseteq \mathscr{I}$ (say, instances with appropriately bounded size). We also assume availability of a finite subset $I \subset \mathscr{I}'$ of instances and let $F = \{ f_\iota \mid \iota \in I \}$ be their feature encodings. We remark that F must be representative of I. Since feature extraction is an intensively studied field, we do not dwell on the specifics here. We also remark that ML methodologies are known to perform well on training sets that are not "overly general" [13, Ch. 5.3]: thus, we assume that I is a set of instances belonging to the same problem, or at least to different variants of a single problem.

In practice, in this paper we focus on a unit commitment problem arising in the energy industry, which is solved hundreds of times per day. The instances all have the same size; the constraints do not vary overmuch; the features are the objective function coefficients. Notwithstanding, our approach is general: the "problem structure" is encoded in the set of features (extracted from the instances), which are certainly class-specific, but need not be size-specific (one can e.g. use dimensionality reduction techniques to achieve feature vectors of the same size even for instances of varying size).

We can then, in principle, compute $p(f, c)$ on each feature vector $f \in F$ with every configuration $c \in C$ by calling the solver configured with c on the instance ι, in order to exploit $(F \times C, p(F, C))$ as a training set for learning estimates $\bar{p}$ or $\bar{C}$ as described above. Hopefully, then, these can be used to generalize our approach to instances outside I, with known encoding and that are in some way similar (in size or otherwise) to those in I.

2.2   Logistic Regression

Logistic Regression (LR) is a supervised ML methodology devised for binary classification of vectors [8].

Let $\mathcal{X} = (\mathcal{X}_1, \ldots, \mathcal{X}_m)$ be a vector of random variables, and let $\mathcal{Y}$ be a Bernoulli distributed random variable depending on $\mathcal{X}$. Following [20], and denoting $\mathsf{P}(\mathcal{X} = x)$ by $\mathsf{P}(x)$ and $\mathsf{P}(\mathcal{Y} = y)$ by $\mathsf{P}(y)$, we have

$$\mathsf{P}(1|x) = \frac{\mathsf{P}(x|1)\mathsf{P}(1)}{\mathsf{P}(x)} = \frac{\mathsf{P}(x|1)\mathsf{P}(1)}{\mathsf{P}(x|1)\mathsf{P}(1) + \mathsf{P}(x|0)\mathsf{P}(0)}$$

$$= \frac{1}{1 + \frac{\mathsf{P}(x|0)\mathsf{P}(0)}{\mathsf{P}(x|1)\mathsf{P}(1)}} = \frac{1}{1 + e^{-z}} = \sigma(z), \tag{2}$$

$$\text{where } z = \ln \frac{\mathsf{P}(x|1)}{\mathsf{P}(x|0)} + \ln \frac{\mathsf{P}(1)}{\mathsf{P}(0)}. \tag{3}$$

We now assume that z depends linearly on x:

$$\exists w \in \mathbb{R}^m, b \in \mathbb{R} \quad z = wx + b. \tag{4}$$

In some cases w, b can be computed explicitly: for example, if we assume that the conditional probabilities $P(x|y)$ are multivariate Gaussians with means μ_y and identical covariance matrices Σ, and use the above expression for $P(1|x)$, we obtain

$$P(1|x) = \frac{1}{1 + e^{-(wx+b)}}$$

where $w = \Sigma^{-1}(\mu_1 - \mu_0)$ and $b = \frac{1}{2}(\mu_0 + \mu_1)^{\top}\Sigma^{-1}(\mu_0 - \mu_1) + \ln(P(1)/P(0))$. In general, however, explicit formulæ cannot always be given, and w, b must be computed from sampled data.

To simplify notation in this section, we let $\tau(x, w, b) \triangleq \frac{1}{1+e^{-(wx+b)}}$, simply denoted $\tau(x)$ when w, b are fixed. Since we only consider two class labels $\{0, 1\}$, we model the probability of $\mathcal{Y} = y$ conditional to $\mathcal{X} = x$ using the function $\tau(x)$ if $y = 1$ and $1 - \tau(x)$ if $y = 0$, i.e.

$$P(y \mid x) = \tau(x)^y (1 - \tau(x))^{1-y}, \tag{5}$$

which evaluates to $\tau(x)$ whenever $y = 1$ and $1 - \tau(x)$ whenever $y = 0$.

We consider a training set $T = (X, Y)$ where $X = (x^i \in \mathbb{R}^m \mid i \leq n)$ and $Y = (y^i \in \{0, 1\} \mid i \leq n)$ consist of independent and identically distributed samples. We then use the Maximum Likelihood Estimation methodology [13] to find the optimal values of the parameters w and b. To this end, we define the likelihood function

$$L_T(w, b) = \prod_{i \leq n} P(y^i \mid x^i) = \prod_{i \leq n} \tau(x^i, w, b)^{y^i}(1 - \tau(x^i, w, b))^{1-y^i}. \tag{6}$$

We want to maximize $L_T(w, b)$. Since the logarithmic function is monotone in its argument, maximizing $\ln(L_T(w, b))$ yields the same optima (w^*, b^*) as maximizing $L_T(w, b)$. The training problem of LR is therefore:

$$\max_{w, b} \sum_{i \leq n} \left[y^i \ln \left(\frac{1}{1+e^{-(wx^i+b)}} \right) + (1 - y^i) \ln \left(1 - \frac{1}{1+e^{-(wx^i+b)}} \right) \right] \tag{7}$$

We recall that the functions

$$\psi_1(z) = \ln \left(\frac{1}{1+e^{-z}} \right) \quad \text{and} \quad \psi_2(z) = \ln \left(1 - \frac{1}{1+e^{-z}} \right)$$

are concave [6, Ex. 3.49(a)]. Since $0 \leq y^i \leq 1$ for each $i \leq n$, (7) maximizes the sum of convex combinations of concave functions, so it is a convex optimization problem which can be solved efficiently.

Once trained, the LR maps input vectors $x \in \mathbb{R}^m$ to an output scalar $y \in [0, 1]$: in this sense, LR approximates a binary scalar; binary values can of course be retrieved by rounding, if necessary. We denote this by $y = \mathsf{LR}(x, w^*, b^*)$.

2.3 The Performance Function

Since any LR output must be in $[0, 1]$ by definition, the performance data set $p(F, C)$ must be scaled to $[0, 1]$. We first measured, on different instance and solver configuration and within a given time limit, the CPLEX integrality gap, which is defined as

$$\frac{|\text{best integer sol.value} - \text{best relaxation value}|}{1e-10 + |\text{best integer sol.value}|}$$

in [17], pg. 263. Unfortunately, CPLEX performance data sometimes include very large values which stand for "infinity" (denoted ∞ below), meaning that CPLEX may not find feasible solutions or valid bounds at every run within the allotted CPU time. Instead of scaling in the presence of these outliers, we employed the following algorithm to obtain our performance function:

1. fix a constant $\gamma > 0$;
2. let $\hat{p} = \max(p(F, C) \setminus \{\infty\})$;
3. for each $v \in p(F, C)$ if $v > \hat{p}$ let $v = \hat{p} + \gamma$;
4. let $\rho = (\rho_1, \ldots, \rho_n)$ be a ranking of $p(F, C) = (v_1, \ldots, v_n)$ so that $\rho_1 \geq \rho_2 \geq \cdots \geq \rho_n$ and ρ_1 ranks the best performance value (equal values in $p(F, C)$ are assigned the same rank);
5. scale ρ to $[0, 1]$ so that ρ_1 is mapped to 1 and ρ_n to 0.

The choice of LR for this work is motivated by the fact that: (a) the parameters chosen for automatic configuration are all binary, and LR is a good method for estimating binary values; (b) the performance function has range $[0, 1]$. In general, LR can be replaced by other ML methodologies: this changes the technical details of the two phases, but it does not change the overall approach.

3 The PMLP

We now describe in details the two announced variants of the PMLP.

3.1 PMLP-PaO

In this variant, the output that we want to produce is an approximation $\bar{p}(f, c)$ of the performance function. Therefore, we interpret the symbols in Sect. 2.2 using the entities defined in Sect. 2.1. We note that the y variables in Eq. (7) are continuous, as noted at the end of Sect. 2.2. We have $X = F \times C \subseteq \mathbb{R}^{t+s}$ and $Y = \rho$; that is, $x = (f, c)$ in (7) encodes the concatenation of features and configurations, and $y = \rho$ is a vector of dimension 1 (i.e., a scalar) encoding the performance (see Fig. 1 (left)). We optimize (7) using some local Nonlinear Programming (NLP) algorithm able to deal with large-scale instances, e.g. stochastic gradient descent (SGD).

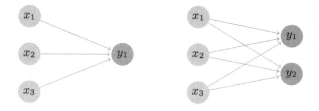

Fig. 1. Standard (left) and multiple (right) logistic regressions.

3.2 PMLP-PaI

In this variant, we want to output an approximation $\bar{C}(f, r)$ of the map that, given an instance and a desired performance in $[0, 1]$, returns the most appropriate configuration. Therefore, $X = F \times [0, 1]$ and $Y = C$ in the training set, i.e., $x = (f, r)$ is a pair (features, performance) and $y \in \mathbb{R}^s$ is a configuration. By the definition of y in PMLP-PaI, the LR requires multiple output nodes instead of a single one (see Fig. 1 (right)), since $s > 1$ in general. This can simply be achieved by considering s standard LRs sharing the same input node set.

Proposition 3.1. *The training problem of a multiple LR with k output nodes consists of k independent training problems for standard LRs, as in Eq. (7).*

Proof. A multiple LR on k outputs is equivalent to k standard LRs with training sets $T^1 = (X, Y^1), \ldots, T^k = (X, Y^k)$ where $Y^h = (y_h^1, \ldots, y_h^n)$ for all $h \le k$ and $t \le n$. Note that all these training sets share the same input vector set X. For each $h \le k$ we define Bernoulli random variables $\mathcal{Y}_h$. Then $P(\mathcal{Y}_h = 1 \mid \mathcal{X} = x)$ (for some $x \in \mathbb{R}^m$) is given by $\tau(x, w^h, b^h)$, where $w^h \in \mathbb{R}^m$ and $b^h \in \mathbb{R}$, for all $h \le k$. The training problem aims at maximizing the log-likelihood functions $\ln L_{T^h}(w^h, b^h)$ Eq. (7) of each output node $h \le k$, which yields the objective function $\max \sum_{h \le k} \ln L_{T^h}(w^h, b^h)$. Now we note that the optimum of $\sum_h \ln L_{T^h}(w^h, b^h)$ is achieved by optimizing each term separately, since each term depends on separate decision variables. $\square$

As anticipated, it is already rather hard to have the LR to produce a bona fide $y \in \{0, 1\}^s$, although this might be easily solved by rounding; what is much harder to obtain is that $y \in C$. Since $s > 1$ in general, the LR requires multiple output nodes instead of one f

4 The CSSP

The CSSP is the problem of computing a good configuration c^* for an input instance $\bar{f}$ and the learnt PMLP map. Clearly, its formulation depends on the output of the learning phase, that is, either $\bar{p}(f, c)$ or $\bar{C}(f, r)$. However, the solution of both the PaO and the PaI variant is guaranteed to be feasible w.r.t. all the dependence/compatibility constraints, i.e. $c^* \in \{c \mid c \in \{0, 1\}^s,\ Ac \le d\}$.

4.1   CSSP-PaO

In this case, the most obvious version of the CSSP would be to just maximize the expected performance over the set of feasible configurations, consistently with the dynamics of the trained LR. This yields the following (nonconvex) Mixed-Integer NonLinear Program (MINLP):

$$\max \left\{ (1 + e^{-(w^*(\bar{f},c)+b^*)})^{-1} \mid Ac \le d \ , \ c \in \{0,1\}^s \right\}, \tag{8}$$

for given $\bar{f}$. Note that Eq. (8) depends on the instance at hand through the input parameters $\bar{f}$, $w^* \in \mathbb{R}^{t+s}$ and $b^* \in \mathbb{R}$. As already remarked in Sect. 2.2, the objective function of (8) is log-concave, which means that

$$\max \left\{ \ln \left(\frac{1}{1 + e^{-(w^*(\bar{f},c)+b^*)}} \right) \ \middle| \ Ac \le d \ , \ c \in \{0,1\}^s \right\} \tag{9}$$

is a MINLP yielding the same optima as (8).

We identified a different interpretation for the CSSP objective, namely that of maximizing the likelihood that any new instance would be matched with a solver configuration and a performance value "as closely as possible" to the associations between (f,c) and $p(f,c)$ established during training. In other words, we maximize the likelihood given in (6) as a function of c and r, r being a specific performance value. In order to have the CSSP pick out a high performance, we add a term $+r$ to the objective:

$$\max_{c,r} r \ln \left(\frac{1}{1+e^{-(w^*(\bar{f},c)+b^*)}} \right) + (1-r) \left(1 - \ln \left(\frac{1}{1+e^{-(w^*(\bar{f},c)+b^*)}} \right) \right) + r \tag{10}$$

$$Ac \le d \ , \ c \in \{0,1\}^s \ , \ r \in [0,1] \tag{11}$$

Note that, while the performance measure r is not binary, it is in $[0,1]$ (where 1 corresponds to maximum (excellent) performance), which is compatible with LR.

Finally, we tested a third CSSP interpretation, where each alternative r and $1 - r$ is weighted by the corresponding conditional probability:

$$\max_{c,r} \left\{ r \left(\frac{1}{1+e^{-(w^*(\bar{f},c)+b^*)}} \right) + (1-r) \left(1 - \left(\frac{1}{1+e^{-(w^*(\bar{f},c)+b^*)}} \right) \right) \mid (11) \right\} \tag{12}$$

While (12) is non-convex, we were still able to (heuristically) solve it efficiently enough using `bonmin`.

4.2   CSSP-PaI

We now consider the multiple LR setting which correlates a given instance feature/performance vector (f,p) to a configuration c. Although p being part of the input means we need not restrict it to $[0,1]$, we chose to replace it with a ranked and scaled version r for better comparing with CSSP-PaO. The most direct interpretation of the CSSP in this case is the nonconvex MINLP

$$\max_{c,r} \left\{ r \mid c_j = \frac{1}{1+e^{-\langle (w^j)^*,(\bar{f},r)\rangle - b_j^*}} \quad \forall j \le s, \ (11) \right\}, \tag{13}$$

where $(w^j)^* \in \mathbb{R}^{t+1}$ is the weight vector of the j-th output. However, this interpretation does not satisfy the feasibility requirements on the c_j.

Proposition 4.1. (13) *is infeasible, even if the constraint* $r \in [0,1]$ *is relaxed.*

Proof. The constraint of the problem implies for all $j \le s$

$$\frac{1}{1+e^{-\langle (w^j)^*,(\bar{f},r)\rangle - b_j^*}} \in \{0,1\}.$$

However, for any given $(w^j)^*$, b_j^* and $\bar{f}$, there is no value of $r \in \mathbb{R}$ which makes the LHS either 0 or 1, hence the result.   □

Because of Proposition 4.1, we consider the same interpretation of the CSSP yielding the best objective function for the PaO case, i.e., the MINLP

$$\max_{c,r} \left\{ \sum_{j \le s} \left[c_j \left(\frac{1}{1+e^{-\langle (w^j)^*,(\bar{f},r)\rangle - b_j^*}} \right) + (1-c_j) \left(1 - \left(\frac{1}{1+e^{-\langle (w^j)^*,(\bar{f},r)\rangle - b_j^*}} \right) \right) \right] \mid (11) \right\}$$

which, through simple rearrangements, can be reformulated as

$$\max_{c,r} \left\{ \sum_{j \le s} \frac{\left(1 - e^{-\langle (w^j)^*,(\bar{f},r)\rangle - b_j^*} \right) c_j - 1}{1+e^{-\langle (w^j)^*,(\bar{f},r)\rangle - b_j^*}} \mid (11) \right\} \tag{14}$$

5   Computational Experiments

We tested both the PaO and PaI variants of our approach in the following general set-up:

- we consider 41 mixed-integer linear programming instances of the Hydro Unit Commitment (HUC) problem [5];
- the MP solver of choice is CPLEX [17];
- the supervised ML methodology used in the PMLP is LR [8,28];
- the CSSP is a MINLP which we heuristically solve—using the bonmin open-source solver [4]—to find good parameter values for CPLEX deployed on 41 instances of the HUC problem.

5.1   Technical Specifications

All experiments were carried out on a single virtual core of a 1.4GHz Intel Core i7 of a MacBook 2017 with 16 GB RAM running under macOS Mojave 10.14.6. Our implementations are based on Python 3.7 [27], AMPL 20200430 [12], and bonmin 1.8.6 [4]. We implemented LR as a Keras+TensorFlow [1,7] neural network with sigmoid activation and a stochastic gradient descent solver minimizing a loss function given by binary cross-entropy (a simple reformulation of the log-likelihood function (7)). The ranking function turning the performance data into ρ was supplied by scipy.stats.rankdata [19], and the scaling to $[0,1]$ by sklearn.preprocessing.minmax_scale [24].

5.2   The Algorithmic Framework

In this section we give a detailed description of the general algorithmic framework we employ.

1. *Feature extraction.* A set of $t = 54$ features was extracted from each of the 41 problem instances, so $|F| = 41$ and $F \subseteq \mathbb{R}^{54}$.
2. *Selection of configuration parameters.* We considered a subset of 11 CPLEX parameters (`fpheur, rinsheur, dive, probe, heuristicfreq, startalgo-rithm, subalgorithm` from `mip.strategy`; `crossover` from `barrier`; and `mircuts, flowcovers, pathcut` from `mip.cuts`), each with a varying number of discrete settings (between 2 and 4), which we combined so as to obtain 9216 configurations. We transformed each of these settings into binary form, obtaining $s = 27$ binary parameters, so $|C| = 27$. These parameters were chosen because, in our experience, they were reasonably likely to have an impact on the problem we considered. Therefore, our dataset is composed of $41 \times 9216 = 377856$ points.
3. *Obtaining the performance data.* For each $(f, c) \in F \times C$ in the dataset, we ran three times, with different random seeds, CPLEX configured by c over the instance described by f for 60 seconds, recording as $p(f, c)$ the second best integrality gap attained (the closer to zero, the better); this allows to mitigate the effect of performance variability issues, by which MIP solvers such as CPLEX have been shown to be affected [21]. We then form the performance value list $\mathbf{p} = (p(f, c) \mid (f, c) \in F \times C)$.
4. *Ranking and scaling.* We ranked $\mathbf{p}$, scaled it to $\hat{\mathbf{p}} \subset [0, 1]$, and let $\rho = 1 - \hat{\mathbf{p}}$ in order for the value 1 to mean "best performance".
5. *Separating in-sample and out-of-sample sets.* We randomly choose 11 out of the 41 instances as "out-of-sample", put them in a set F'', and let $F' = F \setminus F''$ be the "in-sample" set.
6. *Construction of the training sets.* For PMLP-PaO we let $X = F' \times C$ and $Y = \rho$, while for PMLP-PaI we let $X = F' \times \rho$ and $Y = C$.
7. We use the `sklearn.cluster.KMeans` k-means algorithm implementation to cluster the dataset into 5 clusters. We form a training set with 75% of the vectors from each cluster, a validation set with 20%, and a test set with the remaining 5%. By using clustering, we want to ensure that, even after the sampling, the actual distribution of the instances is preserved in all sets.
8. We implement a LR using a `keras.layers.Dense` complete bipartite pair of input/output layers (for PMLP-PaO with $f + s$ input nodes and 1 output node, for PMLP-PaI with $f + 1$ input nodes and s output nodes), with a sigmoid activation function in the output nodes. We train the LR using the corresponding training, validation and test sets, using the `keras` SGD optimizer optimizing the binary cross-entropy loss function (which corresponds to minimizing the negative of Eq. (7)). Then, for further use in the different CSSP (as described in Sect. 4), we save:
   - (w^*, b^*), with $w \in \mathbb{R}^{t+s}$ and $b \in \mathbb{R}$, for PMLP-PaO;
   - $((w^j)^*, b_j^*)$, $\forall j \leq s$, with $w^j \in \mathbb{R}^{t+1}$ and $b_j \in \mathbb{R}$, for PMLP-PaI.

9. For each out-of-sample instance feature vector $g \in F''$ we perform the following actions:
   a. we establish a link from Python to AMPL via `amplpy`;
   b. we solve the CSSP corresponding to the feature vector g with `bonmin`;
   c. we retrieve the optimal configuration c^*;
   d. we retrieve the stored performances $p(g, c^*)$ and $p(g, d)$, where d is the default CPLEX configuration;
   e. if $p(g, c^*) > p(g, d)$ we count an improvement;
   f. if $p(g, c^*) \geq p(g, d) - 0.001$ we count a non-worsening;
   g. we record the performance difference $|p(g, c^*) - p(g, d)|$.
10. We count the number of improvements im and non-worsenings nw over the number of successful `bonmin` runs on the CSSP instances; sometimes `bonmin` fails on account of the underlying NLP solver, which is why some lines of Table 1 consider a total of less than 11 instances.
11. We repeat this process 10 times from Step 5., and report cumulative statistics of improvements, non-worsenings, performance differences, and CPU time.

Table 1. Computational results. Best results are marked in boldface.

Run	im		nw		pd		CPU	
	PaO	PaI	PaO	PaI	PaO	PaI	PaO	PaI
1	0/08	**5/09**	0/08	**7/09**	0.63	**0.30**	41.44	**30.55**
2	0/11	**4/11**	0/11	**6/11**	**0.42**	0.47	41.62	**29.28**
3	**4/09**	4/10	5/09	**8/10**	**0.08**	0.14	43.06	**33.37**
4	0/09	**5/10**	0/09	**8/10**	0.43	**0.12**	42.65	**35.28**
5	**3/10**	1/10	**7/10**	2/10	**0.08**	0.70	43.30	**31.69**
6	**8/09**	3/11	**8/09**	9/11	0.20	**0.18**	43.69	**28.98**
7	**5/10**	1/11	**8/10**	4/11	**0.05**	0.52	45.54	**30.05**
8	**7/08**	3/09	7/08	**8/09**	0.21	**0.02**	45.49	**31.28**
9	0/09	0/10	0/09	0/10	**0.40**	0.88	43.83	**31.88**
10	**8/10**	5/08	8/10	**7/08**	0.21	**0.10**	43.50	**33.88**
sum	**35/93**	34/99	43/93	**59/99**	**2.69**	3.40	434.12	**316.24**
mean	**0.38**	0.32	0.46	**0.60**	**0.26**	0.34	43.41	**31.62**
stdev	0.36	**0.20**	0.39	**0.30**	**0.18**	0.27	**1.30**	1.94

5.3   Results

We first conducted experiments on the simple PaO interpretation (9) of the CSSP. However, this gave very poor results in practice. Equation (11), instead, gave better computational results than those obtained optimizing (9), although each CSSP instance took considerably more time to solve w.r.t. (9) and (12). The PaO formulation (12) is the one which gave the best results and is therefore the only one considered in Table 1. As for the PaI variant, Table 1 shows the results of formulation (14). In the table below, we report improvements im, non-worsenings nw, performance differences pd, and CPU times. We also report cumulative statistics (sum, mean, standard deviation) for the 10 runs of the algorithmic framework in Sect. 5.2 for the PaO and PaI variants. We remark that the "by-run" comparison is only meant for presentation, as the out-of-sample instances involved in each run of PaO and PaI differ.

The results show that the PaO and PaI variants are comparable. PaO improves more, but also worsens more. PaI improves slightly less, but it has three considerable advantages w.r.t. PaO: (i) it does not worsen results more than 60% of the times, which means it can be recommended for usage w.r.t. the default CPLEX configuration; (ii) it is more reliable in terms of standard deviation of improvements and non-worsening; (iii) it is faster.

5.4 Conclusions

We presented a general two-phase framework for learning good mathematical programming solver configurations, subject to logical constraints, using a performance function estimated from data. We proposed two significantly different variants of the methodology, both using Logistic Regression as the Machine Learning technique of choice, but using different configurations for the inputs and outputs of the LR. Tested on a problem arising in scheduling of hydro-electric generators, both variants showed promise, although the PaI one appeared to be preferable for several reasons. We remark that these encouraging results were obtained with a relatively small number of instances.

In future works, we are going to investigate our general framework using different Machine Learning methodologies, such as (deep) Neural Networks and Support Vector Machine/Regression. Each ML technique requires a different definition of both the PMLP and the CSSP (for each of the two PaO and PaI variants); hence, the exploration of the vast landscape of possible versions will offer a vast choice of the trade-offs between computational cost and effectiveness of the obtained configuration, hopefully finally leading to versions that may become actually useful for day-to-day use of MP solvers.

References

1. Abadi, M., et al.: TensorFlow: large-scale machine learning on heterogeneous systems (2015). http://tensorflow.org/
2. Amos, B., Kolter, Z.: OptNet: differentiable optimization as a layer in neural networks. In: Precup, D., Teh, Y.W. (eds.) Proceedings of the 34th International Conference on Machine Learning. Proceedings of Machine Learning Research, 06–11 August 2017, vol. 70, pp. 136–145. PMLR (2017)
3. Battiti, R., Brunato, M.: Reactive search: machine learning for memory-based heuristics. In: Handbook of Approximation Algorithms and Metaheuristics (2007)
4. Bonami, P., Lee, J.: BONMIN user's manual. Technical report, IBM Corporation (2007)

5. Borghetti, A., D'Ambrosio, C., Lodi, A., Martello, S.: A MILP approach for short-term hydro scheduling and unit commitment with head-dependent reservoir. IEEE Trans. Power Syst. **23**(3), 1115–1124 (2008)
6. Boyd, S., Vandenberghe, L.: Convex Optimization. Cambridge University Press, Cambridge (2004)
7. Chollet, F., et al.: Keras (2015). https://keras.io
8. Cox, D.: The regression analysis of binary sequences. J. Roy. Stat. Soc. Ser. B **3**(XX), 215–242 (1958)
9. Doerr, B., Doerr, C., Yang, J.: Optimal parameter choices via precise black-box analysis. In: 2016 Proceedings of the Genetic and Evolutionary Computation Conference, GECCO 2016, pp. 1123–1130. ACM, New York (2016)
10. Eggensperger, K., Lindauer, M., Hutter, F.: Pitfalls and best practices in algorithm configuration. J. Artif. Intell. Res. **64**, 861–893 (2019)
11. Ferber, A., Wilder, B., Dilkina, B., Tambe, M.: MIPaaL: mixed integer program as a layer. CoRR abs/1907.05912 (2019). http://arxiv.org/abs/1907.05912
12. Fourer, R., Gay, D.: The AMPL Book. Duxbury Press, Pacific Grove (2002)
13. Goodfellow, I., Bengio, Y., Courville, A.: Deep Learning. MIT Press, Cambridge (2016)
14. Hoos, H.: Programming by optimization. Commun. ACM **55**(2), 70–80 (2012)
15. Hutter, F., Hoos, H.H., Leyton-Brown, K.: Automated configuration of mixed integer programming solvers. In: Lodi, A., Milano, M., Toth, P. (eds.) CPAIOR 2010. LNCS, vol. 6140, pp. 186–202. Springer, Heidelberg (2010). https://doi.org/10.1007/978-3-642-13520-0_23
16. Hutter, F., Hoos, H., Leyton-Brown, K., Stützle, T.: ParamILS: an automatic algorithm configuration framework. J. Artif. Intell. Res. **36**, 267–306 (2009)
17. IBM: ILOG CPLEX 12.7 user's manual. IBM (2016)
18. IBM: ILOG CPLEX optimization studio parameters reference 12.8. IBM (2017)
19. Jones, E., Oliphant, T., Peterson, P.: SciPy: open source scientific tools for Python (2001). http://www.scipy.org/
20. Jordan, M.: Why the logistic function? A tutorial discussion on probabilities and neural networks. Technical Report. Computational Cognitive Science TR 9503, MIT (1995)
21. Lodi, A., Tramontani, A.: Performance variability in mixed-integer programming. Tutor. Oper. Res. **10**, 1–12 (2013)
22. López-Ibáñez, M., Dubois-Lacoste, J., Pérez-Cáceres, L., Birattari, M., Stützle, T.: The irace package: iterated racing for automatic algorithm configuration. Oper. Res. Perspect. **3**, 43–58 (2016)
23. Mısır, M., Sebag, M.: Alors: an algorithm recommender system. Artif. Intell. **244**, 291–314 (2017)
24. Pedregosa, F., et al.: Scikit-learn: machine learning in Python. J. Mach. Learn. Res. **12**, 2825–2830 (2011)
25. Poirion, P.L., Toubaline, S., D'Ambrosio, C., Liberti, L.: Algorithms and applications for a class of bilevel MILPs. Discrete Appl. Math. **272**, 75–89 (2020)
26. Rice, J.: The algorithm selection problem. Adv. Comput. **15**, 65–118 (1976)
27. Rossum, G.V., et al.: Python language reference, version 3. Python Software Foundation (2019)
28. Schumacher, M., Roßner, R., Vach, W.: Neural networks and logistic regression: part I. Comput. Stat. Data Anal. **21**, 661–682 (1996)
29. Wilder, B., Dilkina, B., Tambe, M.: Melding the data-decisions pipeline: decision-focused learning for combinatorial optimization. In: Proceedings of the AAAI Conference on Artificial Intelligence, vol. 33, pp. 1658–1665 (2019)

Convex Hulls in Solving Multiclass
Pattern Recognition Problem

D. N. Gainanov[1,2], P. F. Chernavin[2], V. A. Rasskazova[1(✉)],
and N. P. Chernavin[2]

[1] Moscow Aviation Institute, Moscow, Russia
damir.gainanov@gmail.com, varvara.rasskazova@mail.ru
[2] Ural Federal University, Ekaterinburg, Russia
chernavin.p.f@gmail.com

Abstract. The paper proposes an approach to solving multiclass pattern recognition problem in a geometric formulation based on convex hulls and convex separable sets (CS-sets). The advantage of the proposed method is the uniqueness of the resulting solution and the uniqueness of assigning each point of the source space to one of the classes. The approach also allows you to uniqelly filter the sourse data for the outliers in the data. Computational experiments using the developed approach were carried out using academic examples and test data from public libraries.

Keywords: Multiclass pattern recognition · Convex hull · Machine learning algorithm

Introduction

The paper deals with multiclass pattern recognition problem in a geometric formulation. Different approaches to solving such a problem could be found in [1,2,5,8,12,15,18,19,21]. Mathematical models for solving applied pattern recognition problems are considered in [1–4,12,13]. In this paper there is proposed a method for solving this problem which is based on the idea of separability of convex hulls of sets of training sample. The convex-hulls and other efficient linear approaches for solving similar problems were also proposed in [2,6,7,17]. To implement this method, two auxiliary problems are considered: the problem of selecting extreme points in a finite set of points in the space $\mathbb{R}^n$, and the problem of determining the distance from a given point to the convex hull of a finite set of points in the space $\mathbb{R}^n$ using tools of known software packages for solving mathematical programming problems. An efficiency and power of the proposed approach are demonstrated on classical Irises Fischer problem [16,22] as well as on several applied economical problems.

Let a set of n-dimensional vectors be given in the space $\mathbb{R}^n$

$$A = \{a_i = (a_{i1}, a_{i2}, ..., a_{in})\} : i = [1, N], \tag{1}$$

© Springer Nature Switzerland AG 2020
I. S. Kotsireas and P. M. Pardalos (Eds.): LION 14 2020, LNCS 12096, pp. 390–401, 2020.
https://doi.org/10.1007/978-3-030-53552-0_35

and let there also be given a separation of this set into m classes

$$A = A_1 \dot{\cup} A_2 \dot{\cup} \ldots \dot{\cup} A_m. \tag{2}$$

You need to construct a decision rule for assigning an arbitrary vector a_i to one of the m classes.

There are a number of methods [14,23] for solving this multiclass pattern recognition problem in a geometric formulation: linear classifiers, committee constructions, multiclass logistic regression, methods of support vectors, nearest neighbors, and potential functions. These methods are related to metric classification methods and are based on the ability to measure the distance between classified objects, or the distance between objects and hypersurfaces that separate classes in the feature space. This paper develops an approach related to convex hulls of subsets A_i, $i = [1, m]$, of the family A.

1 Multiclass Pattern Recognition Algorithm Based on Convex Hulls

The main idea of the proposed approach is as follows.

Let for the given family of points A, which is separated into m classes A_i where $i \in [1, m]$, corresponding convex hulls $conv\ A_i$ contain only points from classes A_i respectively. Then it is natural to assume that any point $x \in conv\ A_i$ represents a vector belonging to the class A_i. Below, we will extend this idea for the general case.

Definition 1. *The set A_i from (2), where $i \in [1, m]$, is named a convex separable set (CS-set, CSS), if the following holds*

$$conv\ A_i \cap A_j = \varnothing, \ \forall\, j \in [1, m] \setminus \{i\}. \tag{3}$$

If the family $A = \{A_1, \ldots, A_m\}$ contains a CSS A_{i_0}, then it is natural to assume that each point $x \in conv A_{i_0}$ belongs to the corresponding set A_{i_0}. In such a case the set A_{i_0} can be excluded from the further process of constructing the decision rule. In other words, the condition $x \in conv A_{i_0}$ must be checked first, and further process on the assigning point x to one of classes from training sample, must continue if and only if $x \notin conv A_{i_0}$.

An interesting case of families (1) is when you can specify a sequence $(i_1, i_2, \ldots, i_m)$, which is a permutation for the sequence $(1, 2, \ldots, m)$, and such that

$$\begin{cases} conv\ A_{i_1} \cap \bigcup\limits_{k=2}^{m} A_{i_k} = \varnothing, \\ conv\ A_{i_2} \cap \bigcup\limits_{k=3}^{m} A_{i_k} = \varnothing, \\ \ldots, \\ conv\ A_{i_{m-1}} \cap A_m = \varnothing. \end{cases} \tag{4}$$

The problem of constructing a decision rule for the family (1) with properties (4) will be called as CSS-solvable.

We denote by $class(x)$ the class number of $[1, m]$, to which the point x belongs. Thus, if $x \in A_i$, $i \in [1, m]$, then $class(x) = i$. For the point $x \notin A$, the problem of pattern recognition in the geometric formulation is to construct a decision rule for determining $class(x)$ for $x \in \mathbb{R}^n \backslash A$.

Let's consider the case of $m = 2$, i.e. $A = A_1 \dot\cup A_2$. Let's construct convex hulls $conv\ A_1$ and $conv\ A_2$. It is natural to assume that if $x \in conv\ A_1 \backslash conv\ A_2$, then $class(x) = 1$.

Similarly, if $x \in conv\ A_2 \backslash conv\ A_1$, then we assume that $class(x) = 2$. If $x \notin conv\ A_1 \cup conv\ A_2$, it is natural to assume that the point x belongs to such a class whose convex hull is located closer to the point x.

Let's denote by $\rho\left(x, conv\ A'\right)$ the distance from the point x to the convex hull of a finite set $A' \subset \mathbb{R}^n$. Then we have $class(x) = arg \min\limits_{i \in \{1,2\}} \{\rho(x, conv\ A_i)\}$.

Finally, let's consider the case of $x \in conv\ A_1 \cap conv\ A_2$.

Let's consider the following two sets.

$$
\begin{aligned}
A_1' &= A_1 \cap conv\ A_1 \cap conv\ A_2, \\
A_2' &= A_2 \cap conv\ A_1 \cap conv\ A_2.
\end{aligned}
\tag{5}
$$

Logically there are possible cases:

$$
\left.
\begin{aligned}
&1.\ A_1' = \varnothing, A_2' = \varnothing \\
&2.\ A_1' \neq \varnothing, A_2' = \varnothing \\
&3.\ A_1' = \varnothing, A_2' \neq \varnothing \\
&4.\ A_1' \neq \varnothing, A_2' \neq \varnothing
\end{aligned}
\right\}
\tag{*}
$$

Following the assumption mentioned above, i.e. $x \in conv\ A_1 \cap conv\ A_2$, we have:

$$
\begin{aligned}
&class(x) \text{ is not defined for the case 1,} \\
&class(x) = 1 \text{ for the case 2,} \\
&class(x) = 2 \text{ for the case 3.}
\end{aligned}
$$

Case 4 leads us to the following situation.

We have a family of two subsets $A' = \left\{A_1', A_2'\right\}$, which locate inside the set $conv\ A_1 \cap conv\ A_2$. You need to construct a decision rule for assigning the vector $x \in conv\ A_1 \cap conv\ A_2$ to one of the two classes A_1', A_2' and, respectively, A_1, A_2.

This problem corresponds to the original one, and therefore the proposed algorithm can be re-applied. Repeating the process we become to situation when for regular sets of the form (5) there holds $conv\ A_1'' \cap conv\ A_2'' = \varnothing$, and thus the process will be completed.

Proposition 1. *If for the sets A_1, A_2 we have $A_1 \cap A_2 = \varnothing$, then algorithm described above converges, i.e. for any point x from $A_1 \cup A_2$ it will lead to the case 1, 2 or 3 (*).*

Proof. Let's consider the following chain of pairs of sets
$A = A_1 \cup A_2, \quad C_1 = conv \ A_1, C_2 = conv \ A_2$:
$$A_1^{(1)} = A_1 \cap C_1 \cap C_2$$
$$A_2^{(1)} = A_2 \cap C_1 \cap C_2$$
$$A^{(1)} = A_1^{(1)} \cup A_2^{(1)}, \quad C_1^{(1)} = conv \ A_1^{(1)}, C_2^{(1)} = conv \ A_2^{(1)}$$
$$A_1^{(2)} = A_1^{(1)} \cap C_1^{(1)} \cap C_2^{(1)}$$
$$A_2^{(2)} = A_2^{(1)} \cap C_1^{(1)} \cap C_2^{(1)}$$
...
$$A^{(k-1)} = A_1^{(k-1)} \cup A_2^{(k-1)}, \quad C_1^{(k-1)} = conv \ A_1^{(k-1)}, C_2^{(k-1)} = conv \ A_2^{(k-1)}$$
$$A_1^{(k)} = A_1^{(k-1)} \cap C_1^{(k-1)} \cap C_2^{(k-1)}$$
$$A_2^{(k)} = A_2^{(k-1)} \cap C_1^{(k-1)} \cap C_2^{(k-1)}$$
$$A^{(k)} = A_1^{(k)} \cup A_2^{(k)}, \quad C_1^{(k)} = conv \ A_1^{(k)}, C_2^{(k)} = conv \ A_2^{(k)}$$
...

Let's show that at some step one of the conditions $A_1^{(k)} = \varnothing$ or $A_2^{(k)} = \varnothing$ will be hold, which means that the proposed algorithm converges.

Let's show that at any step we will have $\left| A_1^{(k+1)} \cup A_2^{(k+1)} \right| < \left| A_1^{(k)} \cup A_2^{(k)} \right|$.
Since $A_1 \cap A_2 = \varnothing$, then $A_1^{(k)} \cap A_2^{(k)} = \varnothing$.
On the other hand,

$$A_1^{(k+1)}, A_2^{(k+1)} \subseteq conv \ A_1^{(k)} \cap conv \ A_2^{(k)}. \tag{6}$$

Let's show that there is a point $x \in A_1^{(k)} \cup A_2^{(k)}$ such that $x \in conv \ A_1^{(k)} \cap conv \ A_2^{(k)}$. Let's assume the opposite:

$$\begin{cases} A_1^{(k)} \subseteq conv \ A_1^{(k)} \cap conv \ A_2^{(k)}, \\ A_2^{(k)} \subseteq conv \ A_1^{(k)} \cap conv \ A_2^{(k)}. \end{cases} \tag{7}$$

Therefore, we have

$$\begin{cases} conv \ A_1^{(k)} \subseteq conv \ A_1^{(k)} \cap conv \ A_2^{(k)}, \\ conv \ A_2^{(k)} \subseteq conv \ A_1^{(k)} \cap conv \ A_2^{(k)}. \end{cases} \tag{8}$$

On the other hand, by the definition of a convex hull, we get

$$\begin{cases} conv \ A_1^{(k)} \supseteq conv \ A_1^{(k)} \cap conv \ A_2^{(k)}, \\ conv \ A_2^{(k)} \supseteq conv \ A_1^{(k)} \cap conv \ A_2^{(k)}. \end{cases} \tag{9}$$

From (8) and (9) there follows that

$$conv \ A_1^{(k)} = conv \ A_2^{(k)}. \tag{10}$$

From (10) there follows that

$$\begin{cases} ext\left(conv \ A_1^{(k)} \right) = ext\left(conv \ A_2^{(k)} \right) \subseteq A_1^{(k)}, \\ ext\left(conv \ A_2^{(k)} \right) = ext\left(conv \ A_2^{(k)} \right) \subseteq A_2^{(k)}. \end{cases} \tag{11}$$

Hence, $A_1^{(k)} \cap A_2^{(k)} \neq \varnothing$, which contradicts the assumption above. Thus, the proposition is proved.

Let's consider the case $m > 2$.

Just as in the case of $m = 2$, the solution of the multiclass pattern recognition problem is reduced to solving a series of similar problems characterized by a sequential decreasing their dimensions. To characterize such a problem, we need to specify the following.

$$
\left.
\begin{aligned}
& X' \subset \mathbb{R}^n \text{ — subset of points for which the problem is solving,} \\
& A' = \left\{ A_i' \subseteq A : i \in J \subseteq [1, m] \right\} \text{ — the family of finite sets,} \\
& \qquad\qquad\qquad\qquad\qquad \text{for which the problem is solving,} \\
& C\left(A'\right) = \left\{ C_i' = conv A_i' : i \in J \subseteq [1, m] \right\} \text{ — the family of convex hulls} \\
& \qquad\qquad\qquad\qquad\qquad\qquad\qquad \text{of the sets of the family } A_i', \\
& J' \subseteq J \text{ — the set of classes, which take part in the problem.}
\end{aligned}
\right\} \quad (12)
$$

Let's denote by $\left\langle x', X', J', A', C\left(A'\right) \right\rangle$ the problem of determining whether a point $x' \in X'$ belongs to one of the classes $J' \subset J$, provided by training sample A' with a set of convex hulls $C\left(A'\right)$.

Further classification of the point $x' \in X'$ will be determined by the value

$$
M' = \left| \left\{ i \in J' : x' \in C_i' \right\} \right|
$$

and will break up into 3 cases: $M' = 0$, $M' = 1$ and $M' > 1$. Let rules of obtaining the problem $\left\langle x'', X'', J'', A'', C\left(A''\right), M'' \right\rangle$ in case $\left| M' \right| > 1$ are as following:

$$
\left.
\begin{aligned}
& x'' = x', \\
& J'' = \left\{ i \in J' : x'' \in C_i' \right\}, \\
& M' = \left| J'' \right|, \\
& X'' = \cap \left\{ C_i' : i \in J'' \right\}, \\
& A'' = \left\{ A_i'' = A_i' \cap X'' : i \in J'' \right\}, \\
& C\left(A''\right) = \left\{ C_i'' = conv A_i'' : i \in J'' \right\}.
\end{aligned}
\right\} \quad (13)
$$

Thus, the decision rule for a multiclass pattern recognition problem based on convex hulls can be represented as a hierarchical tree of basic problems of the form (12). And the root of this tree is the problem of the form $Z = \langle x, \mathbb{R}^n, J = [1, m], A, C(A) \rangle$.

Let's denote by $Z\left(J'\right)$ a problem of the form (13), which is obtained from the problem Z for the set $J' \subseteq J$ such that

$$\bigcap \left\{ C_i : i \in J' \right\} \neq \emptyset. \tag{14}$$

Let $\left\{ J'_1, \ldots, J'_{k_1} \right\}$ be the family of all subsets of $J' \subseteq J$ satisfying (14). Then for the problem Z of the first level, we get k_1 problems of the form $Z\left(J'_i\right)$, $i \in [1, k_1]$, of the second level. For each second-level problem of the form $Z\left(J'_i\right)$, a series of next-level problems of the form $Z\left(J'_i\right)\left(J''_i\right)$ will be obtained, and so on. A vertex in such a hierarchical tree becomes terminal if the subsample A involved in its formulation is included in no more than in one convex hulls involved in its formulation. Thus, to construct a decision rule, you need to construct a hierarchical graph of problems of the form (12) by constructing convex hulls for obtaining subsamples located in at least two convex hulls of the generating problem. To implement such an algorithm for constructing a decision rule, it is necessary to have effective algorithms for solving the following problems.

(1) Let a finite set $A \subseteq \mathbb{R}^n$ be given. You need to find all extreme points of its convex hull $ext\,(conv\,A)$.
    To detect either a point x is an extreme one for a finite set A, you could to solve a following problem LP1 from [20] (see also [24]).
    Let a_j denote an element of A.

$$\min x_j : \sum_{i \in I} x_i a_i = a_j, \sum_{i \in I} x_i = 1, x_i \geqslant 0 \,\forall\, i \in I,$$

    where I denotes the set $\{1, 2, \ldots, n\}$.
    It also should be mentioned that [20] provide an efficient algorithm to solving a problem on the detecting all extreme points of a finite set A by solving a sequence of problems of the form LP1.

(2) Let a point x and a set $ext\,(M)$ of extreme points of the polyhedron M be given. You need to determine whether the point x belongs to the polyhedron M, i.e. is it true that $x \in conv\,exe\,(M)$?
    The LP 2 problem can be used to solve this problem.
    Let $x \in \mathbb{R}^n$ and $A = \{a_1, a_2, \ldots, a_m\} \subseteq \mathbb{R}^n$, and let you need to determine either a point x will belongs to $conv A$.
    Let's consider the following system.

$$\begin{cases} \sum\limits_{i=1}^{m} \alpha_i a_i = x, \\ \sum\limits_{i=1}^{m} \alpha_i = 1, \\ \alpha_i \geqslant 0, i \in [1, m]. \end{cases} \tag{15}$$

It's obvious that $x \in conv$A if and only if a system above is feasible. From the other hand, such a system could be transformed into linear program LP2 of the form:

$$\begin{cases} v + w \longrightarrow \min, \\ \sum_{i=1}^{m} \alpha_i a_i = x, \\ \sum_{i=1}^{m} \alpha_i + v - w = 1, \\ \alpha_i \geqslant 0, i \in [1, m], \\ v \geqslant 0, w \geqslant 0. \end{cases} \tag{16}$$

where v and w are correcting variables in case a system (15) is infeasible. So, a point x will belongs to $conv$A if and only if $g = 0$.

(3) Let a point b and a set $ext\,(M)$ of extreme points of the polyhedron M be given.
You need to find the shortest distance from the point x to M, i.e. $\rho\,(x, M) = \min\,\{\rho\,(x, y) : y \in M\}$.
The following quadratic programming problem can be used to solve this problem:

$$\begin{cases} \sum_{i=1}^{n} (x_i - b_i)^2 \longrightarrow \min, \\ \sum_{j=1}^{m} \alpha_j \cdot a_j = x, \\ \sum_{j=1}^{m} \alpha_i = 1, \\ \alpha_j \; 0, \; j \in [1, m]. \end{cases} \tag{17}$$

Then we get that the required shortest distance from the point b to the convex hull of a finite set A in the space $\mathbb{R}^n$ is equal to the following:

$$\rho\,(b, conv\,A) = \sqrt{\sum_{i=1}^{n} (x_i - b_i)^2}.$$

2   Application of the CSS Machine Learning Algorithm

Let's consider several applied problems, for which proposed CSS machine learning algorithm could be used. Such problems are the problem on the bank scoring [9], analysis of financial markets [10,11], medical diagnostics, non-destructive control, and search for reference clients for marketing activities in social networks.

Problem 1. A Classical Problem of Irises Fisher [16]
There is a training sample of 150 objects in the space $\mathbb{R}^n$, which is divided into 3 classes: class A_1—Setosa, class A_2—Versicolor and class A_3—Virginica, and

each class contains 50 objects. It turns out that this well-known classical problem is CSS-solvable:

$$\begin{cases} conv\ A_1 \cap (A_2 \cup A_3) = \varnothing, \\ conv\ A_2 \cap A_3 = \varnothing. \end{cases}$$

In this case, the $class(x)$ decision rule looks as following:

$$class(x) = \begin{cases} 1, & x \in conv\ A_1, \\ 2, & x \in conv\ A_2, x \notin conv\ A_1 \\ 3, & x \in conv\ A_3, x \notin conv\ A_1. \end{cases}$$

$$arg \min_{i \in \{1,2,3\}} \rho\left(x, conv\ A_i\right), \quad x \notin conv\ A_1 \cup conv\ A_2 \cup conv\ A_3.$$

Problem 2

The proposed approach was used to develop a strategy for trading shares of the Bank of the Russian Federation[1,2]. 5 stock market indicators were selected as input parameters. Table 1 below provides a description of these parameters.

Table 1. Description of features

No.	Indicator	Values range
1	How many days with no break a moving average convergence divergence (MACD) becomes $>$ than 0 or $<$ than 0	Integer
2	Slow stochastic oscillator signal (SSO)	From 0 to +1
3	How many days with no break SSO gives a strong signal	Integer
4	Relative strength index signal (RSI),	From 0 to +1
5	How many days with no break RSI gives a signal	Integer

The following object classes were required to be recognized:

1. Class **Yes**—the set of positions on the trading strategy that were closed with a profit and the profit was greater than the maximum loss for the period of holding the position.

---

[1] When opening a position on the exchange, the position is constantly re-evaluated at current prices. Accordingly, the maximum loss on the position is the maximum amount of reduction in the value of the position relative to the value of the position when opening.

[2] Position hold period is the time from the moment of initial purchase or sale of a certain amount of financial instrument to the moment of reverse in relation to the first trading operation. For more information about the concept of opening and closing positions, see https://www.metatrader5.com/ru/mobile-trading/android/help/trade/positions_manage/open_positions (accessed 01.09.2019).

2. Class **No**—the set of positions on the trading strategy that were closed with a loss or the profit was less than the maximum loss for the period of holding the position.

Corresponding classes were formed based on real data obtained in the period from 26.02.10 until 03.10.19.

Description of cardinality of obtained sets, as well as the number of extreme points and belonging to convex hulls, are shown in the following Table 2.

Table 2. Description of obtained results

	Level 1	Level 2	Level 3
The set **Yes**	125	82	51
An extremal **Yes**	47	38	40
% An extremal	37.60 %	46.34%	78.43%
Yes in the convex hull of the **Yes** only	43	31	4
% **Yes** in the convex hull of the **Yes**	34.40%	37.80%	7.84%
Yes in the convex hull of the **No**	82	51	47
% **Yes** in the convex hull of the **No**	65.60%	62.20%	92.16%
The set **No**	416	290	179
An extremal **No**	83	84	68
% An extremal	19.95%	28.97%	37.99%
No in the convex hull of the **No** only	126	111	167
% **No** in the convex hull of the **No**	30.29%	38.28%	93.30%
No in the convex hull of the **Yes**	290	179	12
% **No** in the convex hull of the **Yes**	69.71%	61.72%	6.70%

From the table above you can conclude that a position needs to be open if and only if current status corresponds to the convex hull of the class **Yes** of the Level 1 or 2. And in other cases the risk is very high.

Problem 3

Convex hulls method was used for solving the problems on the bank scoring. Let's describe the most representative examples of favorable and unfavorable cases we had meet.

Favorable Case. There are 6 input parameters, and all of them are related with financial well-being of the borrower. Data from the first stage of calculations are shown in Table 3.

Further the procedure needs to be repeating for the next 9858 non-default and 242 default items. We will not explain all stages, but it should be mentioned that an acceptable solution was obtained with 7 iterations.

Table 3. Favorable case. First stage

Non-default	15000	Default	300
Including an extreme one	320	Including an extreme one	112
Outside of the Default convex hull	5142	Outside of the Non-default convex hull	58
% Definable in the unique way	34.28	% Definable in the unique way	19.33

Unfavorable Case. There are 5 input parameters (loan amount, loan term, borrower age, loan amount-to-age ratio, loan amount-to-loan term ratio). Data from the first stage of calculations are shown in Table 4.

Table 4. Unfavorable case. First stage

Non-default	62635	Default	1347
Including an extreme ones	612	Including an extreme ones	69
Outside of the Default convex hull	1148	Outside of the Non-default convex hull	29
% Definable in the unique way	1.83	% Definable in the unique way	2.15

In this case, the convex hulls of default and non-default sets are significantly intersected, which is due to the specifics of the problem (the share of default loans is 2.1%), as well as to small number of explanatory features. Further we plan to develop a method for solving similar problems (if one set is fully belongs to the another one and strongly blurred in it). In particular, we plan to consider a problem on the determining the balance between the percentage of points included in the convex hull and the size of this convex hull.

It is naturall that practical situations are much more complicated, but the sequence of actions described above allows you to get an efficient desicion rule.

Conclusion

The paper proposes an approach to solving multiclass pattern recognition problems in geometric formulation based on convex hulls and convex separable sets (CS-sets). Such problems often arise in the field of financial mathematics, for example, in problems of bank scoring and market analysis, as well as in various areas of diagnostics and forecasting. The main idea of the proposed approach is as follows. If for the given family of points A, which is separated into m classes A_i where $i \in [1, m]$, each convex hull $conv\ A_i$ contains only points from class A_i, then we suppose that any point $x \in conv\ A_i$ represents a vector belonging to the class A_i. In the paper is introduced key definition of convex separable set (CSS) for the family of A $= \{A_1, \ldots, A_m\}$ subsets of $\mathbb{R}^n$. Based on this definition another important for this approach definition of CSS-solvable family A $= \{A_1, \ldots, A_m\}$ is introduced. The advantage of the proposed method is the uniqueness of the resulting solution and the uniqueness of assigning each point of the source space to one of the classes. The approach also allows you to uniqelly filter the sourse data for the outliers in the data. Computational experiments using the developed approach were carried out using academic examples and test data from public libraries. An efficiency and power of the proposed approach are demonstrated on classical Irises Fischer problem [16] as well as on several applied ecomonical problems. It is shown that classical Irises Fischer problem [16] is CSS-solvable. Such a fact allows you to expect a high efficiency of the proposed method from the applied point of view.

References

1. Bel'skii, A.B., Choban, V.M.: Matematicheskoe modelirovanie i algoritmy raspoznavaniya celej na izobrazheniyakh, formiruemykh pricel'nymi sistemami letatel'nogo apparata. Trudy MAI (66) (2013)
2. Gainanov, D.N.: Kombinatornaya geometriya i grafy v analize nesovmestnykh sistem i raspoznavanii obrazov, 173 p. Nauka, Moscow (2014). (English: Combinatorial Geometry and Graphs in the Infeasible Systems Theory and Pattern Recognition)
3. Geil, D.: Teoriya linejnykh ekonomicheskikh modelei, 418 p. Izd-vo inostrannoi literatury, Moscow (1963). English: Linear Economics Models
4. Godovskii, A.A.: Chislennye modeli prognozirovaniya kontaktnykh zon v rezul'tate udarnogo vzaimodejstviya aviacionnykh konstrukcii s pregradoi pri avariinykh situaciyakh. Trudy MAI (107) (2019)
5. Danilenko, A.N.: Razrabotka metodov i algoritmov intellektual'noi podderzhki prinyatiya reshenii v sistemakh upravleniya kadrami. Trudy MAI (46) (2011)
6. Eremin, I.I., Astaf'ev, N.N.: Vvedenie v teoriyu lineinogo i vypuklogo programmirovaniya. 191 p. Nauka, Moscow (1970). English: Introduction to the Linear and Convex Programming
7. Eremin, I.I.: Teoriya lineinoi optimizacii, 312 p. Izd-vo "Ekaterinburg", Ekaterinburg (1999). English: Linear Optimization Theory
8. Zhuravel', A.A., Troshko, N., E'dzhubov, L.G.: Ispol'zovanie algoritma obobshhennogo portreta dlya opoznavaniya obrazov v sudebnom pocherkovedenii. Pravovaya kibernetika, pp. 212–227. Nauka, Moscow (1970)
9. Zakirov, R.G.: Prognozirovanie tekhnicheskogo sostoyaniya bortovogo radioelektronnogo oborudovaniya. Trudy MAI (85) (2015)
10. Mazurov, V.D.: Protivorechivye situacii modelirovaniya zakonomernostei. Vychislitel'nye sistemy, vypusk 88 (1981). English: The Committees Method in Optimization and Classification Problems. https://b-ok.xyz/book/2409123/007689
11. Mazurov, V.D.: Lineinaya optimizaciya i modelirovanie. UrGU, Sverdlovsk (1986). English: Linear Optimization and Modelling
12. Nikonov, O.I., Chernavin, F.P.: Postroenie reitingovykh grupp zaemshhikov fizicheskikh licz s primeneniem metoda komitetov. Den'gi i Kredit (11), 52–55 (2014)
13. Syrin, S.A., Tereshhenko, T.S., Shemyakov, A.O.: Analiz prognozov nauchnotekhnologicheskogo razvitiya Rossii, SShA, Kitaya i Evropeiskogo Soyuza kak liderov mirovoi raketnokosmicheskoi promyshlennosti. Trudy MAI (82) (2015)
14. Flakh, P.: Mashinnoe obuchenie. Nauka i iskusstvo postroeniya algoritmov, kotorye izvlekayut znaniya iz dannykh. DMK Press (2015). English: Machine Learning. The Science and Art on Algorithms Construction, which Derive Knowledges from Data

15. Chernavin, N.P., Chernavin, F.P.: Primenenie metoda komitetov k prognozirovaniyu dvizheniya fondovykh indeksov. In: NAUKA MOLODYKH sbornik materialov mezhdunarodnoj nauchnoj konferencii, pp. 307–320. Izd-vo OOO "Rusal'yans Sova", Moscow (2015)

16. Chernavin, N.P., Chernavin, F.P.: Primenenie metoda komitetov v tekhnicheskom analize instrumentov finansovykh rynkov. In: Sovremennye nauchnye issledovaniya v sfere ekonomiki: sbornik rezul'tatov nauchnykh issledovanii, pp. 1052–1062. Izd-vo MCITO, Kirov (2018)

17. Chernikov, S.I.: Linejnye neravenstva, 191 p. Nauka, Moscow (1970). English: Linear Inequalities Theory

18. Fernandes-Francos, D., Fontela-Romero, J., Alonso-Betanzos, A.: One-class classification algorithm based on convex hull. Department of Computer Science, University of a Coruna, Spain (2016)

19. Gainanov, D.N., Mladenović, N., Dmitriy, B.: Dichotomy algorithms in the multiclass problem of pattern recognition. In: Mladenović, N., Sifaleras, A., Kuzmanović, M. (eds.) Advances in Operational Research in the Balkans. SPBE, pp. 3–14. Springer, Cham (2020). https://doi.org/10.1007/978-3-030-21990-1_1

20. Pardalos, P.M., Li, Y., Hager, W.W.: Linear programming approaches to the convex hull problem in $\mathbb{R}^m$. Comput. Math. Appl. **29**(7), 23–29 (1995)

21. Liu, Z., Liu, J.G., Pan, C., Wang, G.: A novel geometric approach to binary classification based on scaled convex hulls. IEEE Trans. Neural Netw. **20**(7), 1215–1220 (2009)

22. http://archive.ics.uci.edu/ml/machine-learning-databases/iris/

23. Koel'o, L.P., Richard, V.: Postroenie sistem mashinnogo obucheniya na yazyke Python, 302 p. DMK Press, Moscow (2016). English: Building Machine Learning Systems with Python. https://ru.b-ok.cc/book/2692544/c6f386

24. Rosen, J.B., Xue, G.L., Phillips, A.T.: Efficient computation of convex hull in $\mathbb{R}^d$. In: Pardalos, P.M. (ed.) Advances in Optimization and Parallel Computing, pp. 267–292. North-Holland, Amsterdam (1992)

Least Correntropic Loss Regression

Mujahid N. Syed[(✉)] [ID]

Department of Systems Engineering,
King Fahd University of Petroleum and Minerals, Dhahran 31261, Saudi Arabia
smujahid@kfupm.edu.sa, snumujahid@gmail.com

Abstract. Robust linear regression is one of the well known areas in data analysis, and various methods to solve the robust regression problems are available in the literature. However, one of the key issues in these methods is the adaptability of the scale/tuning parameter to the data demographics. In this work, a correntropic loss based linear regression model is proposed. An approximation and simplification of the model reduces the model to the well known class of weighted linear regression models. Iterative solution methodology is proposed to solve the proposed formulation. Performance of the proposed approach is evaluated on simulated data. Results of the experiments highlight the usability and importance of the proposed approach.

Keywords: Robust linear regression · Correntropic loss · Weighted least square errors

1. Introduction

Least square error minimization is one of the earliest and commonly known form of linear regression. The method was coined in early 1800's by the individual seminal works of Legendre and Gauss. The survival of linear regression over the past 2 centuries can be attributed to its simplicity and applicability in multitude of pragmatic applications. The literal meaning of word 'regression' is 'return to a formal state'. The linear regression problem can be described as follows. Given Δ independent and identical (iid) records $(\mathbf{x}_r, y_r)$, for $r = 1, \ldots, \Delta$ collected from a system, where $\mathbf{x}_r \in \mathbb{R}^{1 \times D}$ corresponds to the system's input parameters or regressors, and $y_r \in \mathbb{R}$ corresponds to the system's output or response for $r = 1, \ldots, \Delta$; find β such that the following relation holds:

$$y_r = \mathbf{x}_r^T \boldsymbol{\beta}_\bullet + \beta_0 + \varepsilon_r \qquad \forall o = 1, \ldots, \Delta, \tag{1}$$

where $\boldsymbol{\beta} = [\beta_0, \boldsymbol{\beta}_\bullet]$ is unknown $(D+1) \times 1$ vector, and ε_r's are iid errors that are independent of $\mathbf{x}_r$ with $E(\varepsilon_r | \mathbf{x}_r) = 0$. Equation (1) can be written in the compact form as:

$$\mathbf{y} = \mathbf{X}\boldsymbol{\beta} + \boldsymbol{\varepsilon}, \tag{2}$$

© Springer Nature Switzerland AG 2020
I. S. Kotsireas and P. M. Pardalos (Eds.): LION 14 2020, LNCS 12096, pp. 402–413, 2020.
https://doi.org/10.1007/978-3-030-53552-0_36

where $\mathbf{X} \in \mathbb{R}^{\Delta \times (D+1)}$ and the rth row of $\mathbf{X}$ is defined as $[1, \mathbf{x}_r]$, and $\mathbf{y}, \boldsymbol{\varepsilon} \in \mathbb{R}^{\Delta}$ are vectors containing responses and errors respectively. The mathematical formulation of least square linear regression, a.k.a, Ordinary Least Square (OLS) regression can be modeled as follows:

$$minimize:$$
$$\| \mathbf{y} - \mathbf{X}\boldsymbol{\beta} \|_2, \tag{3}$$

where $\| \ \|_2$ is the second norm or the quadratic norm. Although Formulation (3) is nonlinear, it is a convex optimization problem. Furthermore, for a reasonable size of data and computing power, the formulation has a closed form solution. By reasonable size and computation power, we mean that the computer system is capable to inverse or handle $\mathbf{X}^T\mathbf{X}$. The closed form solution for Formulation (3) is an immediate result of the optimality conditions for unconstrained non-linear programs [1]. Since Formulation (3) is convex, the necessary and sufficient conditions for $\widehat{\boldsymbol{\beta}}$ to be optimal is, $\nabla f(\widehat{\boldsymbol{\beta}}) = \mathbf{0}$, where $f(\widehat{\boldsymbol{\beta}}) = \| \mathbf{y} - \mathbf{X}\widehat{\boldsymbol{\beta}} \|_2$. Upon further simplification, the optimality conditions can be stated as:

$$\mathbf{y} - \mathbf{X}\widehat{\boldsymbol{\beta}} = \mathbf{0} \quad or \quad \mathbf{X}\widehat{\boldsymbol{\beta}} = \mathbf{y} \tag{4}$$

If $\mathbf{X}^T\mathbf{X}$ is non-singular, then the solution of Eq. (4) can be written as:

$$\widehat{\boldsymbol{\beta}} = (\mathbf{X}^T\mathbf{X})^{-1}\mathbf{X}^T\mathbf{y} \tag{5}$$

With the development of numerical methods and computing power, $\mathbf{X}^T\mathbf{X}$ of $10,000 \times 10,000$ can be easily inverted in a single go (See [14]). Furthermore, there are many iterative methods to solve Eq. (4), which can extend the usage of OLS to big data. For example, in [5], OLS estimates of 10^{11} regressors are estimated.

2. Relevant Work

One of the critical drawbacks of OLS is its sensitivity to outliers (data points that do not fit in with the majority of the data points). Even a single outlier can have huge impact on the OLS estimate, $\widehat{\boldsymbol{\beta}}$. For example, see Figs. 1 & 2. In Fig. 1, the data is free from outliers. Whereas, in Fig. 2 10% of the observations are replaced with outliers.

To overcome the above limitation for OLS, many approaches have been developed by data scientists (see [7,10,13,20,22,24,29] and the reference there in). Indeed robust regression approaches have been well studied in many research areas originating from various disciplines over the past five decades. The robust approaches typically vary in degree of robustness, type of robustness, and computational complexity. It is out of scope of this work to review or list all the robust regression approaches. Interested readers are directed to see [16,31] in addition to the above references. From the literature, two major ideas for robust linear

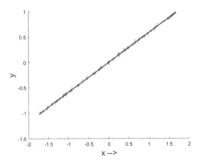

Fig. 1. OLS estimates without outliers

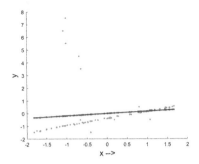

Fig. 2. OLS estimates with 10% outliers

regression can be grouped as: robust approach methods, and robust statistic methods.

Robust Approach: In robust approach methods, the key idea is to use the current OLS method with sampling mechanisms. For example, RANdom SAmple Consensus (RANSAC) is one of the robust approaches that withstood the test of time. In 1981, Fischler and Bolles proposed a generic framework called RANSAC that handles outliers in parameter estimation [4]. Usage of OLS with RANSAC strategy has since then became a popular approach to handle outliers in linear regression. The wide applicability of RANSAC can be attributed to its simple and generic characteristics. Many extensions of RANSAC are also available in the literature [17].

Robust Statistic: In robust statistic methods, the key idea is to use replace the squared error measure with a measures that is insensitive to the outliers. Among myriad of robust methods, some of the well known robust statistic methods used in robust linear regression are: Huber's M-estimates [9,10], MM-estimates [13,30], Generalized M-estimates [2,8], R-estimates [11,15], S-estimates [19], GS-estimates [3,18], LMS-estimate [24], LTS estimates [21], REWLSE estimates [6], and regularized estimates [12,23].

In this work, an adaptive weighted linear regression method that is robust to outliers is proposed. The proposed method uses a robust measure called correntropic loss. Although, weighted methods are available in the literature, the adaptive nature of the weights proposed in this paper improves the quality of the estimates. The rest of the paper is organized as follows: Sect. 3 presents the proposed model, followed by the proposed solution methodology. A numerical study involving simulated data is illustrated in Sect. 4. Some discussion and concluding remarks are depicted in Sect. 5.

3. Methodology

In this section, a mathematical model that is robust and/or insensitive to outliers is presented. An iterative solution methodology for the proposed formulation is developed in the latter part of this section.

3.1 Proposed Model

The following model is proposed for linear regression:

$$minimize:$$

$$\sum_{r=1}^{\Delta} (1 - e^{-\frac{(y_r - \mathbf{x}_r \beta)^2}{2\sigma^2}}),\tag{6}$$

where $\sigma > 0$ is a scale parameter. The exponential objective function (also defined as correntropic loss) in Formulation (6) appears in many data analysis works including [25–28]. From the theory of optimality conditions for unconstrained non-linear programs [1], a local optimal solution to the above formulation should satisfy the following necessary condition:

$$\sum_{r=1}^{\Delta} e^{-\frac{(y_r - \mathbf{x}_r \beta)^2}{2\sigma^2}} \left(\frac{(y_r - \mathbf{x}_r \beta)(x_{rf})}{\sigma^2} \right) = 0 \qquad \forall f \in D \tag{7}$$

Let $\mathbf{w}(\beta) : \mathbb{R}^D \mapsto \mathbb{R}^\Delta$ be defined as $w_r(\beta) = e^{-\frac{(y_r - \mathbf{x}_r \beta)^2}{2\sigma^2}}$ for $r = 1, \dots, \Delta$. The above conditions can be recast as:

$$\mathbf{X}^T \mathbf{W}(\beta)(\mathbf{y} - \mathbf{X}\beta) = \mathbf{0} \tag{8}$$

where $\mathbf{W}(\beta)$ is a diagonal matrix containing $\mathbf{w}(\beta)$ as its diagonal.

3.2 Proposed Solution Approach

In order to find β that satisfies the above necessary conditions, an iterative procedure is proposed. The update rule for the procedure is described as follows:

$$\mathbf{X}^T \mathbf{W}(\beta_{old})(\mathbf{y} - \mathbf{X}\beta_{new}) = \mathbf{0} \tag{9}$$

Since $\mathbf{W}(\beta)$ is a diagonal matrix with positive elements, and if $\mathbf{X}^T\mathbf{X}$ is non-singular, then we have the following closed form solution:

$$\beta_{new} = (\mathbf{X}^T\mathbf{W}(\beta_{old})\mathbf{X})^{-1}\mathbf{X}^T\mathbf{W}(\beta_{old})\mathbf{y}. \tag{10}$$

Notice that the above update rule is similar to the update rule obtained when solving the following weighted linear regression problem:

$$minimize:$$

$$\sum_{r=1}^{\Delta} \omega_r(y_r - \mathbf{x}_r\beta)^2 \tag{11}$$

where ω_r is the non-negative weights assigned to the rth record. The simplification depicted in Eq. (10) drastically reduces the complexity involved in obtaining the solution to Formulation (6). However, the key issue lies in obtaining β_{old} for any value of σ, such that the Hessian of the objective function in Formulation (6) is Positive Semi Definite (PSD) at β_{old}. Obtaining such β_{old} ensures that the necessary conditions stated in Eq. (8) are also sufficient for local optimality. When σ is very large, the Hessian is PSD everywhere. Thus, the main difficulty is to obtain such β_{old} for smaller values of σ. The iterative procedure depicted in Algorithm-1 obtains such β_{old} at each iteration.

Algorithm 1: Proposed Algorithm

Input : $\mathbf{X} \in \mathbb{R}^{\Delta \times (D+1)}, \mathbf{y} \in \mathbb{R}^{\Delta}, \beta^{OLS} \in \mathbb{R}^{D+1}$ and α
Output: $\beta \in \mathbb{R}^{D+1}$

1 Set $\sigma \leftarrow \sqrt{\max\limits_{1 \leq r \leq \Delta}\{(y_r - \mathbf{x}_r\beta^{OLS})^2\}}$;

2 Set $\beta_{new} \leftarrow \beta^{OLS}$;

3 **while** *termination criteria is False* **do**

4 $\quad$ Set $\beta_{old} \leftarrow \beta_{new}$;

5 $\quad$ Construct $\mathbf{W}(\beta_{old})$;

6 $\quad$ Get $\beta_{new} \leftarrow (\mathbf{X}^T\mathbf{W}(\beta_{old})\mathbf{X})^{-1}\mathbf{X}^T\mathbf{W}(\beta_{old})\mathbf{y}$;

7 $\quad$ Update $\sigma \leftarrow \alpha\sigma$;

8 **end**

9 Set $\beta \leftarrow \beta_{new}$

In Algorithm-1, β^{OLS} are the OLS estimates for the given data, $0 < \alpha < 1$ is a tuning parameter. The termination criterion used in the current work is $\sigma < \epsilon$ for some prespecified threshold ϵ. Upon termination at a low value of σ, the algorithm gives a local minimum of Formulation (6). In addition to that, the proposed approach to solve Formulation (6) involves solving Formulation (11) at each iteration (Line-6 in Algorithm-1). Thus, the proposed algorithm may give global minimum of Formulation (6) when $\alpha \longrightarrow 1$.

4.   Experimentation

In order to compare the performance of the proposed method, following existing weighted linear regression methods from the literature of robust linear regression are considered (see Table 1). The first column in Table 1 indicates the commonly known name of the method. The second column describes the mechanism to generate w_r's for each of the methods based on the value of error/residual e_r. Some of these methods are require a tuning parameter, and the third column displays the suggested parameter value.

Table 1. Some robust linear regression methods

Method	Description	Constant
'andrews'	$w_r = \begin{cases} sin(e_r)/e_r & \lvert e_r \rvert < pi, \\ 0 & o/w \end{cases}$	1.339
'bisquare'	$w_r = \begin{cases} (1 - e_r^2)^2 & \lvert e_r \rvert < 1, \\ 0 & o/w \end{cases}$	4.685
'cauchy'	$w_r = 1/(1 + e_0^2)$	2.385
'huber'	$w_r = 1/\max(1, \lvert e_r \rvert)$	1.345
'logistic'	$w_r = tanh(e_r)/r_r$	1.205
'talwar'	$w_r = \begin{cases} 1 & \lvert e_r \rvert < 1, \\ 0 & o/w \end{cases}$	2.795
'welsch'	$w_r = exp(-(e_r^2))$	2.985

Following sequence of experiments are conducted in this section. At first, simulated data containing no outliers is used for checking the validity of the proposed methodology. Next, simulated data containing outliers is used for demonstrating the capability of the proposed methodology to handle outliers. Finally, simulated data that contains outliers in a linear structure is considered.

4.1   Experiment-1

Setup: In this experiment, the data is simulated using the following equation:

$$y = \beta_1 x + \beta_0 + 0.1\varepsilon, \tag{12}$$

where the values of x are uniformly selected from 0 to 1, and ε is a Gaussian noise with zero mean and unit variance. This experiment consists of 30 scenarios, where each scenario contains 100 trials. At the beginning of each scenario, β_0 and β_1 are uniformly randomly selected from the following interval $[1, 10]$. The values for β_0 and β_1 will not be changed during the trials for a given scenario. However, these values will be updated at the start of every scenario. **Validation:** For each scenario, following measure is used for reporting the quality of the estimates:

$$\mu = \frac{1}{|T|} \sum_{t \in T} ||\beta^{act} - \beta_t||_2, \tag{13}$$

where T represents the set of all trials, β^{act} are the actual coefficients used for data generation in the scenario, and β_t represents the estimated coefficients. A two sided hypothesis sign test is utilized for concluding any differences between existing and the proposed method estimates. The null hypothesis (H_0) is that the mean of μ values of the existing method and proposed method are same. The alternate hypothesis (H_a) is that the proposed method's mean μ value is lower than the existing method's mean μ value. **Results:** The results of this experiment are displayed in Table 2. Column labeled Avg μ(Std μ) represents the average(standard deviation) μ value for the method over the 30 scenarios. Column labeled Avg Time(Std Time) represents the average(standard deviation) time in seconds used by the method per trial per scenario. Column labeled Ha contains either 0 or 1. A value of 1 in Ha implies that the sign test supports/favors the alternate hypothesis at 5% significance level. Similarly, a value of 0 in Ha indicates that, at 5% significance level, the test fails to reject the null hypothesis. For Logistic method, based on the sign test, at the 5% significance level, the test favors the alternate hypothesis.

4.2   Experiment-2

Setup: In this experiment, the data is simulated similar to Experiment-1. However, 10% of the response values are modified by updating the response values to $max\{y\} + 10$. **Validation:** The measure, null and alternate hypotheses are similar to Experiment-1. **Results:** The results of this experiment are displayed in Table 3. The columns have similar meaning as described in Experiment-1. For OLS, Logistic and Huber methods, based on the sign test, at the 5% significance level, the test favors the alternate hypothesis.

4.3   Experiment-3

Setup: In this experiment, the data is simulated similar to Experiment-2. However, the number of regressors in this experiment are 10, i.e., $\beta \in \mathbb{R}^{11}$. In addition to that, three cases are considered in this experiment. In Case-1 10% of the data are outliers, in Case-2 20% of the data are outliers, and in Case-3, 30% of the data are outliers. **Validation:** The measure, null and alternate hypotheses are similar to Experiment-1. **Results:** The results of this experiment are displayed in Table 4. The columns have similar meaning as described in Experiment-1. From the results, it can be concluded that as the percentage of outliers increase, the number of existing methods favoring the alternate hypothesis increase (at 5% significance level).

Table 2. Experiment-1 results

Method	Avg μ	Std μ	Avg Time	Std Time	H_a
'OLS'	0.051	0.0015	0	0	0
'proposed'	0.0511	0.0018	0.0001	0.0002	
'andrews'	0.0512	0.0017	0.0008	0.0029	0
'bisquare'	0.0512	0.0017	0.0006	0.0005	0
'cauchy'	0.0512	0.0017	0.0007	0.0002	0
'huber'	0.0511	0.0016	0.0005	0.0002	0
'logistic'	0.0513	0.0018	0.0008	0.0008	1
'talwar'	0.051	0.0015	0.0003	0.0001	0
'welsch'	0.0512	0.0017	0.0006	0.0001	0

Table 3. Experiment-2 results

Method	Avg μ	Std μ	Avg Time	Std Time	H_a
'OLS'	1.9897	0.3105	0	0	1
'proposed'	0.0515	0.0007	0.0069	0.0005	
'andrews'	0.0512	0.0006	0.0006	0.0011	0
'bisquare'	0.0512	0.0006	0.0006	0.0002	0
'cauchy'	0.0513	0.0006	0.0007	0.0001	0
'huber'	0.0575	0.0007	0.0008	0.0001	1
'logistic'	0.0588	0.0008	0.001	0.0003	1
'talwar'	0.051	0.0006	0.0003	0	0
'welsch'	0.0512	0.0006	0.0007	0.0001	0

4.4   Experiment-4

Setup: In this experiment, the data is simulated similar to Experiment-3. However, the outliers form a linear structure. Thus, the methods has to decide the right linear structure based on the majority of the points. **Validation:** The measure, null and alternate hypotheses are similar to Experiment-1. **Results:** The results of this experiment are displayed in Table 5. The columns have similar meaning as described in Experiment-1. From the results, it can be concluded that as the percentage of outliers increase, the number of existing methods favoring the alternate hypothesis increase (at 5% significance level).

Table 4. Experiment-3 results

Case-1: 10% Outliers

Method	Avg μ	Std μ	Avg Time	Std Time	H_a
'OLS'	10.055	0.9696	0	0.0002	1
'proposed'	0.0655	0.0016	0.0139	0.0036	
'andrews'	0.0634	0.0016	0.0014	0.0014	0
'bisquare'	0.0634	0.0016	0.0012	0.0004	0
'cauchy'	0.064	0.0016	0.0015	0.0003	0
'huber'	0.074	0.0015	0.0018	0.0004	1
'logistic'	0.077	0.0015	0.0022	0.0006	1
'talwar'	0.0623	0.0015	0.0006	0.0001	0
'welsch'	0.0636	0.0016	0.0013	0.0002	0

Case-2: 20% Outliers

Method	Avg μ	Std μ	Avg Time	Std Time	H_a
'OLS'	15.253	1.5065	0.0001	0.0006	1
'proposed'	0.0671	0.0022	0.0139	0.0033	
'andrews'	0.0645	0.002	0.0015	0.0013	0
'bisquare'	0.0645	0.002	0.0013	0.0004	0
'cauchy'	0.0651	0.0021	0.0017	0.0003	0
'huber'	0.1072	0.0192	0.0032	0.0011	1
'logistic'	0.1235	0.0294	0.004	0.0014	1
'talwar'	3.1556	0.8304	0.0008	0.0002	1
'welsch'	0.0647	0.002	0.0014	0.0002	0

Case-3: 30% Outliers

Method	Avg μ	Std μ	Avg Time	Std Time	H_a
'OLS'	29.669	20.613	0	0	1
'proposed'	0.0688	0.0017	0.0147	0.0033	
'andrews'	0.9489	0.7712	0.002	0.0014	1
'bisquare'	0.9488	0.7711	0.0017	0.0008	1
'cauchy'	1.5867	1.0927	0.0025	0.0011	1
'huber'	7.4753	5.3805	0.0055	0.0017	1
'logistic'	8.3627	6.1255	0.0086	0.005	1
'talwar'	16.225	12.591	0.0009	0.0004	1
'welsch'	0.9102	0.7206	0.0019	0.001	1

Table 5. Experiment-4 results

Case-1: 10% Outliers

Method	Avg μ	Std μ	Avg Time	Std Time	H_a
'OLS'	14.699	8.5397	0	0	1
'proposed'	0.0652	0.0015	0.0154	0.004	
'andrews'	0.0631	0.0015	0.0015	0.0016	0
'bisquare'	0.0631	0.0015	0.0013	0.0003	0
'cauchy'	0.0636	0.0015	0.0016	0.0003	0
'huber'	0.0658	0.002	0.0018	0.0004	0
'logistic'	0.0681	0.0021	0.0022	0.0005	1
'talwar'	0.062	0.0015	0.0007	0.0001	0
'welsch'	0.0632	0.0015	0.0014	0.0002	0

Case-2: 20% Outliers

Method	Avg μ	Std μ	Avg Time	Std Time	H_a
'OLS'	22.265	16.762	0	0	1
'proposed'	0.067	0.0022	0.0154	0.004	
'andrews'	0.0645	0.0021	0.0016	0.0013	0
'bisquare'	0.0645	0.0021	0.0014	0.0004	0
'cauchy'	0.0649	0.0021	0.0017	0.0003	0
'huber'	0.0887	0.017	0.003	0.001	1
'logistic'	0.1028	0.0306	0.0037	0.0013	1
'talwar'	0.4209	0.4141	0.0008	0.0002	1
'welsch'	0.0647	0.0021	0.0015	0.0003	0

Case-3: 30% Outliers

Method	Avg μ	Std μ	Avg Time	Std Time	H_a
'OLS'	30.865	18.579	0	0.0001	1
'proposed'	0.0682	0.0018	0.0153	0.0037	
'andrews'	1.0296	0.9179	0.002	0.0014	1
'bisquare'	1.0326	0.9182	0.0018	0.001	1
'cauchy'	1.8393	1.4973	0.0025	0.0011	1
'huber'	7.7281	5.1359	0.0055	0.0018	1
'logistic'	8.5068	5.6306	0.0096	0.0042	1
'talwar'	16.947	11.303	0.0009	0.0004	1
'welsch'	0.9615	0.7986	0.0019	0.0009	1

5. Conclusion

In this work, a formulation for robust linear regression related to the correntropic loss minimization is presented. The proposed formulation can be approximated as weighted OLS minimization problem. An iterative solution method for the weighted OLS problem, where the weights are adaptive, has been proposed and implemented. Numerical experiments on the simulated data are presented, that compares the proposed method head-to-head with some of the existing methods from the literature. Based on the numerical study, it can be highlighted that the adaptive nature of weights (or the scale parameter) is the key element in handling outliers.

References

1. Bazaraa, M.S., Sherali, H.D., Shetty, C.M.: Nonlinear Programming: Theory and Algorithms. Wiley, Hoboken (2013)
2. Coakley, C.W., Hettmansperger, T.P.: A bounded influence, high breakdown, efficient regression estimator. J. Am. Stat. Assoc. **88**(423), 872–880 (1993)
3. Croux, C., Rousseeuw, P.J., Hössjer, O.: Generalized s-estimators. J. Am. Stat. Assoc. **89**(428), 1271–1281 (1994)
4. Fischler, M.A., Bolles, R.C.: Random sample consensus: a paradigm for model fitting with applications to image analysis and automated cartography. Commun. ACM **24**(6), 381–395 (1981)
5. Frank, A., Fabregat-Traver, D., Bientinesi, P.: Large-scale linear regression: development of high-performance routines. Appl. Math. Comput. **275**, 411–421 (2016)
6. Gervini, D., Yohai, V.J., et al.: A class of robust and fully efficient regression estimators. Ann. Stat. **30**(2), 583–616 (2002)
7. Hampel, F.R., Ronchetti, E.M., Rousseeuw, P.J., Stahel, W.A.: Robust Statistics: The Approach Based on Influence Functions, vol. 196. Wiley, Hoboken (2011)
8. Handschin, E., Schweppe, F.C., Kohlas, J., Fiechter, A.: Bad data analysis for power system state estimation. IEEE Trans. Power Apparatus Syst. **94**(2), 329–337 (1975)
9. Huber, P.J.: Robust estimation of a location parameter. In: Kotz, S., Johnson, N.L. (eds.) Breakthroughs in Statistics, pp. 492–518. Springer, Heidelberg (1992). https://doi.org/10.1007/978-1-4612-4380-9_35
10. Huber, P.J.: Robust Statistics. Springer, Heidelberg (2011). https://doi.org/10.1007/978-3-642-04898-2_594
11. Jaeckel, L.A.: Estimating regression coefficients by minimizing the dispersion of the residuals. Ann. Math. Stat. 1449–1458 (1972)
12. Lee, Y., MacEachern, S.N., Jung, Y., et al.: Regularization of case-specific parameters for robustness and efficiency. Stat. Sci. **27**(3), 350–372 (2012)
13. Maronna, R.A., Martin, R.D., Yohai, V.J., Salibián-Barrera, M.: Robust Statistics: Theory and Methods (with R). Wiley, Hoboken (2019)
14. MatLab: Largest Matrix Size. https://www.mathworks.com/matlabcentral/. Accessed Sept 2019
15. Naranjo, J.D., Hettmansperger, T.: Bounded influence rank regression. J. Roy. Stat. Soc.: Ser. B (Methodol.) **56**(1), 209–220 (1994)
16. Papageorgiou, G., Bouboulis, P., Theodoridis, S.: Robust linear regression analysis—a greedy approach. IEEE Trans. Sig. Process. **63**(15), 3872–3887 (2015)
17. Raguram, R., Chum, O., Pollefeys, M., Matas, J., Frahm, J.M.: USAC: a universal framework for random sample consensus. IEEE Trans. Pattern Anal. Mach. Intell. **35**(8), 2022–2038 (2012)
18. Roelant, E., Van Aelst, S., Croux, C.: Multivariate generalized s-estimators. J. Multivariate Anal. **100**(5), 876–887 (2009)

19. Rousseeuw, P., Yohai, V.: Robust regression by means of S-estimators. In: Franke, J., Härdle, W., Martin, D. (eds.) Robust and Nonlinear Time Series Analysis, vol. 26, pp. 256–272. Springer, Heidelberg (1984). https://doi.org/10.1007/978-1-4615-7821-5_15

20. Rousseeuw, P.J.: Least median of squares regression. J. Am. Stat. Assoc. **79**(388), 871–880 (1984)

21. Rousseeuw, P.J.: Multivariate estimation with high breakdown point. Math. Stat. Appl. **8**(283–297), 37 (1985)

22. Rousseeuw, P.J., Hubert, M.: Robust statistics for outlier detection. Wiley Interdisc. Rev.: Data Mining Knowl. Discov. **1**(1), 73–79 (2011)

23. She, Y., Owen, A.B.: Outlier detection using nonconvex penalized regression. J. Am. Stat. Assoc. **106**(494), 626–639 (2011)

24. Siegel, A.F.: Robust regression using repeated medians. Biometrika **69**(1), 242–244 (1982)

25. Singh, A., Pokharel, R., Principe, J.: The C-loss function for pattern classification. Pattern Recogn. **47**(1), 441–453 (2014)

26. Syed, M., Pardalos, P., Principe, J.: Invexity of the minimum error entropy criterion. IEEE Sig. Process. Lett. **20**(12), 1159–1162 (2013)

27. Syed, M.N., Pardalos, P.M., Principe, J.C.: On the optimization properties of the correntropic loss function in data analysis. Optim. Lett. **8**(3), 823–839 (2013). https://doi.org/10.1007/s11590-013-0626-5

28. Syed, M.N., Principe, J.C., Pardalos, P.M.: Correntropy in data classification. In: Sorokin, A., Murphey, R., Thai, M., Pardalos, P. (eds.) Dynamics of Information Systems: Mathematical Foundations, pp. 81–117. Springer, Heidelberg (2012). https://doi.org/10.1007/978-1-4614-3906-6_5

29. Tukey, J.W.: Data analysis, computation and mathematics. Q. Appl. Math. **30**(1), 51–65 (1972)

30. Yohai, V.J., et al.: High breakdown-point and high efficiency robust estimates for regression. Ann. Stat. **15**(2), 642–656 (1987)

31. Yu, C., Yao, W.: Robust linear regression: a review and comparison. Commun. Stat.-Simul. Comput. **46**(8), 6261–6282 (2017)

Novelty Discovery with Kernel Minimum Enclosing Balls

Rafet Sifa[1($\boxtimes$)] and Christian Bauckhage[1,2]

[1] Fraunhofer Center for Machine Learning, Sankt Augustin, Germany
{rafet.sifa,christian.bauckhage}@iais.fraunhofer.de
[2] University of Bonn, Bonn, Germany

Abstract. We introduce the idea of utilizing ensembles of Kernel Minimum Enclosing Balls to detect novel datapoints. To this end, we propose a novelty scoring methodology that is based on combining outcomes of the corresponding characteristic functions of a set of fitted balls. We empirically evaluate our model by presenting experiments on synthetic as well as real world datasets.

1 Introduction

The notion of novelty discovery (or detection) [6] can be described as a one-class classification problem (a.k.a *data domain description* [11]) aiming to learn certain characteristics of the analyzed datasets to be able to separate novel datapoints. It finds many applications in numerous scientific and engineering areas such as fraudulent activity detection in financial applications or detecting rare events in medical monitoring [6]. Although reservoir computing based approaches [4] have been proposed to a variety of classification and regression problems, to the best of our knowledge, corresponding methods that are oriented to tackle one-class problems are scarce. The main contribution of this work is about utilizing Minimum Enclosing Balls [1] for novelty discovery. Minimum Enclosing Balls (MEBs) fall into the class of unsupervised representation learning methods that can be used to extract important characteristics about the considered datasets [1]. The main idea behind the MEBs is about determining the smallest ball encapsulating the *entire* dataset in the data- or feature space, which can be found by formulating the problem as an inequality constrained convex minimization problem with a dual allowing for invoking the kernel trick and this dual can be solved using dynamical processes from reservoir computing [1].

Our contribution is based on the decisions of a set of Kernel Minimum Enclosing Balls (KMEBs) by introducing a compound novelty score, which can allow for, for instance, a majority voting based detection as decision based on single balls might be limiting for novelty detection. In addition, our methodology

Supported by the Competence Center for Machine Learning Rhine Ruhr (ML2R) which is funded by the Federal Ministry of Education and Research of Germany (grant no. 01—S18038A).

I. S. Kotsireas and P. M. Pardalos (Eds.): LION 14 2020, LNCS 12096, pp. 414–420, 2020.
https://doi.org/10.1007/978-3-030-53552-0_37

can be easily implemented in neuromorphic architectures and is capable of dealing with nonlinear patterns due to kernelization [1,10]. Figure 1 shows an illustrative example explaining our idea about detecting the novel datapoints (green diamond shaped points in Fig. 1a) given a dataset of normal datapoints (black round points in Fig. 1a), which can neither be detected using euclidean Minimum Enclosing Balls (as seen in Fig. 1b) nor considering probabilistic novelty detection such as the deviation from the sample data mean [6]. Instead by considering the characteristic functions of multiple KMEBs (see example in Fig. 1c) with differently scaled Gaussian kernels we can detect all novel points that the considered balls might not individually be capable of capturing (compare the results of Fig. 1d to the others).

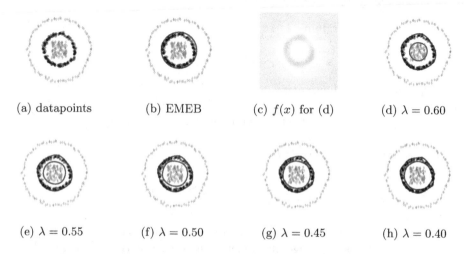

(a) datapoints   (b) EMEB   (c) $f(x)$ for (d)   (d) $\lambda = 0.60$

(e) $\lambda = 0.55$ (f) $\lambda = 0.50$ (g) $\lambda = 0.45$ (h) $\lambda = 0.40$

Fig. 1. A conceptual example illustrating the idea of utilizing Kernel Minimum Enclosing Balls for novelty discovery. (a) shows the data (the inner ring), which is used to compute the ball, and the novel (green diamond) points. It is important to note that neither considering the deviation from the mean vector nor computing the euclidean MEB, which is shown in (b), can in this case isolate the points inside the inner ring. (c) shows a heat-map of the characteristic function from Eq. 7, where colors orange, white and blue respectively indicate positive, zero and negative values. (d–h) shows the dataset and the novel points with the decision boundaries for different Gaussian Kernel scale values λ. Used individually to detect the novel points, the recall values for detecting the novelty are respectively 0.935, 0.995 and 1.000 for the balls in (d), (e) and (f–h). We obtain 1.000 recall when majority-voting over the prediction of the balls (d–h) (i.e. by considering an ensemble of 5 KMEBs with evenly spaced λ values over $[0.4, 0.6]$). (Color figure online)

The remaining of the paper is organized as follows. So as to be self-contained, in Sect. 2 we will formally define the notion of KMEBs, show how we can compute them following a process akin to the ones used in echo state networks and finally show how, once computed, the support vectors of balls can be used to

characterize the interior of the fitted balls. Following that in Sect. 3 we will introduce a new novelty scoring methodology based on the characteristic functions
for novelty discovery. In Sect. 4 we will present empirical results to evaluate our
approach using real world datasets and in Sect. 5 we will conclude our work.

2   An Overview of Kernel Minimum Enclosing Balls

Given a set of m-dimensional data points $\mathcal{X} = \{x_1, \ldots, x_n\}$ (for $x_i \in \mathbb{R}^m$) that
are grouped into a column data matrix $X = [x_1, \ldots, x_n] \in \mathbb{R}^{m \times n}$, we aim to find
the m-ball $\mathcal{B}(c, r)$ containing each of the given data points in $\mathcal{X}$, where $c \in \mathbb{R}^m$
and $r \in \mathbb{R}$ are respectively the center and the radius of $\mathcal{B}$. Finding MEBs can
be cast as a convex optimization problem

$$
c_*, r_* = \operatorname*{argmin}_{c, r} \quad r^2
$$
$$
\text{s.t.} \quad \left\| x_i - c \right\|^2 - r^2 \leq 0 \qquad i \in [1, \ldots, n]. \tag{1}
$$

Upon evaluating the Lagrangian and the KKT conditions, the negated dual
of (1), allows for the kernel trick (as the data *only* occurs in form of inner
products [1]) and can be written as the minimization problem

$$
\mu_* = \operatorname*{argmin}_{\mu} \quad \mu^\mathsf{T} K \mu - \mu^\mathsf{T} k
$$
$$
\text{s.t.} \quad \sum_{i=1}^{n} \mu_i = 1 \quad \wedge \quad \mu_j \geq 0 \ \forall \ j \in [1, \ldots, n], \tag{2}
$$

where $K \in \mathbb{R}^{n \times n}$ is a kernel matrix, k contains its diagonal (i.e. $k = \operatorname{diag}[K]$)
and $\mu \in \mathbb{R}^n$ contains Lagrange multipliers. The kernel matrix K in (2) is built
by considering a Mercer kernel $K : \mathbb{R}^m \times \mathbb{R}^m \to \mathbb{R}$ such that $K_{ij} = K(x_i, x_j)$. An
example kernel function that we considered throughout our work is the Gaussian
kernel that for scale parameter λ is defined as $K(x_i, x_j) = \exp\left(-\frac{\|x_i - x_j\|^2}{2\lambda^2}\right)$.

Considering (2), we note that finding Kernel Minimum Enclosing balls boils
down to finding optimal μ, which resides in the standard simplex Δ^{n-1} and
minimizes a convex function $\mathcal{L}(\mu) = \mu^\mathsf{T} k - \mu^\mathsf{T} K \mu$. Optimization settings of
this kind can be easily solved iteratively using the Frank-Wolfe algorithm [3],
which itself can be implemented as a recurrent neural network (see examples
from [1,2,8,10]). To this end, at each iteration t, the Frank-Wolfe algorithm
evaluates the gradient of the negated dual Lagrangian $\mathcal{L}(\mu)$ from (2), which
amounts to $\nabla \mathcal{L}(\mu) = 2K\mu - k$, and finds the vertex of Δ^{n-1} for the update,
that minimizes

$$
\nu_t = \operatorname*{argmin}_{v_j \in \mathbb{R}^n} v_j^\mathsf{T} \left[2K\mu_t - k\right] \approx g_\beta(2K\mu_t - k), \tag{3}
$$

where $\nu_t \in \mathbb{R}^n$ represent the current solution at t, v_j is the jth standard vector
$v_j = [\delta_{j1}, \delta_{j2}, \ldots, \delta_{jp}]^T\}$ (here δ_{ji} represents the Kronecker delta) and, finally,

$g_\beta(x)$ represents the soft-min operator. This operator is the smooth approximation of argmin., whose the ith entry defined as $\left(g_\beta(x)\right)_i = \frac{e^{-\beta x_i}}{\sum_j e^{-\beta x_j}}$ and has the limit

$$\lim_{\beta \to \infty} g_\beta(x) = \operatorname*{argmin}_{v_j \in \mathbb{R}^n} v_j^\mathsf{T} x = v_i. \tag{4}$$

Given that we can define the *convergent* iterative Frank-Wolfe updates [1] as

$$\mu_{t+1} \leftarrow (1 - \eta_t)\,\mu_t + \eta_t\, g_\beta\big(2\,K\mu_t - k\big), \tag{5}$$

where $\eta_t \in [0, 1]$ is a monotonically decreasing step size. Rearranging the rightmost expression in (5) as $g_\beta(2\,K\mu_t - k) = g_\beta(2\,K\mu_t + \bar{K}\bar{1})$, where $\bar{K} = diag(k)$ and $\bar{1}$ is the vector of -1s defined as $\bar{1} = [-1, \ldots, -1]^T$, allows us to interpret and implement these updates in terms of echo state networks [4]. That is, we can describe this machinery as a structurally constrained echo state network, in which we have the fixed input vector $\bar{1}$ containing -1s, the input weight matrix $\bar{K}$, n reservoir neurons with $g_\beta(\cdot)$ and $2\,K$ respectively being the nonlinear activation function and the reservoir weight matrices and η_t acting as a leaking rate for updating the Lagrange multipliers. Once optimal Lagrange multipliers have been found using the updates from (5), we can determine the kernelized radius and the squared magnitude of the center of the fitted ball $\mathcal{B}$ respectively as $r_* = \sqrt{\mu_*^\mathsf{T} k - \mu_*^\mathsf{T} K \mu_*}$ and $c_*^\mathsf{T} c_* = \mu_*^\mathsf{T} K \mu_*$, which will allow us to define a characteristic function defining the interior of $\mathcal{B}$ [1]. Namely, using these equalities we can represent the inequality $\|x - c_*\|^2 \le r_*^2$ to check whether an arbitrary point $x \in \mathbb{R}^m$ within the ball $\mathcal{B}$ by considering

$$f(x) = \sqrt{K(x, x) - 2\,\bar{k}^\mathsf{T} \mu_* + \mu_*^\mathsf{T} K \mu_*} - \sqrt{\mu_*^\mathsf{T} k - \mu_*^\mathsf{T} K \mu_*}, \tag{6}$$

where $\bar{k} \in \mathbb{R}^n$ is defined as $\bar{k}_i = K(x, x_i)$ [1]. That is, $f(x) > 0$ holds if x is outside of the ball $\mathcal{B}$, whereas, $f(x) \le 0$ is the case when x is inside the ball $\mathcal{B}$. Though, $f(x) = 0$ only holds for the points with nonzero Lagrange multipliers that are the support vectors of $\mathcal{B}$ and can be defined as $\mathcal{S} = \{x_i \mid \forall\, i \in [1, \ldots, n] \wedge \mu_{i*} > 0\}$. It is worth noting that, we can simplify (6) by grouping the $l \le n$ points in $\mathcal{S}$ into a column data matrix $S = [s_1, \ldots, s_l] \in \mathbb{R}^{m \times l}$, putting their corresponding multipliers in $\sigma \in \mathbb{R}^l$, letting $Q \in \mathbb{R}^{l \times l}$ be the kernel matrix for the support vectors (i.e. $Q_{ij} = K(s_i, s_j)$) and $q \in \mathbb{R}^l$ to contain its diagonal (i.e. $q = \operatorname{diag}[Q]$), which yields a simpler characteristic function

$$f(x) = \sqrt{K(x, x) - 2\,\bar{k}^\mathsf{T} \sigma + \sigma^\mathsf{T} Q \sigma} - \sqrt{\sigma^\mathsf{T} q - \sigma^\mathsf{T} Q \sigma} \tag{7}$$

where as in (6), $\bar{k} \in \mathbb{R}^l$ is evaluated as $\bar{k}_j = K(x, s_j)$ and we note that the term $\sqrt{\sigma^\mathsf{T} q - \sigma^\mathsf{T} Q \sigma}$ (which indeed amounts to r_*) is does not depend on x.

3   An Ensemble Approach for Novelty Discovery

Having explained how KMEBs are defined and can be computed so that we can determine their interior, we will now turn our attention to novelty discovery by combining the characteristic functions of a set of balls. We note that, the characteristic function from (7) for a given ball $\mathcal{B}$ can be used to label the points outside of the ball to be the novel points. In this case a query point x is considered novel if $f(x) > 0$ and not novel for $f(x) \leq 0$. Although this approach can capture novel points it might result in very restrictive or too general decision boundaries that respectively might result in detecting every query point to be novel or not novel (see Fig. 1d for the latter case). Both problems, however, can be avoided if we generalize this approach by combining the decisions of multiple balls. One approach for such a combination can be based on uniform voting [7]. That is, given a set of u KMEBs $\mathcal{P} = \{\mathcal{B}_1, \ldots, \mathcal{B}_u\}$, that are trained considering a different setting, and $f_i(\cdot)$ and $[\![\cdot]\!]$ respectively indicating the characteristic function from (7) for ball $\mathcal{B}_i$ and the Iverson bracket, we can assign the *novelty score* of a query point x by evaluating $z(x) = \sum_{i=1}^{u} [\![f_i(x) > 0]\!]$ and, for instance, label x to be novel if $z(x) \geq \lceil \frac{u}{2} \rceil$ (i.e. x is outside of the majority of the balls in $\mathcal{P}$ for an odd u) and not novel if $z(x) < \lceil \frac{u}{2} \rceil$. In the next section, we will empirically evaluate this methodology to detect novelty by showing two conceptual examples on benchmarking datasets.

Table 1. Novelty prediction results in terms of recall (**RC**), precision (**PR**), as well as the harmonic mean and geometric mean of both (respectively referred as **F1** and **GM**) for (a) the CBLC Face and (b) the MNIST datasets to respectively detect non-face images from face images and the images of digit 0 from the ones of 1. We benchmarked methods to detect novelty that consider the deviation from the sample mean (**MDEV**), matrix factorization (**MF**), euclidean MEBs (**EMEB**) and the ensemble of kernelized MEBs (**EKMEB**). The superior prediction results indicate that EKMEB can indeed be used for novelty discovery.

Method	RC	PR	F1	GM
MDEV	0.686	1.000	0.813	0.828
MF	0.711	0.999	0.831	0.843
EMEB	0.790	0.999	0.882	0.889
EKMEB	0.974	0.998	0.986	0.986

(a) MNIST dataset

Method	RC	PR	F1	GM
MDEV	0.043	1.000	0.082	0.207
MF	0.215	0.998	0.354	0.463
EMEB	0.095	1.000	0.173	0.308
EKMEB	1.000	0.949	0.974	0.974

(b) CBLC Faces

4   Empirical Results

We evaluated our method on the MNIST [5] and CBCL-face (bit.ly/2KwOVV6) datasets. For the former we trained models on the digit 1 aiming to obtain the 0s, whereas for the latter we leaned balls on faces to detect non-face novel images. So as to evaluate the precision of the detections, we divided the training data into 90/10 splits and the latter split is combined with the novel points, which resulted in training/evaluation datsets of cardinality values 6067/6598 and 2186/4791 for respectively the MNIST and CBCL-face datasets. We note that for both examples, we constructed ensembles of KMEBs (i.e. distinct $\mathcal{P}$ sets) with the Gaussian Kernel, whose scale values, in our case, were evenly spaced over specified intervals (as in Fig. 1) by considering $u = 5$ KMEBs with λ ranging in $[40, 60]$. We also normalized the datasets to have zero mean and unit variance and always considered $\beta = \infty$ for the softmin function (see (4)).

In Table 1, we compare our method against thresholding the tested points considering the maximum deviation from the sample mean vector [6], euclidean MEBs [1] (where we consider points outside of the ball as novel) and matrix factorization (MF) [8] based reconstruction to validate the use of kernel methods. For the first method, we label points in the test set as novel if the euclidean distance is larger than the furthest point to the sample mean. For the last method, we factorize the matrix with the number of latent factors $k = 50$ using the alternating least squares method [9] and learn a threshold value based on the *worst* reconstruction error ($l2$-norm). Unseen points with reconstruction error exceeding this threshold are considered novel. Table 1a and Table 1b respectively depict the prediction results for the MNIST and CBCL datasets, where we observe the superiority of ensemble KMEBs to detect novel datapoints.

5   Conclusion and Future Work

In this work, we introduced the idea of using ensemble of KEMBs for novelty discovery. We showed how we can construct ensembles of KEMBs and introduced a voting-based approach to detect novel data points. Our empirical evaluation yielded superior results over the use of mean deviation, euclidean MEBs and matrix factorization approaches. Our future work involves studying different ball selection as well as novelty determination strategies and extending the scope of the applications. Another line of future work is related to physical implementation of our methodology and in resource-constrained devices for applications in industrial domains such as for predictive maintenance.

References

1. Bauckhage, C., Sifa, R., Dong, T.: Prototypes within minimum enclosing balls. In: Tetko, I.V., Kůrková, V., Karpov, P., Theis, F. (eds.) ICANN 2019. LNCS, vol. 11731, pp. 365–376. Springer, Cham (2019). https://doi.org/10.1007/978-3-030-30493-5_36
2. Bauckhage, C.: A neural network implementation of Frank-Wolfe optimization. In: Lintas, A., Rovetta, S., Verschure, P.F.M.J., Villa, A.E.P. (eds.) ICANN 2017. LNCS, vol. 10613, pp. 219–226. Springer, Cham (2017). https://doi.org/10.1007/978-3-319-68600-4_26
3. Frank, M., Wolfe, P.: An algorithm for quadratic programming. Naval Res. **3**, 95–110 (1956)
4. Jäger, H., Haas, H.: Harnessing nonlinearity: predicting chaotic systems and saving energy in wireless communication. Science **304**(5667), 78–80 (2004)
5. LeCun, Y., Boottou, L., Bengio, Y., Haffner, P.: Gradient-based learning applied to document recognition. Proc. IEEE **86**(11), 2278–2324 (1998)

6. Pimentel, M.A., Clifton, D.A., Clifton, L., Tarassenko, L.: A review of novelty detection. Sig. Process. **99**, 215–249 (2014)

7. Rokach, L.: Ensemble methods for classifiers. In: Maimon, O., Rokach, L. (eds.) Data mining and Knowledge Discovery Handbook, pp. 957–980. Springer, Boston (2005). https://doi.org/10.1007/0-387-25465-X_45

8. Sifa, R.: An overview of Frank-Wolfe optimization for stochasticity constrained interpretable matrix and tensor factorization. In: Kůrková, V., Manolopoulos, Y., Hammer, B., Iliadis, L., Maglogiannis, I. (eds.) ICANN 2018. LNCS, vol. 11140, pp. 369–379. Springer, Cham (2018). https://doi.org/10.1007/978-3-030-01421-6_36

9. Sifa, R.: Matrix and Tensor Factorization for Profiling Player Behavior. LeanPub, British Columbia (2019)

10. Sifa, R., Paurat, D., Trabold, D., Bauckhage, C.: Simple recurrent neural networks for support vector machine training. In: Kůrková, V., Manolopoulos, Y., Hammer, B., Iliadis, L., Maglogiannis, I. (eds.) ICANN 2018. LNCS, vol. 11141, pp. 13–22. Springer, Cham (2018). https://doi.org/10.1007/978-3-030-01424-7_2

11. Tax, D.M., Duin, R.P.: Data domain description using support vectors. In: Proceedings of ESANN (1999)

DESICOM as Metaheuristic Search

Rafet Sifa[(✉)]

Fraunhofer Center for Machine Learning, Fraunhofer IAIS, Sankt Augustin, Germany
Rafet.Sifa@iais.fraunhofer.de

Abstract. Decomposition into Simple Components (DESICOM) is a constrained matrix factorization method to decompose asymmetric square data matrices and represent them as combinations of very sparse basis matrices as well as dense asymmetric affinity matrices. When cast as a least squares problem, the process of finding the factor matrices needs special attention as solving for the basis matrices with fixed affinities is a combinatorial optimization problem usually requiring iterative updates that tend to result in locally optimal solutions. Aiming at computing globally optimal basis matrices, in this work we show how we can cast the problem of finding optimal basis matrices for DESICOM as a metaheuristic search and present an algorithm to factorize asymmetric data matrices. We empirically evaluate our algorithm on synthetic datasets and show that it can not only find interpretable factors but also, compared to the existing approach, can better represent the data and escape locally optimal solutions.

1 Introduction

Decomposition into Directed Components (DEDICOM) [4] is a popular matrix factorization method to analyze asymmetric data matrices. The method allows for simultaneously extracting the latent structures, that are read-off the columns of the basis matrices, as well as the (asymmetric) relationships among these structures, which can be grouped into lower rank affinity matrices [1,4,5,7]. Formally, DEDICOM represents a given asymmetric matrix $\mathbf{S} \in \mathbb{R}^{n \times n}$ in terms of a basis matrix $\mathbf{A} \in \mathbb{R}^{n \times k}$ and an affinity matrix $\mathbf{R} \in \mathbb{R}^{k \times k}$ such that the data matrix and its arbitrary entry are resp. represented as $\mathbf{S} \approx \mathbf{A}\mathbf{R}\mathbf{A}^T$ and $s_{ij} \approx \mathbf{a}_{i:}^T \mathbf{R}\mathbf{a}_{j:}$, where $\mathbf{A}^T$ represents the transpose of $\mathbf{A}$, k defines the number of latent dimensions (following $k \ll n$) and $\mathbf{a}_{b:}$ represents the bth row of $\mathbf{A}$.

In this work we will focus on aiming to find globally optimal solutions when the Residual Sum of Squares (RSS) is minimized over $\mathbf{A}$ for fixed $\mathbf{R}$ for a variant of DEDICOM, that is called Decomposition into Simple Components [5], and propose a metaheuristic search based algorithm that aims at finding constrained (for interpretability) globally optimal basis matrices $\mathbf{A}$ at each alternating update. DESICOM constrains the basis matrix $\mathbf{A}$ to contain *simple* rows by

Supported by the Competence Center for Machine Learning Rhine Ruhr (ML2R) which is funded by the Federal Ministry of Education and Research of Germany (grant no. 01—S18038B).

I. S. Kotsireas and P. M. Pardalos (Eds.): LION 14 2020, LNCS 12096, pp. 421–427, 2020.
https://doi.org/10.1007/978-3-030-53552-0_38

allowing every row of $\mathbf{A}$ to contain only one non-zero entry while considering a non-constrained affinity matrix. That is, every ith row of $\mathbf{A}$ is defined as $a_{i\hat{b}} \not\oplus 0$, which in our case is tantamount to $a_{i\hat{b}} \neq 0$ and $a_{im} = 0 \ \forall \ m \ \neq \ \hat{b}$, resulting in reconstructing an arbitrary entry of $\mathbf{S}$ as $s_{ij} = a_{i\hat{b}} r_{\hat{b}\hat{c}} a_{j\hat{c}}$ such that $a_{i\hat{b}} \not\oplus 0$ and $a_{j\hat{c}} \not\oplus 0$. Given that we can define the Frobenius norm of the reconstruction matrix (a.k.a. the RSS) as a function of $\mathbf{A}$ and $\mathbf{R}$ for DESICOM as

$$E(\mathbf{A}, \mathbf{R}) = \left\| \mathbf{S} - \mathbf{A}\mathbf{R}\mathbf{A}^T \right\|^2 = \sum_{i,j} (s_{ij} - a_{i\hat{b}} a_{j\hat{c}} r_{\hat{b}\hat{c}})^2 \text{ s.t. } a_{i\hat{b}} \not\oplus 0 \wedge a_{j\hat{c}} \not\oplus 0. \quad (1)$$

In [5] Kiers proposes an alternating least squares (ALS) algorithm (for more information about this method we refer to [1,7]) to minimize (1) by first considering the matrix regression solution as an update for $\mathbf{R}$ as

$$\mathbf{R} \leftarrow \mathbf{A}^{\dagger} \mathbf{S} (\mathbf{A}^T)^{\dagger}, \quad (2)$$

where $\mathbf{A}^{\dagger}$ in this case represents the pseudo-inverse of $\mathbf{A}$. Following that, so as to find optimal and simple loadings minimizing (1), the algorithm alters each row of $\mathbf{A}$ individually and sequentially by considering a greedy update rule while keeping all the other rows of $\mathbf{A}$ as well as the matrix $\mathbf{R}$ fixed. To this end, [5] shows that, for the updated row $\mathbf{a}_{i:}$ this boils down to finding its lth element minimizing

$$E(a_{il}) = \left\| \begin{pmatrix} \hat{\mathbf{s}}_i \\ \hat{\mathbf{s}}_{i:} \end{pmatrix} \begin{pmatrix} \hat{\mathbf{A}}_i \mathbf{R} \\ \hat{\mathbf{A}}_i \mathbf{R}^T \end{pmatrix}_l \right\|^2 + (s_{ii} - a_{il}^2 r_{ll})^2, \quad (3)$$

where $\hat{s}_i$ and $\hat{\mathbf{s}}_{i:}$ represent resp. the ith column and row of the data matrix $\mathbf{S}$ *without* the corresponding diagonal element, similarly $\hat{\mathbf{A}}_i$ represents the matrix $\mathbf{A}$ without the ith row and $(\cdot)_l$ selects the lth column of the given stacked matrix. The norm in (3) can be minimized by solving a fourth degree polynomial (for a_{il}) and the root yielding the lowest reconstruction error value (according to (3)) is set to be a_{il} while all the other elements of $\mathbf{a}_{i:}$ are set to zero. Although this approach is simple to implement and only requires the rank k of $\mathbf{A}$ and $\mathbf{R}$ as the single parameter to be set, it requires solving for a fourth degree polynomial nk times and creating the compound matrices $\hat{\mathbf{A}}_i$ for the update of each ith row on the fly making the algorithm not scalable for large values n and k. We also note that, there exist k^n arrangements of having different basis matrices with zero values (of course with infinitely many possibilities for the value of the nonzero element), which the algorithm in [5] cannot generally explore due to the fact that it only iterates over the rows once.

2   Updating the Loadings as Metaheuristic Search

One way to tackle the above mentioned issues is to formulate the process of finding optimal simple basis matrices $\mathbf{A}$ minimizing (1) for fixed $\mathbf{R}$ as a metaheuristic search, which we will, in the following, cast as a Simulated Annealing (SA) [6] problem. To this end, we propose to treat every $\mathbf{A}$ as a single solution

Procedure UptA($\mathbf{S}$,$\mathbf{R}$,T_0,T_{min},μ,σ,λ,Q,A_{max},A_{min}):

$\quad n, k = shape(\mathbf{A})$

$\quad t = T_0$

$\quad g = \left\| \mathbf{S} - \mathbf{A}\mathbf{R}\mathbf{A}^T \right\|^2$

$\quad$**while** $t > T_{min}$ **do**

$\quad\quad$//Create a copy of A

$\quad\quad \hat{\mathbf{A}} = \mathbf{A}$

$\quad\quad$//Select a random index

$\quad\quad i \sim \mathcal{U}(\{1, 2, \ldots, n\})$

$\quad\quad$//Select a probability for the perturbation-type

$\quad\quad q \sim \mathcal{U}(0, 1)$

$\quad\quad$**if** $Q > q$ **then**

$\quad\quad\quad$//Select a random column index j s.t. $j \neq \hat{b}$

$\quad\quad\quad j \sim \mathcal{U}(\{1, 2, \ldots, k\} \setminus \{\hat{b}\})$

$\quad\quad\quad$//Flip the values between a_{ij} and $a_{i\hat{b}}$

$\quad\quad\quad \hat{a}_{ij} = \hat{a}_{i\hat{b}} \ \wedge \ \hat{a}_{i\hat{b}} = \hat{a}_{ij}$

$\quad\quad$**else**

$\quad\quad\quad$//Select a random perturbation value

$\quad\quad\quad u \sim \mathcal{N}(\mu, \sigma^2)$

$\quad\quad\quad$//Perturb the nonzero element $a_{i\hat{b}}$ and project it back to the defined range

$\quad\quad\quad \hat{a}_{i\hat{b}} = max(min(\hat{a}_{i\hat{b}} + u, A_{max}), A_{min})$

$\quad\quad$//Compute the reconstruction error with the new solution

$\quad\quad \hat{g} = \left\| \mathbf{S} - \hat{\mathbf{A}}\mathbf{R}\hat{\mathbf{A}}^T \right\|^2$

$\quad\quad$//Compute the difference in reconstruction error values

$\quad\quad d = \hat{g} - g$

$\quad\quad$//Select an acceptance probability

$\quad\quad p \sim \mathcal{U}(0, 1)$

$\quad\quad$**if** $d < 0$ *or* $e^{-d/t} > p$ **then**

$\quad\quad\quad$//Update A and g

$\quad\quad\quad \mathbf{A} = \hat{\mathbf{A}} \ \wedge \ g = \hat{g}$

$\quad\quad$//Reducing the temperature

$\quad\quad t = \lambda t$

$\quad$**return A**

Procedure MH-DESICOM($\mathbf{S}$, k,T_0,T_{min},μ,σ,λ,Q,A_{max},A_{min}):

$\quad$Randomly initialize a nonnegative $\mathbf{R}$ and a simple nonnegative $\mathbf{A}$

$\quad$**while** *stopping conditions are not satisfied* **do**

$\quad\quad \mathbf{R} \leftarrow \mathbf{A}^\dagger \mathbf{S}(\mathbf{A}^T)^\dagger$

$\quad\quad \mathbf{A} \leftarrow$ UptA($\mathbf{S}$, $\mathbf{R}$,T_0,T_{min},μ,σ,λ,Q,A_{max},A_{min})

Algorithm 1: The MH-DESICOM algorithm, which incorporates metaheuristic search to the alternating least squares procedure to find optimal basis and affinity matrices for DESICOM. See the text for more details.

(in the metaheuristic search terminology) and at each SA iteration (with a temperature value t) work with a copy $\hat{\mathbf{A}} = \mathbf{A}$ to perturb every of its randomly selected ith *row* $\hat{\mathbf{a}}_{i:}$ s.t. $\hat{a}_{i\hat{b}} \not\equiv 0$ by choosing (with probability $Q \in [0, 1]$) one of the following two strategies. The first strategy is based on randomly selecting a *different* column index j from a uniform distribution (which we represent as $j \sim \mathcal{U}(\{1, 2, \ldots, k\} \setminus \{\hat{b}\})$ s.t. $j \neq \hat{b}$) and flipping the values of $\hat{a}_{ij}$ and $\hat{a}_{i\hat{b}}$. The second strategy is based on creating a random perturbation value u and updating the nonnegative value $a_{i\hat{b}}$ as $a_{i\hat{b}} = a_{i\hat{b}} + u$. Similar to the previous work on neural networks [3], the value of u can be randomly chosen from Gaussian distribution as $u \sim \mathcal{N}(\mu, \sigma^2)$, where μ and σ resp. represent the previously defined mean and the standard deviation of the distribution. Note that, our approach also has the advantage of constraining the matrix $\mathbf{A}$ to have certain properties such as residing in a predefined range $[A_{min}, A_{max}]$ for $A_{min}, A_{max} \in \mathbb{R}$. For the latter we can consider projecting the perturbed value of the considered row $\hat{a}_{i\hat{b}}$ as $a_{i\hat{b}} = max(min(a_{i\hat{b}}, A_{max}), A_{min})$. This can especially be useful when the factorized data matrix $\mathbf{S}$ contains only *nonnegative* entities. In this case, given that $A_{min}, A_{max} \geq 0$ we can safely consider the absolute value of the update of (2) to guarantee having nonnegative factors as this projection cannot *increase* the error in (1).

For consistency with the ALS optimization, we used the RSS from (1) as the fitness function of the current as well as the perturbed solution. Given that, the perturbed (and projected) solution $\hat{\mathbf{A}}$ is accepted either right away if it results in lower error or with probability $e^{-d/t}$ to escape local minima, where e is the irrational base of the natural logarithm and t and d resp. represent the current temperature of the system and the difference value in RSS between the current and the perturbed solutions. Starting with the predefined initial temperature T_0, we consider decreasing the current temperature t value by a cooling parameter $\lambda \in (0,1)$ as $t = \lambda t$ and stop the SA optimization once a given minimum temperature value is reached T_{min}. Algorithm 1 summarizes the whole procedure, that we called Metaheuristic (MH) DESICOM, of finding optimal DESICOM parameters minimizing (1) in an ALS fashion using the update from (2) for $\mathbf{R}$ and our metaheuristic search based approach for finding optimal $\mathbf{A}$.

3 Empirical Evaluation

We consider three sets of experiments to evaluate our approach, where we will evaluate (in order) its performance of obtaining global minimum solution, the interpretability of the resulting factors and its performance on matrix reconstruction on synthetic datasets.

We start by evaluating the ratio of ending up in a local minimum (as in [5]) of (1) for synthetically constructed noise-free (i.e. perfect) asymmetric datasets of two types that have the same individual values for their diagonal, upper diagonal and lower diagonal values resp. x, y, z (i.e. for any $\mathbf{S}$ we will have $s_{ii} = x \; \forall \; i$, $s_{ij} = y \; \forall \; i < j$, $s_{ij} = z \; \forall \; i > j$). The first type will be referred to as mildly asymmetric and have $x = 3$, $y = 1$, $z = 2$ and the second type will be called strong asymmetric containing $x = 3$, $y = 1$, $z = 1$. We created two mildly asymmetric and two strongly asymmetric matrices with $n \in \{3, 4\}$ factorized using the algorithm from and Algorithm 1. For the mildly asymmetric matrices MH DESICOM resulted in local minimum with ratios 0.43 and 0.70 resp. for $k = 3$ and $k = 4$, which for the case of the original row-wise (RW) DESICOM (the algorithm from [5]) were 0.63 and 0.93. Similarly, for the strongly asymmetric matrices our approach's local minima ratios were 0.50 and 0.73 while RW-DESICOM's results were worse (0.67 and 0.97) indicating that updating the loading matrices using our algorithm has a higher probability of finding the global optimum factors.

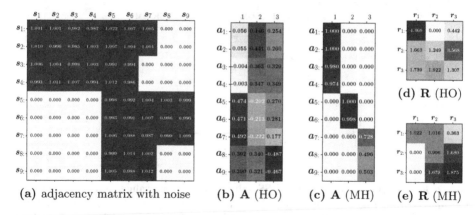

(a) adjacency matrix with noise   **(b) A** (HO)   **(c) A** (MH)   **(e) R** (MH)

Fig. 1. Illustrative example of factorizing a 9×9 noise-added adjacency matrix (shown in (a)) of a directed graph containing three communities with asymmetric relations and different self affinity values with HO-DEDICOM and MH-DESICOM for $k = 3$. (b, d) resp. illustrate the resulting basis and affinity matrices from HO-DEDICOM, whereas, (c, e) show the corresponding results from MH-DESICOM. While the basis matrix in (b) shows soft assignments with negative loadings, that can only be interpreted as a whole, the basis matrices from our approach define hard assignments and can be more interpretable when we can assign every entity to a group.

Following that, we concentrated on factorizing a small matrix to illustrate the interpretability of the basis and the affinity matrices. To this end, we constructed an adjacency matrix $\mathbf{S} \in \mathbb{R}^{9 \times 9}$ (from directed network of three communities with different asymmetric interactions and different self affinity values) and added Gaussian noise to each of its entries to show the resilience of the algorithms against noise (shown in Fig. 1a). We compared the results of factorizing $\mathbf{S}$ using the Semi Nonnegative Hybrid-Orthogonal (HO) DEDICOM algorithm from [8], which constraints the basis matrix to be column orthonormal and the affinity matrix to be nonnegative (shown resp. in Fig. 1b and Fig. 1d), and our approach with nonnegative A_{min}, A_{max} values as well as nonnegative projections of $\mathbf{R}$ (see Fig. 1c and Fig. 1e resp. for the basis and the affinity matrices). Compared to soft assignments of HO-DEDICOM (read off the columns of $\mathbf{A}$ Fig. 1b), the basis matrix $\mathbf{A}$ resulted from our approach clearly allow us to distinguish the three communities (i.e. the objects $\{1, 2, 3, 4\}$, $\{5, 6, 7\}$ and $\{8, 9\}$) in terms of hard assignments and the nonnegative asymmetric affinities in $\mathbf{R}$ allow us to *summarize* the pairwise relationships between these communities (for instance showing high affinity *from* $\{1, 2, 3, 4\}$ *to* $\{5, 6, 7\}$ and low affinity for the opposite case). Our final set of experiments is about investigating the RSS values for different k values when factorizing the matrix from Fig. 1a. To this end, we ran both MH- and RW-DESICOM algorithms 10 times for each $k \in [3, 4, \ldots, 9]$ and showed the results in Fig. 2, which again show the superiority of our approach.

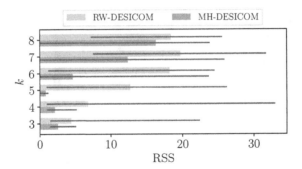

Fig. 2. Empirical evaluation of the RSS values for different values of k for factorizing the matrix from Fig. 1a. The bars and the whiskers resp. show the mean and minimum/maximum values of 10 runs of the benchmarked algorithms (the lower the better). Well aligned with our previous results, our approach can find factors representing the analyzed data better than the baseline approach.

4   Conclusion and Future Work

In this work we presented a metaheuristic search based algorithm to update the loadings of DESICOM in an ALS fashion. Compared to the existing approach, our method aimed at finding globally optimal solutions and our experimental results validated this as we overall obtained more stable and better reconstruction results. Our future work involves evaluating the performance of population-based metaheuristic search models for finding optimal **A** as well as importing our optimization paradigm to other data factorization methods, whose parameter estimation involve discrete optimization problems. For the latter, for instance, one can consider finding archetypal data points using metaheuristic search (see [2,9]) or generalize MH-DESICOM to factorize asymmetric data tensors (as in [1,7]).

References

1. Bader, B.W., Kolda, T.G., Harshman, R.A.: Temporal analysis of social networks using three-way dedicom. Technical report, Sandia National Lab. (SNL-NM), Albuquerque, NM, US (2006)
2. Bauckhage, C., Sifa, R., Wrobel, S.: Quantum computing for max-sum diversification. In: SIAM International Conference on Data Mining (2020)
3. Fogel, D.B., Fogel, L.J., Porto, V.: Evolving neural networks. Biol. Cybern. **63**(6), 487–493 (1990)
4. Harshman, R.: Models for analysis of asymmetrical relationships among N objects or stimuli. In: Proceeding of the Joint Meeting of the Psychometric Society and the Society for Mathematical Psychology (1978)
5. Kiers, H.: DESICOM: decomposition of asymmetric relationships data into simple components. Behaviormetrika **24**(2), 203–217 (1997)
6. Kirkpatrick, S., Gelatt, C.D., Vecchi, M.P.: Optimization by simulated annealing. Science **220**(4598), 671–680 (1983)

7. Sifa, R.: Matrix and Tensor Factorization for Profiling Player Behavior. LeanPub (2019)

8. Sifa, R., Ojeda, C., Bauckhage, C.: User Churn Migration Analysis with DEDICOM. In: Proceedings ACM RecSys (2015)

9. Sifa, R., Bauckhage, C.: k-Maxoids Via Adiabatic Quantum Computing. Research gate technical report (2018)

Author Index

Printed in the United States
By Bookmasters